Biology of Cladocera

Developments in Hydrobiology 71

Series editor

H. J. Dumont

Biology of Cladocera

Proceedings of the Second International Symposium on Cladocera,
Tatranska Lomnica, Czechoslovakia, 13–20 September 1989

Edited by
V. Kořínek & D. G. Frey

Reprinted from Hydrobiologia, vol. 225 (1991)

Kluwer Academic Publishers
Dordrecht / Boston / London

Library of Congress Cataloging-in-Publication Data

International Symposium on Cladocera (2nd : 1989 : Tatranská Lomnica, Czechoslovakia)
Biology of Cladocera : proceedings of the Second International Symposium on Cladocera, Tatranska Lomnica, Czechoslovakia, September 13-20, 1989 / edited by V. Kořínek, and D.G. Frey.
p. cm. -- (Developments in hydrobiology ; 71)
ISBN 0-7923-1460-3 (alk. paper)
1. Cladocera--Congresses. I. Kořínek, V. II. Frey, David G. (David Grover), 1915- . III. Title. IV. Series.
QL444.B83I58 1989
595.3'2--dc20 91-31548

ISBN 0-7923-1460-3

Published by Kluwer Academic Publishers,
P.O. Box 17, 3300 AA Dordrecht, The Netherlands.

Kluwer Academic Publishers incorporates
the publishing programmes of
D. Reidel, Martinus Nijhoff, Dr W. Junk and MTP Press.

Sold and distributed in the U.S.A. and Canada
by Kluwer Academic Publishers,
101 Philip Drive, Norwell, MA 02061, U.S.A.

In all other countries, sold and distributed
by Kluwer Academic Publishers Group,
P.O. Box 322, 3300 AH Dordrecht, The Netherlands.

Printed on acid-free paper

Printed in Belgium

Contents

Preface

This volume contains most of the papers presented at the Second International Symposium on Cladocera, held at Tatranska Lomnica (Czechoslovakia) between 13 and 20 September 1989. The symposium continued the format and tradition begun at the First Symposium in Budapest (Hungary) in 1985.

Fifty-seven participants from 20 countries presented a total of 35 papers and 18 posters. The scheduled program included two field trips, one a sight-seeing trip to nearby Lomnický Peak, and the other to mountain lakes at an elevation of roughly 2000 m. Some of the participants substituted a shorter trip to Strbske Lake at a lower elevation.

The symposium was organized by a group of freshwater scientists from Charles University in Prague, chaired by V. Kořínek. The Tatratour Travel Agency helped with logistical arrangements. The symposium was largely self-financing, with some support from the Faculty of Science at Charles University, which herewith is gratefully acknowledged.

Besides the informal discissions at coffee breaks and during poster sessions, two special workshops were held, one on the 'Biogeography of Cladocera' and the second on 'Feeding in Cladocera'. A summary of the workshop discussions was published in The Cladocera Newsletter No. 3, 1990.

The broad programme of topics was opened by an address of the chairman, followed by two invited papers, one about the limnology of lakes in the Tatras by E. Stuchlík and the other on the distribution of Cladocera in the region by J. Hrbáček. The session on systematics, taxonomy, and comparative morphology contained papers on the genera *Pleuroxus*, *Daphniopsis*, *Daphnia*, *Bosmina*, *Diaphanosoma*, *Ilyocryptus*, and *Alona*, and on the superfamily Sidoidea. Various aspects of cladoceran biology, such as osmoregulation and the organs involved, gamogenesis, size structure and predation pressure, growth rate, survival, and faunistic considerations were presented in the following session. A special session considered feeding behavior and its relationship to morphology, in which two high-speed films by M. Gophen (Israel) and R. Shiel (Australia) documented the movement of the thoracic limbs of *Daphnia* and the capture of food particles. H. Hartmann supported the hypothetical mechanism of food selection by excellent SEM photographs. As in the first symposium, an interesting session was devoted to cladoceran genetics. The poster by S. Giessler provided new insight into the hybridization within the *Daphnia galeata-hyalina* complex. Several papers interpreted cladoceran remains in Austrian, German, and Polish lakes.

In the closing session the participants expressed their enthusiasm for having a broad variety of topics in these international symposia, rather than only a couple selected topics. The next symposium on Cladocera will be held in Norway in 1993.

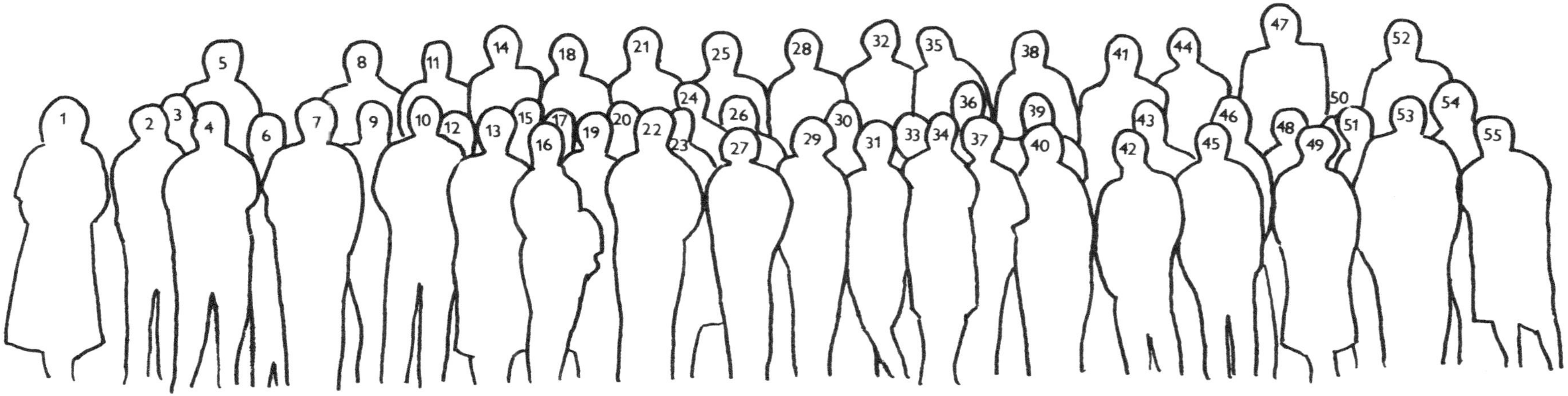

1. Zuzana Stuchlíková
2. Ireneo Ferrari
3. Evžen Stuchlík
4. Martin Černý
5. Milan Lebeda
6. Clara Cotten
7. David G. Frey
8. László Forró
9. Tomáš Herold
10. Nick Aladin
11. Rama Chengalath
12. Wolfgang Hofmann
13. Anna Abaffy-Bothár
14. Pavel Štifter
15. Vladimír Kořínek
16. Dorothy B. Berner
17. Jürgen Jacobs
18. Marian Vranovský
19. Krystyna Szeroczynska
20. Arnold Nauwerck
21. Michal Pop
22. Miguel Alonso
23. Mrs. Røen
24. Dieter Ebert
25. Russel J. Shiel
26. Ulrik I. B. Røen
27. Valeria Rossi
28. Moshe Gophen
29. Evelyn Zoppi de Roa
30. Igor Hudec
31. Antonella Bachiorri
32. Petter Larsson
33. Jaroslav Hrbáček
34. Lilian Matveeva
35. Sergey M. Glagolev
36. Hans J. Hartmann
37. Leonard V. Polishchuk
38. Damia Jaume
39. Vladimir Matveev
40. Sabine Giessler
41. Junpeng Cai
42. Jotaro Urabe
43. Ole T. Kleiven
44. Anders Hobaek
45. Wolfgang Horn
46. Luc De Meester
47. Nikolay M. Korovchinsky
48. Marie Prchalová
49. Miroslava Pražáková
50. Monika Wittingerová
51. Zdeněk Brandl
52. Jiří Macháček
53. Myneni K. Durga Prasad
54. Dietrich Flössner
55. Irina K. Rivier

Hydrobiologia **225**: x, 1991.
V. Kořínek & D. G. Frey (eds), Biology of Cladocera.

List of participants

AUSTRALIA
Shiel R. J. (Albury)

AUSTRIA
Nauwerck A. (Mondsee)

BELGIUM
De Meester L. (Gent)

CANADA
Chengalath R. (Ottawa)

CZECHOSLOVAKIA
Brandl Z. (České Budějovice)
Černý M. (Praha)
Herold T. (Praha)
Hrbáček J. (Praha)
Hudec I. (Košice)
Kořínek V. (Praha)
Macháček J. (České Budějovice)
Pop M. (Praha)
Pražáková M. (Praha)
Stuchlík Mrs. & Mr. E. (Praha)
Štifter P. (Benátky n. Jiz.)
Volfová A. (Praha)
Vranovský M. (Bratislava)
Wittingerová M. (České Budějovice)

DENMARK
Røen Mrs. & Mr. U.I.B. (Kopenhagen)

FRANCE
Hartmann H. J. (Aubiere Cedex)

GERMANY
Cai J. (München)
Flössner D. (Jena)
Giessler S. (München)
Hofmann W. (Plön)
Horn W. (Lengefeld)
Jacobs J. (München)
Lieder U. (Berlin)*

HUNGARY
Abbafy-Bothár A. (Göd)
Forró L. (Budapest)
Ponyi Mrs. & Mr. J. (Tihany)

INDIA
Durga Prasad Myneni K. (Nagarjuna Nagar)

ISRAEL
Gophen M. (Tiberias)

ITALY
Bachiorri A. (Parma)
Ferrari I. (Ferrara)
Rossi V. (Parma)

JAPAN
Urabe J. (Chiba)

NORWAY
Hobaek A. (Bergen)
Kleiven O. T. (Bergen)
Larsson P. (Bergen)

POLAND
Szeroczynska K. (Warszawa)

SPAIN
Alonso M. (Barcelona)
Jaume D. L. (Barcelona)

SWITZERLAND
Ebert D. (Basel)

USA
Berner D. B. (Philadelphia)
Cotten C. (Bloomington)
Frey Mrs. & Mr. D. G. (Bloomington)
Williams-Howze J. (Charlotte)**

USSR
Aladin N. (Leningrad)
Glagolev S. M. (Moscow)
Korovchinsky N. M. (Moscow)
Matveev Mrs. & Mr. V. (Moscow)
Polishchuk L. V. (Moscow)
Rivier I. K. (Borok)
Yampolsky L. (Moscow)

VENEZUELA
Zoppi de Roa E. (Caracas)

---

* Presented by V. Kořínek.
** Presented by P. Štifter.

Hydrobiologia **225**: 1–8, 1991.
V. Kořínek & D. G. Frey (eds), Biology of Cladocera.

A review of the genus *Ilyocryptus* (Crustacea: Anomopoda) from Europe

P. Štifter
Okresní muzeum, Dept. of Nature, Benátky nad Jizerou, Czechoslovakia

Abstract

Anomopoda belong to one of the best known orders of Crustacea, even though there are some genera that need to be revised. The taxonomy of the genus *Ilyocryptus* is not clear at this time. This paper gives information about European species, explains key characters in tabular form together with a pictorial key of European species. Also presented are common problems of working with *Ilyocryptus* material, and a process of further study is explained.

Introduction

The genus *Ilyocryptus* has been known for more than 150 years, the first species *I. sordidus* (described as *Acantholeberis sordida*) was described by Liévin (1848) from the vicinity of Danzig, now Gdańsk (Poland). Unfortunately the description and figures are brief and non-realistic, so that we cannot exactly say which species was really described.

In 1862, G.O. Sars established the genus *Ilyocryptus* and described the second species, *I. acutifrons* from the southern part of Norway. The other European species *I. agilis* was described by Kurz in 1878 from a small pond near Kutná Hora (Czechoslovakia). Kurz also gave a very good redescription of *I. sordidus* with exact figures. I think that this work has had great influence on the conception of the species *I. sordidus* in Lilljeborg's and Sars's papers.

For nearly 100 years the three European species of the genus *Ilyocryptus* were known from only a few monographs and keys: Lilljeborg (1901); Stingelin (1908); Keilhack (1909); Scourfield & Harding (1958); Šrámek-Hušek *et al.* (1962); Manujlova (1964); Flössner (1972); Margaritora (1983) and Negrea (1983).

Romijn (1919) had studied the morphological variability of *I. sordidus* from the Netherlands, and he separated a different form as a new species, *I. silvaeducensis*. Similarly, Hankó (1926) described *I. balatonicus* due to its slight differences from *I. sordidus*. These two species were synonymized with *I. sordidus* not from a detailed study of real material, but from comparisons of literature data. After 1970, Chirkova described two new species – *I. cornutus*, together with Mordukhai-Boltovskoi (1972) from the Rybinsk reservoir (U.S.S.R.) and in 1982, *I. vitali* from tundra lakes on Kola peninsula (U.S.S.R.). Smirnov's well known monograph (1976) lists five European species: *I. sordidus*, *I. acutifrons*, *I. agilis*, *I. silvaeducensis*, and *I. cornutus*. After a study of Dutch material, I redescribed the species *I. silvaeducensis* (Štifter, 1984). Later, large material of *I.* cf. *sordidus* was studied and then divided into three separate species: *I. sordidus*, *I. cuneatus* and *I. spinosus* (Štifter, 1988). Taxonomical status of *I. alexandrinae* Negrea, 1987 is not clear to me at this time. Its description and figures resemble the ephippial females of *I. sordidus*.

The following is a list of European species of the genus *Ilyocryptus* with remarks about their current distribution:

1.	*I. agilis*	Europe
2.	*I. acutifrons*	Europe
3.	*I. cornutus*	USSR, Norway (Štifter, unpubl. data)
4.	*I. cuneatus*	Europe (northerly of Alps)
5.	*I. silvaeducensis*	Holland, Spain (Alonso, pers. commun.)
6.	*I. sordidus*	Europe
7.	*I. spinosus*	Europe
8.	*I. vitali*	USSR (tundra, Kola peninsula).

Some characters used in taxonomy of the genus *Ilyocryptus*

External morphology of the specimens belong to the genus *Ilyocryptus* is unique in many aspects and often very specialised. Every species has its own set of characters. Some of them are variable and depend on geographical distribution, habitat, and its conditions (temperature, oxygen content, food conditions, predators). These characters have low weight in the grouping of taxa compared with characters often used in keys: numbers of teeth, denticles, or spines. They are closely depen-

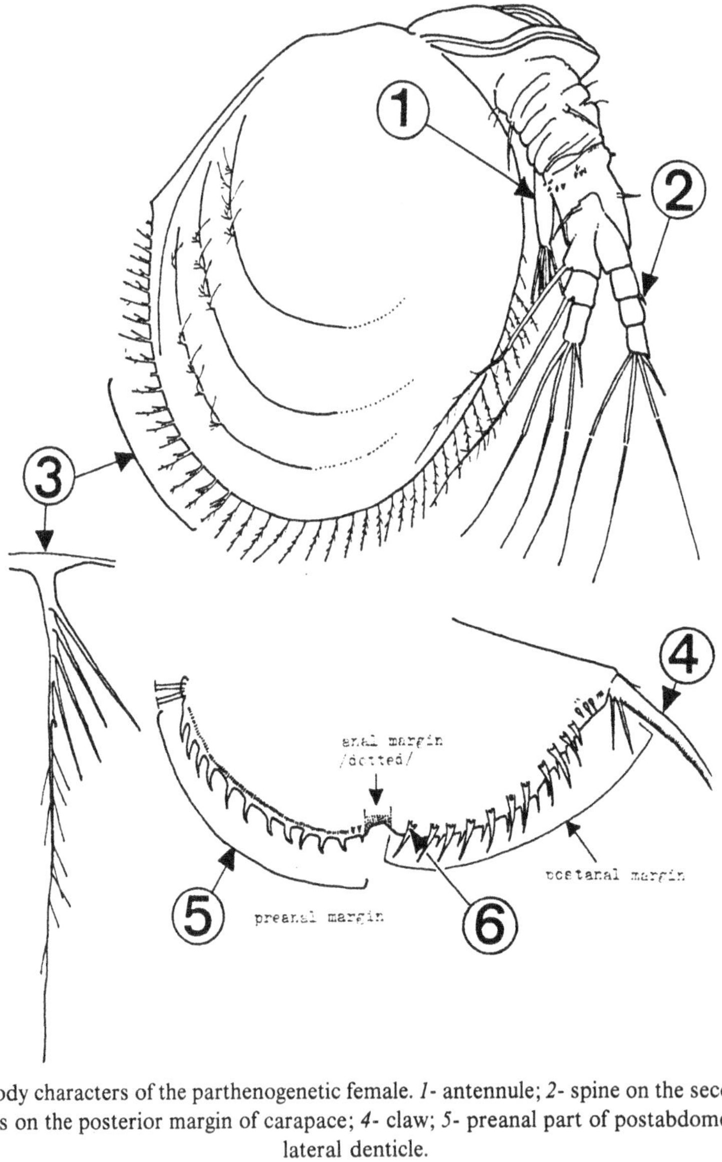

Fig. 1. Location of the body characters of the parthenogenetic female. *1*- antennule; *2*- spine on the second segment of antennal exopod; *3*- defence spines on the posterior margin of carapace; *4*- claw; *5*- preanal part of postabdomen; *6*- the first moveable lateral denticle.

dent on body length of specimens. It is better to use not only numbers but shape, specific form and proper arrangement of, for example, spines. On the other hand, there are some important characters that can be used very successfully in *Ilyocryptus* taxonomy. Unfortunately we use these characters but we know nothing about their purposes, advantages, and their role in the life of animal.

In the taxonomy of *Ilyocryptus* the following characters can be used:

1. Antennule – two segmented. Distal segment nearly globular with one fingershaped projection (in some species). Proximal segment varies widely in the genus: from short and robust (*I. acutifrons*) to very long and thin (*I. silvaeducensis*). (See character number 1 on Figs. 1 and 2.)
2. Spine on the second segment of antennal exopod (see character 2 on Figs. 1 and 2) – this spine can be shorter, equal to, or longer than one-half of the third exopodial segment length.

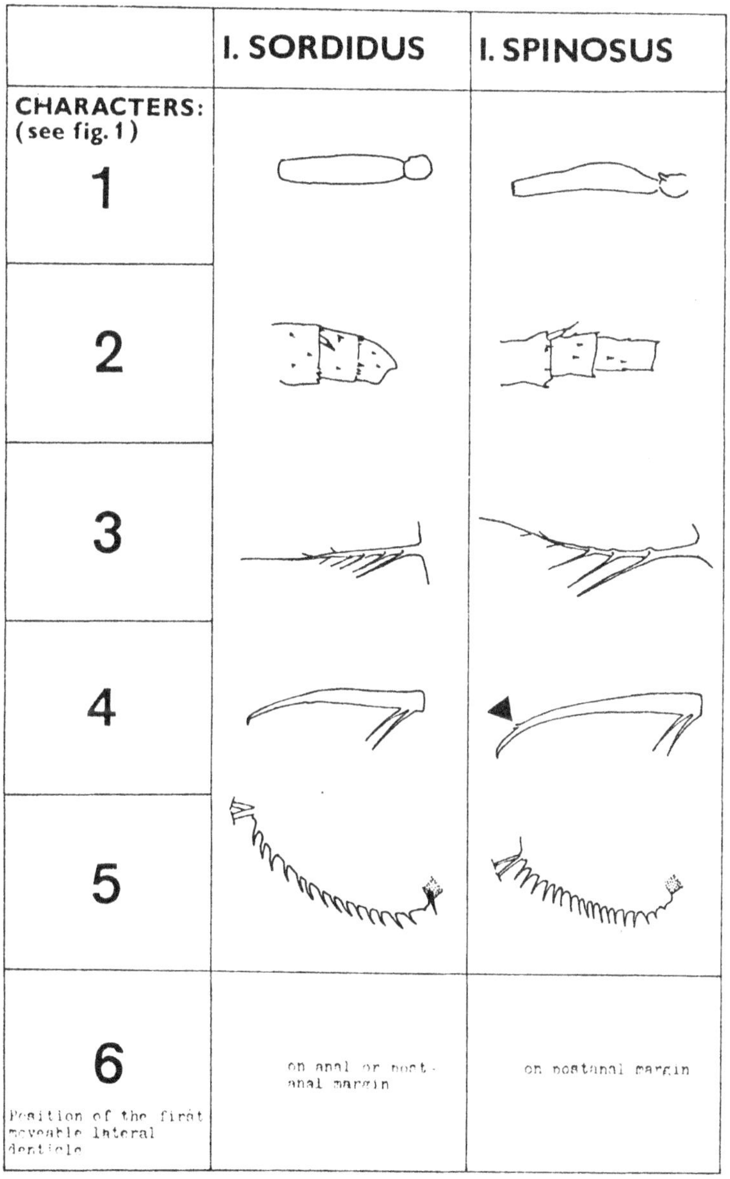

Fig. 2a.

Fig. 2, parts a, b, c: Tabular comparison of some characters in European species of the genus *Ilyocryptus* (for adult parthenogenetic female only).

3. Shape of defense spines on the posterior margin of carapace (see character 3 on Figs. 1 and 2) – is a very important character. Fully developed and the most useful spines are in the middle of their row.
4. Claw (see character 4 on Figs. 1 and 2) – its general shape, form of two basal spines, and spine(s) near distal tip of claw are valuable for taxonomy of the genus.
5. Preanal part of postabdomen (see character 5 on Figs. 1 and 2) – always armed with typical spines. Their specific form, morphology, and sometimes number can be useful for determination of species.
6. Moveable marginal denticles (see character 6 on Figs. 1 and 2) – can be present only on postanal margin or on anal and preanal respectively. The position of the first pair (the one farthest from the claw) can be used for separating species.

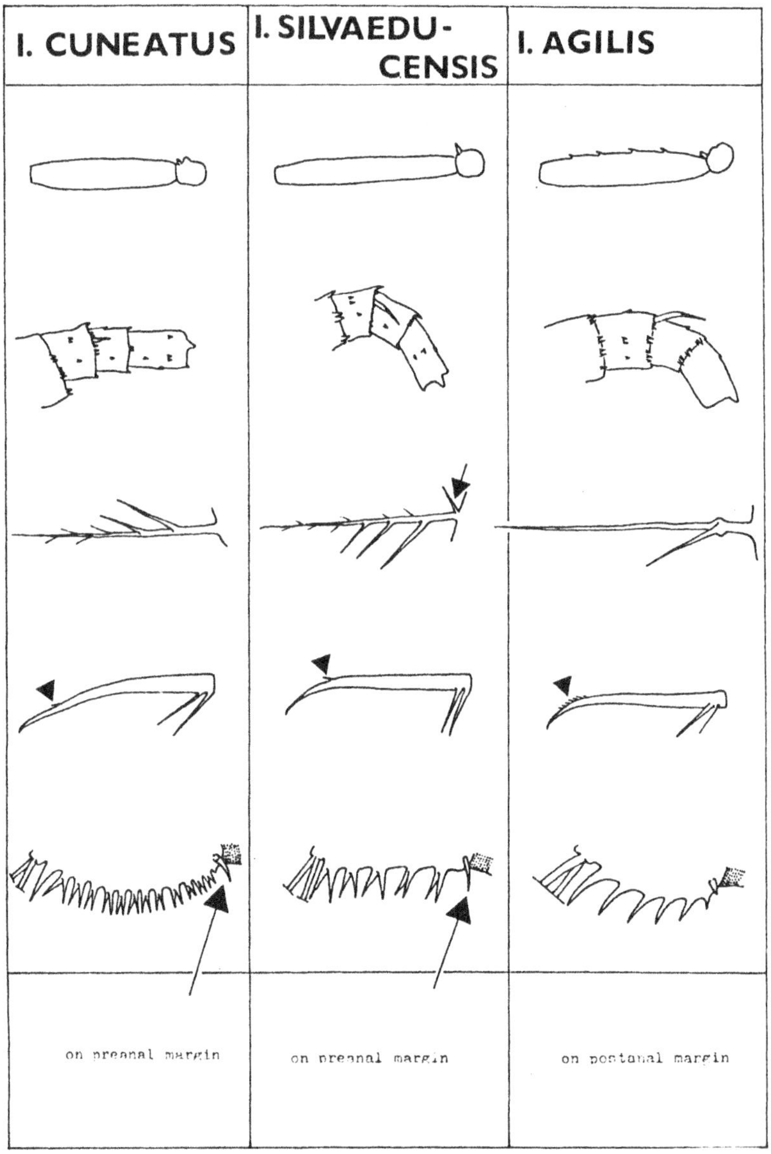

Fig. 2b.

Discussion

It is true that the genus *Ilyocryptus* is one of the poorest known genera of Anomopoda. There are many reasons for this situation. For example:

1. Most of the samples taken in the different parts of the world are planktonic samples, and hence the majority of the benthic species are not found.
2. It often is not possible to collect large samples because of the diffuse spread of *Ilyocryptus* populations.
3. Males and ephippial females are very rare.
4. On initial observation, many species are similar in appearance because of a cover of detritus.
5. Material from museum collections is often not determined correctly and often is fragmentar.
6. The theory of cosmopolitanism, which has been used for a long time, cannot be commonly accepted for many species (Frey 1986, 1984; Williams-Howze, in press). Every species, especially the 'cosmopolitan' ones, must be studied and revised. The supposed world-wide distribution of common European species of *I. sordidus* is questionable.

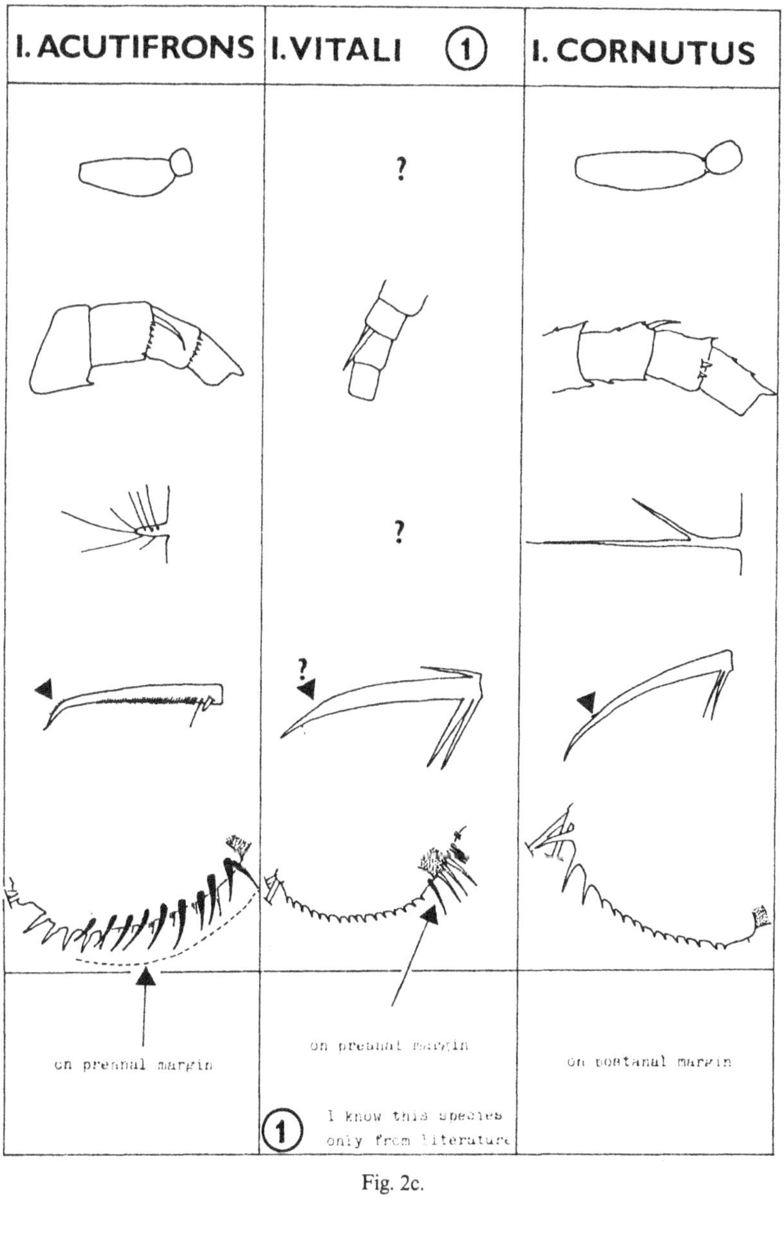

Fig. 2c.

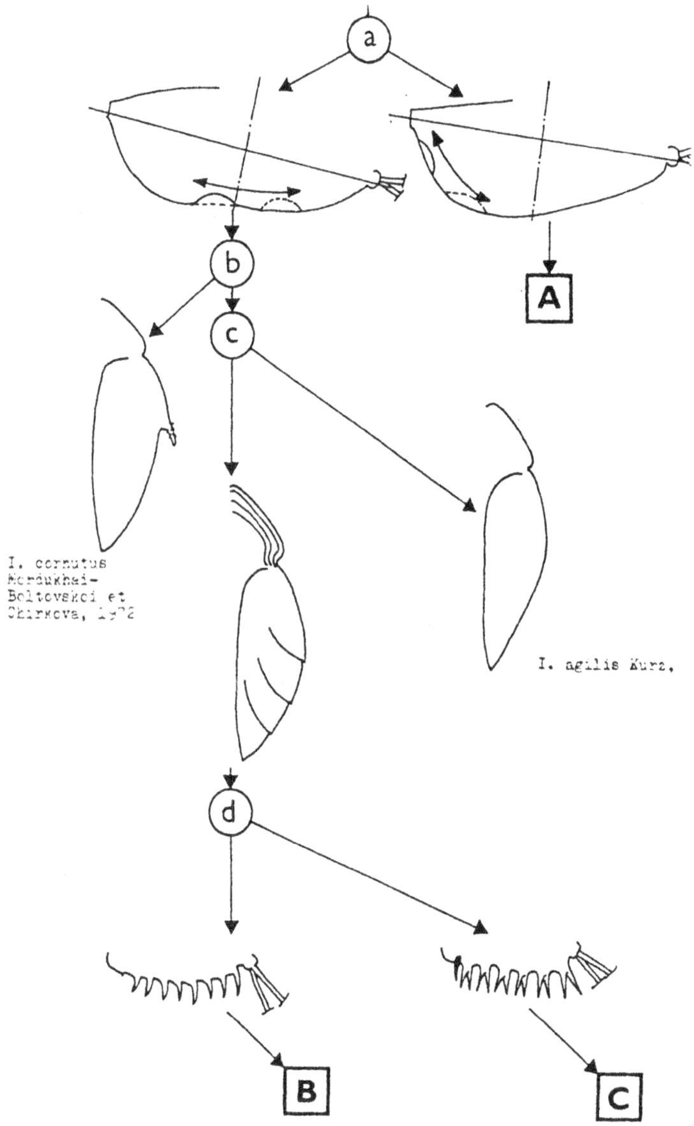

Fig. 3a.

Fig. 3, parts a, b: Pictorial key of European species of the genus *Ilyocryptus* (for adult parthenogenetic female only). Letters in circles mean steps in the key: *a*- position of anus; *b*- horns on carapace; *c*- keeping of old carapaces; *d*- shape of spines on preanal part of postabdomen; *e*- length of moveable lateral denticle's row; *f*- shape and number of spines on preanal part of postabdomen; *g*- shape of preanal part of postabdomen, together with differences in defense spine and length of spine on the second segment of antennal exopod.

Detailed morphological studies, ecological data, information about coexistence (common in *Ilyocryptus*) are necessary for resolving taxonomic and evolutionary relationships. The question is how to work and study these animals when so many problems exist (as mentioned above). I think that there is only this way:

1. to collect more material;
2. to redescribe and revise species (especially 'cosmopolitan' ones from old literature) and study them in the whole area to obtain variability information;
3. to describe new forms (to date there are 8 forms that might be new species – Williams-Howze and Štifter, pers. commun.);
4. to publish new key for *Ilyocryptus* species;
5. to prepare and publish a revision of the genus *Ilyocryptus*.

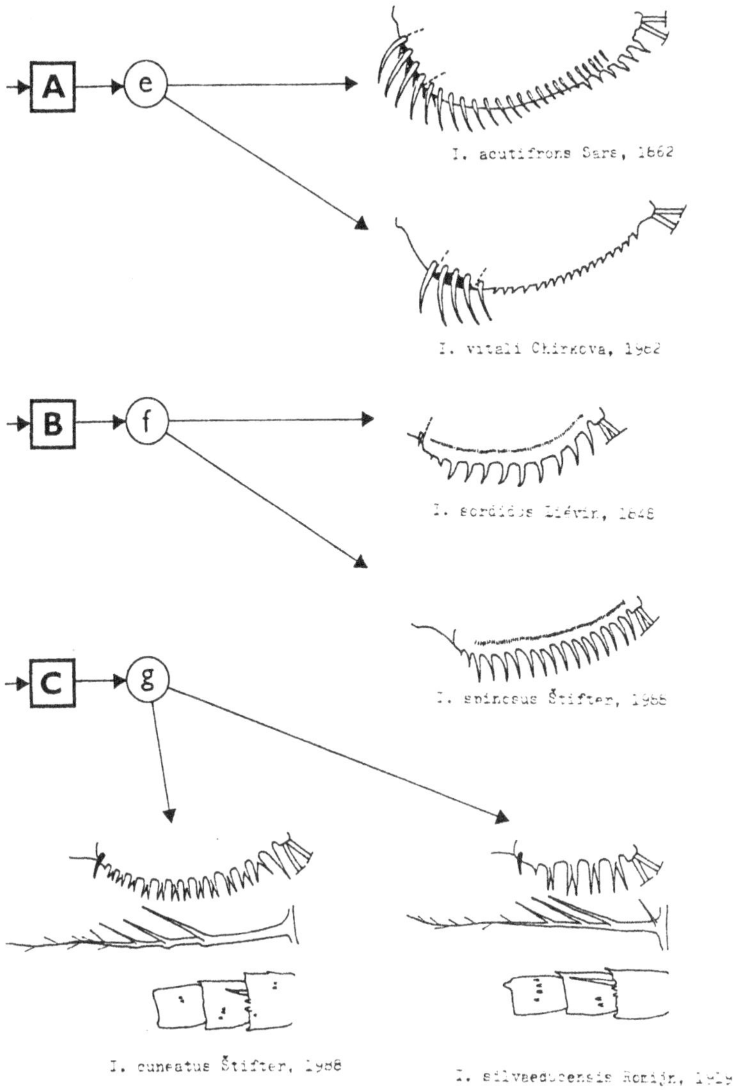

Fig. 3b.

It is clear that this work is long-term and it depends on help (especially with material) of other specialists. This work can bring together much useful information and help to complete part of our knowledge of this very interesting, primitive, and highly specialized genus of Anomopoda.

Acknowledgements

My thanks belong to all who have helped me with material. Thanks are also due to Judy Williams-Howze and Prof. D. G. Frey for reading of manuscript and for language corrections.

References

Chirkova, Z. N., 1982. *Ilyocryptus vitali* sp.n. (Cladocera, Macrothricidae) iz ozer tundry Kolskogo poluostrova. Zool. zhurn. 61(5): 699–706.

Flössner, D., 1972. Krebstiere, Crustacea. Kiemen- und Blattfüsser Branchiopoda. Fischlause, Branchiura. Die Tierwelt Deutschlands, Vol. 60, Fischer Verlag Jena, 501 pp.

Frey, D. G., 1986. The non-cosmopolitanism of chydorid Cladocera: Implications for biogeography and evolution. A. A. Balkema, Rotterdam, Boston: 237–256.

Frey, D. G., 1987. The taxonomy and biogeography of the Cladocera. Hydrobiologia 145: 5–17.

Hankó, B., 1926. *Iliocryptus balatonicus*, eine neue Cladocere aus der Gyttja des Balaton-sees. Arch. Balatonicum 1: 97–115.

Keilhack, L., 1909. Phyllopoda. Die Süsswasserfauna Deutschlands. Heft 10, Jena, 112 pp.

Kurz, W., 1878. Über limicole Cladoceren. Z. wiss. Zool. 30B (Suppl): 396–410.

Liévin, F., 1848. Die Branchiopoden der Danziger Gegend, ein Beitrag zur fauna der Provinz Preussen. Neueste Schriften der Naturforsch. Gesellsch. in Danzig. 4(2): 34–37.

Lilljeborg, W., 1901. Cladocera Sueciae. Nova Acta Upsal. ser. 3(19): 1–701.

Manujlova, E. F., 1964. Vjetvistousyje rachki (Cladocera) fauny SSSR. Opr. po faune SSSR. Zool. Inst. AN SSSR., 88. M.-L.: 1–327.

Margaritora, F., 1983. 22. Cladoceri (Crustacea, Cladocera). In: S. Ruffo (ed.), Guide per il riconoscimento delle specie animali delle acque interne Italiane, Verona, 169 pp.

Mordukhai-Boltovskoi, F. D. & Z. N. Chirkova, 1972. Novyj vid *Ilyocryptus* (Cladocera, Macrothricidae) iz Rybinskogo vodochranilishcha. Zool. zhurn. 51(5): 647–653.

Negrea, S., 1983. Cladocera. Fauna republicii socialiste Romania. Vol. 4, Fasc. 12, Bucurest; 399 pp.

Negrea, S., 1987. *Scapholeberis erinaceus* Daday, 1903 (syn. *Scapholeberis echinulata* Sars, 1903) and *Ilyocryptus alexandrinae* sp. nov. in Romania. Hydrobiologia 145: 105–111.

Romijn, G., 1919. Das Geschlecht *Iliocryptus* G. O. Sars. Int. Revue ges. Hydrobiol. Hydrogr. 8(5): 529–539.

Sars, G. O., 1862. Forsatte sit Fordrag over de af ham i Omegnen af Christiania iaggtagne Crustacea Cladocera. Forhandl. Vidensk. Selsk. Christiania (1861): 283–284.

Scourfield, D. J. & J. P. Harding, 1958. A key to the British species of freshwater Cladocera with notes on their ecology. 2nd edn. Freshwater. Biol. Ass. Brit. Emp., Sci. Publ. 5, 55 pp.

Smirnov, N. N., 1976. Macrothricidae i Moinidae fauny mira. Fauna SSSR, Rakoobraznyje, T.1, Vyp.3, Izd. Nauka, Leningrad, 236 pp.

Štifter, P., 1984. Redescription of *Ilyocryptus silvaeducensis* (Cladocera, Macrothricidae). Věst. čs. Společ. zool. 48: 132–136.

Štifter, P., 1988. Two new species of the genus *Ilyocryptus* (Crustacea, Cladocera) confused with *I. sordidus* Lievin. Věst. čs. Společ. zool. 52: 290–301.

Stingelin, T., 1908. Phyllopodes. Cataloque de Invertébrés de la Suisse II. Genève, 156 pp.

Williams-Howze, J. The macrothricid Cladocera *Ilyocryptus spinifer* Herrick from the North American continent – a review of the species using scanning electron microscopy. Hydrobiologia (in press).

Hydrobiologia **225**: 9–22, 1991.
V. Kořínek & D. G. Frey (eds), Biology of Cladocera.

A comparison of populations from the *Daphnia similis* group (Cladocera: Daphniidae)

Igor Hudec
Institute for Ecology and Agricultural Landscape, SAS, Hydrobiol. Lab., Kukorelliho 10, 04000 Košice, Czechoslovakia

Key words: Crustacea, Cladocera, Daphniidae, Daphnia, Ctenodaphnia, taxonomy, biology

Abstract

European populations of *Daphnia similis* Claus have been compared with populations from tropical Asia. *Daphnia similoides* n. sp. from tropical Asia was described as the sibling species of *D. similis*. The female has characteristic neonate, postabdomen, head, and ephippium. The male has characteristic rostrum, antennules, and the distalmost part of the postabdomen. The *D. similis* group was compared with *D. carinata* s. str. Australian populations.

Daphnia similis was described by Claus (1876) from dried mud from the vicinity of Jerusalem. On the other hand *D. psittacea* Baird, 1850 was described from a pond in Blackheath (England).

Only two species from the subgenus *Ctenodaphnia* (*D. magna* Straus and *D. atkinsoni* Baird) have been found in the British Isles to the present (Scourfield, 1903; Scourfield & Harding, 1941; Johnson, 1952; Fryer, 1985).

The name *D. psittacea* Baird is probably synonymous with *D. atkinsoni* Baird (from the British Isles), and the name *D. psittacea* Baird is a *nomen dubium*.

The variability of *D. similis*, some misinterpreted synonymy of this species with *D. carinata* (Brooks, 1957), and unsatisfactory description and redescription of more species of the *carinata* group from Australia (Hebert, 1977) complicate the taxonomic question: What is the *D. similis* group, and what is *D. carinata* group in tropical Asia.

The geographical distribution of *D. similis* goes from Europe (Hrbáček *et al.*, 1978) throughout Middle East Asia (Claus, 1876) and probably north of the Pamir and Himalayan mountains (Sars, 1903a, 1903b; Manuylova, 1964), Mongolia (Flössner, 1986), North China (Spandl, 1925) and north Japan (Uéno, 1927). Its occurrence in the Central Asia portion of the Soviet Union based on Manuylova's (1964) drawings is not clear. There are probably more related species.

The North American populations described by Brooks (1957) as *D. similis* were not available for study.

Two species have been reported from tropical and subtropical Asia: 1/ *D. psittacea* or *D. similis*; 2/ *D. carinata s. str.* and *D. cephalata*. *D. cephalata* is evidently present in Sri Lanka (Bär, 1924; Fernando, 1980) and in South India (Daday, 1911) as *D. hypsicephala*. It may be that *D. cephalata* is an introduced species, because we know it only from this isolated region in Asia. Benzie (1987) gives the geographical distribution of *D. carinata* – *D. similis* group, but it seems that these conclusions are not quite satisfactory.

***Daphnia similis* Claus, 1876**

Material examined:

The type material has probably been destroyed (personal information from Dr. Hrbáček). It was deposited in the Natural History Museum in Vienna (Richard, 1896). Only the following European populations were compared (Table 1):

Bulgaria: Lovec, Swamp, 25 June 1961, lgt. Naidenov; Russe, pool, 16 June 1961, lgt. Naidenov from collection of Dr. Hrbáček.

Czechoslovakia: 319/Nový Tekov, rural pool, 3 July 1956, lgt. Brtek; 451/Tvrdošovce ditch, 17 July 1969, lgt. Brtek (parthenogenetic, ephippial females, and males); 519/Žarnov – Janík, small village pond, 17 July 1969, lgt. Brtek (juvenile, parthenogenetic, and ephippial females strongly

Table 1. Measurements of *Daphnia similis* Claus, *D. similoides* n. sp., and *D. carinata* s. strr.

Locality	Length (mm) up to	Length (mm) $X \pm s_X$	Spine (mm) $X \pm s_X$	Ratio Spine/Length	N	Sex
D. similis						
319/ Nový Tekov	3.90	1.90 ± 0.66	1.00 ± 0.35	0.53 ± 0.10	13	parth. females
451/ Tvrdošovce	2.90	2.49 ± 0.34	1.33 ± 0.42	0.59 ± 0.03	14	parth. females
	2.10	2.02 ± 0.06	0.67 ± 0.03	0.33 ± 0.01	10	ephip. females
	1.60	1.40 ± 0.13	0.45 ± 0.03	0.31 ± 0.04	10	males
Šamorin	2.40	2.05 ± 0.41	0.85 ± 0.10	0.45 ± 0.03	15	parth. females
D. similis var. *exilis* /like/						
519/ Žarnov-Janík	2.70	2.57 ± 0.09	1.28 ± 0.02	0.49 ± 0.03	9	parth. females
		2.60	1.30	0.50	1	ephip. female
		2.45	1.25	0.51	1	ephip. female
		0.85	0.68	0.80	1	juv. female
		1.26	1.02	0.81	1	juv. female
		1.65	0.45	0.27	1	male
D. similoides						
Ooty Lake	2.40	1.81 ± 0.51	0.52 ± 0.15	0.29 ± 0.11	21	parth. females
		1.35	0.35	0.26	1	male
		2.30	0.75	0.32	1	ephip. female
		2.10	0.55	0.26	1	ephip. female
Singapore	2.10	1.83 ± 0.13	0.40 ± 0.11	0.23 ± 0.04	15	parth. females
		1.85	0.35	0.19	1	ephip. female
		1.80	0.45	0.25	1	ephip. female
		1.80	0.25	0.14	1	ephip. female
		1.25	0.35	0.28	1	male
		1.15	0.35	0.30	1	male
		1.20	0.35	0.29	1	male
D. carinata						
Gloucester	3.3	2.45 ± 0.39	1.7 ± 0.28	0.72 ± 0.1	14	parth. females
		2.30	1.30	0.56	1	ephip. female
		2.45	1.35	0.55	1	ephip. female
		1.14	0.78	0.68	1	male
		0.85	0.77	0.83	1	male
		1.07	0.78	0.73	1	male
		1.05	0.87	0.82	1	male
		1.04	0.73	0.70	1	male

resemble *var. exilis*); 1568/Hamuliakovo – Šamorín, pool, 24 September 1972, lgt. Brtek (parthenogenetic, and ephippial females, and males); Šamorín, pool, 14 August 1956, lgt. Vranovský (juvenile, parthenogenetic, and ephippial females, juvenile and adult males); Kapušany, pool, 11 November 1957, lgt. Straškraba (1st and 3rd generation from laboratory culture).

More information about the distribution in Slovakia is reported in Hudec (1989).

Romania: Jagelia Lake, 11 July 1962, lgt. Negrea (3 parthenogenetic females).

Soviet Union: Region of Leningrad; 76th and 81st generation from laboratory culture in Moskow (from collection of Dr. Hrbáček).

Female (Figs. 1–8; Plate 1 (parts 1 & 2))

The head is large and variable, from widely rounded and semilunar in shape (Fig. 1) to a somewhat quadrate shape (Fig. 1a). Some populations may have a wide forward prolongation of the middle dorsal part of the head (Figs. 2–2c), which resemble *var. exilis*. The rostrum (Figs. 4–4b) is triangular, slightly blunt and pointed. Some natural populations and all examined laboratory cultures have a pointed rostrum. The compound eye is large and the ocellus is present as a minute spot. The triangular fornices are pointed at the posterior end.

Antennules (Figs. 4–4b) are short but not rudimentary, roughly parallel to the anterior margin of the head. Tips of aesthetascs reach the tip of the rostrum.

Postabdomen (Figs. 5–6). The dorsal margin is slightly expanded, with row of 9–12 slightly submarginal anal teeth. Clusters of fine and short hairs are located over the bases of the anal denticles. The praeanal region is covered with clusters or singled minute spicules to near the abdominal setae. The first two abdominal processes are the largest. The 3rd and 4th abdominal processes are strongly reduced. The terminal claw (Fig. 6) has two pectens of larger and stouter spicules that extends basally, distad from which is a long row of minute and fine spicules nearly to the tip of the claw.

The first thoracic limb (Fig. 7) bears two ejector hooks. The first endite has four 2-segmented setae and one unsegmented submarginal seta. The 2nd endite has two 2-segmented setae and one unsegmented seta that is likewise submarginal. The 3rd endite has two 2-segmented setae and one unsegmented seta. The 4th endite has one long unsegmented seta. The exopodite has a basal stalk with one 2-segmented seta that is slightly feathered along the distal segment. A second small seta arises near the base of the long seta. Our examination agrees with the drawing by Wolski (1932).

The second thoracic limb (Fig. 8). The exopodite has a long cylindrical basal segment with two long 2-segmented and feathered setae. The exopodite is 4-parted. The proximalmost part bears one 2-segmented seta, and the second part has none. The 3rd part has one long feathered seta, and the 4th part has two feathered seta and one unfeathered seta (only dense bristles on the one margin (Fig. 26a)). The gnathobase has 3 sections. The one nearest to the endopodite has one long seta feathered with long hairs along one margin, and one somewhat shorter seta that is slightly submarginal, of which the distalmost part is double feathered with minute hairs. Then follow 10–13 2-segmented setae on the second section and one unsegmented seta on a minute hump in the middle part. The 3rd section bears three bent setae feathered on the distal segment. The general setulation agrees that given by Wolski (1932).

We did not find any differences in the 3rd to 5th thoracic limbs, which agree with the drawings by Wolski (1932).

The ephippium (Fig. 1a; Plate 1 (part 2)) is elliptical, with a sadle-like depression in the anterior part. It is heavy sclerotinised and is slightly compressed in the middle between the two eggs.

Valves. The dorsal and ventral margins of the valves are curved. The dorsal margin has two

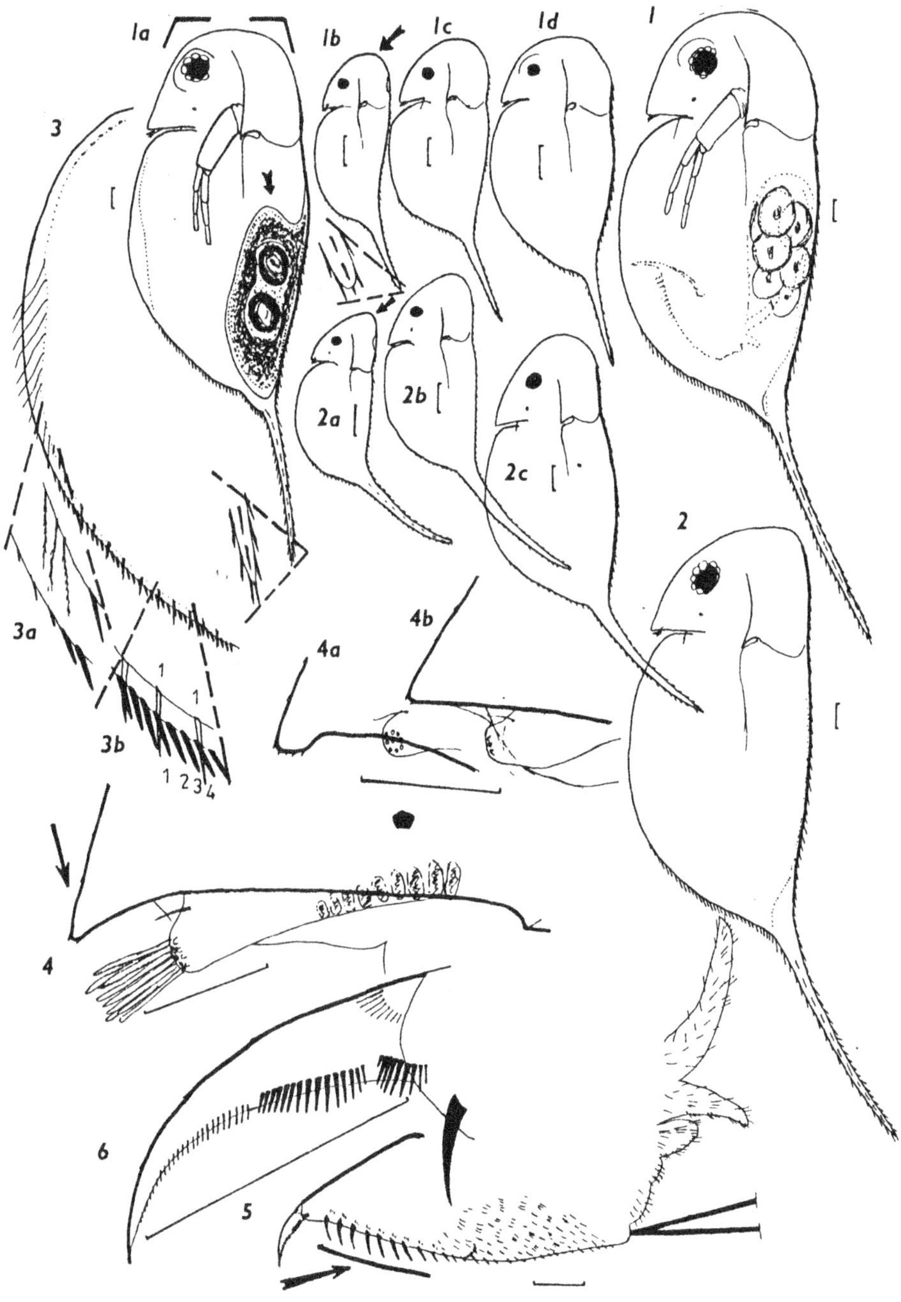

Figs. 1–6. Daphnia similis Claus female.
Fig. 1 adult female; Fig. 1a ephippial female with detail of terminal part of spine; Fig. 1b neonate; Figs. 1c–d juveniles; Figs. 2–2c population resembling var. *exilis*; Fig. 2 adult female; Fig. 2a neonate; Figs. 2b–c juveniles; Figs. 3–3b setulation on ventral margin of valves; Fig. 4 shape of rostrum and antennule; Figs. 4a–b variability of rostrum; Fig. 5 post-abdomen; Fig. 6 terminal claw of postabdomen. Figs. 1-1d, 3–3b, 4–6: 1568/Hamuliakovo – Šamorín. Figs. 2–2c, 4b: 519/Žarnov – Janík; Fig. 4a: Romania.

rows of spicules extending forward to the anterior margin of the compound eye. The posterior third of the ventral margin is provided with marginal spicules, with a submarginal spicula at every fourth marginal spicula (Fig. 3b). The submarginal spicules become transformed in the middle part of the valves to fine row of long feathered hairs (Figs. 3, 3a). The length of shell spine is

Figs. 7–8. Daphnia similis Claus female
Fig. 7. 1st thoracic limb; Fig. 8. 2nd thoracic limb

Figs. 9–11. Daphnia similoides n. sp. female
Fig. 9 ephippial female with detail of setulation on posterioventral margin; Fig. 9a neonate; Figs. 9b–c juveniles; Fig. 10 shape of rostrum and antennule; Figs. 10a–b variability of rostrum; Fig. 11 postabdomen.

Figs. 12–15. Daphnia carinata s. str. female
Fig. 12 parthenogenetic female; Fig. 12a neonate; Fig. 13 shape of rostrum and antennule; Fig. 14 postabdomen; Fig. 15 ephippium.

Figs. 7–8: 1568/Hamuliakovo – Šamorín. Figs. 9–11 Ooty Lake. Fig. 10b: Malaysia, Singapore. Figs. 12–15: Australia, Gloucester.

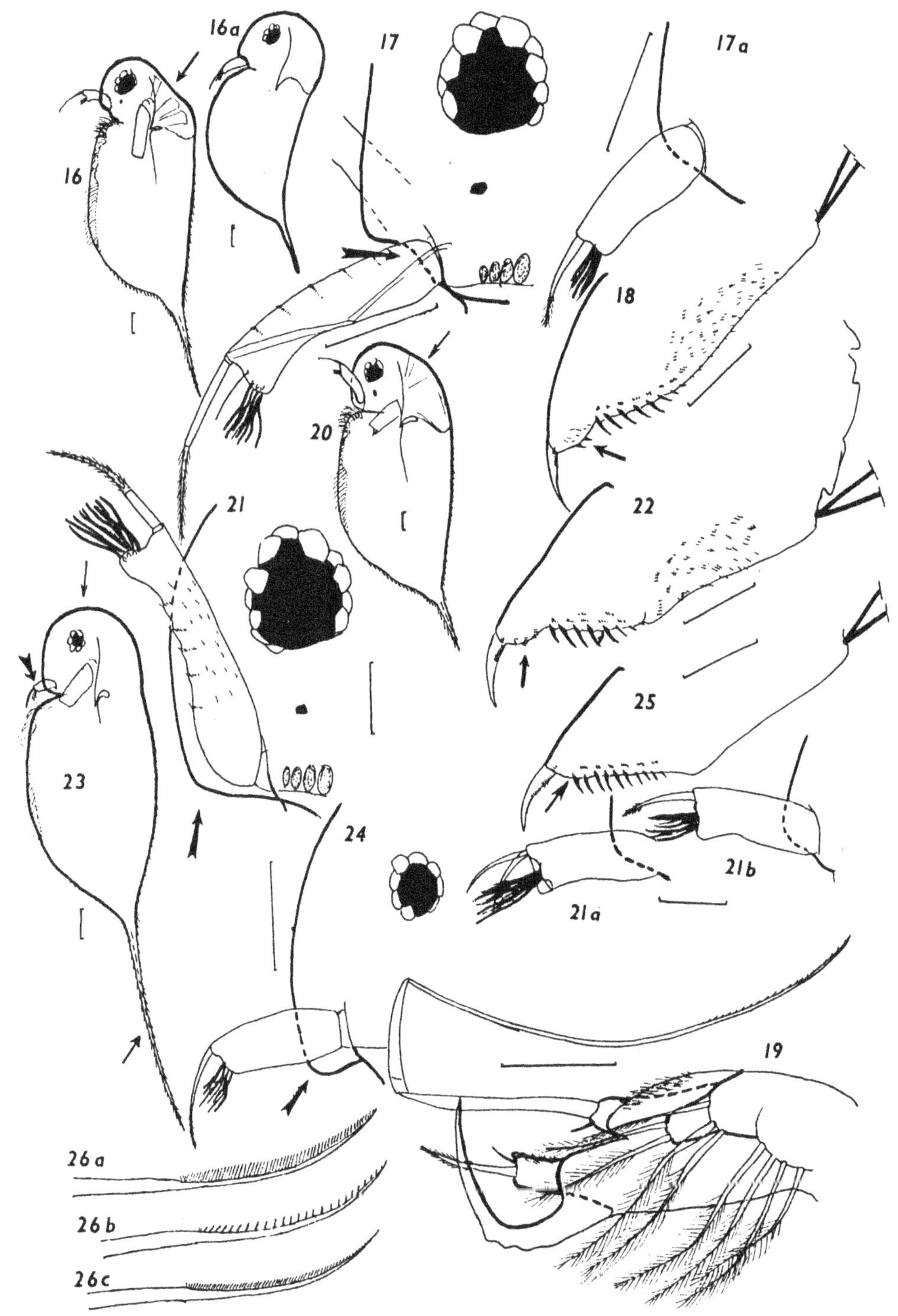

Figs. 16–19. Daphnia similis Claus male

Fig. 16 adult; Fig. 16a juvenile; Fig. 17 rostrum and antennule of adult; Fig. 17a rostrum and antennule of juvenile; Fig. 18 postabdomen; Fig. 19 1st thoracic limb.

Figs. 20–22. Daphnia similoides n. sp. male

Fig. 20 adult; Fig. 21 rostrum and antennule; Fig. 21a antennule with two flabela; Fig. 21b antennule of juvenile; Fig. 22 postabdomen.

moderate from $^1/_5$ to almost $^1/_2$ of total body length.

Juveniles (Figs. 1b–1d, 2a–2c). The neonate and all instars of typical populations have the head oblique over the compound eye but never pointed. Only populations that resemble var. *exilis* have juveniles with a triangular shaped head, but this is never pointed.

Male (Figs. 16–19; Plate 1 (parts 3 & 4))

The male is comparatively smaller than the female. The rostrum (Fig. 17) is widely rounded, bluntly pointed and has a moderate quadrate depression just before the junction of the antennules. The antennules are long, reaching $^1/_3$ of the total height of the head. The flagellum is as long as $^1/_2$ of the length of the antennule. There are 9 aesthetascs on the flagellum. The dorsal surface of the antennules is covered with minute hairs in transverse lines. The supraocular depression is present only in adult males.

The shape of the valves (Fig. 16; Plate 1 (part 3)) corresponds to that of the subgenus.

The first thoracic limb (Fig. 19) corresponds in structure with the drawing by Wolski (1932). The copulatory hook is long and has a pointed tip. The distalmost lobe of the endopodite has one very long seta that is feathered from the middle to the tip, and near its base is a minute seta and a small process.

The 2nd to 5th thoracic limbs are structured like those of the female.

The postabdomen (Fig. 18) is sinuate in the posterior margin as seen from the side. In the postanal region there is a small sinuate knob with 2–3 small denticles. Around the anal grove there are 5–7 marginal denticles. The terminal claw resembles that of the female, but it is straighter and its two basal pectens are more delicate.

Daphnia similoides n. sp.

D. carinata: Brehm, 1914: 511–512, Figs. 1–3; Brehm, 1933: 661–666, Figs. 4–6; Brehm, 1950: 22–23, Figs. 5a–f, 6a–d; Brehm & Wolterek, 1939: 6–8, Fig. 7; Mashiko, 1953: 53–55, Figs. 2a–e; Murugan & Job, 1981: 7, Pl. 1a; Murugan & Venkataraman, 1977: 129–134, 2 Plts.; Spandl, 1925: 186, Figs. 1a–b; Uéno, 1939: 334, Figs. 13–28 (not *D. carinata gibba*); Uéno, 1944: 406–407, Figs. 2a–d.

D. psittacea: Arora, 1931: 66, Table 6, Figs. 3–4; Uéno, 1935: 212–213, Figs. 1a–d; Uéno, 1937: 200–201, Figs. 1–6.

D. similis: Fernando *et al.*, 1987: 112, 124, Figs. 35–47; Idris, 1983: 33, Figs. 9a–f; Uéno, 1966: 288–290, Figs. 1-1 to 1-9.

Etymology: The specific name refers to the similarity with *D. similis*.

Type locality: India – Ootacamund or Udagamandalam: Ooty Lake in Nilgiris Mountains at an elevation of 2288 m; 34 ha, mean depth 3 m. It is highly eutrophic lake, heavy polluted from Ootacamund.

Holotype: parthenogenetic female mounted in Canada balsam is deposited in British Museum Nat. Hist. (BMNH) London 1991. 13.

Figs. 23–25. Daphnia carinata s. str. male
Fig. 23 adult; Fig. 24 rostrum and antennule; Fig. 25 postabdomen.

Fig. 26. Unfeathered seta on 4th part of the exopodite of the second thoracic limb of female: 26a – *D. similis* 26b – *D. similoides* 26c – *D. carinata* s. str.

Figs. 16–19, 26a: 1568/Hamuliakovo – Šamorín; Fig. 17a: Šamorín; Figs. 20–22: Ooty Lake; Figs. 23–25: Gloucester.
All figures were made using the projecting screen of the Fluoval microscope WEB Karl-Zeiss, Jena. Scale prepresent 0.01 mm.

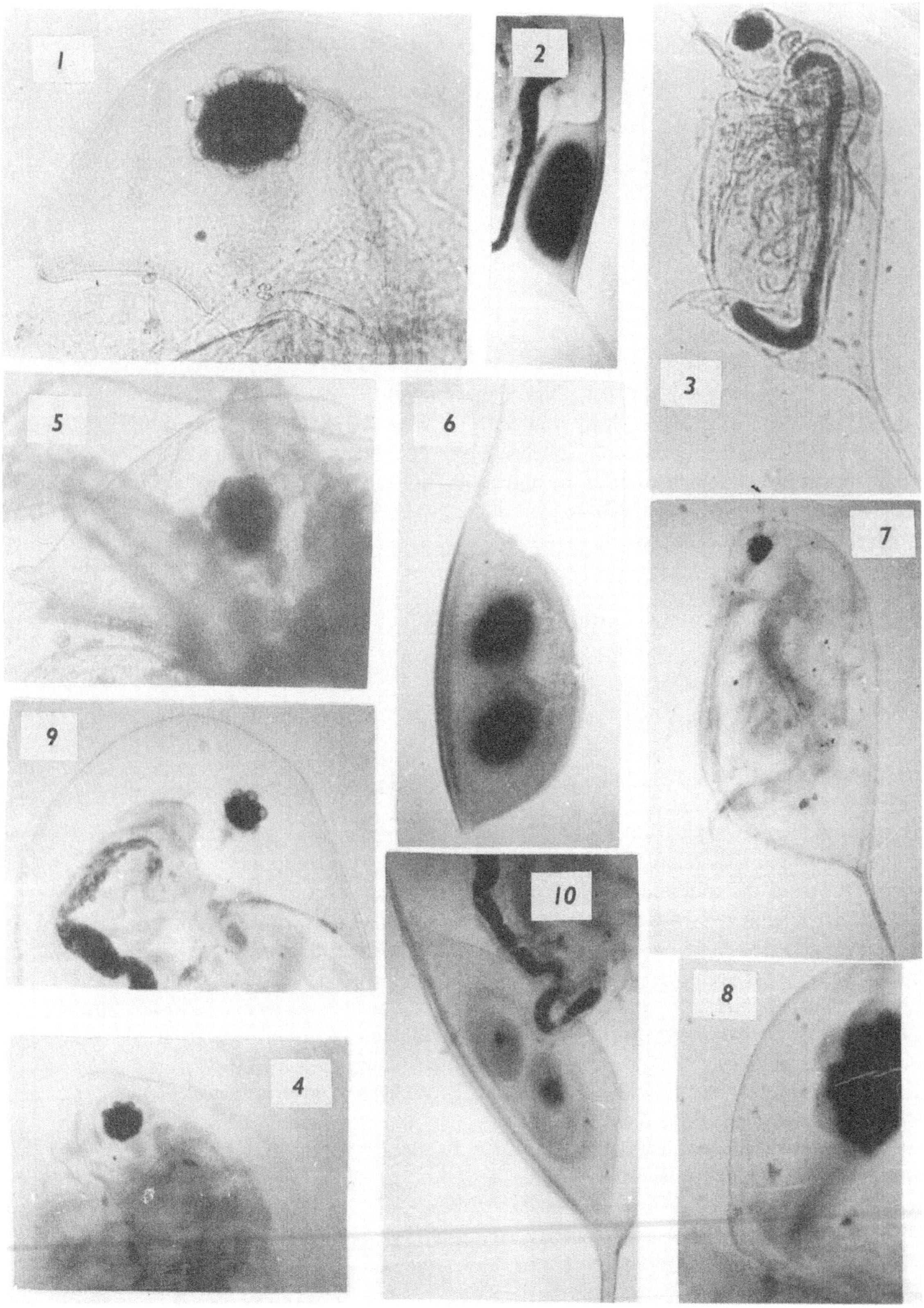

Plate 1

Parts 1–4 *Daphnia similis* Claus: Part 1 frontal part of head of female; Part 2 ephippium; Part 3 male; Part 4 head of male. Parts 5–8 *Daphnia similoides* n. sp.: Part 5 frontal part of head of female; Part 6 ephippium; Part 7 male; Part 8 rostrum of male. Parts 9–10 *Daphnia carinata* s. str.: Part 9 head; Part 10 ephippium.

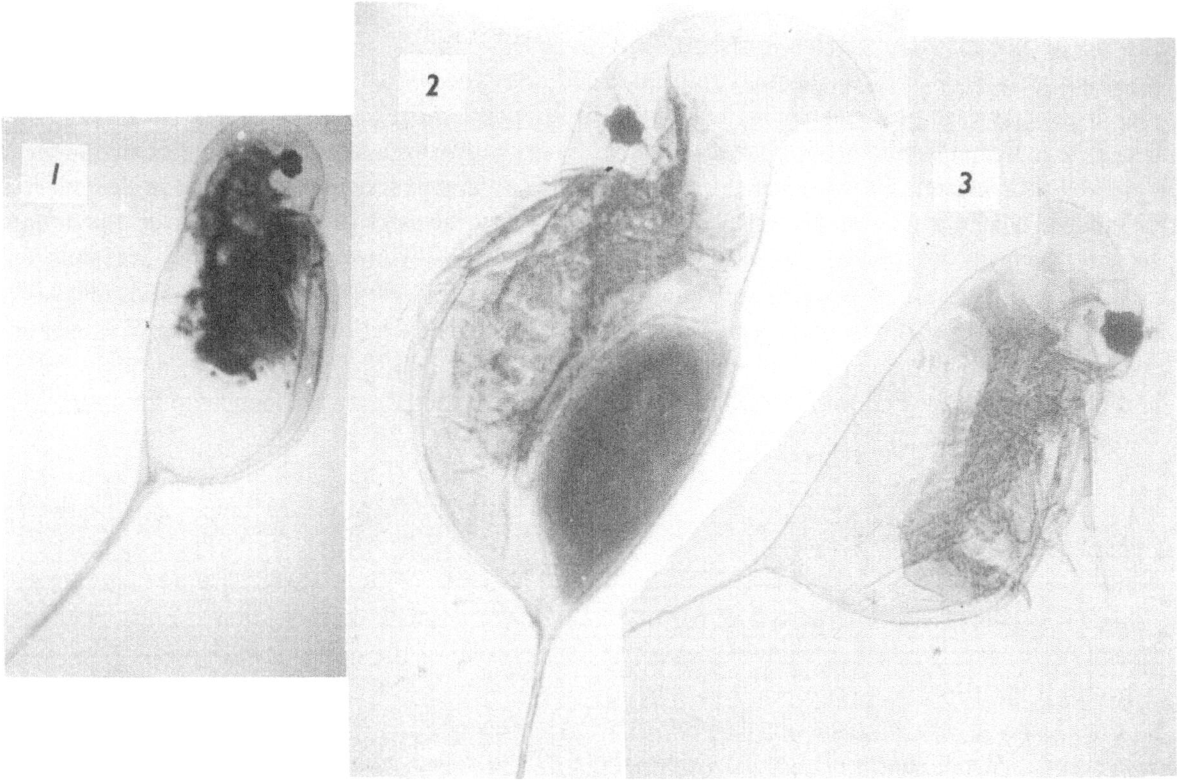

Plate 2
Part 1 *Daphnia carinata* s. str. male. Parts 2–3 *Daphnia* sp. ? China – Wuhan: East Lake ? and Donghu Lake. Part 2 ephippial female; Part 3 male.

Paratypes: series of parthenogenetic females mounted in Canada balsam stained with lignin pink and chlorasol black on slides BMNH 1991. 15–17, 18–22, 25–28 and parthenogenetic females in a tube BMNH 1991. 29–38.

Allotype – male and paratypes sexual and parthenogenetic female stained with lignin pink and chlorasol black in Canada balsam on one slide BMNH 1991. 14. Two slides of dissected females stained with lignin pink and chlorasol black in Canada Balsam BMNH 1991. 23, 24.

A few individuals parthenogenetic females in a tube are in the collection of Dr. Glagolev, USSR, and rest are in my collection.

The type serie is from the author's sample 6 June 1984.

Other material examined:

India: Madurai, pond at the college, 13 June 1984, col. author: one ephippium and one parthenogenetic female; 10 parthenogenetic females labelled *D. carinata*, received from Dr. Murugan, no date, no locality; Calcutta, sewage pond, 10 February 1973, col. Fernando, from the collection of Dr. Hrbáček: juvenile and parthenogenetic females labelled *D. carinata*.

Malaysia: Singapore, Sembawang experimental station, material imported from Japan, 11 September 1973 col. Fernando, from the collection of Dr. Hrbáček: series of parthenogenetic and ephippial females and males labelled *D. carinata*.

Diagnosis
Female: head evenly rounded, rostrum slightly tapering in terminal part, with characteristic

slightly upturned appearance. Antennules short, not rudimentary; sensory setae (aesthetscs) reach the tip of the rostrum. Ocellus present. Postabdomen conical with three slightly compressed marginal denticles in postanal region. Anal region has 7–9 marginal teeth. 3/4 of praeanal region covered with clusters of spicules. Ephippium 'D'-shape. Neonate with small sharply pointed crest.

Male: widely rounded rostrum, no quadrate depression on tip. Supraocular depression not present. Postabdomen conical, with minute knob in postanal region covered with clusters of minute setules.

Description

Female (Figs. 9–11; Plate 1 (parts 5 & 6))
Head (Fig. 9) evenly rounded without crest. The rostrum (Fig. 10–10b) is pointed and slightly tapering, with a characteristic slightly upturned appearance (type material) that resembles *D. obtusa*. Other material examined lacks the upturned rostrum. The compound eye is large; the ocellus is pigmented. The fornices are triangular and have pointed posterior ends.

Antennules (Fig. 10) short, but not rudimentary; terminal part is nearly parallel with the ventral margin of the head. The tips of the aesthetascs do reach the tip of the rostrum.

Valves are curved in ventral and dorsal margin. The dorsal margin has a ridge with a double row of spicules that extends forward to the level of the posterior margin of the compound eye. The posterior third of the ventral margin has marginal spicules, and a row of longer submarginal spicules (Fig. 9 detail), with one submarginal spicula for every two marginal spicules. It agrees with a drawing by Fernando *et al.* (1987). The middle part of the ventral margin has a row of submarginal feathered hairs.

Thoracic limbs. we did not found differences from *D. similis* in general morphology or in the number of setae. The drawing of the first thoracic limb is in Fernando *et al.* (1987).

Postabdomen (Fig. 11) is conical in the anal region. Near the terminal claw are three marginal slightly compressed denticles. The anal region has 7–9 marginal denticles. There are clusters of fine hairs laterally over the bases of the marginal denticles. Three-quarters of the praeanal region is covered with single or clustered spicules and hairs. The first two abdominal processes are the largest, and 3rd and 4th are strongly reduced. The terminal claw is identical to that of *D. similis*.

Ephippium (Fig. 9; Plate 1 (part 6)) has the 'D' shape. It is heavily sclerotinised and has two eggs.

Juveniles (Figs. 9a–9c). The neonates (Fig. 9a) have a small sharply pointed crest, with is not present in the next instars.

Male (Figs. 20–22; Plate 1 (parts 7 & 8))
The male is comparative smaller then the female. The rostrum is widely rounded (Fig. 21; Plate 1 (part 8)) without a depression at the tip. The antennules are long being about 1/3 of the total head height. The flagellum is about half as long as the anthennule. There are 9 aesthetascs and one single seta between the aesthetascs and the flagellum. The surface of the antennules is covered with minute hairs. The general shape of the valves (Fig. 20; Plate 1 (part 7)) is the same as in *D. similis*. We did not find differences between the thoracic limbs of *D. similoides* and *D. similis*. The first thoracic limb of *D. similoides* is as in Fernando *et al.* (1987). The postabdomen (Fig. 22) is conical in anal region. A small hump present near the terminal claw is covered with clusters of minute setules. The terminal claw is identical with that of *D. similis*. The abdominal processes are strong reduced.

Distribution

Despite extensive information about the distribution of *D. similis* (*psittacea*) or *D. carinata* from tropical Asia, it was very difficult to define the zoogeographical distribution of the new species, *D. similoides*.

We restricted ourselves to references with drawings that evidently refer to *D. similoides* based on: 1/ adult females with upturned and tapering rostrum (were determined as *D. psittacea* or *D. similis* the synonymy see); 2/ drawings of juvenile female with pointed crest (all were determined as *D. carinata*, the see synonymy).

Based on this evidence the distribution of D. similoides extends from Baluchistan (Brehm, 1914) through all of India (Arora, 1931; Brehm, 1950; Brehm & Wolterek, 1939; Murugan & Job, 1981; Murugan & Venkataraman, 1977; Nayar, 1971), Sumatra (Brehm, 1933) Malaysia (Idris, 1983) China (Mashiko, 1953; Spandl, 1925; Uéno, 1927, 1937, 1939, 1933).

The distribution of *D. similoides* in Central China is not clear. There is at least one other undescribed species (Plate 2 (parts 2 & 3)) from Wuhan territory (Donghu Lake and East Lake – no date, no characteristic of locality, only few individuals). In Central China *D. similis* is also present.

Table 2. Morphological differences between *Daphnia similis* Claus, *D. similoides* n. sp., and *D. carinata* s. str.

Character	*D. similis*	*D. similoides*	*D. carinata*
Female			
Rostrum	triangular, short	tapering in distalmost part	triangular, long
Aesthetascs	Reach tip of rostrum	Reach tip of rostrum	never reach tip of rostrum
Ocellus	pigmented	pigmented	inconspicuous
Postabdomen: compressed postanal denticles	no	3–2	no
Postabdomen: anal denticles	9–12	7–9	10–14
Ephippium (shape)	"O"	"D"	"double lens"
Neonates: crest on head	oblique	pointed	oblique
Male			
Supraocullar degression	present	absent	absent
Quadrate depresion under rostrum	present	absent	absent
Antennal length to heigh of head	1/3	1/3	1/4
Postabdomen: postanal knob	present	present	absent
Postabdomen: Postanal denticles	2–3	absent, cluster of spicules	absent
Postabdomen: anal denticles	5–7	5–7	8–10
Length of shell spine to body length	1/4–1/2	1/4–1/2	1/2–1

***Daphnia carinata* King, 1853**

Material examined:

Australia: Gloucester, Relfs irrigation dam, 15 July 1968, col. Fernando (large series of juvenile, parthenogenetic, and ephippial females and males); Kentucky, Creck dam - limnetic, via Uralla, N.S.W., 16 August 1967, col. Fernando (parthenogenetic and ephippial females and males). These samples are from the collection of Dr. Hrbáček,

Diagnosis (Figs. 12–15, 23–25; Plate 1 (parts 9 & 10); Plate 2 (part 1))

Female: head is oblique with an elongated pointed rostrum. Antennules reduced with aesthetascs never reaching tip of rostrum. Ocellus inconspicuous. Postabdomen resembles that one of *D. similis*, but has 10–14 marginal teeth. Ephippium resembles a 'double lens'.

Male: head semilunar, with a blunt widely rounded rostrum. Ocellus inconspicuous. Antennular body small (up to 1/4 of height of head). Postabdomen without postanal knob and with 8–10 marginal denticles. Long shell spine.

Discussion

Daphnia similis Claus and *D. similoides* n. sp. are similar in many characters (length, general shape, postabdomen).

Some morphological differences of the head and the shape and length of carapace spine (mainly in juveniles) were found in individuals from Slovakia, which resembles *D. similis* var. *exilis* Uéno (Figs. 2–2c). We assume that both characters are probably the consequence of the specific environmental conditions.

The differences in the terminal portion of the rostrum, the shape of the ephippium, and the density of denticles on the postanal part of the postabdomen is demonstrated by comparison of populations from Europe (Figs. 1, 4–5) and from tropical Asia (Figs. 9–11).

The sharply pointed crest (Fig. 9a) in neonates of *D. similoides* was probably the decisive character for considering that the mentioned populations in tropical Asia were *D. carinata*. For example, Ooty Lake by Brehm & Woltereck (1939).

A sharply pointed crest of the neonate is not present either in *D. similis* (Fig. 1a) or in *D. carinata* (Fig. 12a) (see Sars, 1914).

The males of *D. similis* and *D. similoides* are characterized by the general shape of the head and postabdomen. The adult males of *D. similis* has a quadrate depression in the rostral part just before the base of the antennules (Fig. 17) and a supraocular depression (Fig. 16). The supraocular depression is not present in juveniles (Fig. 16a), and the depression under the rostrum is weakly developed (Fig. 17a). The males of *D. similoides* (adult and juveniles) have neither a quadrate depression (Fig. 21) on the rostrum nor a supraocular depression (Fig. 20).

One of the male of *D. similoides* from Ooty Lake has a double flagellum on the one antennule (Fig. 21b).

Males of *D. similis* and *D. similoides* both have a small knob on the postanal margin of the postabdomen. The knob of *D. similis* usually has 2 small denticles (Fig. 18), while *D. similoides* has clusters of small spicules (Fig. 23).

The specificity of *D. similis* – *D. similoides* and *D. carinata* s. str. is best characterised by males (Plate 1 (parts 3, 4, 7 & 8)); Plate 2 (part 1)). The postabdomen of the female of *D. carinata* (Fig. 14) resembles that of *D. similis* and *D. similoides*, but it has more marginal denticles.

The ephippium of *D. similoides* (Fig. 9; Plate 1 (part 6)) is between that of *D. similis* (Fig. 1a; Plate 1 (part 2)) and *D. carinata* s. str. (Fig. 15; Plate 1 (part 10)).

All three species have an identical general shape of thoracic limbs, only the unfeathered seta on 4th part of the exopodite of the second thoracic limb is different (Fig. 26). *D. similis* has dense bristles on the one margin (Fig. 26a). *D. similoides* has a row of thin bristles on the one margin

(Fig. 26b). *D. carinata* has dense and very short bristles on the one margin (Fig. 26c).

In the respect of the general morphology of thoracic limbs, it is possible to speak of a *D. similis-carinata* group. But on the other hand, other characters (Table 2) separate this group as two different subgroups.

D. similis subgroup: antennule aesthetascs of females reach the tip of the rostrum; contrasting ocellus; long antennular body of males (characteristic for littoral species); small knob on the postanal region of the male postabdomen. *D. similis* Claus, *D. similoides* n. sp., *Daphnia sp.* from region of Wuhan in Central China.

D. carinata subgroup: antennule aesthetascs of females never reach the tip of the rostrum; ocellus inconspicuous; no knob on the postanal region of the male postabdomen; short antennular body of males (characteristic for limnetic species).

Acknowledgement

I thank Dr. J. Hrbáček (Prague) for the most of samples from tropical Asia and Australia, for most of references from his library, and also for helpful suggestions.

References

Arora, G. L., 1931. Entomostraca of Lahore. Bull. Dept. Zool., Panjab. Univ. 1: 62–100.

Bär, G., 1924. Über Cladoceren von der Insel Ceylon. Jenaische Z. Naturwis., 60, N.F., 53: 83–126.

Benzie, J. A. H., 1987. The biogeography of Australian *Daphnia*: clues of an ancient (>70 m.y.) origin for the genus. Hydrobiologia 145: 51–65.

Brehm, V., 1914. Cladoceren und Ostracoden aus Balutschistan. Zool. Anz. 43: 511–515.

Brehm, V., 1933. Die Cladoceren der Deutschen Limnologischen Sunda Expedition. Arch. Hydrobiol. Supp. 2: 631–771.

Brehm, V., 1950. Contribution to the freshwater fauna of India. Part II. Rec. Ind. Mus. 48: 9–28.

Brehm, V. & R. Woltereck, 1939. Die Daphniidae der Yale-North-indische-Expedition. Int. Rev. ges. Hydrobiol. 39: 1–19.

Brooks, J. L., 1957. The systematics of North American *Daphnia*. Mem. Conn. Acad. Arts Sci. 13: 180 pp.

Claus, C., 1876. Zur Kenntniss der Organisation und des feiner Baues der Daphniiden und verwandten Cladoceren. Zeitschr. wiss. Zool. 27: 362–402.

Daday, E., 1911. Eine neue Cladoceren-Art aus Ostindien. Allatt. Közlem. 10: 63–68.

Fernando, C. H., 1980. The specific and size composition of tropical freshwater zooplankton with special references to the oriental region South East Asia. Int. Revue ges. Hydrobiol. 65: 411–426.

Fernando, C. H., J. C. Paggi & R. Rajapaksa, 1987. *Daphnia* in tropical lowlands. In: R. H. Peters and R. de Bernardi /eds./ 'Daphnia'. Mem. Ist. Ital. Idrobiol. 45: 107–141.

Flössner, D., 1986. Beitrag zur Kenntnis der Branchiopoden und Copepodenfauna der Mongolei. Mitt. Zool. Mus. Berl. 62: 3–40.

Freyer, G., 1985. The ecology and distribution of the genus *Daphnia* (Crustacea: Cladocera) in restricted area: the pattern in Yorkshire. J. nat. Hist. 19: 97–128.

Hebert, P. D. N., 1977. A revision of the taxonomy of the genus *Daphnia* (Crustacea: Daphnia) in south-east Australia. Aust. J. Zool. 25: 371–398.

Hrbáček, J., V. Kořínek & D. G. Frey, 1978. Cladocera. In: J. Illies /ed./. Limnofauna Europaea. pp. 189–195. 2nd Edition. Gustav Fischer Verlag: Stuttgart, New York.

Hudec, I., 1989. Distribution and biology of species of the genus *Daphnia* subgenus *Ctenodaphnia* (Cladocera, Daphniidae) in Slovakia. Biológia, Bratislava 44: 489–498. /in Slovak/

Idris, B. A. G., 1983. Freshwater zooplankton of Malaysia. (Crustacea, Cladocera). Penerbit University Pertanian Malaysia, Serdang, Selangor. vii, 153 pp.

Johnson, D. S., 1952. The British species of the genus *Daphnia* (Crustacea, Cladocera). Proc. zool. soc. London, 122: 435–461.

Manuylova, E. F., 1964. Vetvistousye rachki (Cladocera) fauny SSSR. Opred. po fauna SSSR, izd. Zool. Inst. Akad. Nauk SSSR, 88. Nauka, Moskva, Leningrad: 326 pp. /in Russian/

Mashiko, K., 1953. Cladocera and Rotatoria of central China. Sci. Rep. Kanazawa Univ. 2: 49–73.

Murugan, N. & S. V. Job, 1981. Contribution of the biology of Cladocera: III. seasonality, distribution and association of Cladocera. Salsearch 3: 69–82.

Murugan, N. & K. Venkataraman, 1977. Study of the in vivo development of the parthenogenetic eggs of *Daphnia carinata* King (Cladocera, Daphniidae). Hydrobiologia 52: 129–134.

Nayar, K. G., 1971. Cladocera of Rajastan. Hydrobiologia 52: 509–519.

Richard, J., 1896. Revision des Cladocères. Deuxième partie. Ann. Sci. Nat. zool. 8: 187–363.

Sars, G. O., 1903a. On the crustacean fauna of Central Asia. Ann. Mus. Zool. Acad. Imp. Sc. St. Pétersbourg 8: 157–187.

Sars, G. O., 1903b. On the crustacean fauna of Central Asia. Appendix: Local fauna of Central Asia. Ann. Mus. Zool. Acad. Imp. Sc. St. Pétersbourg 8: 233–264.

Sars, G. O., 1914. *Daphnia carinata* King and its remarkable varietes. Arch. Math. Naturv. B. 34: 1–16.

Scourfield, D. J., 1903. Synopsis of known species of British freshwater Entomostraca. Pt. 1. Cladocera. J. Quekett. Micr. Club 2/8: 431–454.

Scourfield, D. J. & J. P. Harding, 1941. A key to the British species of freshwater Cladocera with notes on their ecology. Sci. Publ. Freshw. Biol. Ass. 5: 50 pp.

Spandl, H., 1925. Hydrobiologische Beiträge aus China nach Samlungen Dr. H. Weigolds. Cladocera. Int. Revue ges. Hydrobiol. 13: 185–196.

Uéno, M., 1927. The freshwater Branchiopoda of Japan. I. Mem. Coll. Sci. Kyoto Imp. Univ. B. 2: 259–311.

Uéno, M., 1935. Cladocera from Shanghai. Trans. Nat. Hist. Soc. Formosa 25: 212–215.

Uéno, M., 1937. Cladocera of Manchoukuo. Int. Revue ges. Hydrobiol. 35: 199–216.

Uéno, M., 1939. Manchurian freshwater Cladocera. Ann. Zool. Jap.: 219–231.

Uéno, M., 1943. Cladocera of the Yantze Delta. Report on the limnological survey of Central China. 23. Bull. Shanghai Sci. Res. Inst. 14: 399–418.

Uéno, M., 1966. Crustacea, Cladocera, Copepoda and Amphipoda collected by the Kyoto University Pamir-Hindukush Expedition 1960. Results of the Kyoto University Sciencific Expedition to the Karakoram and Hindukush 8: 287–298.

Uéno, M., 1967. On the occurrence of *Daphnia similis* in Japan. Jap. J. Zool. 15: 249–253.

Wolski, T., 1932. Studien über die Cladoceren Gattung *Daphnia*. 1. *Daphnia psittacea* Baird aus Jugoslavien. Ann. Mus. zool. Pol. 9: 267–285.

Hydrobiologia **225**: 23–35, 1991.
V. Kořínek & D. G. Frey (eds), Biology of Cladocera.

The genus *Diaphanosoma* (Ctenopoda: Sididae) in Spain

Damia Jaume
Departament d'Ecologia, Facultat de Biologia, Universitat de Barcelona, Diagonal 645, 08028 Barcelona, Spain

Key words: cladocera, *Diaphanosoma*, biogeography, taxonomy, ecology, Spain

Abstract

During a 1987–88 extensive limnological study of Spanish reservoirs, euplanktonic cladocerans were sampled all around the country. Two species of *Diaphanosoma* Fischer, 1850 were recorded, viz *D. mongolianum* Uéno, 1938 and *D. brachyurum* (Liévin, 1848). Morphological descriptions of both species are presented, especially of the thoracic limbs, together with notes on their ecology. Both species present a clear allopatric distribution, *D. brachyurum* inhabiting waters with a low Total Dissolved Salt (TDS) content, which equilibrated relative ionic composition, characteristic of a narrow strip in the northern part of the Iberian Peninsula; on the contrary, *D. mongolianum* inhabits waters with high TDS values and high content of sulphate or chloride, characteristic of the central and southern regions.

Introduction

The euplanktonic crustacean fauna of the Iberian Peninsula is very scanty, mainly as a result of the scarcity of true limnetic habitats; there are only two important natural lakes (Sanabria, 108 hm^3; Banyoles, 17 hm^3). Nevertheless, more than 700 reservoirs have been constructed during 1940–70 (Dirección General de Obras Hidráulicas, 1973), and an up-to-date census would probably yield and estimate of about 900. This has given the chance for colonization, and a checklist of the crustaceans of the Spanish reservoirs (Armengol, 1978), which can be considered the first extensive survey of true limnetic habitats carried out in Spain, reported 28 euplanktonic species – 15 cladocerans and 13 copepods. This represent only 2/3 of the expected number of species found in European lakes (Margalef, 1983), an impoverishment which is more a result of the youth and environmental fluctuations of reservoirs than of a deficient habitat prospection.

The existence of a very rich steppic cladoceran fauna, with a high degree of endemism (Alonso, 1985) has switched the attention of local taxonomists far from such euplanktonic groups as *Diaphanosoma* Fischer, 1850, *Daphnia* s. gen. *Daphnia* (O.F. Müller, 1776) Johnson, 1952, and *Bosmina* Baird, 1845. These genera are composed of species with a high phenogenetic plasticity, which require accurate morphological characterization and delimitation of variability in order to elucidate their taxonomic status.

The aim of this paper is to present accurate morphological descriptions and to study the variability of the *Diaphanosoma* species inhabiting the Iberian Peninsula, in order to relate them to some extensive works on the genus (Kořínek, 1987; Korovchinsky, 1987). Some previous references concerning the genus in Spain are limited to the enumeration of localities from which the genus has been reported (Margalef, 1953; Miracle, 1976; Armengol, 1978; Toja, 1981). The pattern used has been to refer all the representatives of the genus in Spain to *D. brachyurum* (Liévin, 1948) (Armengol, 1978; Toja, 1981). Nevertheless, Alonso (1985) distinguished between *D. brachyurum* and *D. birgei* Kořinek, 1981 in his work on Cladocera of Spain. Korovchinsky (1987) studied two populations from Spain and assigned them to *D. mongolianum* Uéno, 1938. Our own collections have confirmed the presence of two *Diaphanosoma* species in the Iberian Peninsula, viz *D. brachyurum* and one representative of the *D. mongolianum* group, with an allopatric distribution determined mainly by ecological factors.

Materials and methods

Material were collected mainly during an extensive limnological study of the Spanish reservoirs carried out during 1987–88. 104 reservoirs were visited two times during this period, corresponding to the maximum stratification and mixture of the water column. Samples from small natural lakes and ponds deposited in Dr M. Alonso's collection (Barcelona, Spain) were also used. Samples were taken with a Nytal net of 200 μm mesh size and 25 cm diameter, and fixed with 4% formalin. Camera lucida drawings were made with a BH-T Olympus microscope. Measurements were made by means of an eyepiece micrometer, the scale factor of which was determined by a stage micrometer, and following Korovchinsky (1978). All measurements and observations were done on adult parthenogenetic females and adult males.

Results

Diaphanosoma mongolianum Uéno, 1938 emend. Korovchinsky, 1987

Synonymy:
Diaphanosoma brachyurum (non Liévin, 1848) Armengol, 1978; Toja, 1981.
Diaphanosoma birgei (non Kořinek, 1981) Alonso, 1985.

Material examined:
Reservoirs (code numbers as in Armengol (1978: 4) and in Fig. 6): 26 (2.8.88), 27 (13.12.87; 1.8.88), 33 (18.12.87; 15.7.88), 34 (19.12.87; 16.7.88), 35 (16.7.88), 36 (14.6.88), 37 (14.6.88), 39 (15.6.88), 44 (21.6.88), 45 (21.6.88), 46 (14.2.88; 22.6.88), 48 (16.2.88); 23.6.88), 49 (23.6.88), 50 (13.2.88), 53 (11.2.88; 9.6.88), 55 (24.6.88), 56 (13.6.88), 57 (2.1.88), 61 (1.9.88), 63 (1.9.88), 64 (25.5.88; 3.9.88), 65 (3.9.88), 66 (27.2.88; 4.9.88), 67 (29.11.88; 6.9.88), 68 (29.2.88; 6.9.88), 70 (14.9.88), 71 (16.9.88), 72 (15.9.88), 73 (11.3.88; 17.9.88), 75 (13.3.88; 25.6.88), 76 (25.6.88), 77 (19.9.88), 82 (1.3.88; 8.9.88), 83 (3.3.88; 3.3.88), 84 (10.9.88), 85 (10.9.88), 86 (11.9.88), 87 (12.9.88), 88 (6.3.88; 12.9.88), 89 (6.3.88; 13.9.88), 90 (28.2.88; 5.9.88), 96 (24.11.87), 98 (1.8.88), 100 (17.7.88), 101 (22.6.88), 102 (14.9.88), 103 (9.3.88; 15.9.88), 104 (3.2.88; 12.6.88).

Lakes and lagoons (materials deposited in the collection of Dr M. Alonso, Barcelona, Spain): Laguna Chica de Villafranca (La Mancha, 26.7.84).

Description

Parthenogenetic female

Head large (38.32% in terms of body length; see Table 1) with a more or less protruded dorsal part, varying in appearance from conical to more or less roundish rectangular (Fig. 1a, b). Ventral side of head with depression (Fig. 1a). Antennules with flagellum twice as long as basal segment; one individual bearing an abnormal thin seta halfway on the margin (Fig. 1e). Labrum reaching anterior margin of valves (Fig. 1b, d). First maxilla with 6 curved biarticulated setae, bearing bristles only on their distal segment, and one triangular spine fringed with a row of bristles (Fig. 1f). Swimming antennae massive (84% in terms of body length), never reaching the posterior margin of the valves (Fig. 1a, 2a), its basipodite longer than the biarticulated branch, with a long seta inserted on the inner side of its flexible basis, and distally with one small subapical feathered seta on inner side

Table 1. Biometrical data (lengths, mm) of 5 spanish populations of *Diaphanosoma mongolianum* Uéno, 1938 (see also Fig. 6 for place localities).

Locality	45 (21.6.88)	73 (17.9.88)	100 (17.7.88)	103 (15.9.88)	MANCHA (26.7.84)
Body length (x ± 1 S.E.) (variation)	1.07 ± 0.18 (0.95–1.23) (*n* = 37)	1.11 ± 0.02 (1.00–1.23) (*n* = 30)	1.19 ± 0.02 (0.98–1.40) (*n* = 49)	1.08 ± 0.12 (0.78–1.23) (*n* = 49)	0.88 ± 0.04 (0.80–1.02) (*n* = 9)
Head length: body length (%)	38.15 ± 0.61 (*n* = 37)	38.17 ± 0.48 (*n* = 31)	38.59 ± 0.30 (*n* = 49)	37.72 ± 0.27 (*n* = 49)	38.97 ± 1.32 (*n* = 9)
Head height: body length (%)	22.73 ± 0.84 (*n* = 10)	23.97 ± 1.05 (*n* = 11)	24.12 ± 0.51 (*n* = 28)	22.65 ± 0.41 (*n* = 29)	22.44 ± 0.84 (*n* = 9)
Shell denticles	21–38 (*n* = 37)	30–39 (*n* = 30)	19–39 (*n* = 30)	21–34 (*n* = 30)	18–26 (*n* = 9)

and one spine on outer side (Fig. 2a, b). Distal part of proximal segment of biarticulated antennal branch with one spine and one rounded prominence (Fig. 2c); distal part of distal segment with one spine, 2 rounded and one pointed prominences (Fig. 2d). One subapical spine on distal part of the intermediate segment of triarticulated branch; 2 spines, one of them very small, on distal segment (Fig. 2e). Formula of setae: 4-8/0-1-4. Carapace convex in the middle part of dorsal side, with a laterally protruding brood chamber, but narrower in its posterior part (Fig. 1a, c). Infero-posterior margin of valves merging smoothly with the flap-like ventral margin of valves (Fig. 1a, g); up to 39 denticles implanted on the infero-posterior margin of valves, in groups of 3–4 denticles with one thin seta implanted between adjacent groups (Fig. 1g); this row of denticles extends somewhat into the posterior margin of the valves. 12(-14) long feathered setae implanted on the edge of the ventral flap, the 3 proximal ones smaller and implanted on the inner side of the valves; the distalmost seta is shortened also, and implanted in the fist group of denticles of the infero-posterior margin of valves (Fig. 1g). There are 7–8 small setules between successive long setae (Fig. 1h). Posterior margin of valves with a row of marginal cilia on inner side, the ventralmost ones thicker; one small and thick spine implanted on the inner side of the supra-posterior margin of the valves (Fig. 1i).

Postabdomen with 7–8 groups of small denticles and two rows of small setules above them on the lateral side. Proximal basal spine of claw smaller than the other two (Fig. 1j). Inner ventral side of claw with 5–8 regularly distributed flat denticles; 2 rows of spinules on its outer lateral side, one of them implanted over basal claw spines (Fig. 1j).

First thoracic limb (Fig. 3a) with exopodite narrowed distally, and bearing 5 apical and 5 laterally implanted feathered setae, the proximal one with its basis thicker than the rest, and directed proximally. Endopodite of the same length as exopodite and subdivided into 4 not very distinct segments; the subterminal one bears a row of biarticulate filtering setae and one shorter, nonarticulate feathered stout seta like those of the exopodite; outer basal side of endopodite with a lobe having as many as 13 unfeathered, spine-like setae (Fig. 3b). Margin of gnathobasic prominence (Fig. 3b) with one long and stout biarticulate seta, one nonsegmented feathered seta, not surpassing the proximal segment of the former, and one biarticulated seta of the same size as the last one described; on the outside of these

Fig. 3. Diaphanosoma mongolianum Uéno, 1938 from Iznájar reservoir (73). Thoracic limbs of female: *a*, first limb; *b*, gnathobasic prominence of first limb; *c*, second limb; *d*, detail of gnathobasic prominence of second limb; *e*, detail of apical part of endopodite of second limb; *f*, third limb; *g*, gnathobasic prominence of third limb; *h*, detail of apical part of endopodite of third limb; *l*, fourth limb; *j*, fifth limb; *k*, sixth limb (rows of filtering setae not represented in d, e, g and h).

there is a row of 8 small biarticulate setae, only feathered on their distal part, the first one slightly thicker and longer than the rest.

Second thoracic limb (Fig. 3c) with exopodite expanded distally; 7 terminal setae, roughly of the same size, and 4 lateral setae of the same size also, shorter than the former; the proximal lateral seta has its basis thicker than the rest of the setae, and is directed proximally. Endopodite shorter than exopodite; distal segment with a row of biarticulate filtering setae and one thick nonarticulate feathered seta (Fig. 3c), roughly of the same length as the biarticulate ones; second segment with a row of biarticulate setae and one feathered nonarticulate seta (Fig. 3e), thinner and shorter than that of the distal segment. Margin of gnathobasic prominence bearing a long and thick biarticulate seta (Fig. 3d) and a row of up to 30 biarticulate filtering setae (Fig. 3c).

Third thoracic limb (Fig. 3f) with terminal part of exopodite more broadly expanded than that of the second thoracic limb, and bearing 7 long, feathered, unsegmented setae, all of them roughly the same size; 4 lateral setae, similar to those of the distal part, but the proximal one with a thicker basis, and directed proximally. Endopodite shorter than exopodite, with structure resembling that of the second thoracic limb (Fig. 3h). Gnathobasic prominence bearing a long biarticulate seta resembling that of the second thoracic limb, a small unsegmented seta shorter than the basal segment of the biarticulate one, bearing some denticles on its distal part (Fig. 3g), and a row of biarticulate filtering setae (Fig. 3f).

Fourth thoracic limb (Fig. 3l) resembling the third one, but smaller.

Fifth thoracic limb (Fig. 3j) with general structure resembling those of the third and fourth ones, but smaller and bearing 6 setae on the terminal part of the exopodite and 3 on its lateral side.

Sixth thoracic limb (Fig. 3k) very small. Exopodite with 5 setae on terminal plate and a lateral seta. Endopodite reduced to an unsegmented, broad plate with 7 nonarticulate feathered setae of the same size, and one small prominence inserted between the third and second proximal setae. Gnathobasic prominence with 2 nonarticulate feathered setae, 2 prominences, and one thick lateral feathered seta.

Gamogenetic female
Identical to the parthenogenetic female. One resting egg, brownish, oval, with a smooth cuticle.

Male
Diagnostic features as in female. Antennae surpassing the posterior margin of valves. Antennules long, with flagellum covered by minute setules, and olfactory setae inserted halfway on basis (Fig. 1m). Hook of first thoracic limb endopodite with minute setules along its distal half (Fig. 1k). Tube-like copulatory appendage (penis) narrower towards end (Fig. 1l).

Remarks

Korovchinsky (1987) has presented an emended redescription of two *Diaphanosoma* species, viz *D. mongolianum* Uéno, 1938 and *D. lacustris* Kořinek, 1981. Both are essentially Palearctic species, very similar in morphology, and constitute the so-called 'mongolian' group. The diagnostic character for their correct assignment are mainly biometrical (length and height of head in terms of body length, number of denticles on the infero-posterior margin of valves), and also morphological (shape of postero-dorsal margin of valves). The Spanish populations can be assigned to *D. mongolianum* s.st., in accordance with the morphological and biometrical data recorded (Table 1). Nevertheless, some peculiarities have been detected: The ratio head length: body length falls in the lower range recorded for *D. mongolianum* (Table 1) (Korovchinsky, 1987), as is also evident comparing our drawings (Fig. 1a) with those presented by Korovchinsky, loc.cit.; nevertheless, Korovchinsky had noted this characteristic in one of the two populations he studied from Spain, namely, the one roughly assigned to the province La Mancha. A sample from the same locality, i.e. Laguna Chica de Villafranca, a temporary athalassohaline pool, was kindly offered to me by Dr M. Alonso. Only

9 well-preserved, mature parthenogenetic females were found, the biometrical data of which appear in Table 1. Head length : body length mean values of the latter population and those from reservoirs 45, 73, 100 and 103 (Table 1) were subjected to One Factor ANOVA in order to test the biometrical homogeneity of Spanish populations; no significative differences were found (F = 1.571, N.S). Hence, Spanish populations are characterized by a small head, a feature also reported from the population inhabiting Lake Issyk-Kul (Kirghizia, USSR) (Korovchinsky, 1987).

Some meristic characters have presented a wider range of variation than that previously reported. The number of denticles of the infero-posterior margin of valves can be up to 39 (maximum reported by Korovchinsky (1987) was 36). The number of feathered setae along the ventral margin of the valves can be up to 14 (up to 12 in Korovchinsky (1987), who considers this feature characteristic of the 'mongolian' group). Both these results are probably a consequence of big individuals involved in our measurements. The thin lateral seta of the female's antennule has not been reported in previous works on the 'mongolian' group (Korínek, 1981; 1987; Korovchinsky, 1987). We have observed a small, hyaline, spine-like prominence on the distal part of distal segment of the triarticulated antennal branch (Fig. 2e). The natatory antennae do not surpassed the posterior margin of the valves in all the females examined; in males, on the contrary, the length of the antennae has always surpassed the posterior margin of valves with the exception of males from La Tranquera reservoir (33). The first maxilla has 6 curved setae (only 5 reported by Kořinek, 1987).

Spanish reservoirs' populations of *D. mongolianum* differ from others studied elsewhere (Korovchinsky, 1987) by producing males in early winter (males are known from 27(13.12.87), 33(18.12.87) and 96(24.11.87)) instead than in summer; this is perhaps related to the midl autumn conditions. In Laguna Chica de Villafranca, on the contrary, males are known from April 1985 (Korovchinsky, 1987) and 26.7.84; Korovchinsky (in litt.) considers this population as peculiar because of its small head, the nature of the habitat (temporary athalassohaline pool), and the occurrence of gamogenesis in spring. No differences, either morphological or biometrical (Table 1), have been detected in comparison with other Spanish populations. Moreover, the variability in time of gamogenesis perhaps is related to the unpredictability of the wet period.

***Diaphanosoma brachyurum* (Liévin, 1848)**

Material examined:
Reservoirs (code numbers as in Armengol (1978 : 4) and in Fig. 6):
1 (27.11.87), 6 (9.7.88), 10 (4.12.87; 22.7.88), 11 (4.12.87; 21.7.88), 12 (4.12.87; 23.7.88), 13 (21.7.88), 14 (23.7.88), 16 (7.12.87; 25.7.88), 17 (9.12.87), 18 (8.12.87; 25.7.88), 19 (8.12.87; 26.7.88), 20 (10.12.87; 28.7.88), 21 (9.12.87; 27.7.88), 22 (11.12.87; 30.7.88), 24 (12.12.87; 31.7.88), 32 (15.7.88), 95 (25.11.87), 97 (28.7.88), 99 (30.7.88).

Lakes and lagoons (materials deposited in the collection of Dr M. Alonso, Barcelona, Spain): Laguna de Navalhornos, Laguna de Matisalvador (Cantalejo, Segovia, July 1980); Laguna de Guardafuegos (Marzagón, Huelva, 15.4.79); Lago de Isoba (León, July 1982); Lago Ubales (Asturias, 18.9.89); Llac de Banyoles (Girona, July 1983).

Description

Parthenogenetic female

General profile more slender than that of *D. mongolianum*; head without protruded dorsal part (Figs. 4a, b). Antennae never reaching posterior margin of valves, and bearing the same setation as in *D. mongolianum*; there is no spine on the distal part of proximal segment of the biarticulated antennal branch (Figs. 4a, b).

Ventral margin of valves with a row of long, feathered setae with minute setules intercalated between them (Fig. 4c). Infero-posterior margin

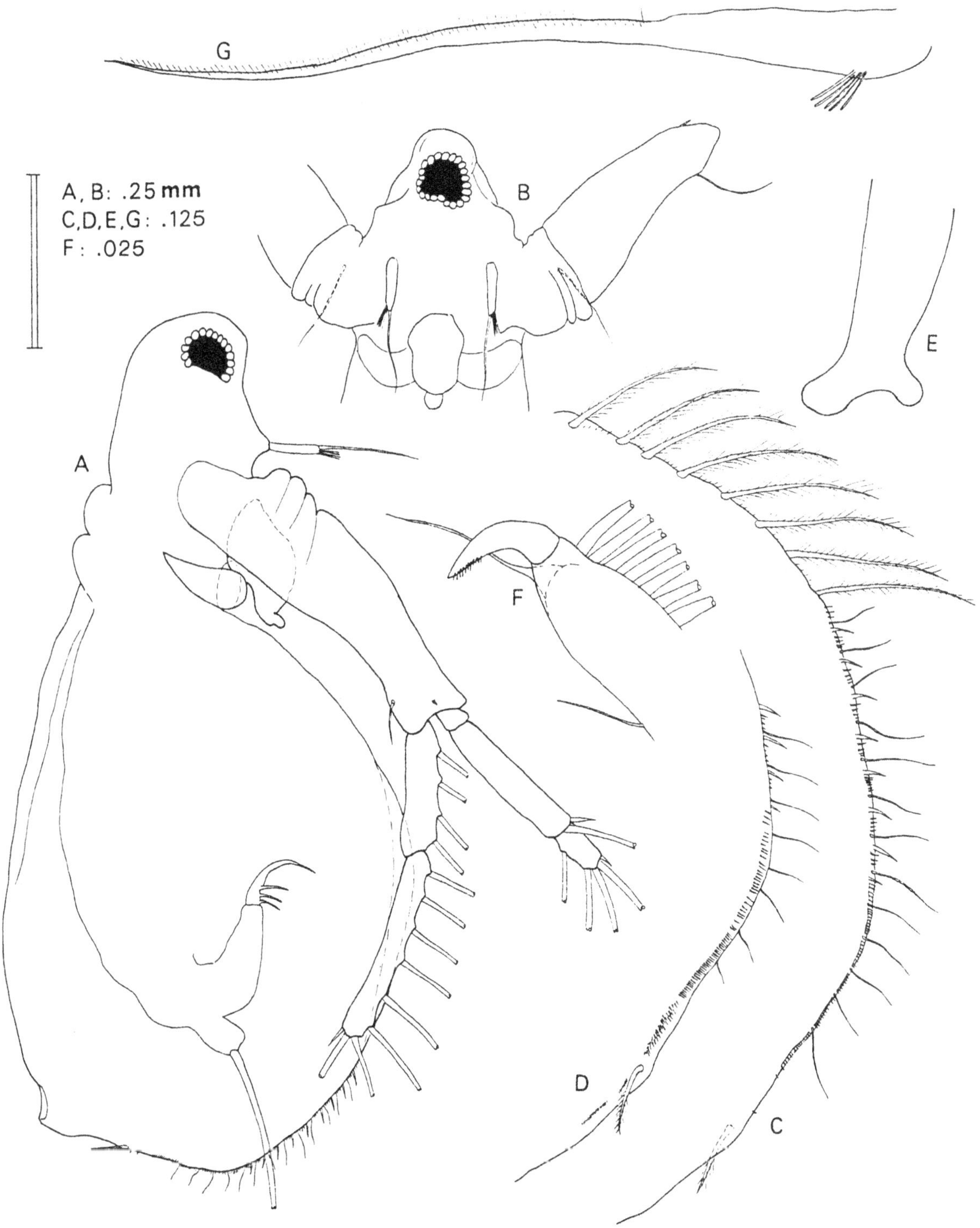

Fig. 4. Diaphanosoma brachyurum (Liévin, 1848) female from Belesar reservoir (18): *a*, lateral view; *b*, frontal view of head; *c*, posterior margin of valves, outer side; *d*, posterior margin of valves, inner side. Male from las Conchas reservoir (20): *e*, distal part of penis; *f*, detail of distal part of first limb endopodite; *g*, antennula.

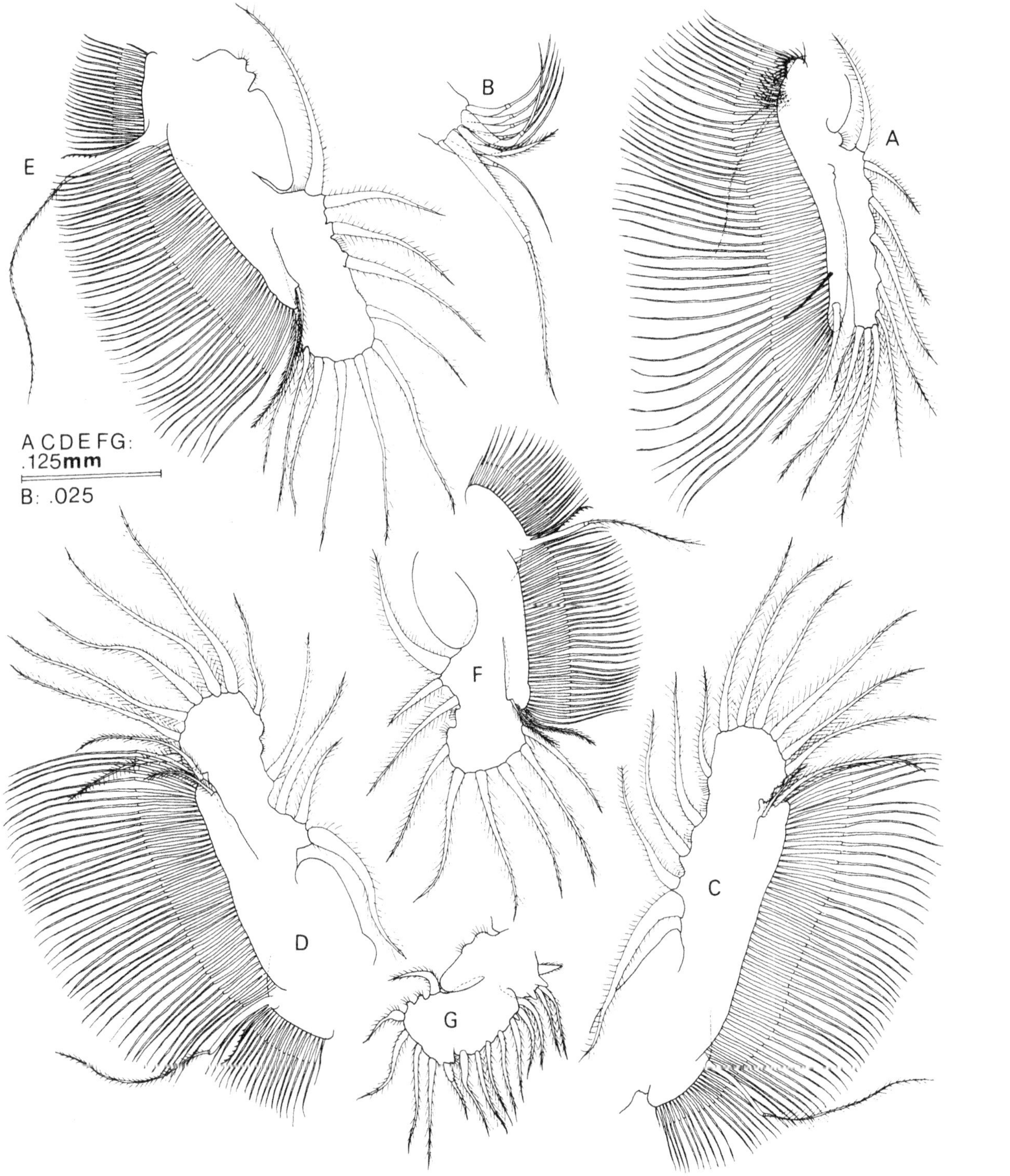

Fig. 5. Diaphanosoma brachyurum (Liévin, 1848), thoracic limbs of female from Belesar reservoir (18): *a*, first limb; *b*, gnathobasic prominence of first limb (row of filtering setae not represented); *c*, second limb; *d*, third limb; *e*, fourth limb; *f*, fifth limb; *g*, sixth limb.

with a row of 7–8 marginal spines, with groups of 3–5 small spinules and one hair in between them (Fig. 4c). Posterior margin of valves with a row of marginal cilia and one plumose, thick spine on inner side (Fig. 4d), and 4–5 groups of 10–11 cilia and one long hair on outer side (Fig. 4c).

Thoracic limbs somewhat smaller than in *D. mongolianum* but resembling them in general

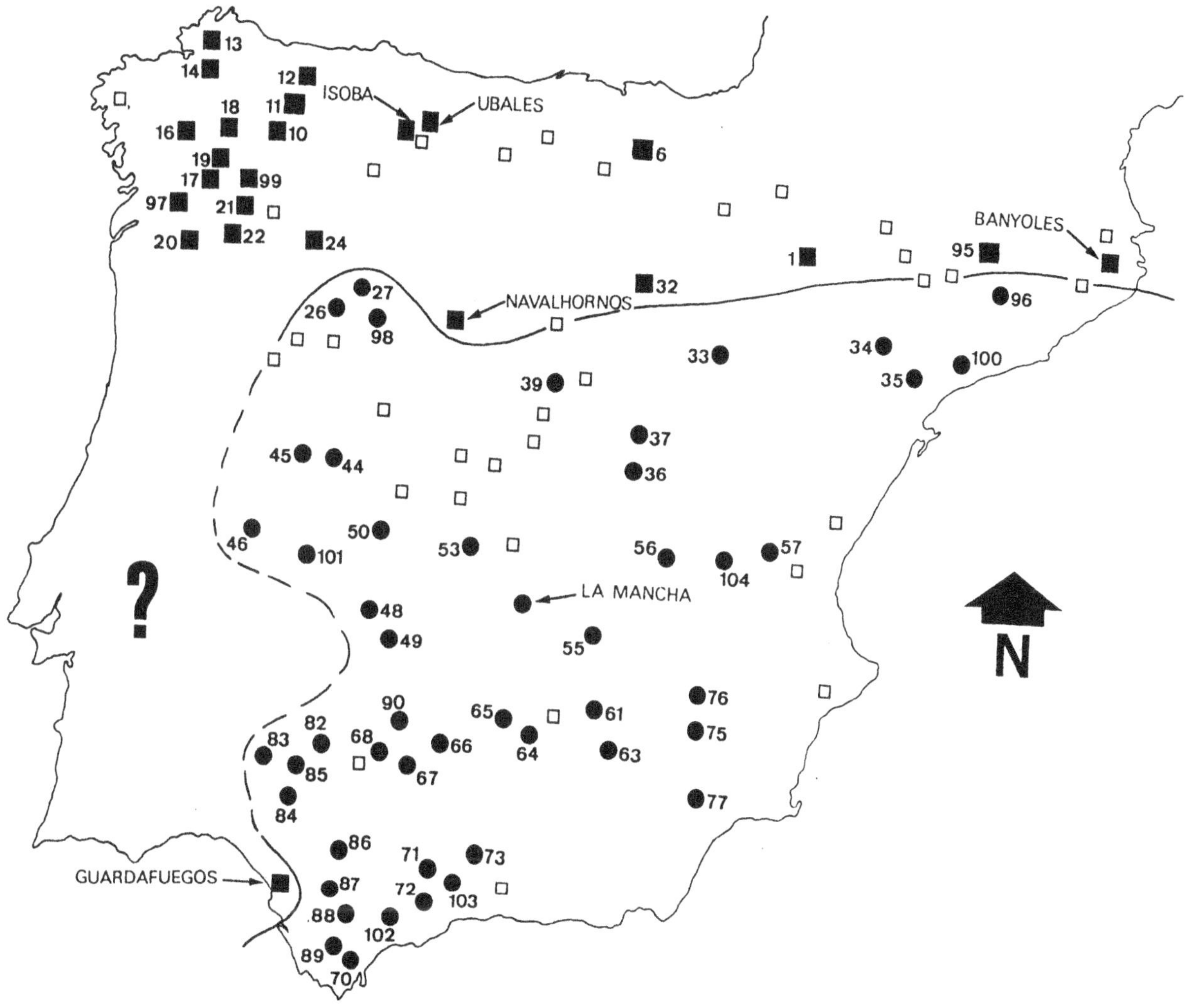

Fig. 6. Distribution of the genus *Diaphanosoma* Fischer, 1850 in the Iberic peninsula. Dark squares represent localities where *D. brachyurum* (Liévin, 1848) has been recorded, dark circles those for *D. mongolianum* Uéno, 1938. Open squares represent reservoirs without *Diaphanosoma*. Reservoirs are, 1 : Sotonera (Huesca); 6 : Ebro (Santander); 10 : Salime, 11 : Doiras, 12 : Arbón (Asturias); 13 : Forcadas, 14 : Ribeira, 16 : Portodemouros (Coruña); 17 : Velle (Orense); 18 : Belesar, 19 : Peares (Lugo); 20 : Conchas, 21 : San Esteban, 22 : Chandreja (Orense); 24 : Cernadilla; 26 : Villalcampo; 27 : Ricobayo (Zamora); 32 : Cuerda del Pozo (Soria); 33 : Tranquera, 34 : Mequinenza (Zargoza); 35 : Flix (Tarragona); 36 : Buendía, 37 : Entrepeñas (Guadalajara); 39 : El Atazar (Madrid); 44 : Gabriel y Galán (Cáceres); 45 : Borbollón, 46 : Alcántara (Cáceres); 48 : Orellana, 49 : Zújar (Badajoz); 50 : Valdecañas (Cáceres); 53 : Guajaraz, 55 : Peñarroya (Ciudad Real); 56 : Alarcón (Cuenca); 57 : Benaixebe (Valencia); 61 : Guadalmena, 63 : Tranco de Beas, 64 : Rumblar, 65 : Jándula (Jaén); 66 : Guadalmellato, 67 : Breña, 68 : Bembézar (Córdoba); 70 : Guadarranque (Cádiz); 71 : Guadalteba; 72 : Conde Guadalhorce (Málaga); 73 : Iznájar (Granada); 75 : Cenajo; 76 : Talave (Albacete); 77 : Puentes (Murcia); 82 : Pintado (Sevilla); 83 : Aracena (Huelva); 84 : Minilla, 85 : Cala, 86 : Torre del Aguila (Sevilla); 87 : Bornos; 88: Guadalcacín, 89 : Celemín (Cádiz); 90 : Puente Nuevo (Córdoba); 95: Oliana, 96 : Sant Ponç (Lleida); 97 : Frieira (Orense); 98 : San Román (Zamora); 99: Mao (Orense); 100 : Riudecanyes (Tarragona); 101 : Guadiloba (Cáceres); 102 : Concepción, 103 : Guadalhorce (Málaga); 104 : Contreras (Valencia).

structure and setation (Fig. 5); minor differences occur in the row of small biarticulate setae on the outside of the gnathobasic prominence of the first thoracic limb (Fig. 4b), which is composed by (6)-7 setae instead of the 8 in *D. mongolianum*. No gamogenetic females were found in the populations studied.

Male

Diagnostic features as in female. Copulatory appendages with characteristic bilobated distal part (Fig. 4e). Hook of first thoracic limb endopodite with minute denticles along its distal third (Fig. 4f). Antennula long, with olfactory setae inserted on proximal third part of basis (Fig. 4g).

Remarks

Some differences in morphology and denticulation of the hook of the male first thoracic limb endopodite are appreciable in comparison with that of a Czechoslovakian male represented in Kořinek (1984). Nevertheless, *Diaphanosoma brachyurum* probably constitutes a group of closely related species as does the 'mongolian' group, and is now under revision (Korovchinsky, pers.com.).

Ecology and distribution

The euplanktonic preferences of both *Diaphanosoma* species in the Iberian Peninsula are clearly manifested by their extensive distribution along reservoirs, and their absence from shallower water bodies (Alonso, 1985).

Both species are characteristic of warm waters, and are preponderant elements of the summer zooplankton communities of Spanish reservoirs. Nevertheless, as has already been pointed out by Alonso (1985), *D. brachyurum* is less thermophilic than *D. mongolianum*: it is not rare in the winter plankton of the northern reservoirs, but occurs at low densities. Both species are absent from the high altitude lakes of the Pyrenees (Miracle, 1978) and Sierra Nevada (Cruz, 1981).

D. mongolianum and *D. brachyurum* present a clearly allopatric pattern of distribution in the Iberian Peninsula (Fig. 6). This seems ecological in nature: *D. brachyurum* inhabits waters in the NW part of the Peninsula with low levels of Total Dissolved Salts (TDS < 250 mg l^{-1}) and a relative ionic composition of $CO3^{2-} > Cl^- = SO4^{2-}$, and a group of reservoirs in rivers draining the Cantabrian mountains and the Pyrenees having bicarbonate waters ($CO3^{2-} > SO4^{2-} > Cl^-$) and intermediate TDS values (150–450 mg l^{-1}). *D. mongolianum* inhabits the central and southern parts of the Iberian Peninsula, in waters characterized by preponderance of sulphate or chloride, and high TDS value (Armengol *et al.*, in press).

Acknowledgements

I am grateful to Dr N. Korovchinsky and Dr M. Alonso for valuable comments and materials. This research has been supported by the Comisión Asesora para la Investigación Científico y Técnica, Program PB85-0166.

References

Alonso, M., 1985. Las lagunas de la España peninsular: taxonomía, ecología y distribucíon de los Cladóceros. Tesis doctoral. Univ. Barcelona. 795 pp.

Armengol, J., 1978. Los crustáceos del plancton de los embalses españoles. Oecol. aquat. 3: 3–96.

Armengol, J., J. L. Riera & J. A. Morguí, 1989. Major ionic composition in the Spanish reservoirs. Verh. int. Ver. Limnol. 24, in press.

Cruz, L., 1981. Estudio de la comunidad zooplanctónica de un lago de alta montaña. Tesis Doctoral. Univ. Granada. 180 pp.

Dirección General de Obras Hidráulicas, 1973. Inventario de presas españolas. Ministerio de Obras Públicas, Servicio de Publicaciones, Madrid. 393 pp.

Kořinek, V., 1981. *Diaphanosoma birgei* n. sp. (Crustacea, Cladocera). A new species from America and its widely distributed subspecies *Diaphanosoma birgei* ssp. *lacustris* n. ssp. Can. J. Zool. 59: 1115–1121.

Kořinek, V., 1984. Cladocera. In J. J. Symoens (ed.), Hydrobiological survey of the Lake Bangweulu Luapula River Basin. Bruxelles, Belgium: 1–107.

Kořinek, V., 1987. Revision of three species of the genus *Diaphanosoma* Fischer, 1850. Hydrobiologia 145: 35–45.

Korovchinsky, N. M., 1978. Izmenchivost *Sida cristallina* i *Diaphanosoma* cf. *brachyurum* (Crustaces, Cladocera) ozera Glubokogo. Zoologichesky J. 57: 1330–1341.

Korovchinsky, N. M., 1987. A study of *Diaphanosoma* species of the '*mongolianum*' group. Int. Revue ges. Hydrobiol. 72: 727–758.

Margalef, R., 1953. Los crustáceos de las agaus continentales ibéricas. Biología de las aguas continentales, 10. Minist. Agricultura, Inst. Forest. Invest. y Exper., Madrid. 243 pp.

Margalef, R., 1983. Limnología. Editorial Omega, Barcelona. 1010 pp.

Miracle, M. R., 1976. Distribución en el espacio y en el tiempo de las especies del zooplancton del lago de Banyoles. Monografías del Instituto Nacional para la Conservación de la Naturaleza 5. 270 pp.

Miracle, M. R., 1978. Composición específica de las comunidades zooplanctónicas de 153 lagos de los Pirineos y su interés biogeográfico. Oecol. aquat. 3: 167–192.

Toja, J., 1981. Zooplancton de los embalses de Aracena y la Minilla durante 1977. In N. Prat (ed.), Actas del primer congreso español de Limnologia: 105–114. Barcelona.

Hydrobiologia **225**: 37–43, 1991.
V. Kořínek & D. G. Frey (eds), Biology of Cladocera.

Review of Iberian Cladocera with remarks on ecology and biogeography

M. Alonso
Departament d'Ecologia, Facultat de Biologia, Universitat de Barcelona, Barcelona, Spain

Key words: Iberian Cladocera, taxonomy, ecology, biogeography

Abstract

A checklist of 88 freshwater Cladocera from the Iberian Peninsula is given, based on the examination of approximately 1 500 samples collected from all parts of the peninsula from 1976 to 1989. Ecology and species assemblages are considered. Distribution of the species versus regional limnology of the Iberian Peninsula is discussed.

Introduction

Taxonomic studies on Iberian freshwater Cladocera started with Arevalo (1916). Later, in Margalef (1953) a comprehensive list of 46 Spanish cladocerans was published; this author also carried out several regional studies on limnology, some of which (Margalef, 1947, 1955, 1958) made a significant contribution to our knowledge of Cladocera. Since then, extensive surveys of cladocerans were carried out in the Spanish reservoirs (Armengol, 1978), in the Pyrenees (Miracle, 1978), and in small permanent and temporary waterbodies scattered over Spain (Alonso, 1987a).

The present paper is a review of the current knowledge of Cladocera in the Iberian Peninsula. The results presented here are not definitive, because the status of some taxa is uncertain, but may make a further contribution to our understanding and also provide a useful tool for further investigations. Remarks on the autoecology and distribution of the species are included.

Materials and methods

Collections were made in 105 reservoirs (Data from Armengol 1978 and own data), alpine lakes in all the Iberian mountain systems (Data of Miracle 1978 and own data), all Spanish karstic lakes (about 20), a large number of permanent and temporary small water bodies (about 1 200), almost all Spanish saline lakes (about 100), and 9 slow-moving streams. Sampling trips were carried out from 1976 to 1989 and covered all the Spanish provinces including the Balearic Islands and the southern part of Portugal. Driest areas were sampled in winter and spring, mountain lakes in summer and autumn, and reservoirs, karstic lakes, and streams throughout the whole year. Samples were obtained from the most representative environments in each waterbody (littoral areas, offshore water, among vegetation, etc.) using a hand net of mesh size 100 μm. Specimens were preserved in 4% formaldehyde.

Results

A total of 88 species of cladocerans have been identified in the Iberian Peninsula (Table 1).

Table 1. List of freshwater cladoceran taxa from the Iberian Peninsula.

Family Sididae (Baird, 1850) Sars, 1865
Sida crystallina (O.F. Müller, 1776)
Diaphanosoma brachyurum (Liévin, 1848) emend. Negrea, 1983
D. mongolianum Uéno, 1938 emend. Korovchinsky 1987

Family Holopediidae
Holopedium gibberum Zaddach, 1885

Family Daphniidae (Straus, 1820) Sars, 1865
Daphnia magna Straus, 1820
D. similis Claus, 1876, emend. Brooks, 1957
D. hispanica Glagolev & Alonso, 1990
D. mediterranea Alonso, 1985
D. atkinsoni Baird, 1859
D. bolivari Richard, 1888
D. curvirostris Eylmann, 1887, emend. Johnson, 1952
D. obtusa Kurz, 1874, emend. Scourfield, 1942
D. pulicaria Forbes, 1893, emend. Hrbácek, 1959
D. longispina O.F. Müller, 1785
D. rosea Sars, 1862, emend. Richard, 1896, Brooks, 1957
D. galeata Sars, 1864, emend. Richard, 1896, Brooks, 1957
D. cucullata Sars, 1862
D. parvula Fordyce, 1901
Simocephalus vetulus (O.F. Müller, 1776)
S. exspinosus (Koch, 1841)
Ceriodaphnia reticulata (Jurine, 1820)
C. dubia Richard, 1894
C. quadrangula (O.F. Müller, 1785)
C. pulchella Sars, 1862
C. cornuta Sars, 1886
C. laticaudata P.E. Müller, 1867
Megafenestra aurita (Fischer, 1849)
Scapholeberis mucronata (O.F. Müller, 1776)
S. rammeri Dumont & Pensaert 1983

Family Moinidae Goulden, 1968
Moina brachiata (Jurine, 1820), emend. Goulden, 1968
M. micrura Kutz, 1874, emend. Goulden, 1968
M. macrocopa (Straus, 1820)
M. salina Daday, 1888, emend. Negrea, 1983

Family Macrothricidae Norman & Brady, 1867
Ilyocryptus sordidus (Liévin, 1848)
I. acutifrons Sars, 1862
I. silvaducensis Romijn, 1919, emend. Stifter, 1984
Streblocerus serricaudatus (Fischer, 1849)
Macrothrix laticornis (Fischer, 1848)
M. hirsuticornis (Norman & Brady, 1867)
Macrothrix rosea Liévin, 1848
Drepanothrix dentata (Eurén, 1861)

Family Bosminidae Sars, 1865
Bosmina longirostris (O.F. Müller, 1776)

Family Chydoridae Stebbing, 1902

Subfamily Eurycercinae Kurz, 1875
Eurycercus lamellatus (O.F. Müller, 1776)

Subfamily Chydorinae Stebbing, 1902
Pleuroxus aduncus (Jurine, 1820)
P. laevis (Sars, 1862)
P. uncinatus Baird, 1850
P. truncatus (O.F. Müller, 1785)
P. letourneuxi (Richard, 1888)
P. denticulatus Birge, 1879
Alonella excisa (Fischer, 1854)
A. exigua (Lilljeborg, 1853)
A. nana (Baird, 1850)
Ephemeroporus margalefi Alonso, 1987
E. epiaphantoii Alonso, 1987
E. phintonicus Margaritora, 1969
Estatheroporus gauthieri Alonso, 1990
Phrixura rostrata (Koch, 1841) emend. Michael & Frey, 1984
P. leei (Chien, 1970) emend. Michael & Frey, 1984
Chydorus sphaericus (O.F. Müller, 1776) emend. Frey, 1980
C. pizarri Alonso, 1988
C. piger Sars, 1862
Dunhevedia crassa King, 1853

Subfamily Aloninae Frey, 1967
A. quadrangularis (O.F. Müller, 1776)
A. rectangula Sars, 1862
A. elegans Kurz, 1875
Alona sp 1 gr *elegans*
Alona sp 2 gr *elegans*
A. phreatica Dumont, 1983
A. azorica Frenzel & Alonso, 1988
A. nuragica Margaritora, 1971
A. rustica Scott, 1895
A. costata Sars, 1862
A. guttata Sars, 1862
A. affinis Leydig, 1860
A. iberica Alonso & Pretus 1989
A. intermedia Sars, 1862
A. verrucosa Sars, 1901
Acroperus neglectus Lilljeborg, 1901
A. harpae (Baird, 1834)
Camptocercus rectirostris Schoedler, 1862
Leydigia leydigi (Schoedler, 1863)
L. acanthocercoides (Fischer, 1854)
Tretocephala ambigua (Lilljeborg, 1901)
Rynchotalona falcata (Sars, 1862)
Oxyurella tenuicaudis (Sars, 1862)
Graptoleberis testudinaria (Fischer, 1848)
Monospilus dispar Sars, 1862

Family Polyphemidae
Polyphemus pediculus (Linné, 1761)

Daphniids and chydorids are the most diversified families, and the two genera with the highest number of species are *Daphnia* (13 species) and *Alona* (15 species). The majority of the species are currently considered widespread (cosmopolitan, holarctic, palaearctic, or tropicopolitan in distribution), and 11 species are circummediterranean, including five that have so far been reported only from the Iberian Peninsula (*Daphnia hispanica*, *Ephemeroporus epiaphantoii*, *E. margalefi*, *Chydorus pizarri*, and *Alona iberica*). *Alona* sp 1 and 2 are close to *A. elegans* but certainly different species.

Cladoceran assemblages: characteristic environments and distribution

Regional limnology is highly diverse in the Iberian Peninsula, and the different types of waterbodies have an associated characteristic fauna of cladocerans (Alonso & Comelles, 1984). Some species are eurioic and occur almost anywhere, but others (the majority) appear preferentially in definite kinds of aquatic environments. The most important parameters that have influence on the structure of the cladoceran community are: persistence of water, salinity, and trophic conditions. Limnetic and littoral communities are usually different in deep waterbodies with no water level fluctuations.

Permanent waters

Natural permanent waterbodies are scarce in the Iberian Peninsula. There are two types: alpine lakes and tarns in the mountains, and karstic lakes at lower altitudes. The rest of permanent waters are reservoirs, which have been built during the last 50 years. Another interesting habitat for cladocerans is slow-moving streams.

The greatest assemblages of alpine lakes occur in the Pyrenees, Cantabric ridge, Central system, and Sierra Nevada. The limnetic community of such lakes is quite uniform and poor in species (Miracle, 1978; Cruz, 1981): the most characteristic are *Daphnia longispina* and *Holopedium gibberum*. This last species is so far known only from the Pyrenees. The littoral community is more diverse, mainly if macrophytes are present; the most characteristic species are *Scapholeberis mucronata*, *Eurycercus lamellatus*, *Chydorus piger*, *Alonella nana*, *Alona quadrangularis*, *A. affinis*, *A. rustica* (only in the Pyrenees), *A. intermedia*, *Acroperus harpae*, *Rynchotalona falcata* (only in the Cantabric ridge), *Monospilus dispar*, *Polyphemus pediculus*, and *Streblocerus serricaudatus* (both only in the Pyrenees).

Karstic lakes appear in limestone, and the water has high levels of total hardness (4–6 meq/l^{-1} of total alkalinity); they are distributed throughout the eastern part of Spain. The most characteristic planktonic species are *Daphnia longispina* and *Diaphanosoma brachyurum*. *Sida crystallina*, *Simocephalus vetulus*, *Pleuroxus truncatus*, *Acroperus neglectus*, *Alona affinis*, and *Graptoleberis testudinaria* occur in the littoral, and *Ilyocryptus sordidus* in the benthos.

Reservoirs are widespread all over Spain. They may be divided into two main groups, those built in headwaters that are oligotrophic or mesotrophic and with low-mineralized water, and those built downstream, in which the water is more eutrophic and mineralized. Littoral communities of cladocerans are poor in the reservoirs because of the water-level fluctuations; the most frequent species are *Macrothrix laticornis*, *Leydigia leydigi*, *Chydorus sphaericus*, and *Alona rectangula*. The planktonic community of the headwater reservoirs is quite similar to that of alpine lakes, *Daphnia longispina* being the most characteristic species. In eutrophic reservoirs *Bosmina longirostris* and *Ceriodaphnia pulchella* are very frequent; besides these species, such environments have been colonized by species that supposedly were not present in the Iberian Peninsula before construction of the reservoirs such as *Diaphanosoma mongolianum*, *Daphnia galeata*, *D. parvula*, *D. cucullata*, and *Ceriodaphnia cornuta* (Armengol, 1978).

Streams have a very interesting cladoceran fauna. *Alona phreatica* (groundwater species) and *Ephemeroporus epiaphantoii* are so far exclusive to running waters (Dumont, 1983, 1987; Sabater,

1987; Alonso, 1987b), and *Phrixura leei* and *Alona verrucosa* have been collected only in rivers in the Iberian Peninsula. Species characteristic of lakes also occur in streams. *Eurycercus lamellatus*, *Chydorus piger*, *Alona intermedia*, and *Monospilus dispar*, typical species of alpine lakes, occur in the rivers of the western Palaeozoic shield with weakly mineralized waters. *Pleuroxus truncatus*, *P. aduncus*, and *Acroperus neglectus*, species characteristic of karstic lakes, occur in streams draining calcareous regions with more mineralized waters. In addition to these, other species are also quite frequent in streams, namely *Alona quadrangularis*, *A. guttata*, and *Chydorus sphaericus*.

Temporary waters

Temporary lentic waters are widespread in semi-arid regions (less than 600 mm rainfall). They are especially abundant in the flatlands of the Duero, Ebro, Guadiana, and Guadalquivir basins. The chemical composition of temporary waters is very variable and is conditioned by the nature of bedrock and the kind of drainage. As far as the cladoceran composition is concerned three habitat types can be considered: Fresh waters on weakly soluble substratum (granite, terrace gravels, siliceous sand), fresh waters in Miocenic clay, and mineralized waters.

The first type of waterbodies occur mainly in the western part of the Iberian Peninsula. Due to the silicic nature of the substratum the waters are soft (less than 100 $\mu s\ cm^{-1}$ of conductivity). The water is usually clean, allowing the existence of macrophytic vegetation. In this kind of waters, the cladoceran fauna is characterized by species richness and diversity; many species occurring in other habitats (alpine lakes, reservoirs, streams) appear here also, but the most significant are *Ilyocryptus silvaducensis*, *Scapholeberis rammneri*, *Macrothrix rosea*, *Ephemeroporus margalefi*, *Estatheroporus gauthieri*, *Alona azorica*, and *A. iberica*.

Fresh waters in Miocenic clay are abundant in steppes. The water is usually turbid because of suspended clay particles, and macrophytes are scarce. The community is very characteristic and, unlike the previous group of lagoons, does not include many species from other habitats. The most typical species are *Daphnia similis*, *D. atkinsoni*, *D. hispanica*, *D. bolivari*, *D. obtusa*, *Ceriodaphnia quadrangula*, *Moina brachiata*, *Macrothrix hirsuticornis*, *Pleuroxus letourneuxi*, *Ephemeroporus phintonicus*, *Chydorus pizarri*, *Alona* sp 2, and *A. nuragica*.

Mineralized waters also occur in steppes, but in drier and endorheic regions. These waters are characterized by low species richness. In saline waters of less than 40‰ salinity, halophyle species occur; *Daphnia magna*, *Simocephalus exspinosus*, and *Alona* sp 1 being the most characteristic. At levels of salinity of between 40‰ and 60‰ only *Daphnia mediterranea* and *Moina salina* are found (Comin & Alonso, 1988).

Discussion

The cladoceran fauna of the Iberian Peninsula appears to be rather differentiated. Species richness is slightly lower than in the neighbouring European countries and higher than in North Africa; 88 species in the Iberian Peninsula compared to 94 in Italy (Margaritora 1985), 92 in France (Amoros, 1984), and 53 in Algeria, Tunisia, and Morocco (Gauthier, 1928, 1929; Ramdani, 1986). Many genera occurring in Europe have not yet been reported in the Iberian Peninsula, such as *Latona*, *Eubosmina*, *Lathonura*, *Bunops*, *Anchistropus*, *Pseudochydorus*, *Alonopsis*, *Kurzia*, *Leptodora*, and *Bythotrephes*. Some of these genera (*Leptodora*, *Eubosmina*, and *Bythotrephes*) are typical of large, deep lakes, and their absence from the Iberian Peninsula is probably due to the lack of suitable habitats. Iberian reservoirs, especially those of the northern part, could partly reproduce the ecological characteristics needed for such genera; however, their colonization by European cladocerans has been rather poor; apparently, only *Daphnia galeata* and *D. cucullata* are recent successful colonizers. Differences between the Iberian Peninsula and North Africa are lower; cladoceran genera of both countries are the same, and only nine species occurring in North Africa have not been recorded in the Iberian Peninsula (*Ceriodaphnia setosa*, *C. rotunda*, *Daphnia lumholtzi*, *D. chevreuxi*, *D. barbata*, *Alona elegans lebes*, *Alona pulchella*, *Alona karua*, and *Leydigia ciliata*).

Taxonomic groups especially involved in faunistic differences between countries are the subgenus *Ctenodaphnia* and some genera of chydorids, which suggests that dispersal capability of such groups is low. *Daphnia* (*Ctenodaphnia*) has six species in the Iberian Peninsula, one of which (*D.(C.) hispanica*) is endemic (Glagolev and Alonso, 1990). *D.(C.) magna*, *D.(C.) similis*, and *D.(C.) atkinsoni* occur also in Europe and in North Africa; *D.(C.) mediterranea* in Romania (Negrea, 1983), Sardinia (Margaritora, 1972), and in North Africa (Ramdani, 1986); finally, *D.(C.) bolivari* also appears in North Africa (Gauthier, 1937) and in Sardinia (Margaritora, 1967). *D.(C.) similis* has two different forms in the Iberian Peninsula; northern populations are similar to the European populations, with an evenly rounded head and a medium-sized carapace spine, and populations from the south have a helmeted head and very long carapace spine. Iberian chydorids are quite differentiated. Four endemic species have so far been found in the Iberian Peninsula namely *Ephemeroporus margalefi*, *E. epiaphantoii*, *Chydorus pizarri*, and *Alona iberica* (Alonso, 1987b, 1988; Alonso & Pretus, 1989). Both species of *Ephemeroporus* belong to the *E. barroisi* complex, whose North African populations are somewhat close to *E. epiahantoii* (pers. obs.). *Alona iberica* belongs to the tropicopolitan *A. karua* complex. *Chydorus pizarri* is apparently little related with other *Chydorus* species.

Some of the remaining chydorid species deserve more comments. *Pleuroxus truncatus* has two forms that are probably different species (L. Frey *in litt.*), one in the Pyrenees and another in the southern lowlands. *Alonella excisa* should also be revised because there are at least two different forms in the Iberian Peninsula, one with a rounded tip to the rostrum and another with pointed tip. *Phrixura rostrata* was reported to be widespread in Europe; however, neither the description of Negrea (1983) nor the majority of the Iberian populations correspond to this species. On the contrary, the commonest taxon in the Iberian Peninsula is *P. leei* which was considered exclusive to North America (Michael & Frey, 1984). Only three species of *Chydorus* have been identified: *C. pizarri*, *C. sphaericus*, and *C. piger*; surprisingly, none of the remaining species of this genus recorded in Europe (*C. ovalis*, *C. gibbus*, *C. latus*) have been found in the Iberian Peninsula. Iberian *Chydorus sphaericus* probably comprise more than one species, judging by the high size variability between different populations, but detailed morphological analysis has not yet been carried out. *Alona* sp 1 and *Alona* sp 2 are close to *A. elegans* but differ from this species in the carapace sculpture and in the shape of the postabdomen; besides, *Alona* sp 2 has the headshield completely lined with longitudinal striae. *Alona azorica* has two forms, one in the Azores and western Iberian Peninsula (Frenzel & Alonso, 1988) and another in the central Spanish steppes, in the Balearic islands, and in North Africa. Both forms differ in the head-pore structure; in the typical form, the posterior pore grows away from the middle one in the course of the growth, while in the form from the steppes, the distance between such pores remains constant. This last form was named *A. esteparica* (Alonso, 1987a; Ramdani, 1986), however, since intermediate forms were subsequently found in the same locality (same population?), the validity of this taxon must be revised. As set out above, the Iberian fauna of cladocerans is really far from being well known. In accordance with the idea that cladocerans are not cosmopolitan (Frey, 1982a), comparisons between Iberian populations and type material from other countries are needed for the majority of the Iberian taxa. Besides, gamogenetic populations of some of the species should be intensively searched in order to produce complete descriptions.

Distribution of cladocerans in the Iberian Peninsula arises from ecological and historical factors. Ecological factors explain the distribution of widespread species inhabiting particular environments. Such species have high facilities for dispersal and colonization ('r' strategists). Colonization of reservoirs, which are recently created new habitats, provides good examples of easily dispersible species such as *Daphnia parvula* (New World species), *D. cucullata* and *D. galeata* (Eurasian species). The process of colonization of a cooling reservoir in central Spain by the warm-water stenotherm *Ceriodaphnia cornuta* is illustrative in the matter. This species occurred only in the summer plankton of two south Spanish reservoirs (Armengol, 1978), and became dominant during the whole year in the cooling reservoir from the beginning of the exploitation of the power plant (Alonso, 1989). The pattern of distribution of the cladocerans reflects in large measure the Iberian regional limnology. Two examples: chydorid species richness is greater in the western part and in the mountains, where soft and permanent water is more frequent (similar results have been obtained in other countries by Frey (1982b) and Chengalath (1987)), and the subgenus *Ctenodaphnia* is typical from the semi-arid regions of the central and south Iberian regions where waters are temporary and more mineralized (Alonso & Comelles, 1984).

But dispersal is not a general adaptation for all of the cladocerans; on the contrary, some species are specially adapted to continue in a particular site (Frey, 1982a). It is for the distribution of these species for which historical factors are important. Three patterns of distribution can be taken into account: a) Boreo-Alpine species. Colonization of the mountains could have been in two ways (Miracle, 1982), one across the Pyrenees and another across the Cantabric ridge. Species that suggest the Pyrenean pathway are *Holopedium gibberum*, *Alona rustica*, *Polyphemus pediculus*, and *Streblocerus serricaudatus*. *Rynchotalona falcata* suggests penetration from the north-west (Atlantic or Lusitanian species). Typical Boreo-Alpine species have not so far been reported either in the Central system nor in the mountains of the south (Sierra Nevada). b) Species from the steppes. On the whole, the fauna of the central and southern Iberian steppes reminds us of the fauna of Sicilia and Sardinia (Stella & Margaritora, 1976; Stella *et al.*, 1972; Margaritora *et al.*, 1982), and that of North Africa (Gauthier, 1928; Ramdani, 1986). Differences between these countries and regions of the Peninsula itself suggest that wetlands in steppes are rather isolated. This could have produced the regional differentiation of the *Daphnia (C.) chevreuxi* complex (one of whose elements seem to be *D. (C.) hispanica*, *D. (C.) similis* and *Alona azorica*. c) Tropicopolitan species. There are species belonging to tropicopolitan (or South Hemisphere) complexes as are *Ephemeroporus margalefi*, *E. epiaphantoii*, *Alona iberica*, and *A. verrucosa*, which occur preferentially in the westernmost part of the Peninsula. Three of them are Iberian endemics, and it is reasonable to think that their differentiation has been produced in this country. If, as Frey states (1987), cladocera are very old and conservative, these species must have been isolated for a long time; ancestors probably colonized the emerged part of the Peninsula when a tropical climate prevailed during the Tertiary period. Subsequently, persistence of these species throughout the time could have been possible, because this part of the country was not affected by Pleistocene glaciers and has a mild Atlantic climate.

Acknowledgements

This work has been supported by DGICYT grant ID CSIC 88IA081.

References

Alonso, M., 1987a. Las lagunas de la España peninsular: Taxonomía, ecología y distribución de los cladóceros. Col. lecció Tesis Doctorals Microfitxades 139. Universitat de Barcelona, 795 pp.

Alonso, M., 1987b. Morphological differentiation of two new *Ephemeroporus* species (Cladocera, Chydoridae) belonging to the *barroisi* complex: *E. margalefi* and *E. epiaphantoii*, in Spain. Hydrobiologia 145: 131–146.

Alonso, M., 1988. *Chydorus pizarri* sp. nov., a new chydorid (Cladocera) from western Spain. Limnetica 4: 27–40.

Alonso, M., 1989. Temperature increase effects on zooplankton structure in a cooling reservoir. In J. F. Talling, Z. Brandl & V. Straskrabová (eds), Proceedings of the International Conference on Reservoir Limnology and Water Quality. 2. Chemical limnology, primary productivity, plankton, benthos and fish interaction. Arch. Hydrobiol. Beih., Ergebn. Limnol. 33: 503–511.

Alonso, M. & M. Comelles, 1984. A preliminary grouping of the small epicontinental water bodies in Spain and distribution of crustaceans and Charophyta. Verh. int. Ver. Limnol. 22: 1699–1703.

Alonso, M. & J. L. Pretus, 1989. *Alona iberica* n. sp.: First evidence of non-cosmopolitanism within the *A. karua* complex (Cladocera, Chydoridae). J. Crust. Biol. 9 (3): 459–476.

Amoros, C., 1984. Crustacés cladocères. Introduction pratique a la systematique des organismes des eaux continentales françaises. Bull. Soc. Linn. Lyon 3/4: 1–63.

Arévalo, C., 1916. Introducción al estudio de los Cladóceros del plancton de la Albufera de Valencia. An. Inst. Gen. Tec. Valencia 1: 1–65.

Arévalo, C., 1916. Introdución al estudio de los Cladóceros del plancton de la Albufera de Valencia. An. Inst. Gen. Tec. Valencia 1: 1–65.

Armengol, J., 1978. Los crustáceos del plancton de los embalses españoles. Oecologia aquatica 3: 3–96.

Chengalath, R., 1987. The distribution of chydorid Cladocera in Canada. Hydrobiologia 145: 151–157.

Comin, F. A. & M. Alonso, 1988. Spanish salt lakes: Their chemistry and biota. Hydrobiologia 158: 237–245.

Cruz, L., 1981. Estudio de la comunidad zooplanctónica de un lago de alta montaña (La Caldera, Sierra Nevada, Granada). Ph. D. dissertation, University of Granada, Granada, Spain, 180 pp.

Dumont, H. J., 1983. Discovery of groundwater-inhabiting Chydoridae (Crustacea: Cladocera) with the description of two new species. Hydrobiologia 106: 97–106.

Dumont, H. J., 1987. Groundwater Cladocera: A synopsis. Hydrobiologia 145: 169–173.

Frenzel, P. & M. Alonso, 1988. *Alona azorica* sp. nov., a new Chydorid (Cladocera) from the Western Palaearctis. Arch. Hydrobiol. 11 (3): 449–465.

Frey, D. G., 1982a. Questions concerning cosmopolitanism in Cladocera. Arch. Hydrobiol. 93 (4): 484–502.

Frey, D. G., 1982b. Relocation of *Chydorus barroisi* and related species (Cladocera, Chydoridae) to a new genus and description of two new species. Hydrobiologia 86: 231–269.

Frey, D. G., 1987. The taxonomy and biogeography of the Cladocera. Hydrobiologia 145: 5–17.

Gauthier, H., 1928. Recherches sur la faune des eaux continentales de l'Algerie et de la Tunisie. Imp. Minerva, Alger, 419 pp.

Gauthier, H., 1929. Cladocères et ostracodes du Sahara Central. Bull. Soc. hist. nat. Afr. 20: 143–162.

Gauthier, H., 1937. Euphyllopodes et cladocères continentaux récoltés par M. Monod au Sahara Occidental et en Mauritanie. Bull. Soc. Sc. Nat. Maroc 17: 75–98.

Glagolev, S. M. & M. Alonso, 1990. *Daphnia (Ctenodaphnia) hispanica* sp. nov. a new daphniid (Cladocera) from Spain. Hydrobiologia 194: 149–162.

Margalef, R., 1947. Estudios sobre la vida en las aguas continentales de la región endorréica manchega. P. Inst. Biol. Apl. 4: 5–51.

Margalef, R., 1953. Los crustáceos de las aguas continentales ibéricas. Biología de las aguas continentales 10. Minist. Agricultura, Inst. Forest. Invest. y Exp., Madrid, 243 pp.

Margalef, R., 1955. Contribución al estudio de la fauna de las aguas dulces del noroeste de España. P. Inst. Biol. Apl. 21: 137–171.

Margalef, R., 1958. Distribución de los crustáceos en las aguas continentales españolas. Grado de asociación entre las especies con factores ecológicos e históricos. P. Inst. Biol. Apl. 27: 17–31.

Margaritora, F. G., 1967. Su una forma di *Daphnia atkinsoni* Baird nuova per l'Italia e considerazioni sulla variabilita' della specie. Ist. Lomb. (Rend. Sci.) B 101: 369–376.

Margaritora, F. G., 1972. Sulla presenza di *Ctenodaphnia* Dybowski e Grochowski 1895 nelle acque astatiche della Sardegna. Ist. Lomb. (Rend. Sci.) B 103: 36–49.

Margaritora, F. G., 1985. Cladocera. Fauna d'Italia 23, Ediz. Calderini, Bologna: 399 pp.

Margaritora, F. G., L. Mastrantuono, D. Grosetti & F. Lombardi, 1982. Contributo allo studio della fauna ad entomostraci delle acque interne della Sicilia. Animalia 9 (1/3): 87–102.

Michael, R. G. & D. G. Frey, 1984. Separation of *Disparalona leei* (Chien, 1970) in North American from *D. rostrata* (Koch, 1841) in Europe (Cladocera, Chydoridae). Hydrobiologia 114: 81–108.

Miracle, M. R., 1978. Composición específica de las comunidades zooplanctónicas de 153 lagos de los Pirineos y su interés biogeográfico. Oecologia aquatica 3: 167–192.

Miracle, M. R., 1982. Biogeography of the freshwater zooplanktonic communities in Spain. J. Biogeogr. 9: 455–467.

Negrea, St., 1983. Cladocera. Fauna Republicii Socialiste Romania 4, 12, Edit. Academiei, Bucaresti: 399 pp.

Ramdani, M., 1986. Ecologie des Crustacés (Copépodes, Cladocères et Ostracodes) des Dayas Marocaines. Ph. D. dissertation, University of Marseille, Marseille, France: 217 pp.

Sabater, F., 1987. On the interstitial cladocera of the river Ter (Catalonia, NE Spain), with a description of the male of *Alona phreatica*. Hydrobiologia 144: 51–62.

Stella, E. & F. G. Margaritora, 1976. Contributi alla conoscenza della fauna ad Entomostraci di acque astatiche della Sardegna (zone Nord occidentale e centrale). Considerazioni ecologiche e biogeografiche. Rend. Acc. Naz. 40: 145–155.

Stella, E., F. G. Margaritora & V. Cottarelli, 1972. La fauna ad Entomostraci di acque astatiche della Sardegna nord orientale. Ren. Acc. Naz. 22: 163–210.

Hydrobiologia **225**: 45–62, 1991.
V. Kořínek & D. G. Frey (eds), Biology of Cladocera.

Additional cladoceran records for Mantecal and new for Venezuela

Evelyn Zoppi de Roa & Wilman Vasquez
Instituto de Zoologia Tropical, Facultad de Ciencias, Universidad Central de Venezuela

Key words: Cladocera, taxonomy, distribution, Venezuela

Abstract

24 cladoceran species were found in collections from several fresh waterbodies, including environmental samples and gut contents of small micro-characid fishes from Mantecal (Apure state). Nine of these were new records for Venezuela and 18 for the Mantecal area. Brief descriptions of locations and species, including range-distribution, are given.

Introduction

A review of collections (1979–1988) from several freshwater locations of Venezuela was done in order to get new records for the cladoceran fauna, after Rey & Vasquez (1986, Orinoco basin), and Zoppi de Roa *et al.* (1985, Mantecal, Apure State). New material from Mantecal includes environment samples as well as gut contents of small micro-characid fishes from flooded savannas (south western Llanos). Other locations are San Antonio de Caparo (pool at Andes cordillera lowlands), Agua Fria (reservoir at central cordillera), Yaracuy River (central region), Alpargaton stream (central coastal region), Atapirire stream (northeastern Orinoco basin), Provincial lagoon and Atabapo river (southwestern Orinoco basin), and Churun-Merun stream (Auyantepuy, Bolivar state) (Fig. 1).

Material and methods

Collections are catalogued at Coleccion Laboratorio de Zooplancton, Instituto de Zoologia Tropical, Facultad de Ciencias, Universidad

Table 1. Some environmental (max-min) values of studied locations.

Locations	Altitude M.A.S.L.	Water Temp. (°C)	Conductivity mhos/cm	Depth (m)	PH
Mantecal grasslands	79	26–33	110–140	0.3–2.0	4.8–6.5
Yaracuy river	2	29	900	4.0	7.0–7.3
Alpargaton stream	2	30	1400	0.4–2.2	4.3–5.0
Atapirire stream	303	27	300	0.6–1.5	6.7–7.8
Provincial lagoon	40	31–35	900–1400	0.6–0.8	4.5–6.8
S. Antonio Caparo pool	400	27–28	...	1.0	5.8–7.2
Agua Fria reservoir	1720	18–22	210–270	25.0	6.5–7.2
Churun-Merun (Auyantepuy) stream	2000	12–16	180	0.5–2.0	...
Atabapo river	40	31–35	180	8.0	4.2–5.0

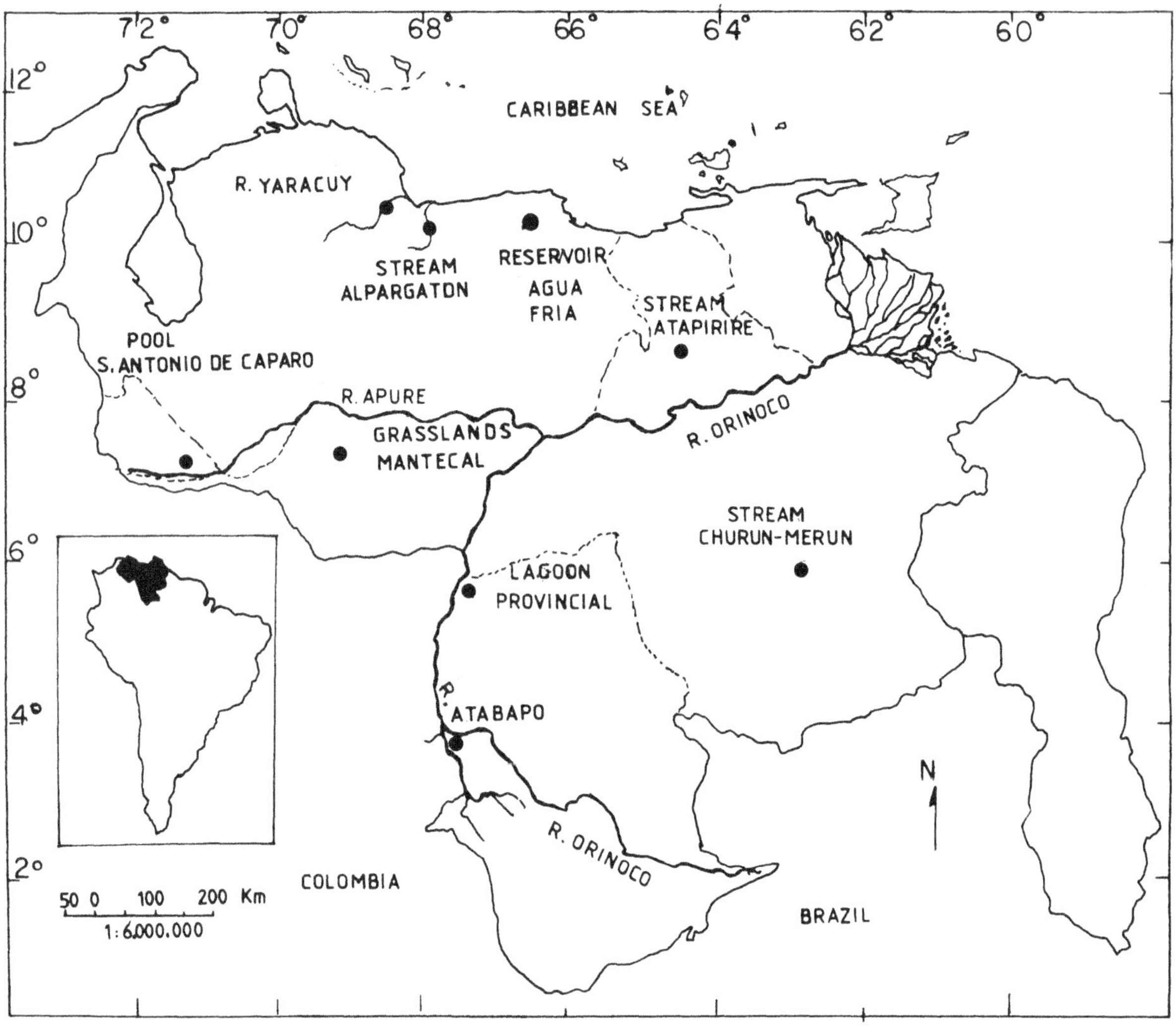

Fig. 1. Map of Venezuela showing locations studied.

Central de Venezuela. Samples were taken from several habitats at each location (flooded grassland, close to emerging plant associations, and/or open waters), using 2 l and 1 l Van Dorn samplers and a plankton net of 69 μ mesh.

Results

Table 1 shows some environmental conditions of the locations, with generally similar values corresponding to tropical conditions. Low water temperature in atypic oligotrophic Agua Fria reservoir and in Churun-Merun stream (Auyantepuy) result from their altitude of more than 1700 m above sea level. Other locations below 400 m, had temperatures of 27–33 °C, with the exception of 31–35 °C in Provincial lagoon. The pH is acidic, except the Yaracuy River (7.0–7.3) near the sea mouth, which is polluted with domestic wastes. The Alpargaton stream has a very acid pH of 4.3–5.0 and high conductivity, both probably resulting from industrial pollution.

Description of species

Sididae

Latonopsis occidentalis **Birge, 1891 (Fig. 2)**
Valves with pyramidal granulations, and ventral rim with 2 long setae, instead of 3 as described for

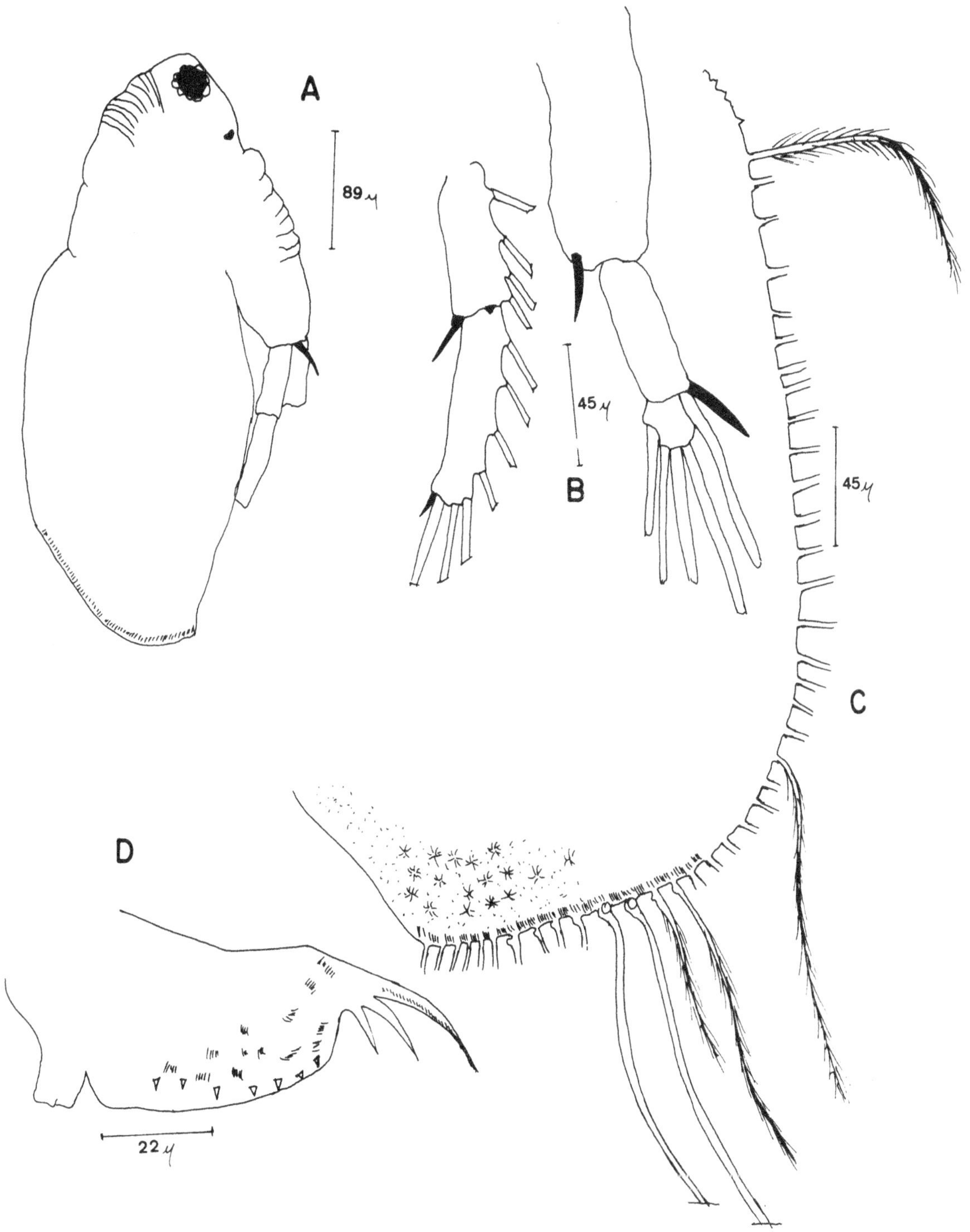

Fig. 2. Latonopsis occidentalis Birge, 1891. – A: Lateral view. B: Antenna. C: Postero-ventral angle of valve. D: Postabdomen.

the Orinoco by Rey & Vasquez (1986). Postabdomen with 7 submarginal spines, instead of 8 and 9 described for other locations. Claw with 2 unequal basal spines. Female with 12 eggs.

First record for Venezuela. Relatively low abundance in grasslands and in stomach contents of fishes in Mantecal.

Size: 1.65 mm length; 0.79 mm height.

Daphniidae

Daphnia laevis ***Birge, 1878***

Agrees with the description of Paggi (1976). Three-fourth of the rim of the valves covered by spines. Postabdomen with 17 basal spines; curved claw with 2 saw-like processes at superior edge. Parthenogenetic female with up to 3 eggs, in contrast with 1 or maximum 2 reported earlier.

Highly abundant in pelagic and littoral zones of Agua Fria reservoir, in contrast with low abundance observed by Infante (1984) for the same location and for high-mountain Andean lagoons.

Size: 0.84 mm body length; 0.37 mm height; 0.44 mm spine length.

Scapholeberis ***cf.*** *kingi* ***Sars, 1903 (Fig. 3)***

'Normal' form (without apical horn) was first reported by Zoppi de Roa *et al.* (1985) for grasslands at Mantecal. Its diagnosis is closely related

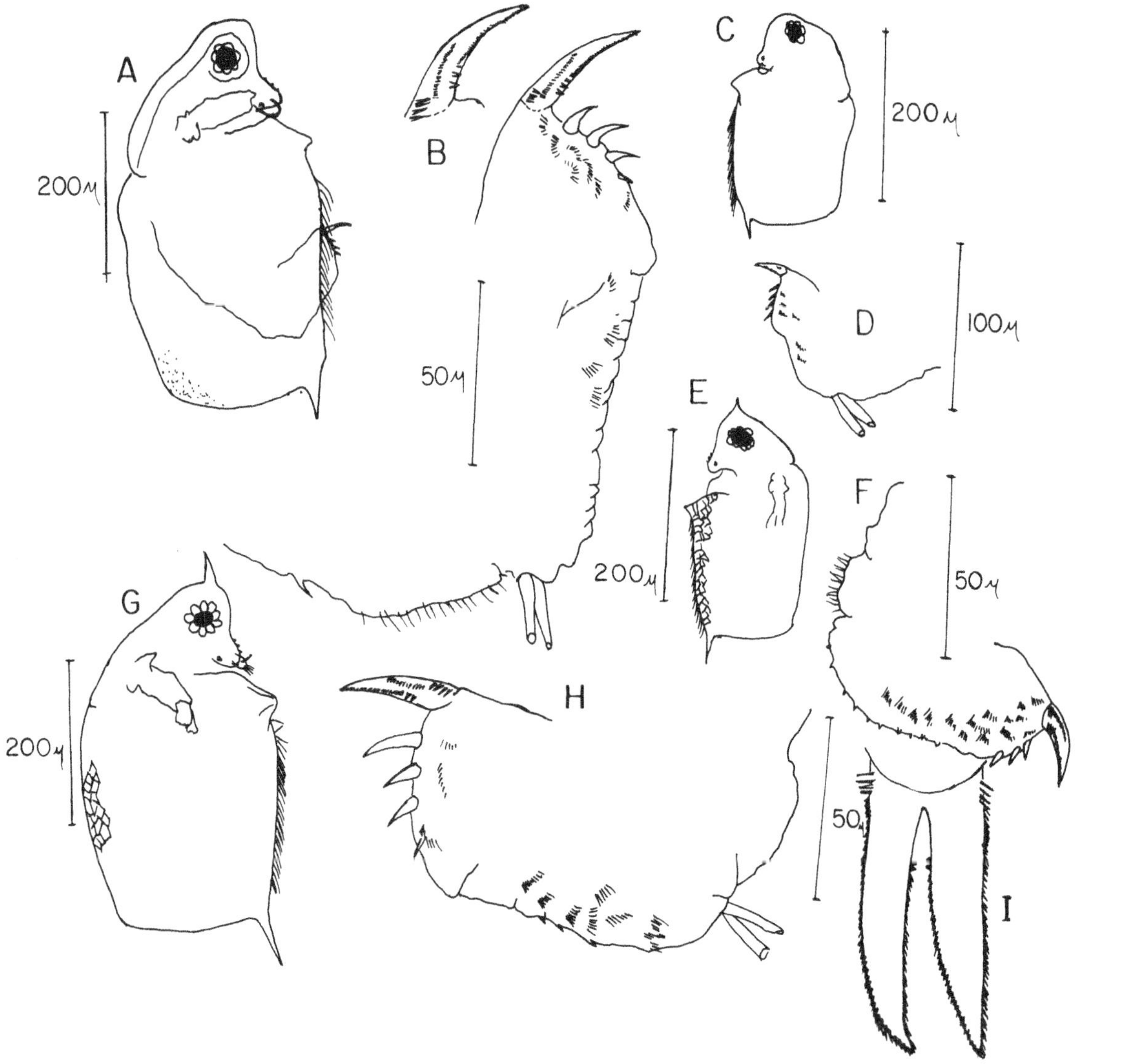

Fig. 3. Scapholeberis cf. *kingi* Sars, 1903. – A: Lateral view of adult without apical horn. B: Postabdomen and detail of claw of adult (A). C: Lateral view of juvenile without apical horn. D: Postabdomen of juvenile (C). E: Lateral view of juvenile with apical horn. F: Postabdomen of juvenile (E). G: Lateral view of adult with apical horn. H: Postabdomen of adult (G). I: Ventral view of postabdominal claw of adult (A).

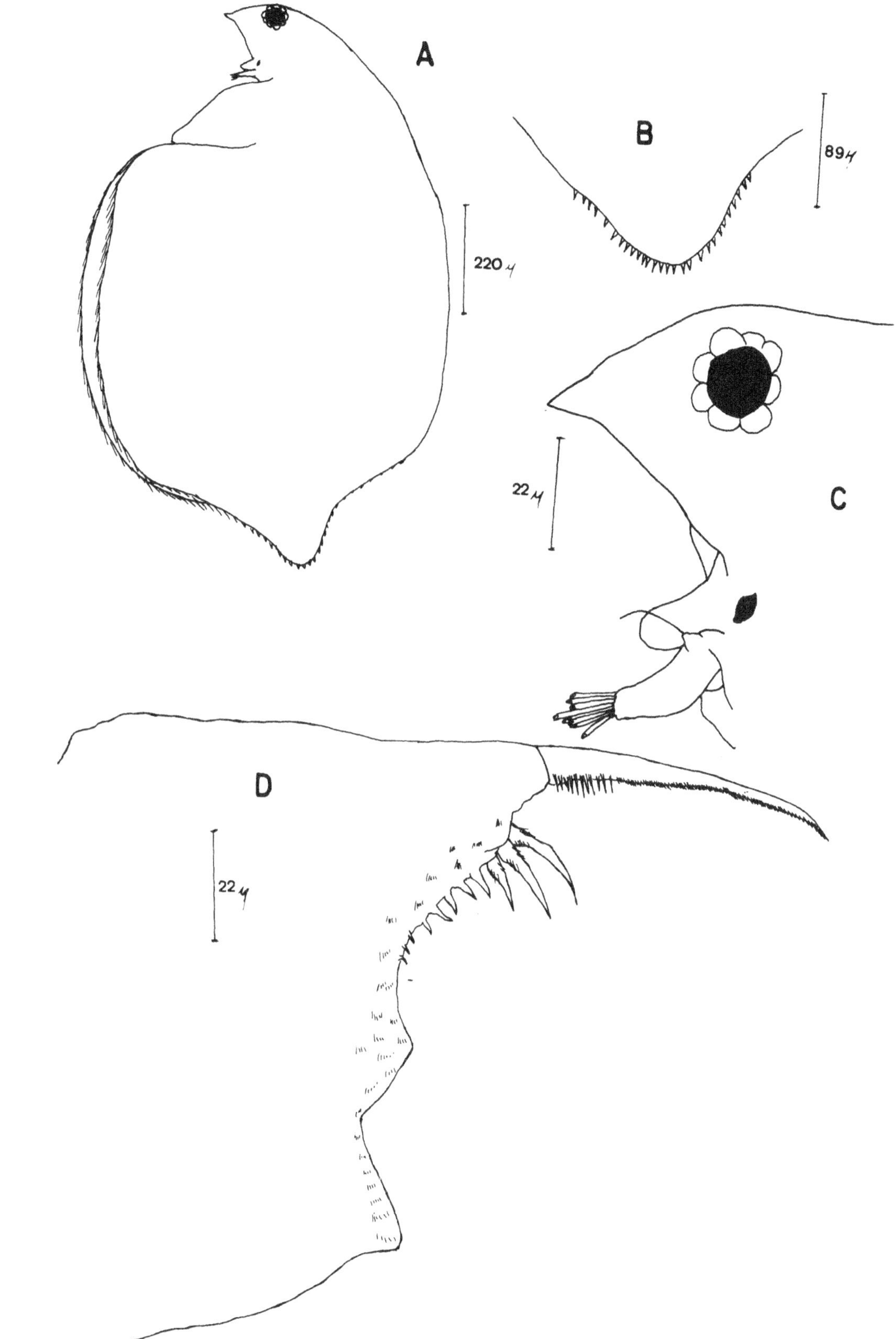

Fig. 4. Simocephalus acutirostratus (King, 1853). – A: Lateral view. B: Postero-ventral angle of valve. C: Detail of head and rostrum. D: Postabdomen.

to that described by Dumont & Pensaert (1983). Polymorphic 'cornuta' form with apical horn has trilobate and terminally-rounded head, 3 and 4 spines in postabdomen of juvenile and adult stages, respectively. Dorsal external pecten of claw with 2 thick and small basal spines, followed by several very small spines, and dorsal upper pecten of claw with 4 long spines.

These two varieties were founded together in the Atabapo River, near its confluence with the Orinoco River during the rainy months of June and August. Juvenile specimens of the 'cornuta' form were found during March (dry season) at Provincial lagoon near the Atabapo River. Both varieties from Orinoco basin have the same size (0.42–0.44 mm adults and 0.26 mm juveniles). Specimens from Mantecal were larger (0.45–0.68 mm adults and 0.25 juveniles). Mucron size of valves was 0.028–0.031 mm for Mantecal 'normal' form, and 0.06–0.08 mm for

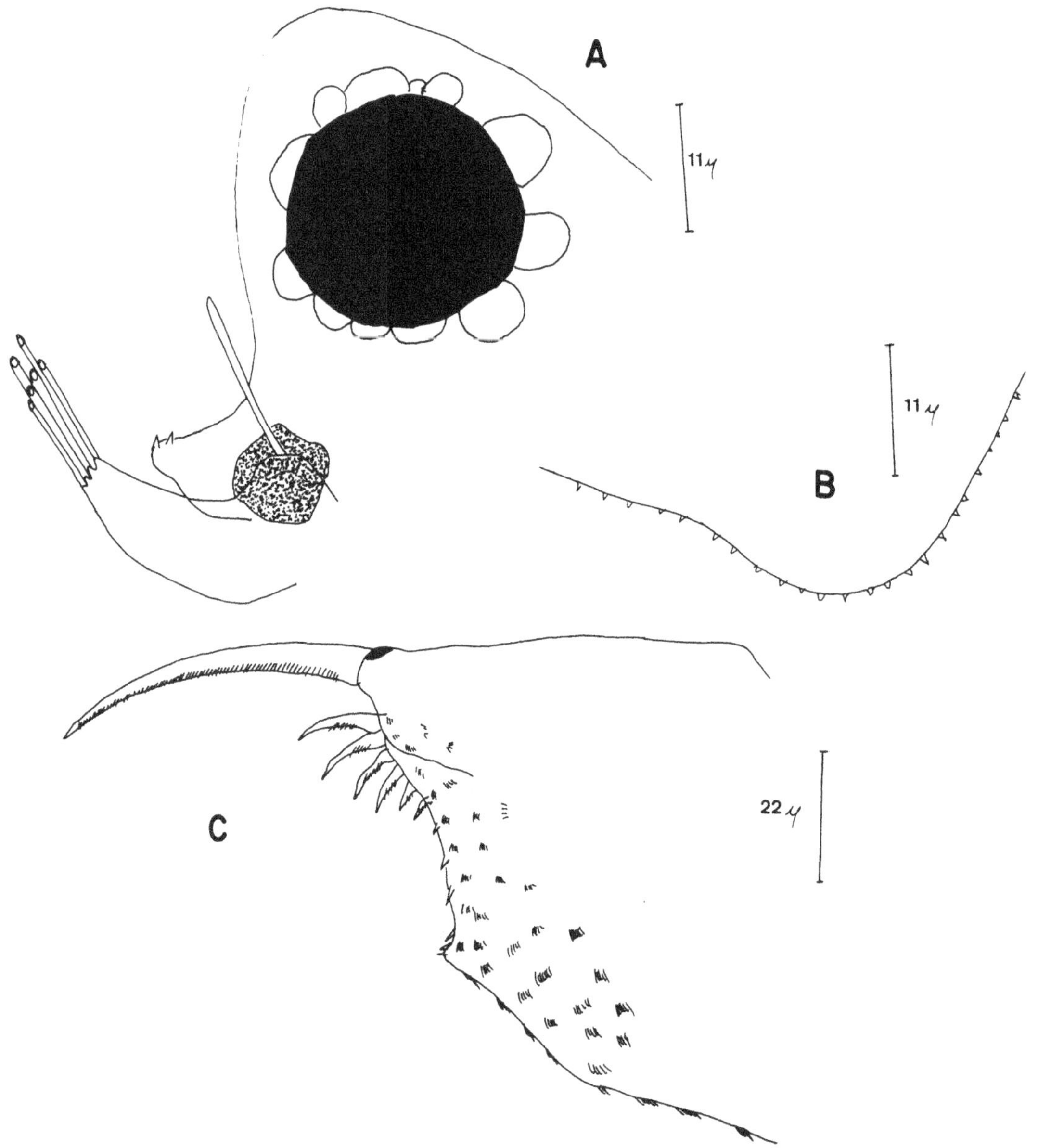

Fig. 5. Simocephalus latirostris (Stingelin, 1906). – A: Lateral view of head and rostrum. B: Detail of valve structure. C: Postabdomen.

Fig. 6. Simocephalus cf. *vetulus* (O.F. Müller, 1776). – A: Lateral view. B: Detail of head and rostrum. C: Postero-ventral angle of valve. D: Postabdomen.

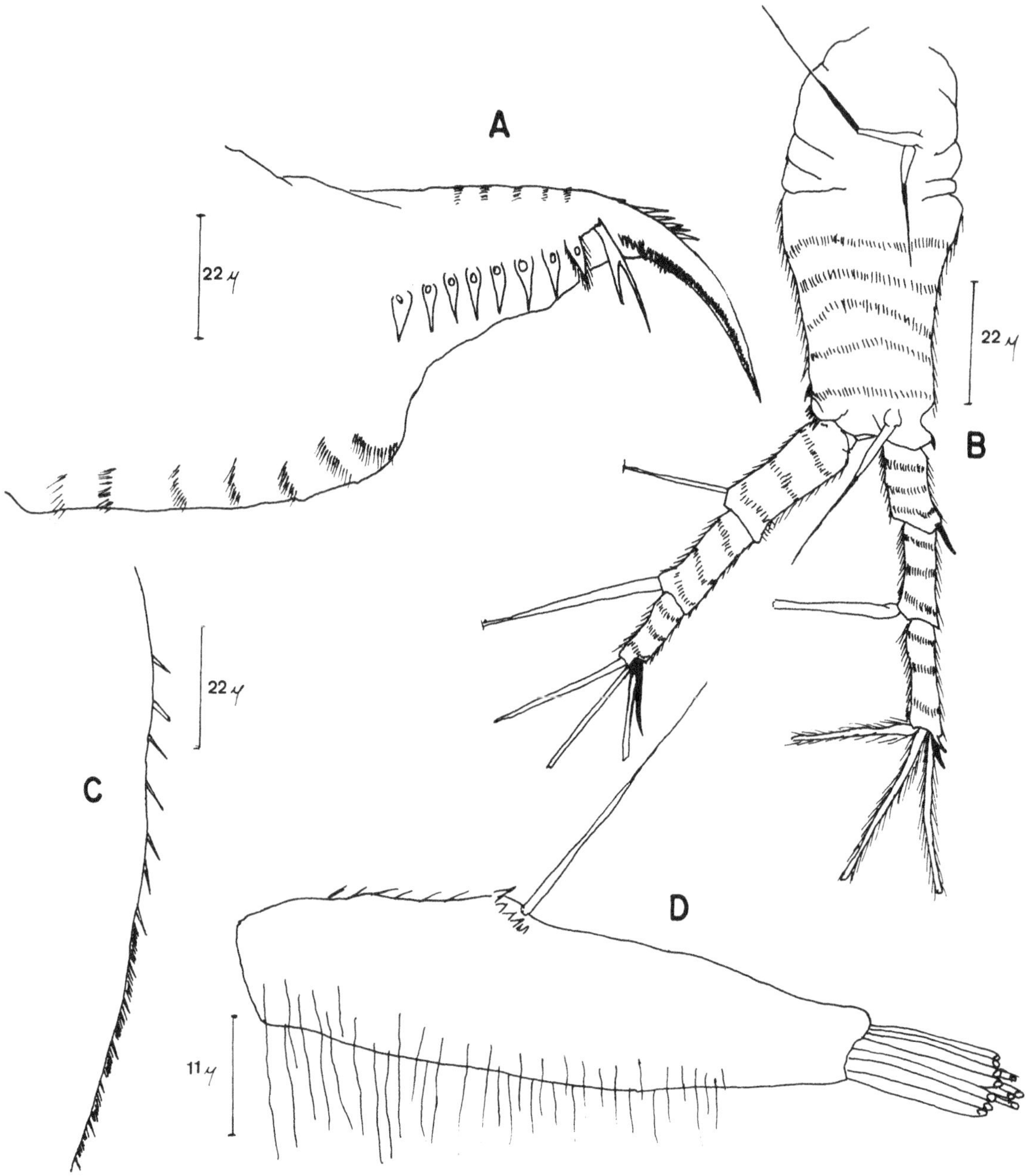

Fig. 7. Moina micrura ciliata (Daday, 1905). – A: Postabdomen. B: Antenna. C: Detail of postero-ventral angle of valve. D: Antennula.

adult 'cornuta' form (0.04 mm for juvenile). Apical horn size was 0.02–0.04 mm.

Simocephalus acutirostratus ***(King, 1853) (Fig. 4)***
Body shape with marked protuberance at posterior end of valves and a pointed rostrum, both of which are very characteristic of this species. Rhomboid ocellus at the base of the antennula. Wide postabdomen, with two fissures and 11 spines decreasing in length proximally, the two distalmost ones considerably longer than the others. Claw with 11 tooth-like spines, basally coincident with the description of Brehm (1937), who reported 14 spines on the postabdomen, in

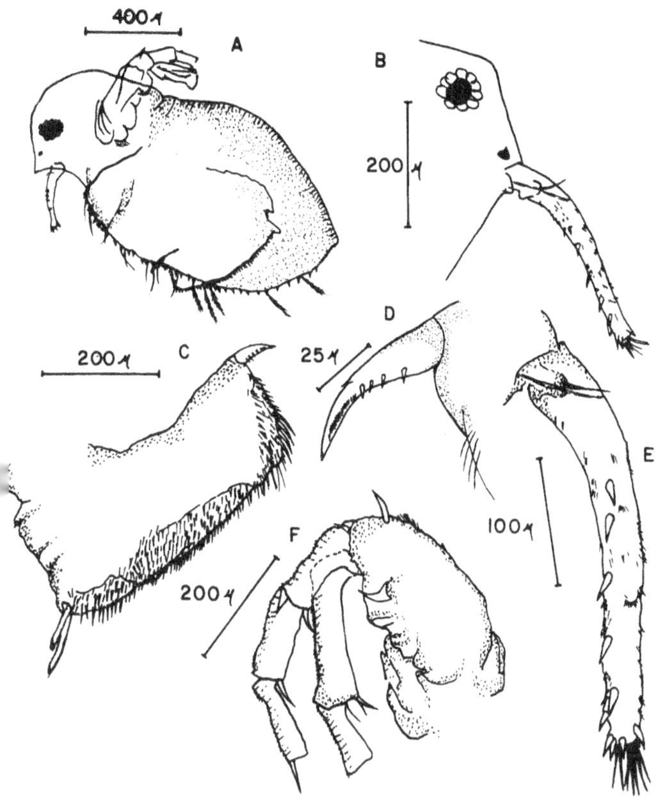

Fig. 8. Iheringula paulensis Sars, 1900. – A: Lateral view. B: Detail of head and rostrum. C: Postabdomen. D: Detail of postabdominal claw. E: Antennula. F: Antenna.

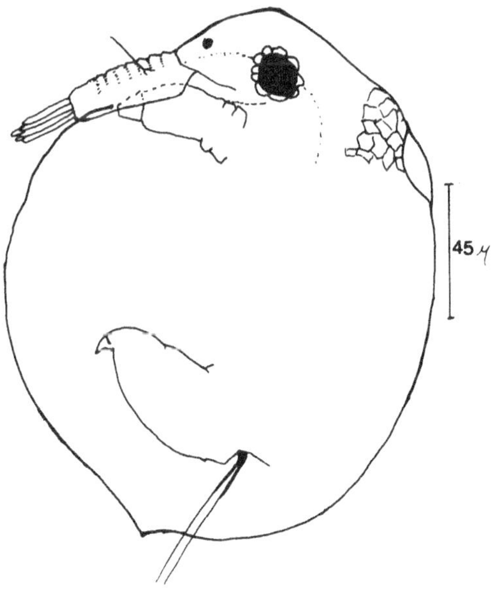

Fig. 9. Guernella rafaelis Richard, 1892. – Lateral view.

contrast with the 5–6 spines reported by Mamaril & Fernando (1978).

First record for Venezuela. From grasslands, free water, and stomach contents of fishes at Mantecal, as well as from the Alpargaton stream.

Size: 2.11 mm length; 1.72 mm height.

Simocephalus latirostris ***Stingelin, 1906 (Fig. 5)***

Coincident with description of Dumont & Van de Velde (1977). 2 small spines on front edge of rostrum; postabdomen with 6 spines of decreasing length, first 3 longer than others. Female with 12 eggs.

First record for Venezuela. Low abundance in flooded grassland and free water at Mantecal and Yaracuy River; highly abundant in Agua Fria reservoir.

Size: 1.08 mm length; 0.66 mm height.

Simocephalus ***cf.*** *vetulus* ***(O.F. Müller, 1776) (Fig. 6)***

Coincident with description of Negrea (1983). Elongated ocellus; postabdomen with 9 spines, first 2 for the juvenile and first 4 for the adult longer than the following ones, all progressively decreasing in length. Female with 6 eggs.

Highly abundant in Agua Fria reservoir.

Size: 1.39 mm length; 0.92 mm height.

Moinidae

Moina micrura ciliata ***(Daday, 1905) (Fig. 7)***
Coincident with description of Goulden (1968). Margin of antennula with bristles; 2 basal sensorial long setae on antenna, with several small basal denticles and bristles on proximal anterior margin, instead of the short setae described by Goulden (1968); 6–8 basal spines on postabdomen, 5 small spines on ventral base of claw.

First record for Venezuela. Abundant in San Antonio de Caparo pool for fish culture, and in free water and in stomach of fishes of Mantecal.

Size: 0.75–0.84 mm length; 0.51 mm height.

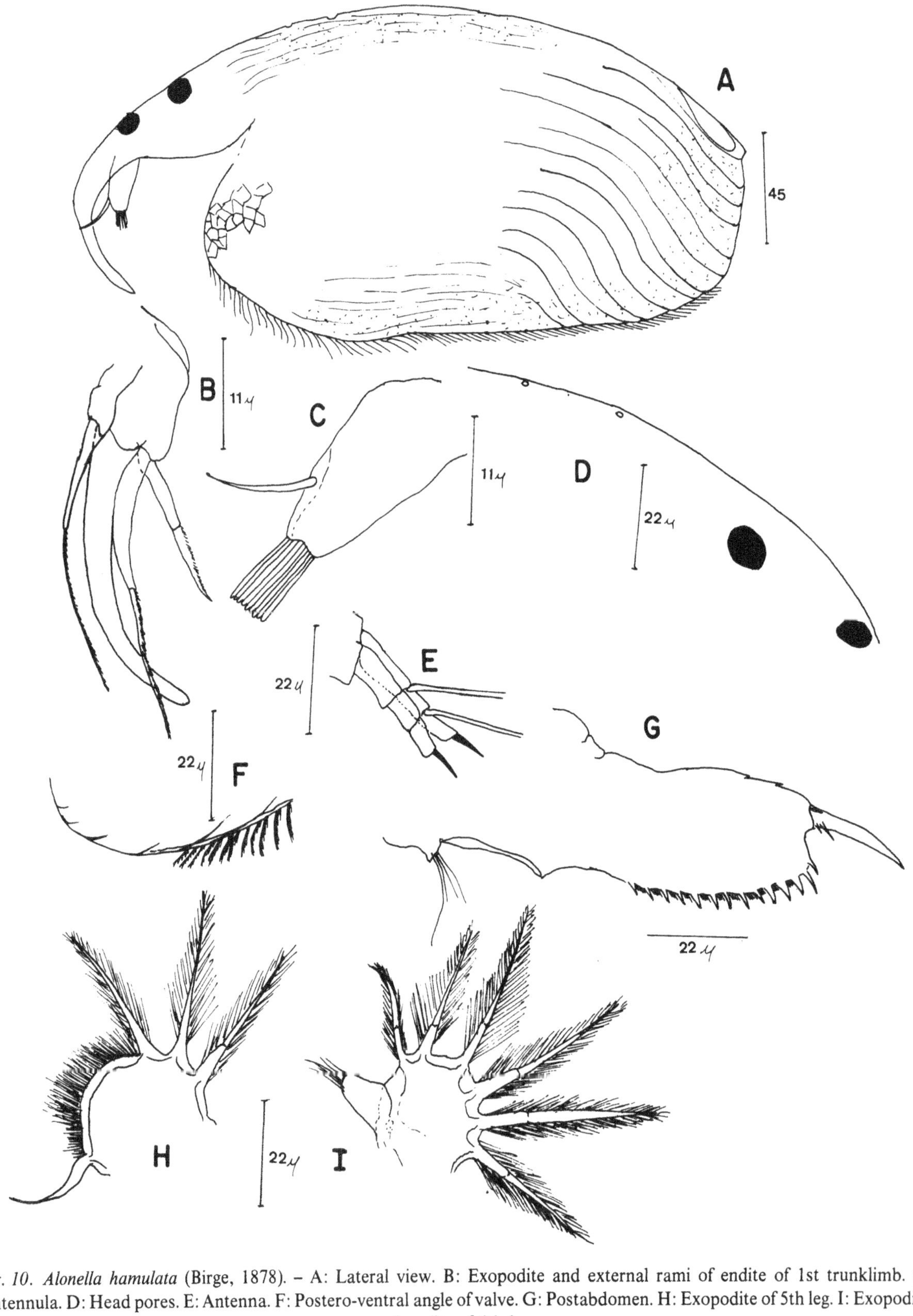

Fig. 10. Alonella hamulata (Birge, 1878). – A: Lateral view. B: Exopodite and external rami of endite of 1st trunklimb. C: Antennula. D: Head pores. E: Antenna. F: Postero-ventral angle of valve. G: Postabdomen. H: Exopodite of 5th leg. I: Exopodite and endite of 4th leg.

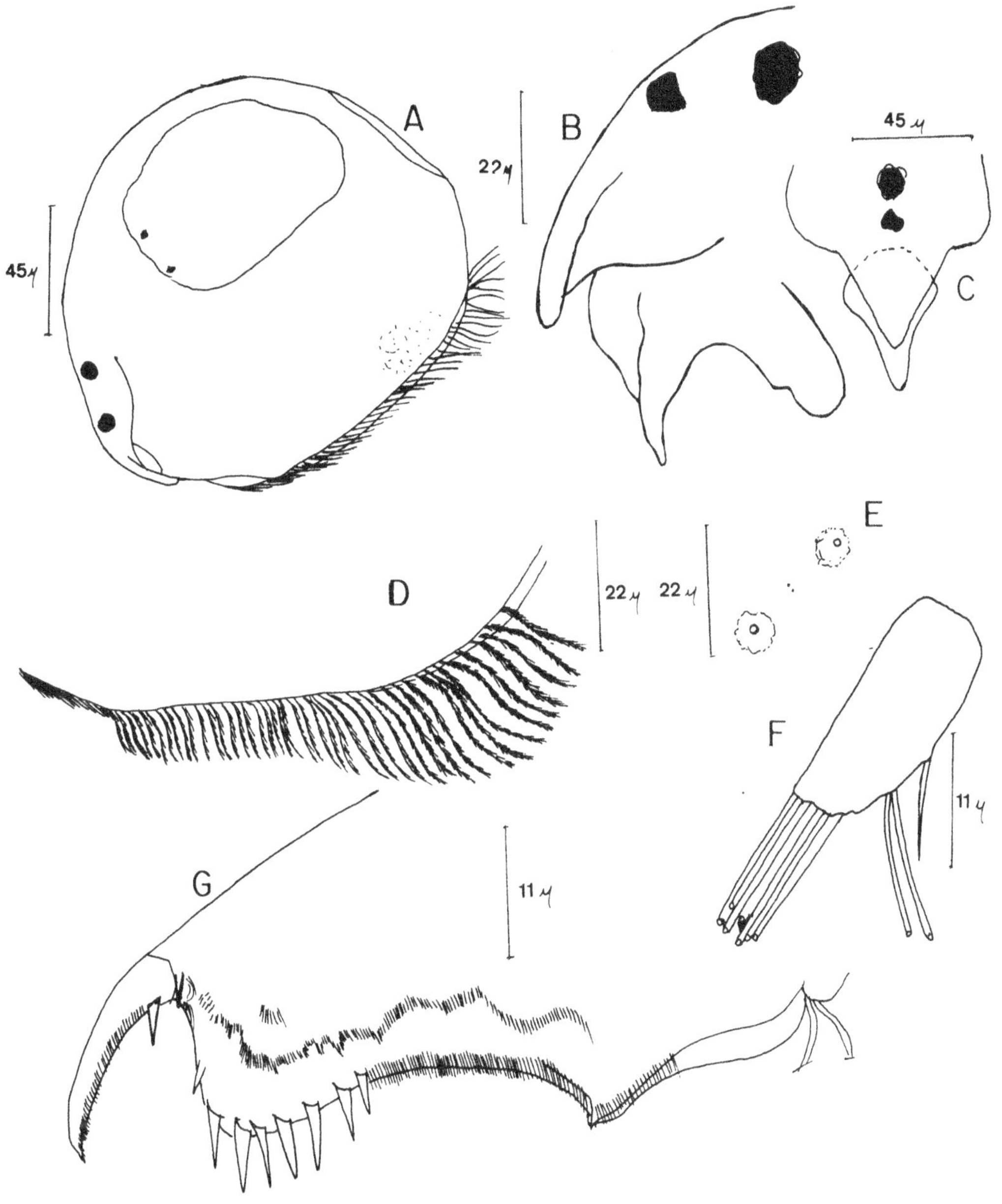

Fig. 11. Chydorus piger Sars, 1862. – A: Lateral view. B: Detail of head and labrum. C: Dorsal view of head. D: Detail of postero-ventral angle of valve. E: Head pores. F: Antennula. G: Postabdomen.

Macrothricidae

Iheringula paulensis ***Sars, 1900 (Fig. 8)***
Coincident with description of Sars (1900). 6 strong spines on margin and 4 on base of antennula. 4 basal spines and a small protuberance on distal superior part of postabdominal claw.

First record for Venezuela. Scarce and rare in grassland of Mantecal and in Provincial lagoon.

Size: 2.18 mm length, 1.54 mm height.

Guernella raphaelis ***Richard, 1892 (Fig. 9)***
Globular superior organ at end of head. Strong antennules with denticles.

First report for Venezuela. Low abundance in grassland and in stomach of fishes at Mantecal; abundant in Alpargaton stream.

Size: 0.33 mm length, 0.28 mm height.

Chydoridae - Chydorinae

Alonella hamulata ***(Birge, 1878) (Fig. 10)***
This species received several names and synonymies since its original description as *Pleuroxus hamatus* Birge, 1878. There is a complicated questioning about synonymies of *Alonella hamulata* (= *Pleuroxus chappuisi* Brehm, 1933 = *P. striatoides* Šrámek-Hušek, 1946), these are not yet resolved. Reported for Venezuela by Rey & Vasquez (1986).

Thoracic 5th leg with 3 posterior feathered bristles and 1 anterior, slightly curved smooth bristle on exopodite; margin of exopodite is rounded, covered by thin long hairs (Fig. 10-H). 4th leg with 6 identical feathered bristles on exopodite; endite reduced, with apical clusters of thin hairs (Fig. 10-I).

Low abundance in free water at Mantecal.

Size: 0.57 mm length, 0.27 mm height.

Chydorus pubescens ***Sars, 1901***
Only one specimen, with valve structure covered by large, strong spinules, found in a very abundant population of *Ch. pubescens* 'smooth' form in a free water sample from Mantecal. Paggi (1972) reported mixed populations of 'smooth' and pubescent forms, with several intermediate shapes.

Chydorus piger ***Sars, 1862 (Fig. 11)***
Coincident with description by Margaritora (1983), with the exception of a row of spines on base of postabdominal claw. Peculiar labrum, point terminated; large and rounded headpores with punctuated borders forming a halo.

First report for Venezuela from Churun Merun stream.

Size of parthenogenetic female: 0.32 mm length, 0.22 height.

Dunhevedia odontoplax ***Sars, 1901 (Fig. 12)***
Coincident with descriptions of Sars (1901) and Valdivia Villar (1987). Small spine and large teeth on rear inferior angle of valve.

Low abundance in grassland of Mantecal, Yaracuy River, and Alpargaton stream.

Size: 0.95 mm length; 0.66 mm height.

Ephemeroporus hybridus ***(Daday, 1905) and*** *Ephemeroporus tridentatus* ***(Bergamin, 1939)***
Coincident with descriptions of Rey & Vasquez (1986). Very abundant in grassland and stomach of fishes at Mantecal.

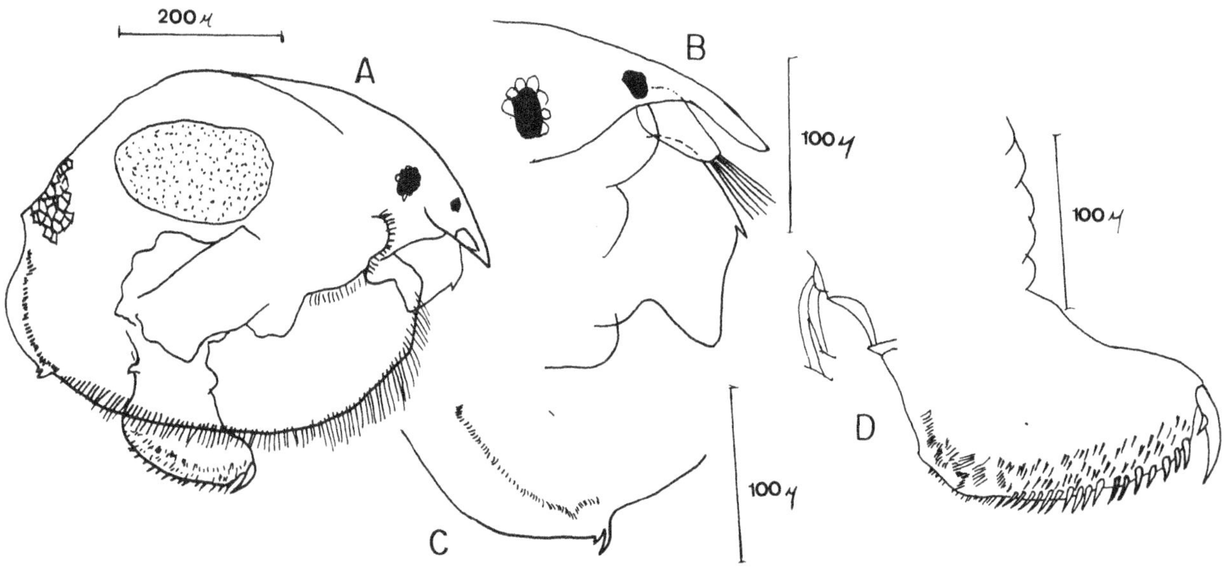

Fig. 12. Dunhevedia odontoplax Sars, 1901. – A: Lateral view. B: Detail of head and labrum. C: Detail of postero-ventral angle of valve. D: Postabdomen.

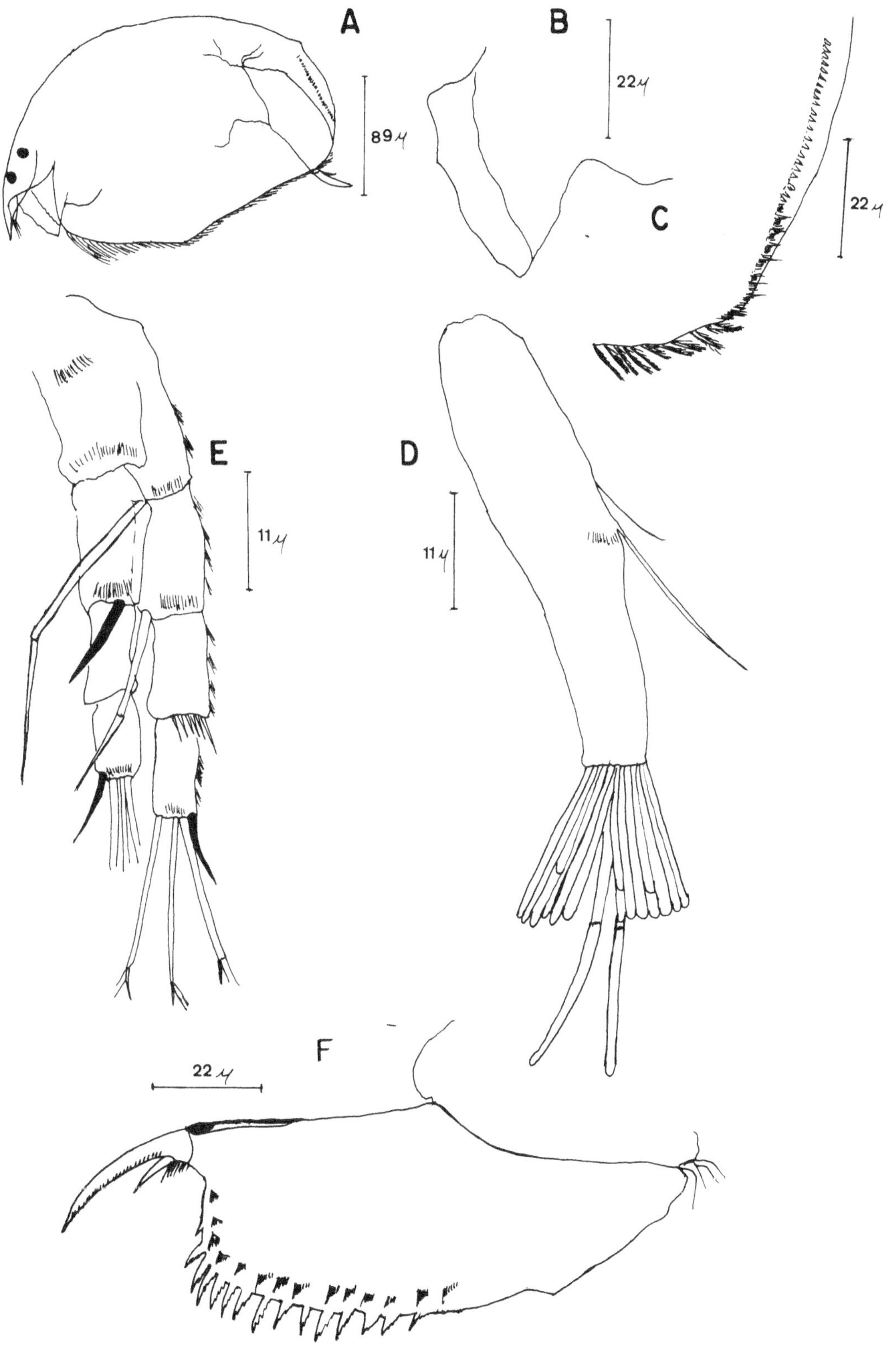

Fig. 13. Alona affinis affinis (Leydig, 1860). – A: Lateral view. B: Labrum. C: Postero-ventral angle of valve. D: Antennula. C: Antenna. F: Postabdomen.

Size: *E. hybridus*: 0.26 mm length, 0.22 mm height.
E. tridentatus: 0.27 mm length, 0.23 mm height.

Aloninae

Alona affinis affinis ***(Leydig, 1860) (Fig. 13)***
Reported by Rey & Vasquez (1986). Coincident with description of Margaritora (1983).

Low abundance at grassland in Mantecal, Churun-Merun River, and Alpargaton stream.

Size: 0.95 mm length; 0.51 mm height.

Alona cf. affinis f. ornata ***(Stingelin, 1895) (Fig. 14)***
Coincident with description of Negrea (1983). 2 series of small bristles on lower side of labrum; head pore with rounded border, instead of triangular as in *A. affinis affinis*. Saw-like processes in upper margin of postabdomen, near claw.

First for Venezuela, from Churun-Merun stream.

Size: 0.95 mm length, 0.32 mm height.

Alona rustica ***Scott, 1895 (Fig. 15)***
Concentric, reticulated lines on valves; postero-ventral setae longer, diminishing in length towards anterior end. Oval labrum with two crescents of bristles on posterior margin; typical three head-pores, with rounded lateral pores. Typical post-abdomen, with 2 small marginal spinules at distal tip, followed proximally by 9 longer spines.

Reported for Orinoco by Rey & Vasquez (1986). Abundant in Churun Merun stream (Auyantepuy).

Size of parthenogenetic female: 0.40 mm length, 0.31 mm height.

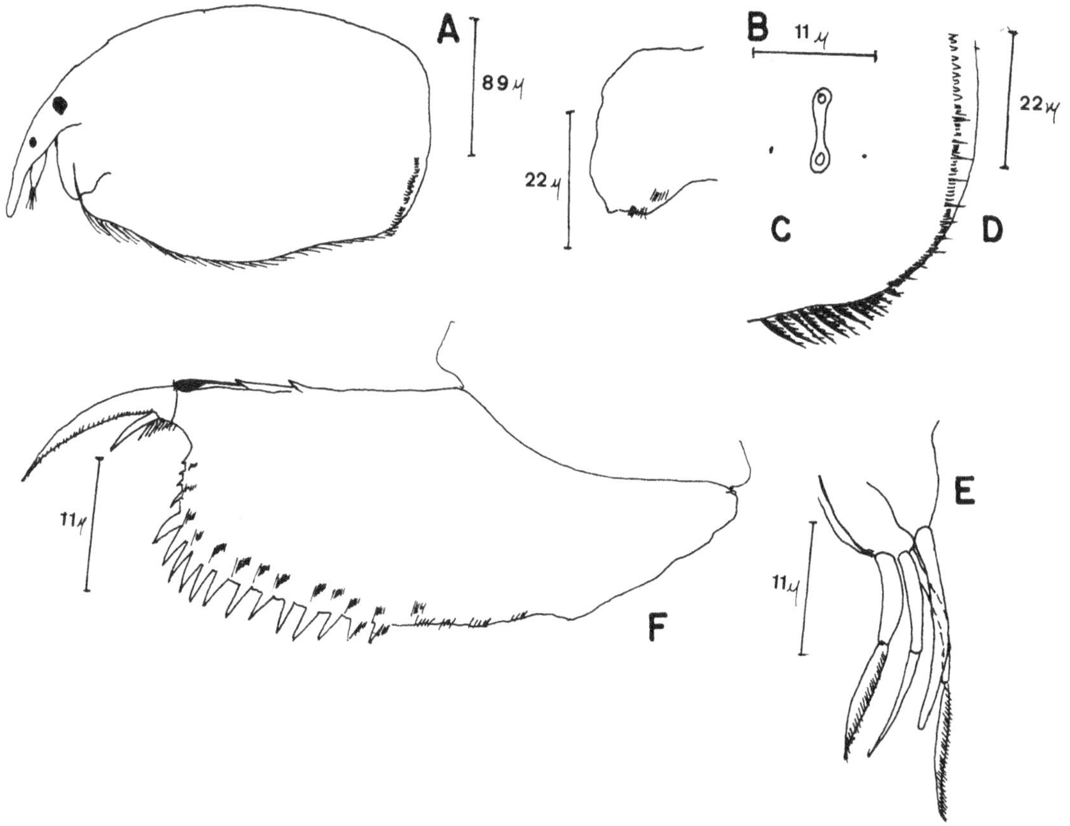

Fig. 14. Alona cf. *affinis* f. *ornata* (Stingelin, 1895). – A: Lateral view. B: Labrum. C: Head pores. D: Postero-ventral angle of valves. E: Leg 1. F: Postabdomen.

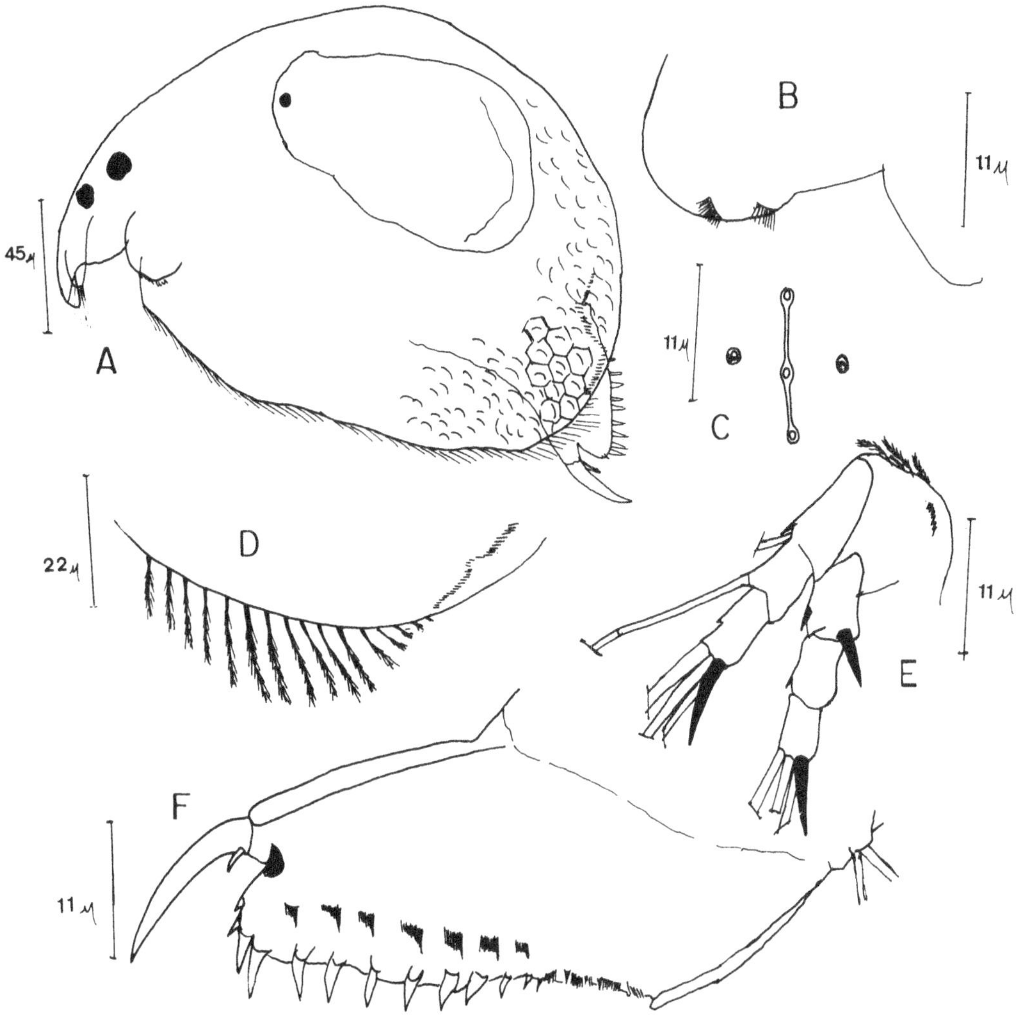

Fig. 15. Alona rustica Scott, 1895. – A: Lateral view. B: Labrum. C: Head pores. D: Detail of postero-ventral angle of valve. E: Antenna. F: Postabdomen.

Kurzia latissima **(*Kurz, 1875*)**
Coincident with description of Rey & Vasquez (1986).

Wide distribution and low abundance. In grasslands and free water at Mantecal, Yaracuy River, and Alpargaton stream.

Size: 0.314 mm length, 0.21 mm height.

Leydigiopsis curvirostris **Sars, *1901 (Fig. 16)***
Coincident with description of Sars (1901). 1 spine and transverse row of denticles on middle upper portion of labrum.

First record for Venezuela. Rare in grassland at Mantecal and Atapirire stream.

Size: 1.41 mm length, 0.86 mm height.

Discussion

24 cladoceran species were founded, including 9 new records for Venezuela and 18 additional records for flooded grasslands of Llanos at Mantecal, where Zoppi de Roa *et al.* (1985) reported 33 species, for a total of 52 species for that area, second in richness in Venezuela to the 59 species reported by Rey & Vasquez (1986) for the Orinoco Basin.

Some of the species were found also in other localities, showing a wide distribution. For example *S. latirostris* was found in the Agua Fria reservoir, the Yaracuy River (near its mouth at sea level), and in the Mantecal grasslands (very low

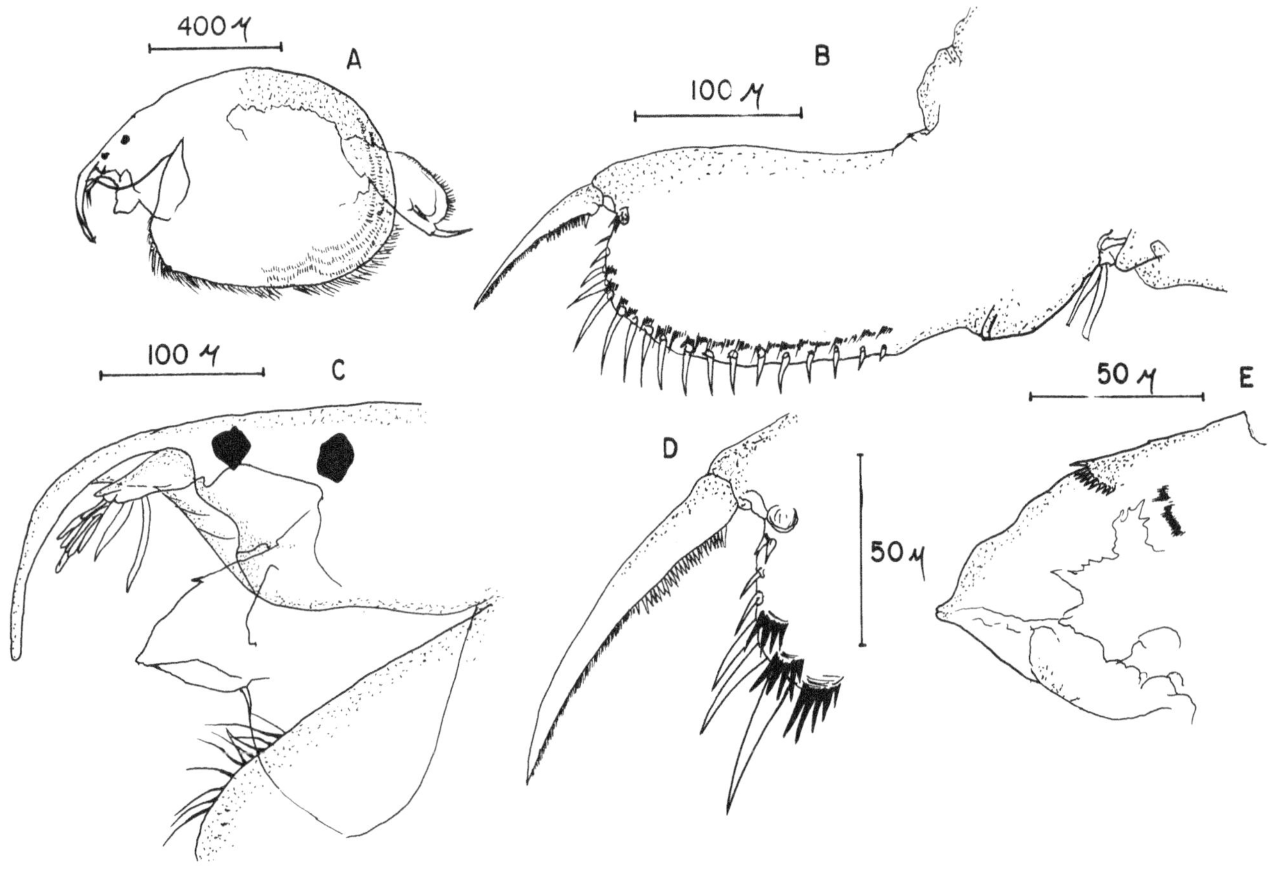

Fig. 16. Leydigiopsis curvirostris Sars, 1901. – A: Lateral view. B: Postabdomen. C: Detail of head and labrum. D: Detail of postabdominal claw. E: Labrum.

altitude, far away from the sea). In contrast, *D. laevis* and *S. vetulus* seem restricted to the Agua Fria reservoir, and *B. affinis f. ornata* to the Churun-Merun River (from the first zooplankton samples taken in Auyantepuy highlands), both at more than 1800 m above sea level. Van de Velde *et al.* (1978) reported a nearctic distribution for *D. laevis*, that seems to be endemic and restricted to high mountain lakes of Neotropical region.

It is difficult to relate *S. kingi* to previous diagnoses because of the scarcity of material, restricted only to parthenogenetic females and juveniles. The 'normal' form (without an apical horn) was found only in Mantecal flooded grasslands. It is possible that those habitats provide protection against predators, which could be one of the factors that helps explain polymorphism in the Tropics.

Predation by fish and other organisms seems to be a factor influencing the composition and abundance of cladocerans in tropical environments. The finding of some uncommon species in the stomachs of small fishes suggests the importance of including studies of the diets of zooplankton predators along with normal environmental samples.

Acknowledgements

Thanks to Consejo de Desarrollo Cientifico y Humanistico and Instituto de Zoologia Tropical of Universidad Central de Venezuela for financial support. Thanks to G. Colomine and M. J. Pardo (I.Z.T., U.C.V.) for technical assistance, Dr. D. G. Frey (Indiana University, U.S.A.) and Dr. I. Hudec (PBH, OOAV-hydrobiol. Lab., Kosice,

Table 2. Species list and relative abundance in each location.

Species	Locations									
	Mantecal		Agua fria	Yara cuy	Alpar gaton	Atapi rire	SN Ant caparo	Provin cial	Churun merun	Ata bapo
	GL	FW								
*Latonopsis occidentalis**		L								
Pseudosida ramosa		L								
Diaphanosoma brevireme	L									
Daphnia laevis			H							
Scapholeberis cf. kingi	L					L		L		L
*Simocephalus acutirostratus**	L	L			L					
*S. latirostris**	L	L	H	L						
*S. cf. vetulus**			H							
*Moina micrura ciliata**	L						H			
*Iheringula paulensis**		L						L		
*Guernella raphaelis**		L			H			L		
Alonella hamulata		L								
Chydorus pubescens f. pubescente	L									
*Chydorus piger**									H	
Dunhevedia crassa		L		L						
D. odontoplax		L		L	L					
Ephemeroporus hybridus		H								
E. tridentatus		H								
Alona affinis affinis		L			L				H	
*A. cf. affinis f. ornata**									L	
A. rustica									H	
Kurzia latissima	L	L		L	L					
Leydigiopsis curvirostris		L				L				
Oxyurella ciliata		L								

GL: Grassland
FW: Free water
L: Low abundance
H: High abundance
*: New record for Venezuela

Czechoslovakia) for comments and suggestions about some of the species.

References

Brehm, V., 1937. Brasilianische Cladoceren. Int. Revue ges. Hydrob. Hydrogr. 35: 497–512.

Dumont, H. J. & I. Van de Velde, 1977. Cladocères et Conchostracés récoltés par le professeur Th. Monod dans la moyenne vallée du Niger en décembre 1972 et janvier 1973. Bull. l'I.F.A.N. T. 39, sér. A, No. 1: 75–93.

Dumont, H. J. & J. Pensaert, 1983. A revision of the Scapholeberinae (Crustacea: Cladocera). Hydrobiologia 100: 3–45.

Goulden, C. E., 1968. The systematics and evolution of the Moinidae. – Trans. Am. Phil. Soc., New Series, Vol. 58, Part VI: 5–98.

Infante, A., 1984. A note about *Daphnia* in Venezuela waterbodies. Hydrobiologia 119: 81–81.

Mamaril, A. C. & C. H. Fernando, 1978. Freshwater zooplankton of the Philippines (Rotifera, Cladocera and Copepoda). Nat. Appl. Sci. Bull. 30: 109–221.

Margaritora, F., 1983. Cladoceri (Crustacea: Cladocera). Guide per il riconoscimento delle specie animali delle acque interne italiane. Consiglio Nazionale delle Ricerche, AQ/1/197. Stamperia Valdonega, Verona: 169 pp.

Negrea, A., 1983. Fauna Republicii Socialiste Romania, Crustacea, Volumul IV, Fascicula 12, Cladocera. Editura Academiei Republicii Socialiste Romania, Bucuresti: 399 pp.

Paggi, J. C., 1972. Nota sistematica acerca de algunos cladoceros del genero *Chydorus* Leach 1843, de la Republica Argentina. Physis, T.31, No. 82: 223–236.

Paggi, J. C., 1976. Cladoceros Macrothricidae nuevos para la fauna Argentina. Physis, Seccion B, T. 35, No. 91: 103–112.

Rey, J. & E. Vasquez, 1986. Cladocères de quelques corps d'eaux du bassin moyen de l'Orénoque (Vénézuéla). Ann. Limnol. 22: 137–168.

Sars, G. O., 1900. Description of *Iheringula paulensis*, G. O. Sars. A new generic type of Macrothricidae from Brazil. Arch. Math. Naturvidens. 22(6): 1–27.

Sars, G. O., 1901. Contributions to the knowledge of the freshwater entomostraca of South America. Archiv Math. Naturvidens 23(3): 1–102.

Valdivia Villar, R. S., 1987. Estudio de la fauna Cladocera Neotropical (Cladocera, Chydoridae): I. Redescripción de *Dunhevedia odontoplax* Sars, 1901 del Peru. Amazoniana 10: 137–147.

Van de Velde, I., H. J. Dumont & P. Grootaert, 1978. Report on a collection of Cladocera from Mexico and Guatemala. Arch. Hydrobiol. 83: 391–404.

Zoppi de Roa, E., F. Michelangelli & L. Segovia, 1985. Cladocera (Crustacea, Branchiopoda) de sabanas inundables de Mantecal, Edo. Apure, Venezuela. Acta Biol. Venez., 12: 43–55.

Hydrobiologia **225**: 63–70, 1991.
V. Kořínek & D. G. Frey (eds), Biology of Cladocera.

Comparative-morphological and comparative-ecological analysis of the Sidoidea (Crustacea: *Daphniiformes*): the main moments of the topic

N. M. Korovchinsky
A. N. Severtsov Institute of Animal Evolutionary Morphology and Ecology of the USSR Academy of Sciences, Leninsky Prospect 33, Moscow 117071, USSR

Key words: comparative morphology, *Sidoidea*, predation

Abstract

Comparative-morphological analysis of the structure of the *Daphniiformes* of the superfamily *Sidoidea* (= *Ctenopoda*), comprising the families *Sididae* and *Holopediidae*, compared with representatives of other groups – *Chydoroidea* (= *Anomopoda*), *Limnadiiformii* (= *Conchostraca*), *Polyphemiformii* (= *Onychopoda*, *Haplopoda*), and *Branchipodiones* (= *Anostraca*, *Notostraca*) – is made. The main parameters and structures important in functional-morphological and ecological aspects are considered. In contrast with all other crustaceans under consideration, in the *Sididae* and the *Polyphemiformii*, the organs contributing to rapid maneuverable swimming, soaring, and good orientation in the environment, have received the most development. All this predetermined elaboration of the life strategy in these forms mainly aimed, especially in the *Sididae*, at active avoidance of predation, combined with elements of passive protection (imperceptibility attained by various methods). All others, including the *Holopediidae*, have emphasized passive protection.

Introduction

Many representatives of the superfamily *Sidoidea* are still poorly described, there being no data on some important morphological structures or such data available are erroneous. No attempts at detailed consideration of these crustaceans in a wide comparative morphological aspect have been made.

While the extremely peculiar appearance of *Holopedium* is well known to hydrobiologists, the notion of specificity of the body structure of the *Sididae* is far more vague. Consequently in the mind of experts the *Sididae* are not duly distinguished from other cladocerans (= *Daphniiformes*) having a general superficial similarity, primarily the *Daphniidae*, and are fused together with them into some general image concerned with some common morpho-functional and ecological essence.

Only in a recent paper by Fryer (1987), attention for the first time is drawn to significant general morphological distinctions between the *Sidoidea* and the *Chydoroidea*, this being done, however without a detailed consideration of some important functionally active organs and of the evolutionary-ecological aspect of the formation of these morphological features.

This study is aimed at elusidation of the evolutionary-morphological and evolutionary-ecological development of the *Sidoidea*, mainly those that inhabit continental waters. This can be done by comparison of the principal morpho-functional and ecological characters of their representatives with those of other *Branchipodiodes* (= *Branchiopoda*) living in similar biotopes. Of course the small size of this paper does not permit the description of all in detail. This will be published elsewhere.

In the present paper mainly the well-elaborated classification and terminology of Starobogatov (1986) will be used.

Comparative-morphological analysis

The body length of adult parthenogenetic females of different species of the *Sididae* and the *Holopediidae* is very different, ranging from 0.6 to 4.0 mm. Size differences are well-expressed in the species of the same genus. Intra- and interpopulational size differences may be rather high too.

It is quite clear that the *Sidoidea* belong among largest *Daphniiformes* along with the *Daphniidae* and some *Moinidae* and *Macrothricidae*. Even those that inhabit the pelagic zone of waterbodies attain a rather large size of 1.4–2.2 mm (*Diaphanosoma dubia*, *D. mongolianum*, *Limnosida frontosa*, *Holopedium gibberum*). It is the *Sidoidea* along with *Daphnia* which provide the commonest large-sized component of planktonic communities of various regions of the World, even surpassing *Daphnia* in the total number of species in planktonic communities.

Data on the main morphological traits of the *Sididae* and other crustaceans under consideration are summarized in Table 1.

In all *Sididae* the body is elongated oblong-oval. During swimming the long axis is oriented horizontally or almost horizontally. A large head is also characteristic of representatives of this group, making up 30–43% of the body length. The form of the head is highly variable. It is important that in many *Sididae* it depends much on the extent of the development and the character of attachment of the system of muscles situated in its dorsal and frontal parts and those moving the swimming antennae. This is especially true of species permanently living in the open water (*Limnosida*, *Diaphanosoma*). But there are some species with these muscles developed comparatively weakly, and the head, especially its dorsal part, is accordingly smaller, lower and more rounded (*D. sarsi*, *D. brevireme*, *D. volzi*, etc.). In part, this is associated with a relatively smaller body size in the latter species and probably with lesser locomotive activity of the antennae.

In contrast to the *Sididae*, most of *Chydoroidea*, including those that are constantly planktonic, possess mainly a rather short body, that is high and wide, rounded or cuneiform. Sometimes the habitus of these crustaceans is determined by various outgrowths, expressed reticulation, and spines on the exoskeleton. The form and size of their head are not mainly related to the development of antennal muscles. Formation of a massive helmets is attended by an increase in the muscles moving the antennae (Hebert, 1978). Yet the latter is most likely a secondary feature and is unavoidably results from the increased body weight. Often the thick headshield and the rostrum of various size and form are strongly developed.

Comparison of the visual possibilities of the eye of various *Sidoidea* yields very interesting results as indicated by the number of ommatidia (Table 2). Most representatives of this group have numerous ommatidia (44–81), especially the active planktonic forms *Diaphanosoma* and *Limnosida*. The same is true of *Sida crystallina*, which is capable of rapid shottime swimming oriented at the substrat for attachment using vision (Szlauer, 1973; Fairchield, 1981). Populating of a substratum by these crustaceans was found to be 2–4 times less intensive at night than in the daytime (Korovchinsky, 1981). It is remarkable that bottom-living *Latona*, *Latonopsis* and *Sarsilatona* also have a well-developed eye, peculiarly shifted to the dorsal side of the head and situated far from the ocellus, in contrast with other sidids.

At the same time in the *Chydoroidea* the eye is developed much less, as indicated by the number of ommatidia (Table 2), especially in most *Chydoridae*. In some of the latter having a bottom or underground mode of life (*Chydorus gibbus*, *Rhynchotalona falcata*, *Leydigia*, *Leydigiopsis*, *Alona*

Table 1. Morphological features of different groups of *Branchipodiodes (= Branchiopoda)*

	Features	*Sididae*	*Polyphemiformii*	*Chydoroidea*	*Holopediidae*	*Limnadiiformii*	*Branchipodiformes*	*Triopiformes*
1. Body form (with shell if it occurs)	Elongate, relatively narrow in most representatives	+	+				+	
	Mostly massive and broad, more or less rounded or oval			+	+	+		+
2. Form and size of the head	Relatively large, more or less separate from the trunk	+	+				+	
	Relatively small, usually merged with trunk			+	+	+		+
3. Eye(s)	Developed, with numerous ommatidia	+	+			+	+	+
	Poorly developed, with few ommatidia			+	+			
4. Swimming antennae	Strongly developed, with diverse structure and armament	+	+					
	Poorly developed, with relatively uniform structure and armament. In some cases antennae may be absent (Triopiformes)			+	+	+	+	+
5. Shell coverings	Relatively thin, transparent or pigmented, without outer defensive structures	+	+				+	
	Relatively thick, may be strong sclerotized or with different outer defensive structures			+	+	+		+
6. Cyclomorphic variability	Present			+				
	Absent	+	+		+	+	+	+
7. Long posterior part of the body, long posterior setae or prominences	Present	+	+				+	+
	Absent			+	+	+		

phreatica), the eye is much reduced and disappears completely in *Monospilus* and *Bryospilus*.

In all *Sidoidea* without exception swimming antennae are very long and powerful. In different *Sididae* the relative length of these antennae (excluding the setae) is noticeably different.

In representatives of different genera of *Sididae* the structure of the antennal branches, number of sulted in the development of various external protective and disguising formations, morphoadaptations related to better buoyancy and stabilization of the body in comparatively slow swimming, crawling, or feeding on substrata. The outer structures (helmets, crests, spines, mucro) of the *Daphniidae* and *Bosmina* may be rapidly formed and variously transformed under the influence of various factors (e.g., Grant & Bayly, 1981; Krueger & Dodson, 1981; Hebert & Grewe, 1985). Most representatives of the *Chydoroidea*, the *Limnadiiformii* and *Triopiformes* mainly prefer hard substrata, swim little, some have lost the ability to swim. In the *Chydoroidea* especially, the system of thoracic limbs is diversely constructed, manifesting multifaceted specializations, including locomotory. Sometimes the postabdomen participates in locomotion (in some cases only the postabdomen is used, its structure and armament being very specialized (Fryer, 1968; Smirnov, 1971).

To elucidate reasons for such diverse morphoadaptive development of crustaceans of continental waters, attention must be given first of all to interrelations of these crustaceans with predators. At present the great role of predators, especially fishes, in the formation of the structure of freshwater zoocenoses is well defined (Hrbacek, 1962; Brooks & Dodson, 1965; Greene, 1983; DeBernardi *et al.*, 1987; Ghilyarov, 1987). The pressure of fishes renders a longterm evolutionary impact (Kerfoot & Lynch, 1987). Predation to a great extent controls the range of body size, many traits of morphology and behaviour and of life cycles of crustaceans of continental water bodies (Kerfoot *et al.*, 1980; Lynch, 1980; Greene, 1983).

Already in the late Paleozoic era and in the first half of the Mesozoic era because of the increasing influence of fishes the especially large and evolutionary primitive *Branchipodiformes*, *Triopiformes* and *Limnadiiformii* were mainly driven out from most freshwater habitats into small, often temporary, waters and some other refugia (Potts & Durning, 1980; Kerfoot & Lynch, 1987). At the same time most of the *Chydoroidea* and all the *Sidoidea*, including the largest representatives (*Sida*, *Limnosida*, *Latona*, etc.) commonly populated habitats with potentially high pressure from invertebrate or vertebrate predators and had to adapt themselves to it.

According my point of view the previously outlined morphogenetic and behavioural differences are directly related to different ways that crustaceans have evolved to avoid predation. In the first case (*Sididae*) the active avoidance of predators by prey animals is most significant. The chief peculiarity of the *Sididae*, radically differentiates them from other cladocerans, is that their avoidance tactics are reminiscent of those used by the *Copepoda*, especially the *Calanoida*. Pelagic, phytophilous, and bottom-living representatives of this family possess, as already noted, a very strong locomotory apparatus and are capable of rapid jumps to avoid predators, followed by rather prolonged periods of inactivity, when they become essentially invisible to predators that detect their prey by mechanoreceptors or eyes. The importants of the latter for *Diaphanosoma* has been recorded (Brooks, 1968; Zaret, 1980; Williamson, 1983). It has been shown experimentally that *Diaphanosoma* are caught by sucking worse than any other cladocerans, being second in this respect only to the *Copepoda* (Szlauer, 1965; Drenner & McComas, 1980). This property is very important in protecting them from planktophagous fishes that feed by filtration.

In the second case various methods of passive protection are basic, including relocation to habitats inaccessible to fish. For example, many of the above mentioned morphological characters of the *Chydoroidea*, especially thick covers, outgrowth, crests, etc., suggest, first of all, adaptations for passive protection. Being not able to swim rapidly they may have a special mode of locomotion that disorients predators (Li & Li, 1979), akinesis which is typical for many of them (Smirnov, 1977; Kerfoot *et al.*, 1980). Mode of passive protection (Hamilton, 1958); Zaret, 1980; Stenson, 1987; etc.) obviously brings together the *Holopediidae* and representatives of the latter group.

Morphological and ethological evidence on the specificity and success of the protective strategy of the *Sididae* against predation can be supported by some hydrobiological data. Thus most groups of planktonic cladocerans exhibit a decrease of species diversity in tropical waters (Fernando, 1980a, b) as a result of much higher fish predation there (Kerfoot & Lynch, 1987), but *Diaphanosoma* manifests the opposite trend. The species of this genus everywhere live continuously in zones of high feeding activity of fish (for example, in littoral or sublittoral zones, in the upper layers of waters of the pelagic zone of large waterbodies where their vertical migration rarely exceeds 5 m in the epi-metalimnion), in comparison with pelagic *Daphnia*, especially the larger ones, which prefer to remain in the meta-hypolimnion during daylight (Stich & Lampert, 1981; Kiselev, 1980; Matveev, 1986), thereby escaping predation. Coefficient of electivity of *Diaphanosoma* preyed on by fish are negative or equal to small positive values (Gophen & Scharf, 1981; Boykova, 1986; etc.).

So the *Sididae* are unique, being an example of the purest strategy of active avoidance of predators. Combined, to a certain extent, with elements of passive protection (imperceptibility attained in various ways), it ensures the successful survival of species of this group at a considerable predation pressure.

The fresh-water *Polifemiformii*, especially *Bythotrephes* and *Leptodora*, are still more skilled and strong swimmers than the *Sididae*, and like *Diaphanosoma* they inhabit the open water of the littoral zone or the upper layers of the pelagic zone and can withstand a slight and moderate pressure of fishes (Zaret, 1980; Ghilyarov, 1987; DeBernardi *et al.*, 1987). It seems that their morphoadaptations are related primarily to their active predatory mode of life. From the viewpoint of protcction of themselves from predation, they are more noticeable and vulnerable to predators of higher trophic level than are other pelagic *Cladocera*, especially *Diaphanosoma*. *Leptodora* and *Bythotrephes* are very large (in the latter the body is also rather intensively pigmented) and have a big pigmented eye. Besides, they are in perpetual movement and tend to swarm, which occurs especially in *Polyphemus* (Butorina, 1982; Mordukhai-Boltovskoi & Rivier, 1987). Selection of the fresh-water *Polyphemiformii* by fish is very high (Boykova, 1986, etc.).

If the variety of the crustaceans considered is expended, the freeliving *Calanoida* and *Cyclopoida* are obviously the closest analog of the *Sididae*.

In contrast, the *Ostracoda* as well as most *Branchipodiodes* (or *Branchiopoda*) did not developed structures related to rapid swimming, and they use mainly the strategy of passive protection. Realization of the same life strategies in the *Branchipodiodes*, the *Copepoda*, and the *Ostracoda* is based in many respects on different morphofunctional foundations.

Acknowledgements

I am deeply indebted to my colleagues O.S. Boykova, V.F. Matveev, and N.N. Smirnov (Institute of Animal Evolutionary Morphology and Ecology of the USSR Acad. Sci.) for their very useful comments made during the discussion of the present paper, which greatly contributed to its refinement.

References

Behning, A., 1941. Kladotsera Kavkaza. Zarja Vostoka, Tbilisi, 383 pp. (*Cladocera* of the Caucasus).

Berg, K., 1929. A faunistic and biological study of Danish *Cladocera*. Vidensk. Medd. Dansk naturh. Foren. 88: 31–112.

Boykova, O. S., 1986. Feeding of fish in Lake Glubokoe. Hydrobiologia 141: 95–111.

Brooks, J. L., 1968. The effects of prey size selection by lake planktivores. Syst. Zool. 17: 272–291.

Brooks, J. L. & S. I. Dodson, 1965. Predation, body size and composition of plankton. Science 150: 28–35.

Butorina, L. G., 1982. O stajah *Polyphemus pediculus* (L.). Trudy Inst. biol. vnutrennyh vod 46/49: 69–88. (On the swarms of *Polyphemus pediculus* (L.)).

DeBernardi, R., G. Giussani & M. Manca, 1987. *Cladocera*: Predators and prey. Hydrobiologia 145: 225–243.

Drenner, R. W. & S. R. McComas, 1980. The roles of zooplankter escape ability and fish size selectivity in the selective feeding and impact of planktivorous fish. In W. C. Kerfoot (ed.) 'Evolution and ecology of zooplankton communities'. The University Press of New England, Hanover (N.H.): Lond. 587–593.

Fairchield, G. W., 1981. Movement and microdistribution of *Sida crystallina* and other littoral microcrustacea. Ecology 62: 1341–1352.

Fernando, C. H., 1980a. The freshwater zooplankton of Sri Lanka, with a discussion of tropical freshwater zooplankton composition. Int. Revue ges. Hydrobiol. 65: 85–125.

Fernando, C. H., 1980b. The species and size composition of tropical freshwater zooplankton with special reference to the Oriental region (South East Asia). Int. Revue ges. Hydrobiol. 65: 411–425.

Fryer, G., 1968. Evolution and adaptive radiation in the *Chydoridae* (*Crustacea: Cladocera*): a study in comparative functional morphology and ecology. Phil. Trans. r. Soc. B 254: 221–385.

Fryer, G., 1987. Morphology and the classification of the so-called *Cladocera*. Hydrobiologia 145: 19–28.

Ghilyarov, A. M., 1987. Dinamica chislennosti presnovodnyh planktonnych rakoobraznyh. Nauka, Moskva, 191 pp. (Quantity dynamics of freshwater planktonic crustaceans).

Gophen, M. & A. Scharf, 1981. Food and feeding habit of *Mirogrex* finderlings in Lake Kinneret (Israel). Hydrobiologia 78: 3–9.

Grant, J. W. G. & I. A. E. Bayly, 1981. Predator induction of crests in morphs of the *Daphnia carinata* King complex. Limnol. Oceanogr. 26: 201–218.

Greene, C. H., 1983. Selective predation in freshwater zooplankton communities. Int. Revue ges. Hydrobiol. 68: 297–315.

Hamilton, J. D., 1958. On the biology of *Holopedium gibberum* Zaddach (*Crustacea: Cladocera*). Verh. int. Ver. Limnol. 13: 785–788.

Hebert, P. D. N., 1978. The adaptive significance of cyclomorphosis in *Daphnia*: more possibilities. Freshwat. Biol. 8: 313–320.

Hebert, P. D. N. & P. M. Grewe, 1985. *Chaoborus*-induced shift in the morphology of *Daphnia ambigua*. Limnol. Oceanogr. 30: 1291–1297.

Hrbacek, J., 1962. Species composition and the amount of the zooplankton in relation to the fish stock. Rozpr. csl. Akad. Ved. 10: 1–116.

Kashin, S. M. & U. A. Rudyakov, 1978. Plavanie leptodory. Biol. vnutrennyh vod 39: 48–52 (The swimming of *Leptodora*).

Kerfoot, W. C., J. Kellog & J. R. Strickler, 1980. Visual observations of live zooplankters: evasion, escape and chemical defenses. In W. C. Kerfoot (ed.) 'Evolution and Ecology of zooplankton communities'. The University Press of New England, Hanover (N.H.); Lond: 10–27.

Kerfoot, W. C. & M. Lynch, 1987. *Branchiopod* communities: associations with planktivorous fish in space and time. In W. C. Kerfoot & A. Sih (eds) 'Predation. Direct and indirect impacts on aquatic communities'. The University Press of New England, Hanover (N.H.); Lond: 367–378.

Kiselev, I. A., 1980. Plankton morey i kontinentalnyh vodoyjomov. Nauka, Leningrad, 439 pp. (Plankton of the seas and continental water-bodies).

Korovchinsky, N. M., 1981. Rasprostranenie pribrezhnykh rakoobraznych v pelagitcheskoi zone ozera Glubokogo. Gidrobiologitcheski zhurnal 17: 25–31 (Distribution of nearshore crustaceans in the pelagic zone of Lake Glubokoe).

Krueger, D. A. & S. I. Dodson, 1981. Embryological induction and predation ecology in *Daphnia pulex*. Limnol. Oceanogr. 26: 219–223.

Li, J. L. & H. W. Li, 1979. Species-specific factors affecting predator-prey interactions of the copepod *Acanthocyclops vernalis* with its natural prey. Limnol. Oceanogr. 24: 613–626.

Lynch, M., 1980. The evolution of cladoceran life histories. Q. Rev. Biol. 55: 23–42.

Matveev, V. F., 1986. Long-term changes in the community of planktonic crustaceans in Lake Glubokoe in relation to predation and competition. Hydrobiologia 141: 33–43.

Mordukhai-Boltovskoi, F. D. & I. K. Rivier, 1987. Khyshnye vetvistousye *Podonidae*, *Polyphemidae*, *Cercopagidae* i *Leptodoridae* fauny mira. Nauka, Leningrad, 182 pp. (Predatory cladocerans *Podonidae*, *Polyphemidae*, *Cercopagidae* and *Leptodoridae* of the World fauna).

Potts, W. T. W. & C. T. Durning, 1980. Physiological evolution in the branchiopods. Comp. Biochem. Physiol. B67: 475–484.

Sars, G. O., 1865. Norges Ferskvandskrebsdyr. Forte afnit, *Branchiopoda*. I. *Cladocera Ctenopoda* (*Sididae* & *Holopedidae*). Efter det Acad. Collegium, Christiania, 79 pp.

Sars, G. O., 1901. Contribution to the knowledge of the Freshwater *Entomostraca* of South America, as shown by artificial hatching from dried material. Arch. Math. naturw. 23: 1–102.

Smirnov, N. N., 1971. *Chydoridae* fauny mira. Fauna SSSR, Novaya seriya 101. Rakoobraznyye, Nauka, Leningrad, 1(2): 531 pp. (*Chydoridae* of the World fauna).

Smirnov, N. N., 1976. *Macrothricidae* i *Moinidae* fauna mira. Fauna SSSR, Novaya seriya 112. Rakoobraznyye, Nauka, Leningrad, 1(3): 237 pp. (*Macrothricidae* and *Moinidae* of the World fauna).

Smirnov, N. N., 1977. Morfologitcheskiye osnovy obraza zhyzny vetvistousyh rakoobraznyh. VIII. Akinez u Kladocera. Zool. J. 56: 571–572. (Morphofunctional basis of life mode of *Cladocera*. VIII. Akinensis in *Cladocera*).

Starobogatov, Y. I., 1986. Systema rakoobraznykh. Zool. J. 65: 1769–1781. (The system of crustaceans).

Stenson, J. A. E., 1987. Variation in capsule size of *Holopedium gibberum* (Zaddach): a response to invertebrate predation. Ecology 68: 928–934.

Stich, H. B. & W. Lampert, 1981. Predator evasion as an explanation of diurnal vertical migration by zooplankton. Nature 293: 396–398.

Szlauer, L., 1965. The refuge ability of plankton animals before models of plankton-eating animals. Pol. Arch. Hydrobiol. 13: 89–95.

Szlauer, L., 1973. Settlements of *Sida crystallina* O.F.M. (Cladocera) on solid objects. Ecol. pol. 21: 185–192.

Williamson, C. E., 1983. Behavioral interactions between a cyclopoid copepod predator and its prey. J. Plankton Res. 5: 701–711.

Zaret, T. M., 1980. Predation and freshwater communities. New Haven and London, Yale University Press: 187 pp.

Zozulya, S. S., 1978. O dvizheniy *Bythotrephes longimanus*. Leydig. Gidrobiologitcheski zhurnal 14: 14–17 (On the movement of *Bythotrephes longimanus* Leydig).

Hydrobiologia **225**: 71–80, 1991.
V. Kořínek & D. G. Frey (eds), Biology of Cladocera.

The *Bosmina kessleri*-like morphotype of *Eubosmina* in Lake Muskoka, Ontario, Canada, as putative interspecific hybrids *

Ulrich Lieder
Mahlsdorfer Straße 54, 0–1170 Berlin, Germany

Key words: zooplankton, Cladocera, *Bosmina*, hybridization

Abstract

Like *Bythotrephes*, *Bosmina* (*Eubosmina*) *coregoni coregoni* Baird appears to have invaded North America from Eurasia, since there are no records from before 1965. All previously known North American populations have a well-rounded ventrocaudal corner of the carapace, suggesting that the founding race is one of the *rotunda* type, the most common eubosminid in European lakes. In Lake Muskoka, immediately to the east of Georgian Bay, Ontario, eubosminids were found that shared features of *B.* (*E.*) *longicornis kessleri* Ulianine. In contrast to other North American populations of *B.* (*E.*) *c. coregoni*, these specimens have a short, broadly based mucro. Morphological features suggest that these are interspecific F_1-hybrids between *B.* (*E.*) *longispina* Leydig and *B.* (*E.*) *coregoni*, without any signs of introgression.

Introduction

Two species of the subgenus *Eubosmina* Seligo, 1900 are presently known to occur in North America: *B.* (*E.*) *longispina* Leydig, 1860 and *B.* (*E.*) *coregoni* Baird, 1857. The Holarctic *B.* (*E.*) *longispina* occurs in many lakes across northern North America; its southern limit is not precisely known but seems to approximate the 40th parallel. In contrast, *B.* (*E.*) *coregoni* seems to have been recently introduced to North America, since it has been found in only a few lakes, and only since the 1960's.

Research on Eurasian eubosminids has shown that interspecific and intersubspecific hybridizations, with and without signs of introgression, occur frequently (Lieder, 1953, 1957, 1958). Lieder (1956, 1958, 1983a, b, 1986) postulated that the differentiation of *Eubosmina* into an enormous number of transitional and morphologically distinct populations can best be explained by assuming that the lakes were originally colonized by genotypically variable, hybrid populations. This hybridization in Eurasia involved 4 species with 10 subspecies (Lieder, 1983a, b).

The occurrence of *B.* (*E.*) *c. coregoni* in North America since approximately 1965 raised the question of whether hybridization between the immigrant and indigenous species is occurring in North America as well, even though 25 years is extraordinarily short for this kind of hybridogenic evolutionary process. As a result of very different environmental conditions, especially with regard to trophic state, there are few lakes with sympatric populations of both *Eubosmina* species. It is thus even more interesting that mucronate *Eubosmina* were found in Lake Muskoka, near the east coast

---

*This paper is dedicated to the memory of Dr. John C.H. Carter.

of Georgian Bay. Morphological features indicate that these must be regarded as interspecific hybrids between *B.* (*E.*) *longispina* and *B.* (*E.*) *c. coregoni*. The hybridization seems to be limited to the formation of the F1 generation. No indications of further hybridization or introgression were found.

Methods

Zooplankton in Lake Muskoka, Moon River, and Georgian Bay were sampled by Dr. J.C.H. Carter using vertical and horizontal hauls of an ordinary plankton net in May and August 1985 and in late autumn 1986. Samples were preserved with formaldehyde and sent to the present author. About 10–20 adults (instars 4 and 5) of each type (*B.* (*E.*) *longispina*, *B.* (*E.*) *c. coregoni*, and intermediate types) were mounted in glycerol for morphological examination, measurement and photography under a Zeiss-Lumipan microscope. Measurements were made in accordance with Rühe (1912). Flössner (1972) and Kerfoot (1980), Fig. 5) and included total body length, carapace length (according to both Rühe and Flössner and to Kerfoot), height of the dorsal part of the carapace, mucron length, lengths of the proximal and distal parts of the antennules, length of the seta kurzi. The number of sutures on the distal part of the antennule was also recorded.

Sketches were made with the aid of a camera lucida. Ages (instars) were determined from scatter diagrams of body length against the height of the dorsal part of the carapace (Woltereck, 1929).

Lake Muskoka (79° 05′ W, 45° N) is one of a group of softwater lakes draining into Georgian Bay via the Moon River. 'This is an immensely complicated system with the river from Lake Muskoka containing soft waters flowing through a number of bays and then into the hard water of Georgian Bay proper' in an area containing a vast number of small islands (J.C.H. Carter, pers. comm.). The mouth of the Moon River is part of Georgian Bay Islands National Park.

Results

The plankton of Lake Muskoka contained 4 species of Bosminidae: *Bosmina* (*Neobosmina*) *tubicen* Brehm, 1953 and 1956; a species which superficially resembles *B.* (*Bosmina*) *longirostris* O.F.M. but differs in important features, including some characteristics of the subgenus; *B.* (*Eubosmina*) *longispina* Leydig, 1860; and, very infrequently, *B.* (*E.*) *coregoni coregoni* Baird, 1857. In addition, large numbers of specimens were found that were morphologically intermediate, and almost certainly hybrids, between *longispina* and *c. coregoni* (see Fig. 3, 8).

B. (*E.*) *c. coregoni* was the dominant species in Georgian Bay; *B.* (*E.*) *longispina* and the intermediate morphotypes were very rare and only found at the mouth of the Moon River. The species that resembles *longirostris* was frequent throughout the whole area. This very interesting and ecologically important species will be de-

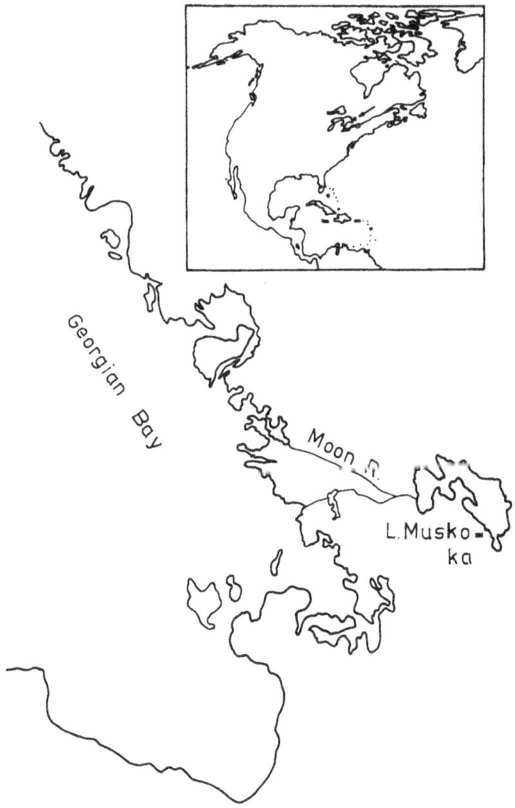

Fig. 1. Study area and Lake Muskoka.

scribed by Kořínek as a new species in the subgenus *Sinobosmina* Lieder, 1957 (Kořínek, pers. comm.).

No further mention will be made here of *B.* (*N.*) *tubicen*, since it is well known from this area (Brandlova *et al.*, 1972).

Bosmina (*E.*) *c. coregoni* occurred mostly in Georgian Bay. It was very rare in Lake Muskoka and absent from Moon River. After examination of specimens from many populations, I have found no morphological differences between American and Eurasian *B.* (*E.*) *c. coregoni.* Populations from Georgian Bay and other lakes, e.g. Lake Erie (samples provides by K. Patalas), have a high completely round carapace and completely rounded ventro-caudal margin without the slightest indication of a mucro ('scharfen Ecke' (Rühe, 1912; Lieder, 1983c)) or a 'seta kurzi'. The American populations are therefore the same as the *coregoni* occurring in the Berlin Lakes area and many central and eastern European lakes, which Schödler (1865, 1866) described as *B. rotunda*, the morphotype that is most frequently placed in the subspecies coregoni. The dorsum of the carapace in the Lake Muskoka population is hexagonally patterned as in other North American populations. In Europe, *B.* (*E.*) *c. coregoni* exhibits all variations from strong hexagonal patterning to weak stria on the dorsum of the carapace. For 'rotunda populations', however,

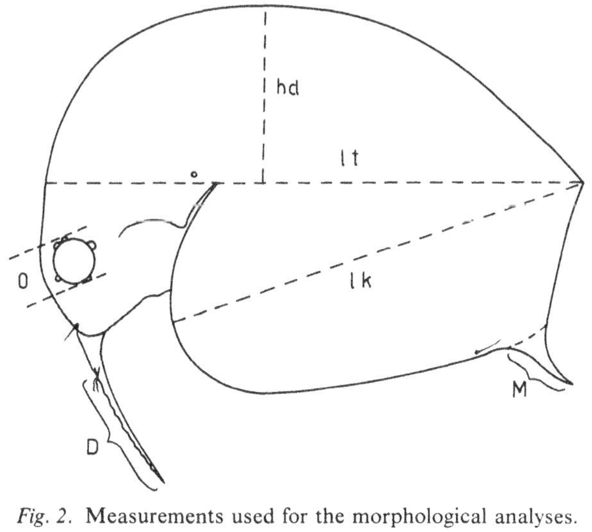

Fig. 2. Measurements used for the morphological analyses.

the hexagonal patterning, as in the American populations, is typical. The forehead of the Georgian Bay and Lake Muskoka *coregoni* is flat, the rostrum relatively long and pointed, and the eye is relatively small (an average of 35 μm). The antennules are long and directed straight downwards or bend slightly backwards. The number of sutures on the distal part of the antennules (from the triangular 'Schildchen' to the tip) is 16–21 (ave. 18.3). According to Lieder (1957), European populations of *B.* (*E.*) *c. coregoni* (with round ventrocaudal edges = rotunda) have 13–20 (ave. 17.2) sutures (188 individuals from 15 populations). The number of spines on the postabdominal claw (8–9) is quite high, but within the range for European populations. The position of the lateral head pore is typical for the subgenus *Eubosmina* Seligo, 1900 (Goulden & Frey, 1962). Kořínek (1971) has noted the same features for the *B.* (*E.*) *c. coregoni* of Lakes Erie and Ontario.

Body length of adult females is between approximately 450 and 700 μm.

B. (*E.*) *longispina* occurred only as single individuals in Georgian Bay but was found frequently in the plankton of Lake Muskoka. It was absent from Moon River. These animals were morphologically close to the populations of the morpho-group obtusirostris-lacustris of European authors. G.O. Sars erected the species *B. obtusirostris* and *B. lacustris* in 1862, two years after the description of *B. longispina* by Leydig from Lake Constance. Today both *obtusirostris* and *lacustris* (as well as several other forms) are regarded as ecological morphs of *longispina* (Lieder, 1957, 1983b c; Nilssen & Larsson, 1980). The length of adult females from Lake Muskoka was 460–655 μm. Body height was approximately 88% of body length. *B.* (*E.*) *longispina* is thus relatively low and elongate. The highest point of the dorsal margin is situated anterior to the middle of the body, so that the dorsal contour is sloped backwards. The mucrones are of intermediate length. The seta kurzi are very clearly developed. On the ventral margin of the mucro there is a point of a sudden change of direction so that there is a 'Mukrobucht' (Rühe, 1912) in the area of the seta kurzi, which becomes more pronounced with increasing age of the animal. The mucro sutures (2–3) are always oriented ventrally (which is always the case in *Eubosmina* in contract to *Neobosmina*); they are, as in all Bosmina, especially obvious in juveniles and become reduced with increasing age. The rostrum is short and blunt. The eye is large (ave. 50 μm) and its crystalline lenses protrude only indistinctly. The antennules are slightly bent and relatively short (about 30% of the body length). The number of sutures on the distal part, 11–13 (ave. 11.9), roughly corresponds with the average count for circumpolar *longispina* populations (8–19, ave. 13.7, n = 268 from 40 populations, Lieder, 1957). There are 4–5 spines in the pecten of the postabdomenal claw, significantly fewer than in *B.* (*E.*) *c. coregoni* from Georgian Bay. The lateral headpore is in the typical *Eubosmina* position. The headshield of Lake Muskoka *longispina* is clearly reticulated, and the body carapace is weakly reticulated dorsally (long polygons).

Morphologically intermediate specimens were found in Lake Muskoka (frequently) and in Georgian Bay (isolated individuals), which are considered by the author to be interspecific hybrids (see Figs. 3, 8) between *B.* (*E.*) *longispina* and *B.* (*E.*) *c. coregoni*. The body form is sleeker than *coregoni* and shorter than *longispina*. The opposite is true with regard to the length of the rostrum (Figs. 3 and 4). The size of the eyes is also about intermediate (ave. 42 μm) (Fig. 7), as is the number of spines in the comb of the postabdominal claw, which is 6 for the presumed hybrids (*coregoni* 8–9, *longispina* 4–5) (Fig. 5). Instead of the typical mucrones of *B.* (*E.*) *longispina*, and the completely amucronate *B.* (*E.*) *c. coregoni*, the hybrids have a 'pseudomucro' (Figs. 3, 4, 8) (Carter, pers. comm.), i.e. a sharply protruding, pointed carapace edge with a wide base. (Fig. 8). The seta kurzi (completely absent in coregoni and strongly developed in longispina) is weakly developed. The number of sutures on the distal position of the antennules falls between the parental species (Fig. 6).

Bosminids with this combination of characteristics have never before been reported from North America. Morphologically they corre-

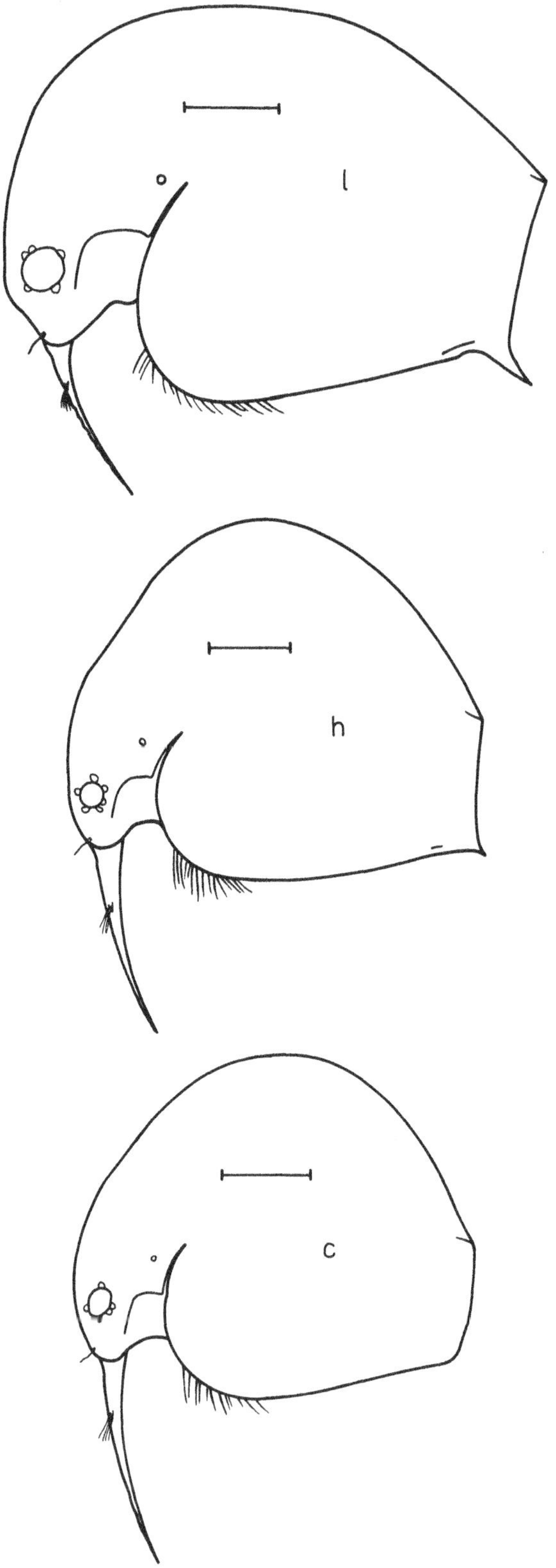

Fig. 3. Parental species and presumed F_1-hybrids (centre) between *B.* (*E.*) *c. coregoni* (bottom) and *B.* (*E.*) *longispina* (top) from Lake Muskoka, August 1985 (bar = 100 μm).

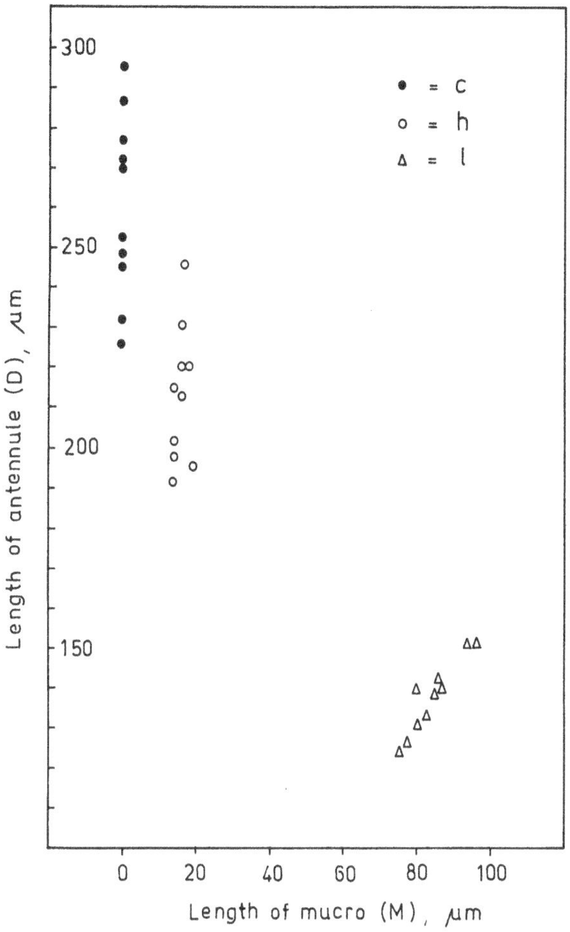

Fig. 4. Plot of the relationship between antennule (distal part) length and mucron length.

spond to *Eubosmina* from Eurasia which were described by Ulianine in 1874 as *B. kessleri.** The type location for *B. kessleri* is L. Onega (Karelia, USSR). This conclusion is based on the auther's examination of *B.* (*E.*) *longicornis kessleri* Ulianine, 1874 (syn. *B. c. kessleri* in Rühe, 1912; *B.* (*E.*) *mixta kessleri* in Lieder 1957, 1983b, c; *B.* (*E.*) *kessleri* in Flössner 1972) from Lake Onega provided by O. Bauer (Leningrad).

In Europe, interspecific hybridization of mucronate *Eubosmina* with the amucronate *B.* (*E.*) *c. coregoni* has produced intermediate populations of the morphotype *B. kessleri* (Lieder, 1953, 1957,

*N.B. Karl Fedorowitsch Kessler (1815–1881) worked on Lake Onega; W. Ulianine was head of the Biological Station in Sevastopol, Russia.

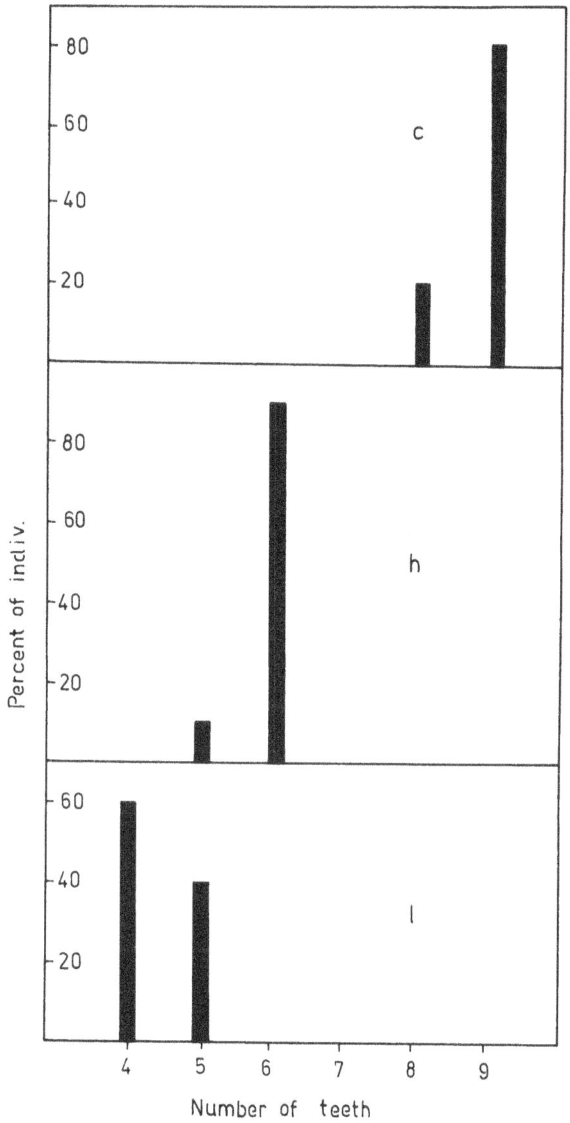

Fig. 5. Frequency distributions of the numbers of spines in the pectaen of th terminal postabdominal claw of *B.* (*E.*) *c. coregoni* (c), *B.* (*E.*) *longispina* (l), and the presumed hybrid (h).

1983a, b, c). The *kessleri* form, however, occurs in Eurasia not only as an interspecific hybrid, but also as a well-defined subspecies of *B.* (*E.*) *longicornis* Schödler, 1865 (= *B.* (*E.*) *mixta* Lilljeborg, 1901), possibly representing old, stabilized hybrid complexes. Analogous situations have been described by Flössner & Kraus (1986) for *Daphnia galeata*, and by Lieder (1986) for *D. cucullata procurva*. Therefore, it also seems likely that where *B.* (*E.*) *longispina* and *B.* (*E.*) *c. coregoni* occur in the same, or in neighboring connected North Ameri-

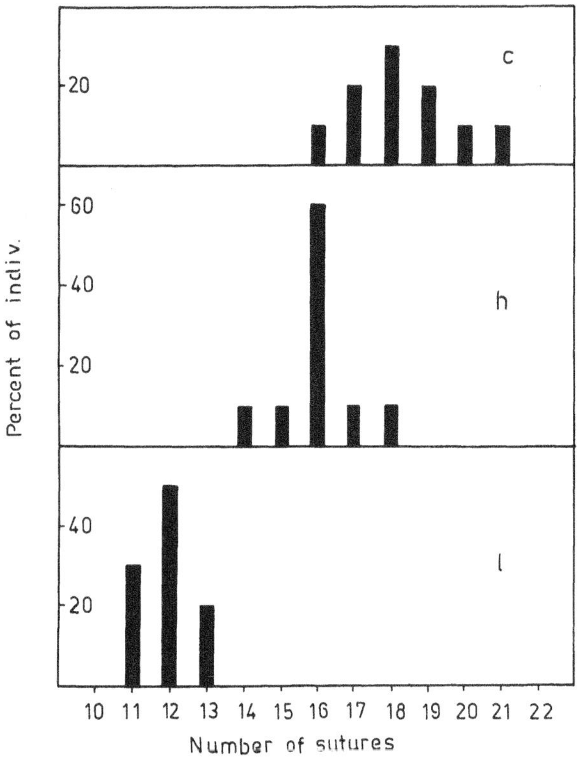

Fig. 6. Frequency distributions of the numbers of sutures on the distal part of the antennules of *B.* (*E.*) *c. coregoni* (c), *B.* (*E.*) *longispina* (l), and the presumed hybrid (h).

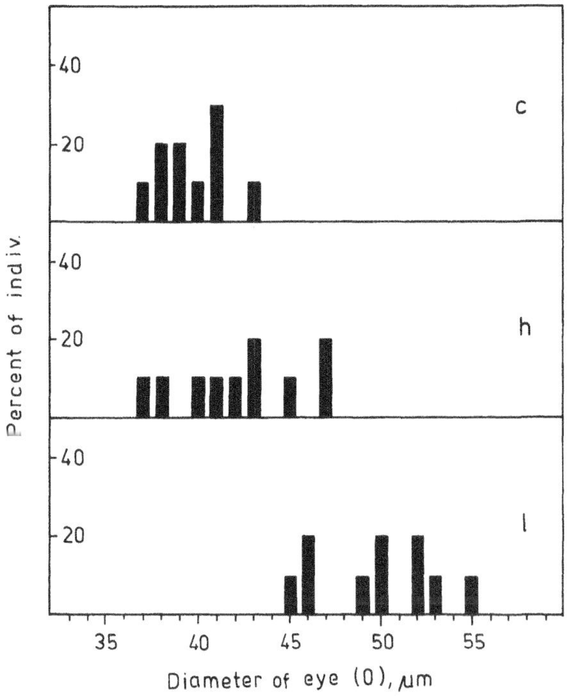

Fig. 7. Frequency distributions of the size of the eye of *B.* (*E.*) *c. coregoni* (c), *B.* (*E.*) *longispina* (l), and the presumed hybrid (h).

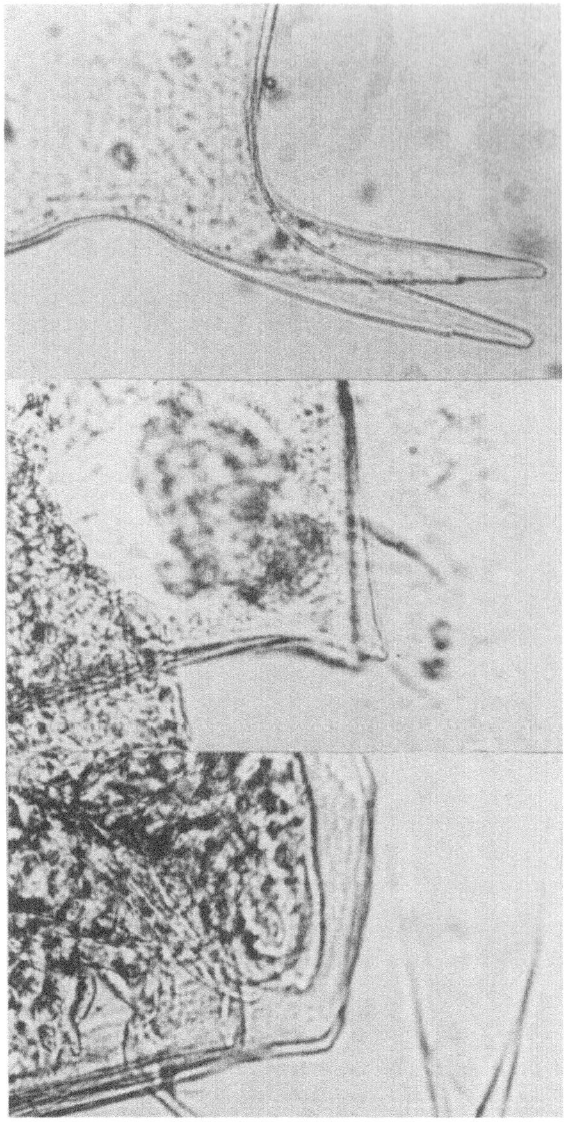

Fig. 8. Ventro-caudal carapace edge of parental species and presumed hybrids (centre) between *B.* (*E.*) *c. coregoni* (bottom) and *B.* (*E.*) *longispina* (top).

can lakes, hybridization could be expected to result in the production of the *kessleri* morphotype. According to the opinion of the author, this is the situation in Lake Muskoka.

Discussion

All North American populations of *B.* (*E.*) *coregoni* belong to the subspecies *coregoni*, which differs from the other (Eurasian) subspecies of the species in not exhibiting any tendency toward hump formation during cyclomorphosis, such as is typical for the subspecies *gibbera* Schödler, 1866 and *thersites* Poppe, 1887.

The subspecies *coregoni* occurs as two morphotypes in Eurasia. These differ primarily with regard to the development of the ventro-caudal carapace edge: the first type has a completely rounded carapaca corner (= *rotunda* Schödler, 1865); the second (= *diaphana* P.R. Müller, 1867) has a sharply protruding ventro-caudal carapace edge, which is, however, hardly a real mucro. There are also certain differences in the patterning of the dorsum of the carapace: the '*rotunda*' populations are strictly hexagonally patterned, the '*diaphana*' populations show signs of a dorsal striated structure.

B. (*E.*) *longispina* is holarctic in distribution (cf. Chengalath, 1987; Patalas, 1971, 1975; Patalas & Salki, 1973). It occurs across northern North America from the Aleutian Islands, to Washington in the west and from Labrador to Maine in the east. The southern limit of its distribution is not exactly known but probably coincides with the 40th parallel. The several ecophenotypes that occur within this huge area may reflect predation pressure (O'Brien & Schmidt, 1979), or the trophic state and humic content of the water (Lieder, 1957, 1983b, c). These may also be genetically different types in the sense of a generic complex within *B.* (*E.*) *longispina* s. lat.

While *B.* (*E.*) *longispina* is an original element of the Nearctic freshwater fauna, *B.* (*E.*) *c. coregoni* seems to have invaded North America only recently. The following points support this: first, in contrast to Europe where it is the most common *Eubosmina* (Lieder, 1957, 1983c), it occurs in only very few lakes in North America, namely in the Great Lakes (except Lake Superior) (Patalas, 1969, 1972; Davis, 1969; Wells, 1970; Carter, 1971, 1980; Brandlova *et al.*, 1972; Kořínek, pers. comm.), a small number of lakes in Ontario, Sanctuary Lake, PA (Deevey & Deevey, 1971a, b), and a lake in the Adirondacks of New York (Cotten, pers. comm.). Secondly, there were no reports of *B.* (*E.*) *c. coregoni* from Nearctic lakes before the 1960's G.O. Sars (1915), an outstanding expert on the European *Bosmina* species, did not list it for Georgian Bay. According to Wells (1960, 1970), it did not occur in Lake Michigan in 1954, a few individuals were found in 1966, and the species was common by 1968. Beeton (1965) attributed this to eutrophication, but Brooks (1969) felt it to be the result of predation pressure. Introduction from Europe during the early 1960's seems more likely. This could be verified through paleolimnological research as well as through enzyme electrophoresis to estimate the genetic distance from the European *B.* (*E.*) *c. coregoni.* The discharge of bilge water from ships involved in international trude has been assumed to be the mechanism responsible for the appearance of *Bythotrephes cederstroemi* (Bur *et al.*, 1986; Lange & Carp, 1986), *Dreissena polymorpha* (Leach & Hebert, 1989) and *Gymnocephalus cernua* (Pratt *et al.*), in the Great Lakes in recent years.

Of the two types of *B.* (*E.*) *c. coregoni*, only the '*rotunda*' type seems to occur in North America, suggesting that individuals of a *rotunda* population founded the American populations. The distribution of both morphotypes in Europe clearly points to genetic differences between '*rotunda*' and '*diaphana*' populations, even though paleolimnological research leaves a different impression (Hofmann, 1977, 1978). Taxonomical investigations have shown that interspecific and intersubspecific hybrids, with and without signs of introgression, are not a rare occurrence in Eurasian lakes. The differentiation of *Eubosmina* (and some 'critical' *Daphnia* spp.) into a large number of races with morphological transitions can be most easily explained by assuming that postglacial colonization was by hybrid populations with great genotypic diversity (Lieder, 1953, 1956, 1957, 1958, 1983a, b c, 1986, 1987). The occurrence of *B.* (*E.*) *c. coregoni* in the last 25 yrs makes it possible to examine this problem in the Nearctic, even though this time span is to be considered as minimal with regard to evolutionary processes. In the Nearctic only *B.* (*E.*) *longispina* can be considered as a sibling species. However, it has disappeared, or is found tential impact of the zebra mussel, *Dreissena polymorpha* (Pallas), in the Great Lakes. 32 Conference on Great Lakes Research, Madison, Wisconsin, U.S.A., in press.

Lieder, U., 1953. Über Bastarde zwischen einigen Formtypen des *Coregoni*-Kreises. Arch. Hydrobiol. 47: 453–469.

Lieder, U., 1956. Introgressive Hybridisation als Evolutionsfaktor bei den Gattungen *Bosmina* und *Daphnia* (Crustacea, Cladocera). Naturwissenschaften 43: 207.

Lieder, U., 1957. Beiträge zur Kenntnis des Genus *Bosmina*. IV. Versuch einer Monographie der Untergattung *Eubosmina* Seligo, 1900. Thesis Humboldt-Univ. Berlin (GDR), p. 1–247.

Lieder, U., 1958. Ablaufsformen und Tempo der Evolution der polytypischen baltischen Planktoncladoceren. Verh. Int. Ver. Limnol. 13: 789–898.

Lieder, U., 1983a. Revision of the Genus *Bosmina* Baird, 1845 (Crustacea, Cladocera). Int. Revue ges. Hydrobiol. 68, 1: 121–139.

Lieder, U., 1983b. Introgression as a Factor in the Evolution of polytypical Plankton Cladocera. Int. Revue ges. Hydrobiol. 68, 2: 269–284.

Lieder, U., 1983c. Die Arten der Untergattung *Eubosmina* Seligo, 1900 (Crustacea, Cladocera, Bosminidae). Mitt. zool. Mus. Berlin 59, 2: 195–292.

Lieder, U., 1986. *Bosmina* und *Daphnia*: taxonomisch kritische Gruppen unter den Cladoceren (Crustacea, Phyllopoda). Limnologica (Berlin) 17, 1: 55–66.

Lieder, U., 1987. The possible origin of *Daphnia cucullata procurva* Poppe, 1887 in the lakes of the Pomeranian Lakeland by hybridization in the past. Hydrobiologia 145: 201–211.

Lilljeborg, W., 1901. Cladocera Sueciae. Nova Acta reg. soc. sci. Upsal. ser. 3, 19: 1–701.

Müller, O. E., 1785. Entomostraca seu Insecta Testacea, qual in aquis Danial et Norvegial reperit, descriprit et iconibus illustravit. Lipsiae et Havnial: 1–135.

Müller, P. E., 1867. Danmarks Cladocera. Naturhist. Tidsskrift (3) 5: 53–240.

Nilssen, J. P. & P. Larsson, 1980. The systematical position of the most common Fennoscandian *Bosmina* (*Eubosmina*). Z. zool. Syst. Evol.-Forsch. 18: 62–68.

O'Brien, W. J. & D. Schmidt, 1979. Arctic *Bosmina* morphology and copepod predation. Limnol. Oceanogr. 24: 564–568.

Patalas, K., 1969. Composition and horizontal distribution of Crustacean plankton in Lake Ontario. J. Fish. Res. Bd Can. 26: 2135–2164.

Patalas, K., 1971. Crustacean plankton communities in forty-five lakes in the Experimental lakes Area in north-western Canada. J. Fish Res Bd. Can. 28, 2: 231–244.

Patalas, K., 1972. Crustacean plankton and eutrophication of St. Lawrence Great Lakes. J. Fish Res. Bd Can. 29: 1451–1462.

Patalas, K., 1975. The crustacean plankton communities of fourteen North American great lakes. Verh. Int. Ver. Limnol. 19: 504–511.

Patalas, K. & A. Salki, 1973. Crustacean plankton and the eutrophication of lakes in the Okanagan Valley, British Columbia. J. Fish Res. Bd Can. 30: 519–551.

Poppe, S. A., 1887. Beschreibung einiger neuer Entomostraken aus norddeutschen Seen. Z. wiss. Zool. 45: 278–281.

Pratt, D., J. Selgeby & J. Spurrier, 1989. The ruffe (*Gymnocephalus cernua*) in Lake Superior: abundance, distribution and prospects for the future. 32nd Conference on Great Lakes Research, Madison, Wisconsin, U.S.A., in press.

Rühe, F. E., 1912. Monographie des Genus *Bosmina*. *Bosmina coregoni* im baltischen Seengebiet. Zoologica 25 (63): 1–141.

Sars, G. O., 1862. Forsatte sit Fordrag over de af ham i Omegnen af Christiania iaggtagne Crustacea Cladocera. Forhandl. Vidensk. Selsk. Christiania (1861): 250–302.

Sars, G. O., 1915. Entomostraca of Georgian Bay. Contrib. Can. Biol. 2: 221–222.

Shan, R. K. & D. G. Frey, 1968. Induced interbreeding between two stocks of a chydorid cladoceran. Bio Science 18: 203–205.

Shan, R. K. & D. G. Frey, 1983. *Pleuroxus denticulatus* and *P. procurvus* (Cladocera, Chydoridae) in North America: distribution, experimental hybridization and the possibility of natural hybridization. Can. J. Zool. 61: 1605–1617.

Shan, R. K., 1971. Genetic stability in the Cladocera. Trans. Am. Microsc. Soc. 90: 116.

Shan, R. K., 1974. Reproduction in laboratory stocks of *Pleuroxus* (Chydoridae, Cladocera) under the influence of photoperiod and light intensity. Int. Revue ges. Hydrobiol. 59: 643–666.

Schödler, J. E., 1865. Zur Diagnose einiger Daphniden. Arch. Naturgeschichte 31: 283–285.

Schödler, J. E., 1866. Die Cladoceren des Frischen Haffs nebst Bemerkungen über anderweitig vorkommende verwandte Arten. Arch. Naturgeschichte 32: 1–56.

Seligo, A., 1900. Untersuchungen in den Stuhmer Seen. Danzig, p. 66.

Ulianine (Uljanin), W. N., 1874. Cladocera und Copepoda einiger Seen in Rußland. Protok. Ver. Kais. Ges. Freunde der Naturw. Moscow, 10: 1–14.

Wells, L., 1960. Seasonal abundance and vertical movement of planktonic crustacea in Lake Michigan. Fish Bull. U.S.A. 172, 60: 343–369.

Wells, L., 1970. Affects of Alewife predation on zooplankton populations in Lake Michigan. Limnol. Oceanogr. 15, 4: 556–565.

Wolf, H. G., 1987. Interspecific hybridization between *Daphnia hyalina*, *D. galeata*, and *D. cucullata* and seasonal abundances of these species and their hybrids. Hydrobiologia 145: 213–217.

Wolf, H. G. & M. A. Mort, 1986. Inter-specific hybridization underlies phenotypic variability in *Daphnia* populations. Oecologia (Berlin West) 68: 507–511.

Woltereck, R., 1929. Technik der Variations- und Erblichkeitsanalyse bei Crustaceen. Handb. biol. Arbeitsmethoden 9: 567–636.

Hydrobiologia **225**: 81–85, 1991.
V. Kořínek & D. G. Frey (eds), Biology of Cladocera.

The late-glacial/holocene *Bosmina* (*Eubosmina*) fauna of Lake Constance (Untersee) (F.R.G.): traces of introgressive hybridization

Wolfgang Hofmann
Max-Planck-Institut für Limnologie, Abt. Mikrobenökologie, Postfach 165, D-2320 Plön, Germany

Key words: *Bosmina* (*Eubosmina*), Lake Constance, sediment analysis, morphological variation, introgressive hybridization

Abstract

Subfossil *Bosmina* (*Eubosmina*) remains were analysed in a sediment core from the Untersee of Lake Constance which covered the Late-Glacial/Holocene period.

During the Late-Glacial and the early Holocene the lake was inhabited only by *Bosmina longispina* Leydig. In the Late Holocene a second species, *Bosmina coregoni* f. *kessleri* Uljanin, appeared. In the uppermost sediment layers the morphological gap between the two taxa disappeared, apparently from introgressive hybridization.

Introduction

Analysis of *Bosmina* remains in lake sediments provides information about the history of the *Bosmina* fauna of lakes, in particular about long-term morphological variation. This method can be used effectively as a tool for testing suppositions about the origin of morphological variation of present populations and its taxonomic significance (Hofmann, 1987a).

The aim of this study was to determine if the same pattern of Late-Glacial/Holocene *Bosmina* development observed in some north German lakes is discernible in the pre-alpine region.

In the pre-alpine Starnberger See, a long-term species succession has been found that basically coincides with the development observed during the Late-Glacial/Holocene in north German lakes. However, because of insufficient resolution of time, confusion remained about what happened during the most recent times (Hofmann, 1987b).

Material and methods

The samples were obtained from a deepwater sediment core of the Untersee (US 8707), cored in April 1987 within a research project on prehistoric settlements in the pre-alpine region (Schlichtherle, 1988). The core was 6.78 m long and reached down into the Late-Glacial. As the datings have not been finished thus far, the samples are related only to sediment depth.

The Untersee is a small, shallow basin (surface area: 63 km^2; maximum depth: 46 m; mean depth: 13.2 m) of Lake Constance separated from the Obersee (surface area: 476 km^2; maximum depth: 252 m; mean depth: 100 m) (Geller & Güde, 1989).

For preparation of the sediment samples and morphometric analysis (mucro length, number of antennule segments, antennule length) refer to Hofmann (1987b). For mucro measurements, only specimens with a carapace length $> 300\ \mu m$

were used. Mucro length was defined as the distance between the end of the mucro and the nearest point on an auxiliary line indicating a rounded, spineless ventral-caudal angle of the carapace.

Results and discussion

In Fig. 1 mucro lengths are plotted against sediment depth. The point clusters indicate two different units: In the lower section of the core (200–600 cm) only specimens with long mucrones were present. The range of lengths increased from 90–140 μm in the lowermost section (556–600 cm) to 130–>200 μm in the middle section (40–400 cm). At a sediment depth of 100 cm specimens appeared with very short mucrones (12–25 μm). In the section 40–100 cm these specimens were separated from the long mucrones by a broad gap. Towards the sediment surface this gap disappeared, and a continuous transition from the short to the long mucrones occurred. In the uppermost section (2–10 cm) specimens with very long and very short mucrones disappeared and intermediate specimens (40–120 μm) predominated.

Figure 2 shows the shape of selected mucrones representing the range of variation of four major phases: (1) only one morph was present with respect to mucro morphology (300 cm); (2) a second morph characterized by very short mucrones appeared, clearly separated from the morph with long mucrones (50 cm); (3) continuous morphological transition from short to long mucrones (20 cm); (4) predominance of intermediate forms (2 cm).

The number of antennule segments exhibited several significant shifts (Fig. 3): In the lowermost section (556–600 cm) numbers ranged from 13 to 19. In the 500 cm sample there was a distinct decrease in the number of segments (range: 9–15), followed by larger numbers again at lesser depths. In the section 100–450 cm most of the specimens had 12 to 17 segments. The appearance of specimens with short mucrones in section above 100 cm was combined with an extension of the

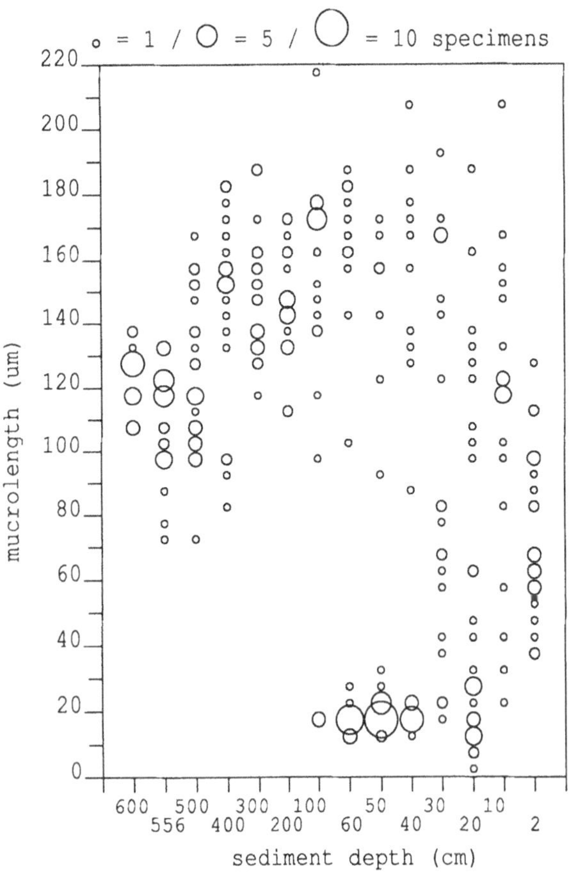

Fig. 1. Lake Constance, Untersee, core US8707, *Bosmina* (*Eubosmina*): mucro length in relation to sediment depth.

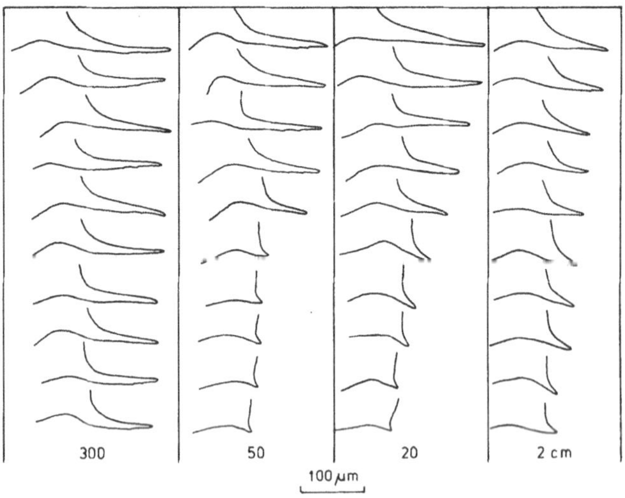

Fig. 2. Lake Constance, Untersee, core US8707, *Bosmina* (*Eubosmina*): variation of mucro shapes at sediment depths of 300, 50, 20, and 2 cm.

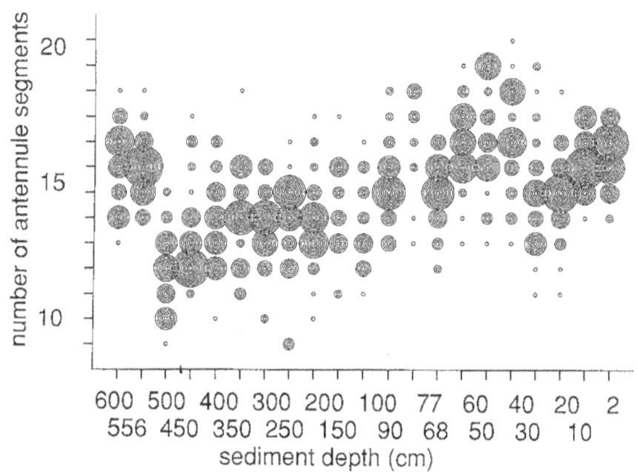

Fig. 3. Lake Constance, Untersee, core US8707, *Bosmina* (*Eubosmina*): number of antennule segments in relation to sediment depth (number of rings indicate number of specimens).

range due to the occurrence of specimens with up to 21 segments. In the youngest sediments the same trend as observed in mucro length was discernible: Specimens with high (> 18) and low (< 14) segment numbers were no longer found, and intermediate morphs (14–18 segments) predominated.

The long-term variation in antennule length corresponds with the pattern observed in the number of segments: (1) A sharp decrease between 556 and 500 cm sediment depth; (2) an increase above this layer; (3) a further increase in the range of variation due to the occurrence of antennules > 250 μm above 100 cm; (4) disappearance of antennules < 200 μm confined the range of variation in the uppermost samples (2–10 cm) (Fig. 4).

On the basis of the data on long-term variation in mucro and antennule morphology, the following *Bosmina* (*Eubosmina*) history of the Untersee is suggested: The morph with long mucrones (70– > 200 μm) and short antennules (< 270 μm) with a low number of segments (mostly < 18) was *Bosmina longispina* Leydig. Lake Constance is *locus typicus* of this species (Leydig, 1860).

The morph with very short mucrones (12–25 μm) and longer antennules (maximum: 400 μm) with higher numbers of segments (maximum: 21) which appeared at 100 cm sediment depth clearly corresponds with *Bosmina coregoni* f. *kessleri* Uljanin (Hofmann, 1984, 1987b), which is considered as *Bosmina mixta kessleri* by Lieder (1983).

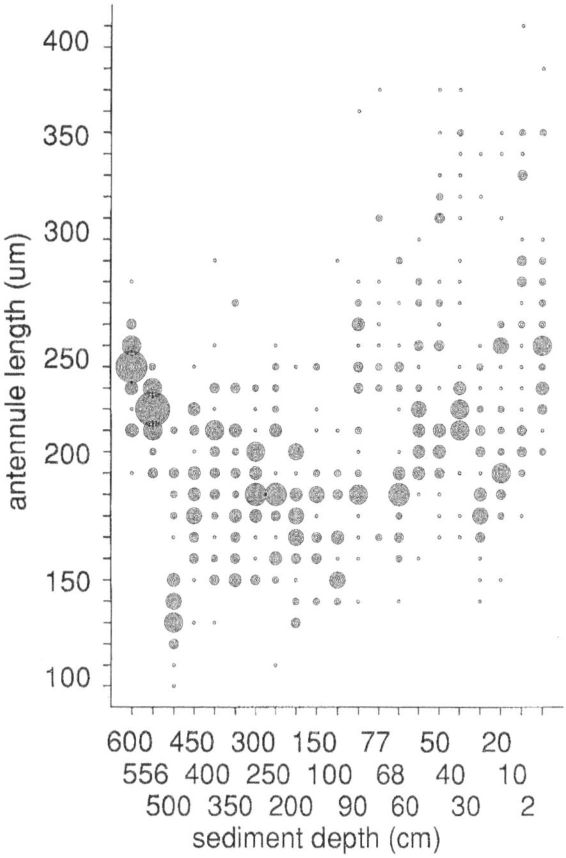

Fig. 4. Lake Constance, Untersee, core US8707, *Bosmina* (*Eubosmina*): antennule length in relation to sediment depth (number of rings indicate number of specimens).

The pattern of species succession found in the Untersee in general coincided with the pattern observed in all the deep, stratified north German and pre-alpine lakes studied: *Bosmina longispina* was the only representative of the subgenus during the Late-Glacial and early (and middle) Holocene. During the Holocene (north Germany: Boreal; Starnberger See: Subatlantic) a second species *Bosmina coregoni* f. *kessleri* invaded the lakes, and the two species have since co-occurred (Hofmann, 1984, 1987a, 1987b).

However, morphological shifts in the Untersee populations during this phase of co-occurrence were basically different from what happened in north German lakes: due to long-term variation the morphological differences disappeared by occurrence of intermediate specimens, which formed a continuous morphological transition between the species *Bosmina longispina* and *Bosmina coregoni*. Finally, these intermediates became predominant in the lake.

In contrast, in north German lakes the morphological gap between the populations of the same two species increased during the phase of co-occurrence due to divergent morphological changes: increase in mucro length in *Bosmina longispina* and mucro reduction in *Bosmina coregoni* f. *kessleri* (Hofmann, 1984).

The most obvious explanation for the morphological shifts in the Untersee is introgressive hybridization. In general, this process leads to the production of morphological intermediates. In the light of this hypothesis, the development was characterized by four major phases (Fig. 5):

1. *Bosmina longispina* occurred alone (600–150 cm).
2. *Bosmina coregoni* f. *kessleri* invaded the lake; the two species co-occurred maintaining their morphological differences (100–40 cm).
3. Hybridization between the populations led to the appearance of intermediates. The morphological gap disappeared (30–20 cm).
4. The *Bosmina* (*Eubosmina*) population consisted almost exclusively of hybrids (10–2 cm).

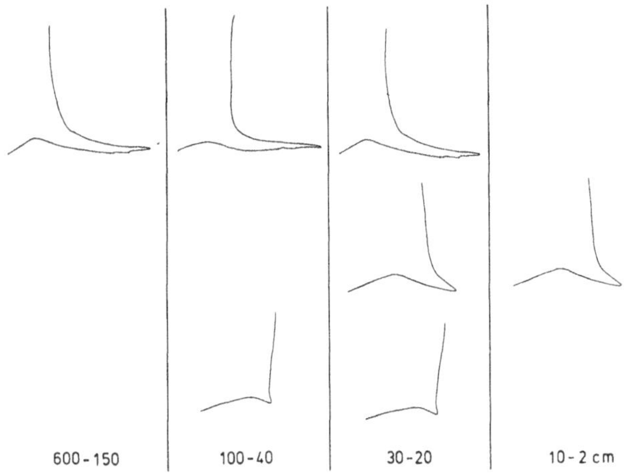

Fig. 5. Lake Constance, Untersee, core US8707, *Bosmina* (*Eubosmina*): scheme of long-term variation based on mucro morphology, showing four developmental stages: (1) 600–150 cm: solitary occurrence of *Bosmina longispina*; (2) 100–40 cm: immigration of *Bosmina coregoni* f. *kessleri*: co-occurrence and morphological gap between the taxa; (3) 30–20 cm: appearance of intermediate morphs from introgressive hybridization; morphological transition between the taxa: (4) 10–2 cm: almost complete disappearance of parent taxa and predominance of hybrids.

The hypothesis of introgressive hybridization has to be tested by further case studies in prealpine lakes including the Obersee of Lake Constance. However, it is known from other planktonic cladocerans of the genus *Daphnia* that morphological intermediates found were indeed products of inter-specific hybridization, as has been shown by electrophoretic patterns (Wolf & Mort, 1986).

Furthermore, Lieder (1983) has postulated introgressive hybridization in the case of the *Bosmina* population from the Untersee. There is still a minor difference between the development suggested by Lieder and thc data presented above. Lieder proposed that in Lake Constance hybridization occurred between the taxa *Bosmina longispina* and *Bosmina coregoni*, thereby establishing the species *Bosmina mixta kessleri*. The subfossil remains showed, however, that the morph *kessleri* of the Untersee was not a product of hybridization but was one of the parent taxa, which had also been suggested by Flößner (Einsle, 1978).

Nevertheless, the long-term morphological changes observed in the Untersee indicate that indeed introgressive hybridization might have contributed to the enormous morphological variation observed in recent *Bosmina* populations.

References

Einsle, U., 1978. Qualitative und quantitative Änderungen im Crustaceenplankton des Bodensee-Obersees. Arch. Hydrobiol. 82: 300–315.

Geller, W. & H. Güde, 1989. Lake Constance – the largest German lake. in: Lampert, W. & Rothhaupt, K.-O. (eds.) Limnology in the Federal Republic of Germany, Plön, 9–17.

Hofmann, W., 1984. Postglacial morphological variation in

Bosmina longispina Leydig (Crustacea, Cladocera) from the Großer Plöner See (north Germany) and its taxonomic implications. Z. zool. Syst. Evolut.-forsch. 22: 294–301.

Hofmann, W., 1987a. Cladocera in space and time: Analysis of lake sediments. Hydrobiologia 145: 315–321.

Hofmann, W., 1987b. The late Pleistocene/Holocene and recent *Bosmina* (*Eubosmina*) fauna (Crustacea: Cladocera) of the pre-alpine Starnberger See (FRG). J. Plankton Res. 9: 381–394.

Leydig, F., 1860. Naturgeschichte der Daphniden (Crustacea Cladocera). Laupp & Siebeck, Tübingen, 252 pp.

Lieder, U., 1983. Die Arten der Untergattung *Eubosmina* Seligo, 1900 (Crustacea: Cladocera, Bosminidae). Mitt. zool. Mus. Berlin 59: 195–292.

Schlichtherle, H., 1988. Siedlungsarchäologische Untersuchungen im Alpenvorland. Jb. histor. Forsch. 1987, Saur, München, 20–23.

Wolf, H. G. & M. A. Mort, 1986. Inter-specific hybridization underlies phenotypic variability in *Daphnia* populations. Oecologia 68: 507–511.

Hydrobiologia **225**: 87–103, 1991.
V. Kořínek & D. G. Frey (eds), Biology of Cladocera.

The history of the genus *Eubosmina* in Lake Mondsee (Upper Austria)

Arnold Nauwerck
Öst. Akad. Wissensch., Institut für Limnologie, Abt. Mondsee, A-5310 Gaisberg 16, Austria

Key words: postglacial sediments, *Bosmina*, successions, morphometric changes

Introduction

Since early work by Frey (1964), Goulden (1964), and Deevey (1969), cladoceran remains in sediments have been used as a tool in paleontological analysis (Boucherle & Züllig, 1983; Ravera & Zarini, 1987) and for the study of species evolution. The genus *Bosmina*, whose remains preserve well in the sediments, has been the object of numerous studies, particularly by Hofmann (1977, 1978, 1983, 1984a, 1984b, 1987a, 1987b).

As a general result, these studies have shown, that *Eubosmina longispina* was first to colonize European and North American postglacial lakes. *Bosmina longirostris* and other eubosminids appear later, and there are indications for immigration of the *E. coregoni* group (Hofmann, 1984a, 1987a) as well as indications for transgressions from *E. mixta kessleri* forms into typical *E. coregoni coregoni* forms, eventually also from *E. longispina* forms into *E. mixta kessleri* forms. Species replacements and morphological alternations have been interpreted as results of successive changes of the environment, namely by eutrophication or by water level changes (Hofmann, 1977, 1983, 1984a). However, only in *Bosmina longirostris* are such interpretations (Deevey, 1969; Boucherle & Züllig, 1983) in accordance with longterm studies (Müller, 1985; Nauwerck, 1988). In case of *Eubosmina*, interpretations mainly rely on the conclusions based on distribution of the species along trophic scales (Pejler, 1965; Patalas & Patalas, 1966). Also the *Eubosmina mixta* group still offers taxonomical problems (Lieder, 1957, 1983, 1986; Flössner, 1972).

The aim of the present paper is to contribute to the knowledge of the history of zooplankton colonization of Mondsee and the role of the genus *Bosmina* in the history of this lake. Results will be compared with results of other authors. Some unpublished results from other lakes in central and northern Europe will also be used for the comparison.

Methods

From sediment cores, taken with a Kullenberg corer, samples of 1 cm^3 were drawn and washed through 100 μm gauze filters. The filtrate then was examined microscopically. *Bosmina* remains were counted and identified to the species as far as possible. Mucrones were measured and antennal segments were counted, normally 50 items of each. Fractionating the sample through different mesh size showed that practically all *Bosmina* remains were retained in the 100 μm filter. No additional cleaning or bleaching was needed.

The material from Mondsee was further analysed as to pollen distribution (M. Handl, Salzburg), carbon, nitrogen and phosphorus content (M. Leichtfried, Lunz), fossil pigments (E. Schultze, Mondsee), and grain composition (J. Schneider, Göttingen). The author wishes to thank all these people for their contributions.

Results

A 7 m core from Mondsee

One core was taken at 48 m depth in the northeastern part of Mondsee. The core did not reach glacial sediments, and its upper part was slightly disturbed. Figure 1 shows an increase in density of *Bosmina* remains downward from about 6 m, together with a decrease of mucro lengths and an increase of antennule sediment numbers. Above 5 m, mucro and antenna curves fluctuate around relatively constant mean values. Contradistinctive peaks indicate that these fluctuations reflect real alternations. In the uppermost parts of the core, mucro lengths decrease considerably.

Figure 2 shows these changes in greater detail. Figure 3 shows a random choice of mucro forms from different depths, sorted according to length. Most specimens can be assigned to *Eubosmina longispina*. Only in superficial sediments is there a fraction of spineless forms, such as also are found in recent plankton samples. Those forms can be classified as *E. coregoni coregoni*. The longest mucrones occur a few decimeters below the sediment surface. They fit *E. longispina ruehei*.

In order to check how far relative size proportions can be affected by selective decay of some features, percentages of antennules in pairs and the relationship between numbers of mucrones and numbers of antennules were estimated. Figure 4 indicates steady and relatively slow changes of these parameters, which means that strong alterations according to Fig. 1 are not likely to be result of selective decay.

Figures 5 and 6 show some variables that may be correlated with the observed changes of *Bosmina* features and might be expected to help to explain them. However, all parameters, except total phosphorus, which does not change at all, provide only one clear point of information: a major change of the lake has taken place at about 2.5 m sediment depth. With the help of the pollen diagram this event can be identified as the result of the colonization of the Mondsee region by man. Clear-cutting of the woods and agriculture increased run-off from the catchment area of the lake and increased primary productivity in the lake. Besides considerable irregularities which in no case support each other and may be regarded as aberrant, the curves in Fig. 5 also show weak indications of a slow and steady increase of pro-

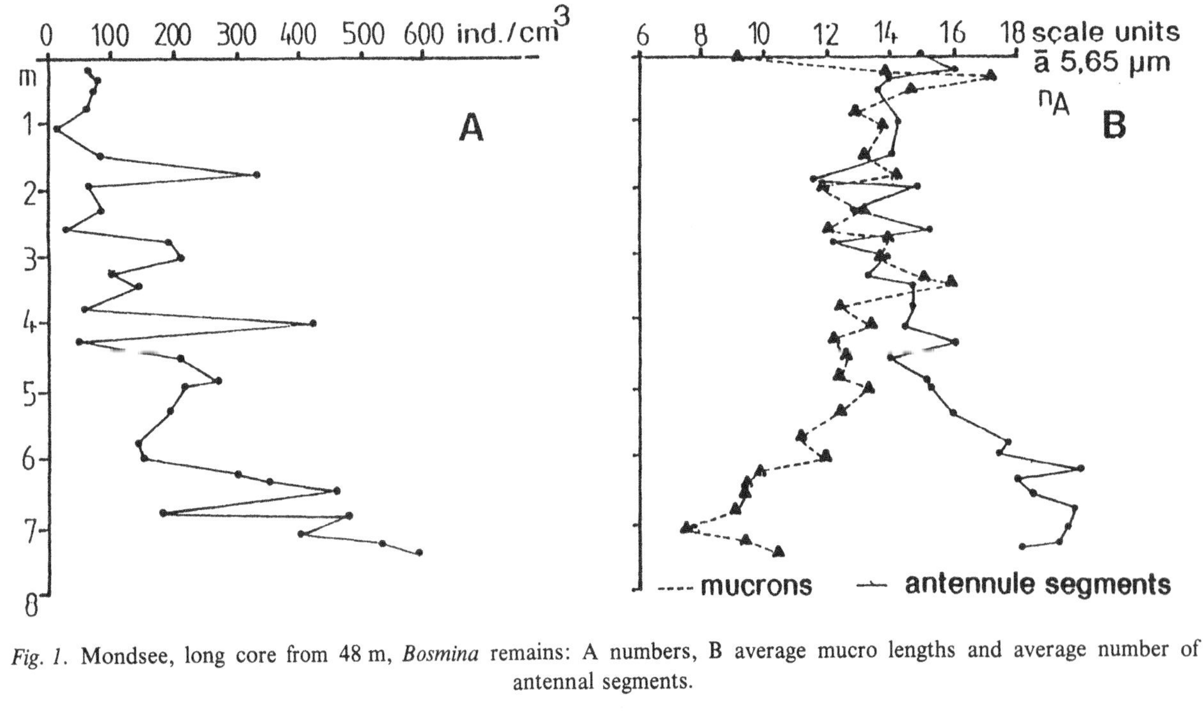

Fig. 1. Mondsee, long core from 48 m, *Bosmina* remains: A numbers, B average mucro lengths and average number of antennal segments.

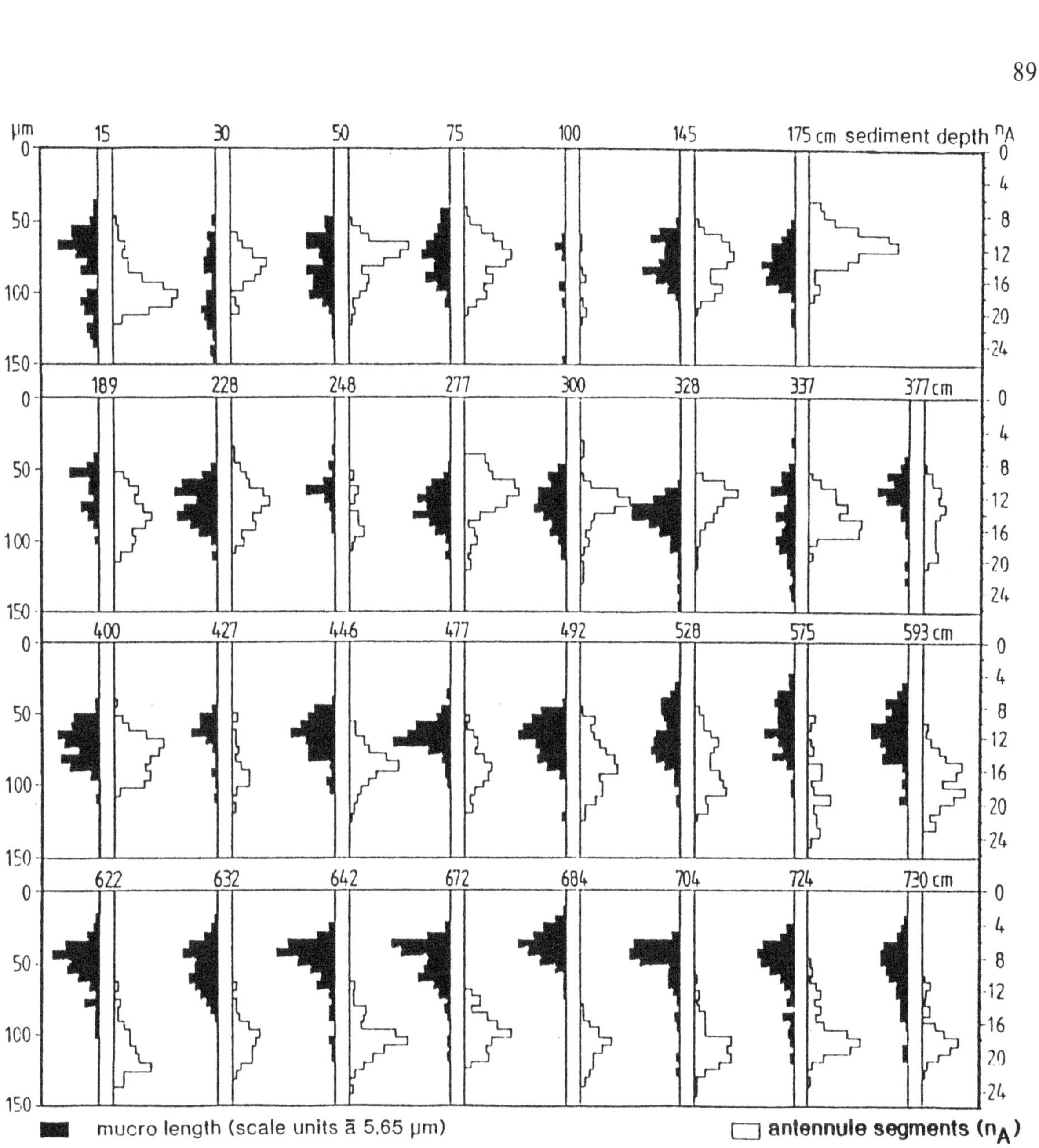

Fig. 2. Mondsee, long core from 48 m, *Bosmina* remains: distribution of mucro length and of number of antennal segments in different sediment depths.

ductivity in the lake even in older times. However, none of the changes occurs in parallel with the development of *Bosmina* densities or *Bosmina* morphology. In particular the changes between 5 and 7 meters sediment depth remain unexplained.

Late sediments and early sediments

It is obvious from the upper part of core one that changes of *Bosmina* features in the lake have accelerated during the last few decades. There are also indications that the trend observed in the lower part of the core changes direction below 7 m depth. More information can be gained from a carefully taken core of the upper sediment layers (Fig. 7 and 8) and from a new long core extending into the glacial deposits from the birth of the lake (Fig. 9). The development of the upper sediments of Mondsee has been studied in detail by several researchers (Schultze, 1985; Klee & Schmidt, 1987). Preserved plankton samples from 1943

Fig. 3. Mondsee, long core from 48 m, *Bosmina* mucrones: random choice from different sediment depths.

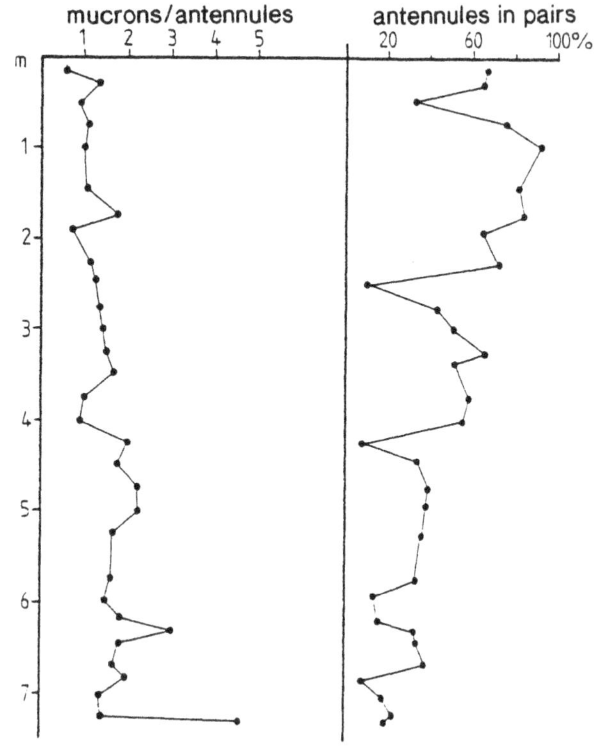

Fig. 4. Mondsee, long core from 48 m, ratio mucrones: antennules and percentage of unbroken pairs of antennules in total antennule numbers.

and onward (Nauwerck, 1988) allow further identifications. From the long core only *Bosmina* analyses have been carried out so far.

In the late sediments (Fig. 7), *Bosmina* features display changes that to some extent seem to be connected to major events in the recent history of the lake. During the 1950s and early 1960s, rubble and gravel from road constructions were deposited in the lake in such amounts that allochthonous fine particles caused plankton kills (Einsele, 1963) and left behind a sand horizon in the sediments (Klee & Schmidt, 1987). During the same period mean mucro lengths of *Bosmina* increase, or, more exactly, the share of the *Eubosmina longispina ruehei* form in the *Bosmina* population increases (Fig. 8).

At the end of the 60's the lake was subjected to accelerating eutrophication, which culminated in heavy *Oscillatoria* blooms at the beginning of the 1970s but later was controlled by a sewage treatment plant (Jagsch & Dokulil, 1989). Signs of recovery are restricted so far to phosphorus load, oxygen conditions, and algal composition and do not include zooplankton composition or biomass (Nauwerck, 1988). According to Fig. 7 and 8 *Bosmina* mucrones become shorter (or short-featured forms become more common) along with

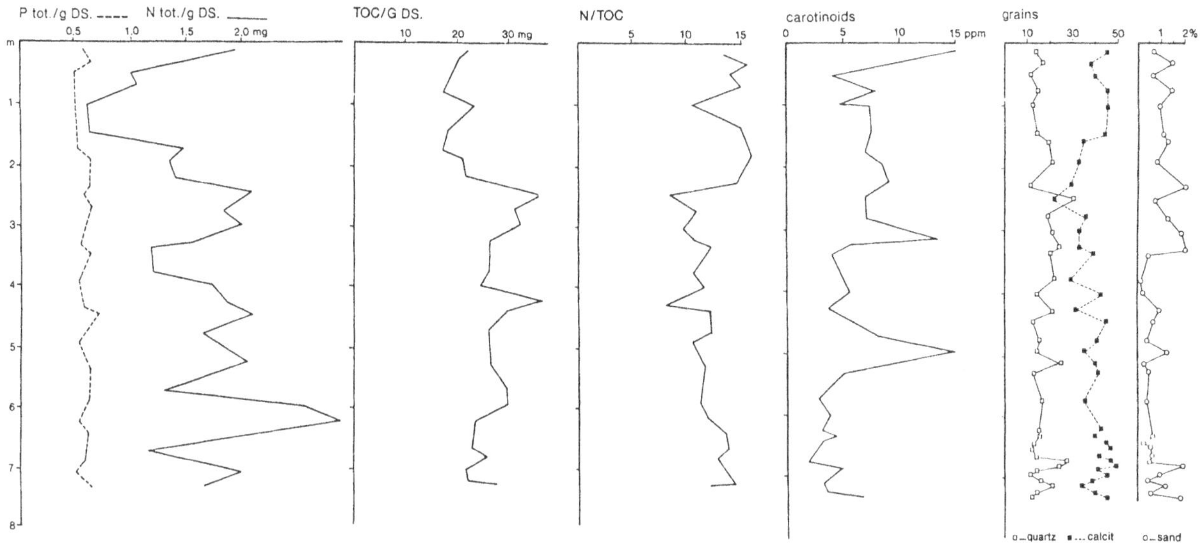

Fig. 5. Mondsee, long core from 48 m. Depth variation of different trophic parameters. From left: total phosphorus/dry substance, total nitrogen/dry substance, total organic carbon/dry substance, total nitrogen/total organic carbon, carotinoids, percentage of quartz, calcite and sand particles.

eutrophication, but long-featured forms increase when eutrophication culminates; since then, short featured forms have taken over successively.

The longer core (Fig. 9), taken from 68 m in the deepest part of the lake, displays the same *Bosmina* pattern as the core from 48 m. Individual densities increase down to 7–8 m, and average mucro lengths show a distinct decrease below 5 m down to about 7 m. Still deeper, both curves turn in the opposite direction, except that densities increase again until a maximum at about 11 m. Mucrones stay at about 130 μ length between 8.5 and 11 m and belong to *E. longispina ruehei* forms.

The inset diagrams in Fig. 9 show developing trends in greater detail. It is clear that the change towards shorter mucrones from 9 to 7 m is due to a jump from one size group to another, but also that populations at a given time may be composed of several size groups (cf. Fig. 2 and 8). Changes in relative proportions of different size groups thus appear to be the crucial mechanism behind size changes of populations.

Even if general trends obviously are the same in the two long cores, a comparison of absolute mucro sizes at comparable depths reveals systematic differences, mucrones from the second core being generally longer than those from the first one. It may be important to point out that the first core stems from a part of the lake that because of affluents and shallow shore regions has, and probably always has had a generally more eutrophic character than the deep southern part of the lake with its steep and rocky shores, where the second long core was taken.

A core from meromictic eutrophied Höllerer See

Figure 10 shows *Bosmina* changes in a small forest lake, some 50 km northeast of Mondsee. Thanks to a very restricted catchment area and wind protection by surrounding woods, Höllerer See in the past had extremely low sedimentation rates. Artificial draining and other agricultural measures eutrophied the lake and increased sedimentation rates enormously (Dokulil & Jagsch, 1987). From its birth and during the period of low sedimentation (oligotrophy) the lake was inhabited by a *Eubosmina longispina* of the *ruehei* type. With eutrophication, *E. longispina* more or less disappeared from the lake and was replaced by *Bosmina longirostris*, which has remained the dominating bosminid in the lake. While the remaining *E. longispina* have decreased in size

Fig. 6. Mondsee, long core from 48 m. Pollen distribution at different depths.

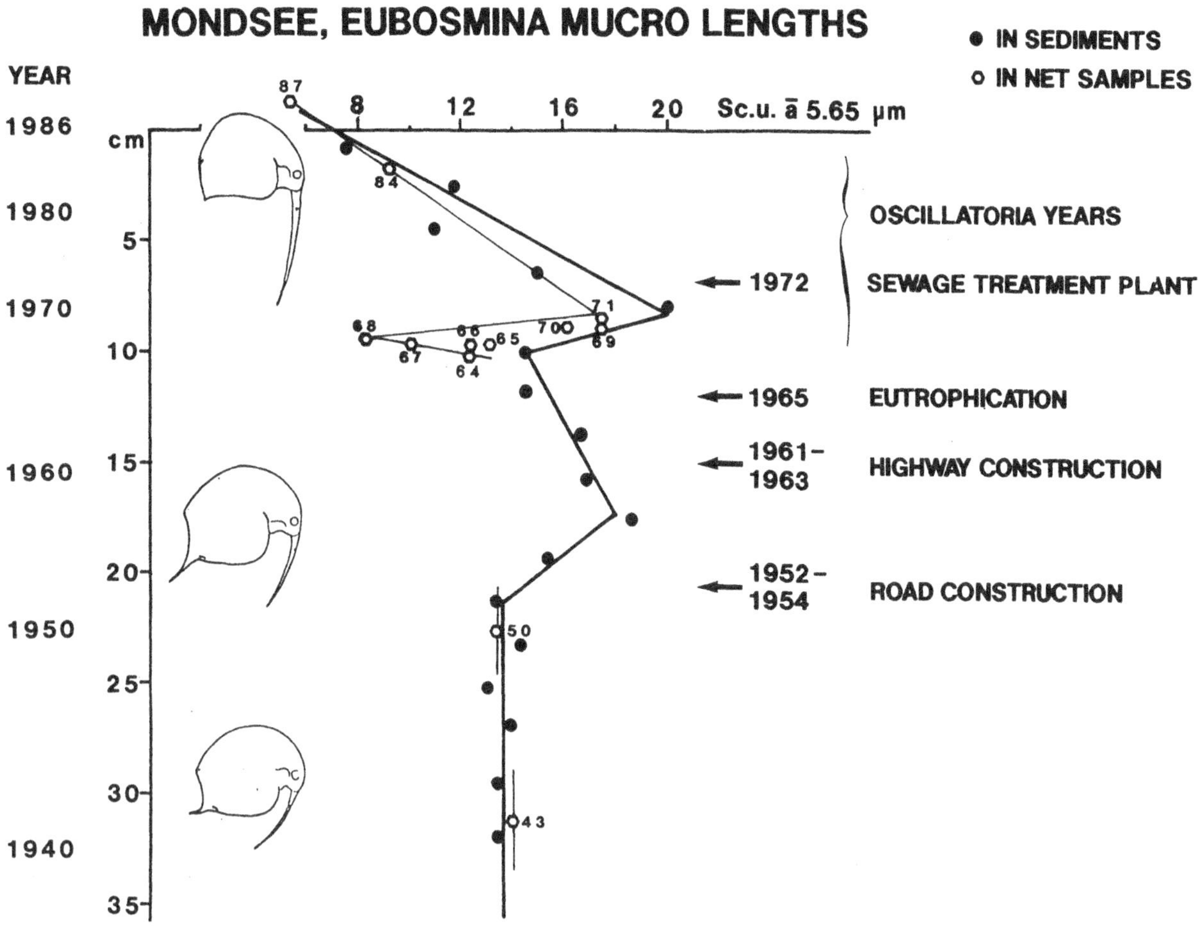

Fig. 7. Mondsee, short core from 48 m, *Bosmina*: distribution of average mucro length distribution in different sediment depths (black symbols) and in preserved net samples from recent years (white symbols). Major antropogenous interferences indicated at right.

and mucro length, *B. longirostris* has not shown any significant changes during the time of its existence in the lake. The case proves a primary postglacial colonization of the water bodies of the region with *Eubosmina longispina*. Further colonization of Höllerer See by *Bosmina longirostris* corresponds to the *Bosmina* development observed by Hofmann (1983) in small Lake Poolsee, and *Bosmina* development in Swiss lakes according to Boucherle & Züllig (1983).

Some comparisons of recent Bosmina types from lakes of the Salzkammergut region

Already Rühe (1914) called attention to the fact that what he then called *Bosmina coregoni longispina*, appeared as different 'races' in different lakes of the Salzkammergut, and he considered the form of the Hallstätter See to be 'the most genuine one' among them. It is the form to which Lieder (1957) assigned the rank of subspecies, and which still is believed to be endemic in the lakes of the river Traun catchment area or at least in the large lakes of the Salzkammergut (Flössner, 1972; Lieder, 1983). According to Rühe, differences between the populations of different lakes have to be explained by peculiarities of the lakes rather than by cyclomorphosis. Lieder (1983) on the other hand, maintains that there is pronounced cyclomorphosis in *Eubosmina ruehei*.

Figure 11 gives a survey over the occurrence of *Bosmina* in some lakes of the Salzkammergut. Arrows in Table 1 indicate tendencies of form

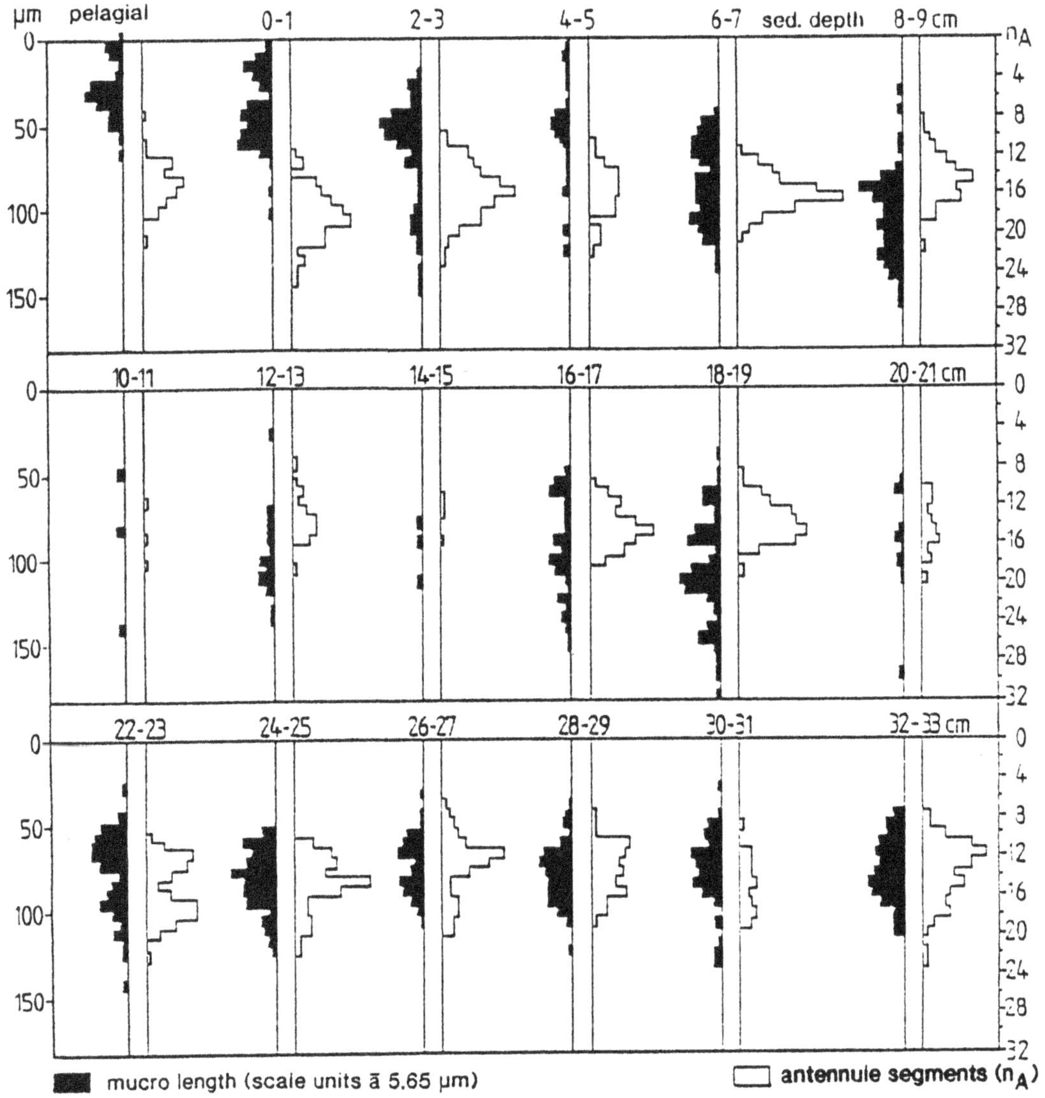

Fig. 8. Mondsee, short core from 48 m, *Bosmina* remains. Distribution of mucro length and of number of antennule segments at different sediment depths.

deviations. Unfortunately, information about the trophic state of the lakes is incomplete. Most of them are mesotrophic or oligotrophic. Only Wallersee is definitely eutrophic. Most lakes have *Eubosmina longispina* forms with variable mucro lengths up to the typical *E. longispina ruehei* form. The latter is found mainly in the large and oligotrophic lakes, but also in some smaller lakes and it is not confined to the catchment area of the river Traun. Hofmann (1987a) also mentions it from Starnberger See. On the other hand, the *E. longispina ruehei* form is not found in some of the most oligotrophic lakes of the region. Those lakes are usually smaller than the *ruehei*-lakes, and some of them are also younger in age.

Most remarkable is the presence of *Eubosmina coregoni coregoni* only in Wallersee and in Mondsee, whereas only Mondsee belongs in the river Traun catchment area. *Bosmina longirostris* as an accompanying species in several lakes may indicate higher degrees of eutrophy than elsewhere. It may be of interest to note, however, that *B. longirostris* is the only bosminid found in meromictic but oligotrophic Krottensee.

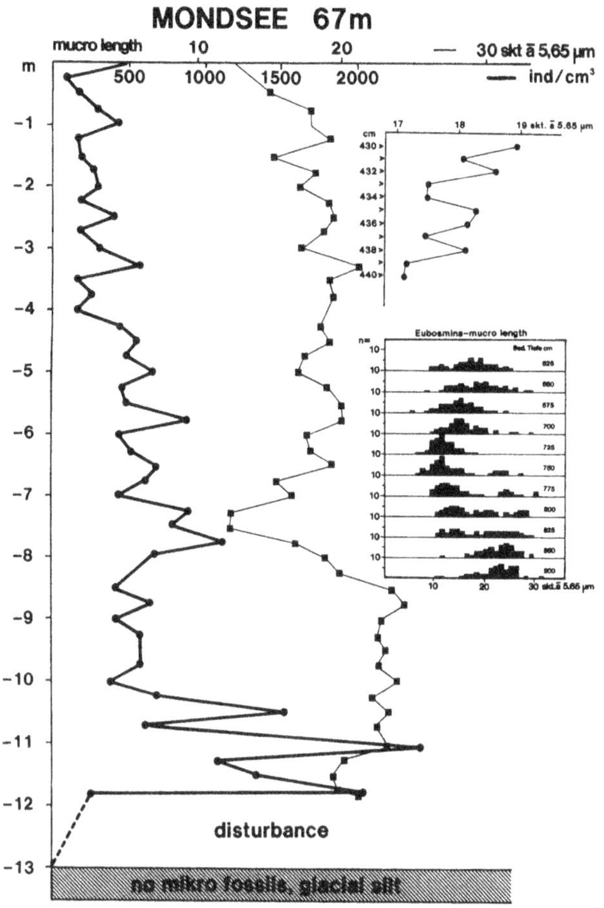

Fig. 9. Mondsee, long core from 67 m, *Bosmina*. Average mucro length (thin line) and individual numbers (thick line) at different sediment depths. Inserted: development details in depth segments 430 cm–440 cm and 625 cm–900 cm.

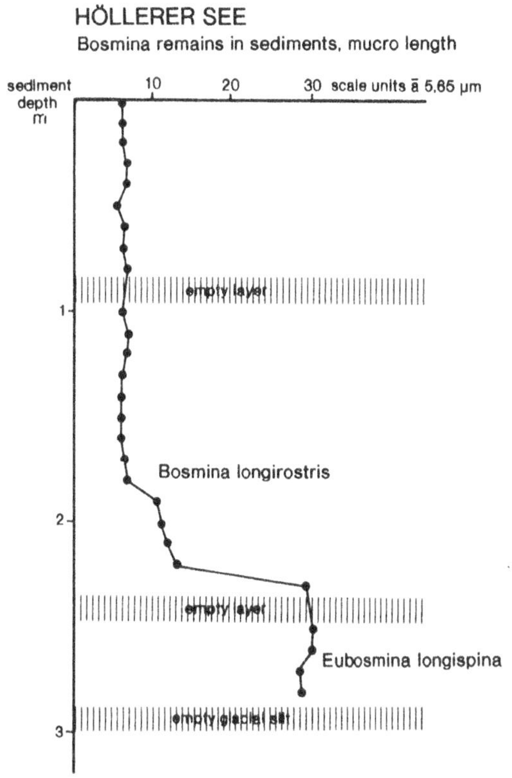

Fig. 10. Höllerer See, core from 20 m, *Bosmina* remains. Mean mucro length at different sediment depths.

Even if the *Bosmina* collections from most lakes are occasional, there is enough evidence for morphometrically rather stable populations. *Eubosmina longispina ruehei* from Hallstätter See and of *E. longispina* from Mondsee are kept in culture at the Limnological institute in Mondsee and have so far shown remarkable morphological stability. This is in accordance with the findings of Kerfoot & Peterson (1980) in *Bosmina longirostris* and also fits the experience of Black & Hairston (1983) and of Gerritsen (1983).

Some examples from Northern Sweden

Lakes in the coastal region of Northern Sweden have turned from bays of the Baltic into fresh-water bodies since the end of deglaciation about 6500 years BC, and the process is still going on. From a number of sediment cores from such lakes, three examples of *Bosmina* development are shown in Fig. 12–14. Further biological and chemical data will be published elsewhere.

Lake Raitajärvi is situated just below the highest postglacial coast line (HPC) and is the easternmost of the three lakes. Lake Lillträsk is situated closest to the coast, and lake Forsträsk is situated at medium level and westernmost.

Bosmina development in the three lakes shows common patterns, which are also found in other lakes below the HPC. In oldest sediments, only *E. longispina* is present. Then, as the lake is cut off from the Baltic, there is a sudden increase of total *Bosmina* remains, usually with a strong dominance of *Bosmina longirostris*. In lakes nowadays distant from the sea, a *Eubosmina mixta* then follows. Later *E. mixta* is accompanied or replaced by *E. coregoni*. In lakes close to the sea, *E. coregoni* also may appear all of a sudden.

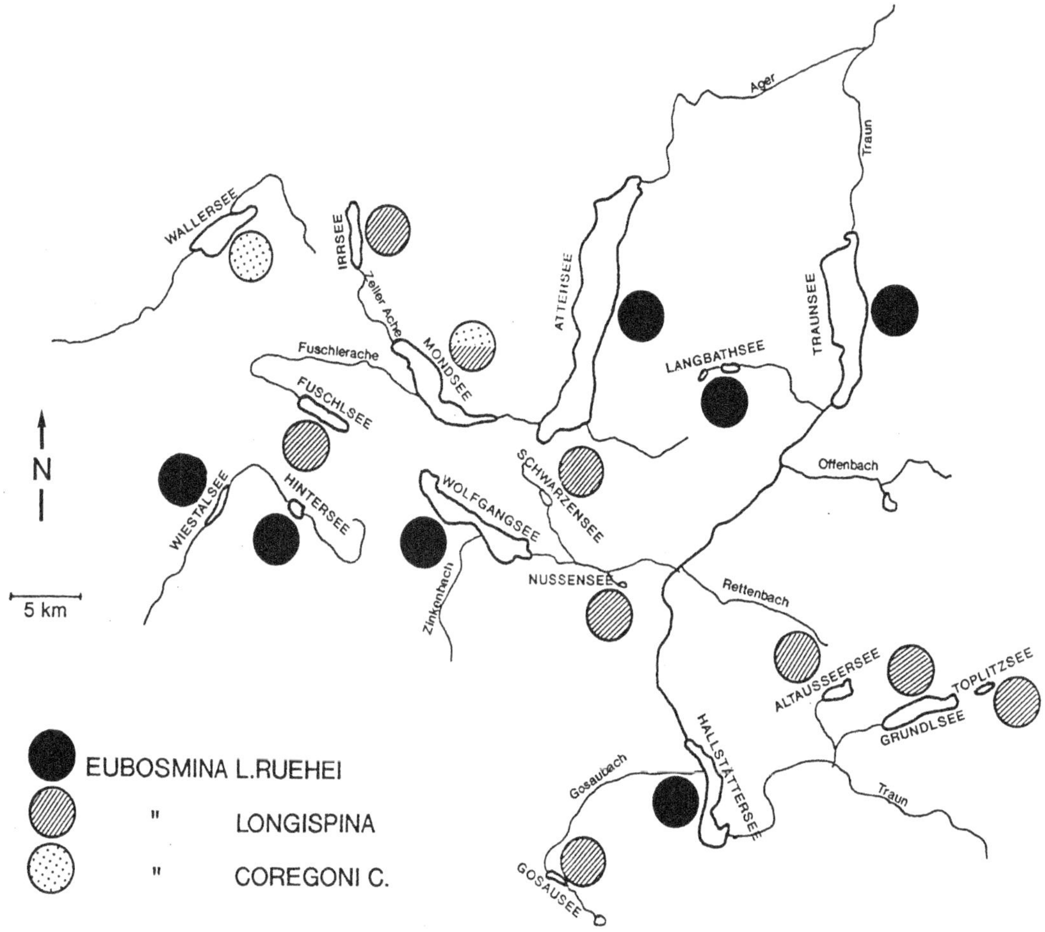

Fig. 11. Eubosmina distribution in Salzkammergut lakes 1987.

E. longispina is almost always found along with the others, but it is the only *Eubosmina* in lakes above the HCP.

There may be some doubts that the *E. longispina* of the lake phase is the same as that of the baltic phase. Here one has to remember that the Baltic after deglaciation was a freshwater body ('Ancyclus lake') for a few thousand years, and that salinity along the coast of northern Sweden after that has also been very low. Lilljeborg (1901) expressed the opinion that *E. longispina* carried by the rivers into the Baltic may have adapted to saline conditions, transforming into *E. longispina maritima*. In transplantation experiments I carried out with different *Eubosmina* species along a gradient from freshwater to brackish water (salinity 3‰), *E. coregoni* and *E. mixta* died quickly, *E. longispina* survived a little longer, but *E. longispina maritima* survived and reproduced in fresh water. This experiment does not necessarily contradict Lilljeborgs concept, but it shows that *E. longispina maritima* is perfectly able to continue its existence in cut-off bays of the Baltic as a freshwater *E. longispina*.

Remarkable in the Swedish examples is the stability of mucro lengths throughout time, except in *E. mixta* which is extremely variable. In lake Forsträsk its mucro length decreases, and the species at last is succeeded by *E. coregoni*. This may look like a transgression from one species into the other. But in lake Raitajärvi, on the contrary, the mucro length increases until *E. mixta*

Table 1. Bosminids in the Salzkammergut lakes 1987. Arrows indicate directions of form varieties.

	Eubosmina l. ruehei	longispina	coregoni	Bosmina longirostris	Lake surface km³	Secci-depth*	Total Phosphorus* µg < l⁻¹
Attersee	×				45.9	10–12	6
V. Langbathsee	×				< 1		
Hallstätter See	×				8.6	6–8	20
Traunsee	×				25.6	5–7	10–18
Wolfgangsee	×				12.8	4–9	10
Wiestalsee	×				< 2		
Hintersee	× ⟶			×	< 2		
Grundlsee	⟵	×			4.1	9–11	10
Altauseer See	⟵	×			2.1	10–11	20
Toplitzsee		×			< 2		
Nussensee		×		×	< 1		
Fuschlsee		× ⟶			2.7	5–7	7–13
Gosausee		× ⟶			< 3		
Schwarzensee		× ⟶		×	< 1		
Zeller See		×			3.5	3–6	20
Mondsee		× ⟶	×	×	14.2	4–7	10–20
Wallersee			×		6.1	2.5–4	25
Krottensee				×	< 1		

* From 'Seenreinhaltung in Österreich' BMLF Wien 1989. Annual averages.

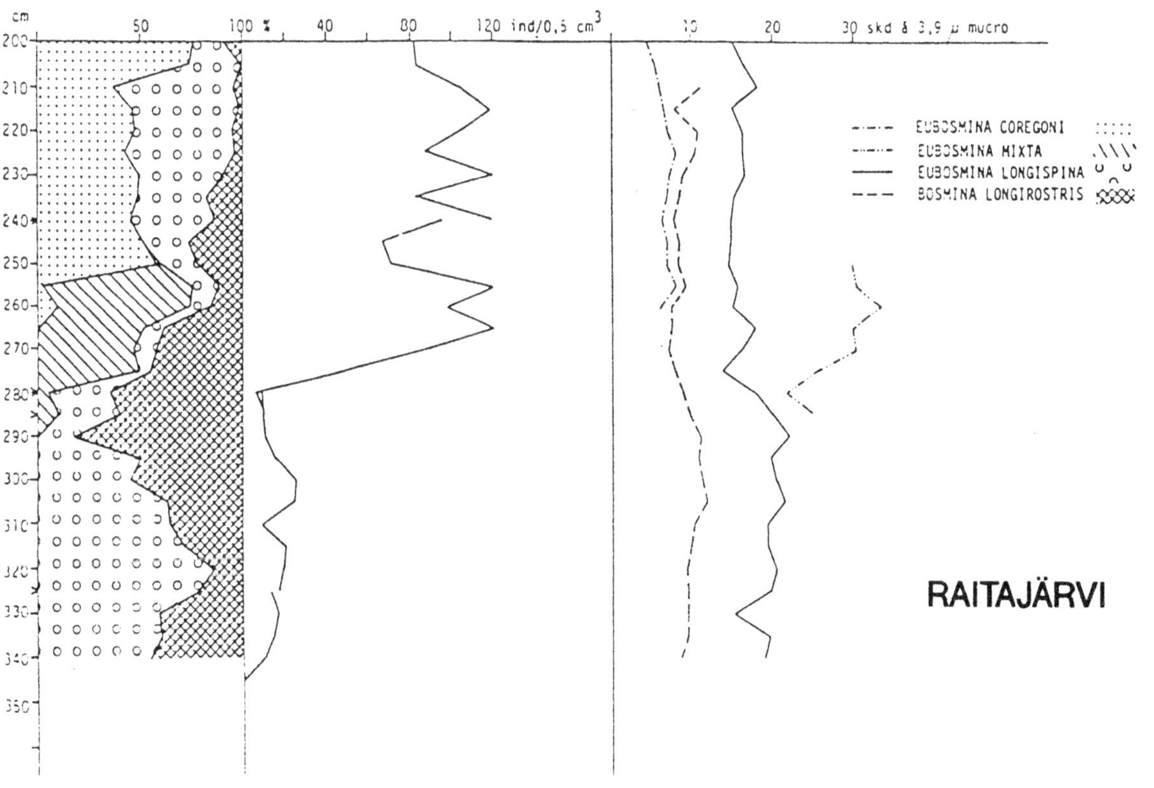

Fig. 12. Raitajärvi, core from 2.0 m, *Bosmina* remains. From left: species composition, numbers/0.5 cm³, average mucro lengths (scale units à 3.9 µ).

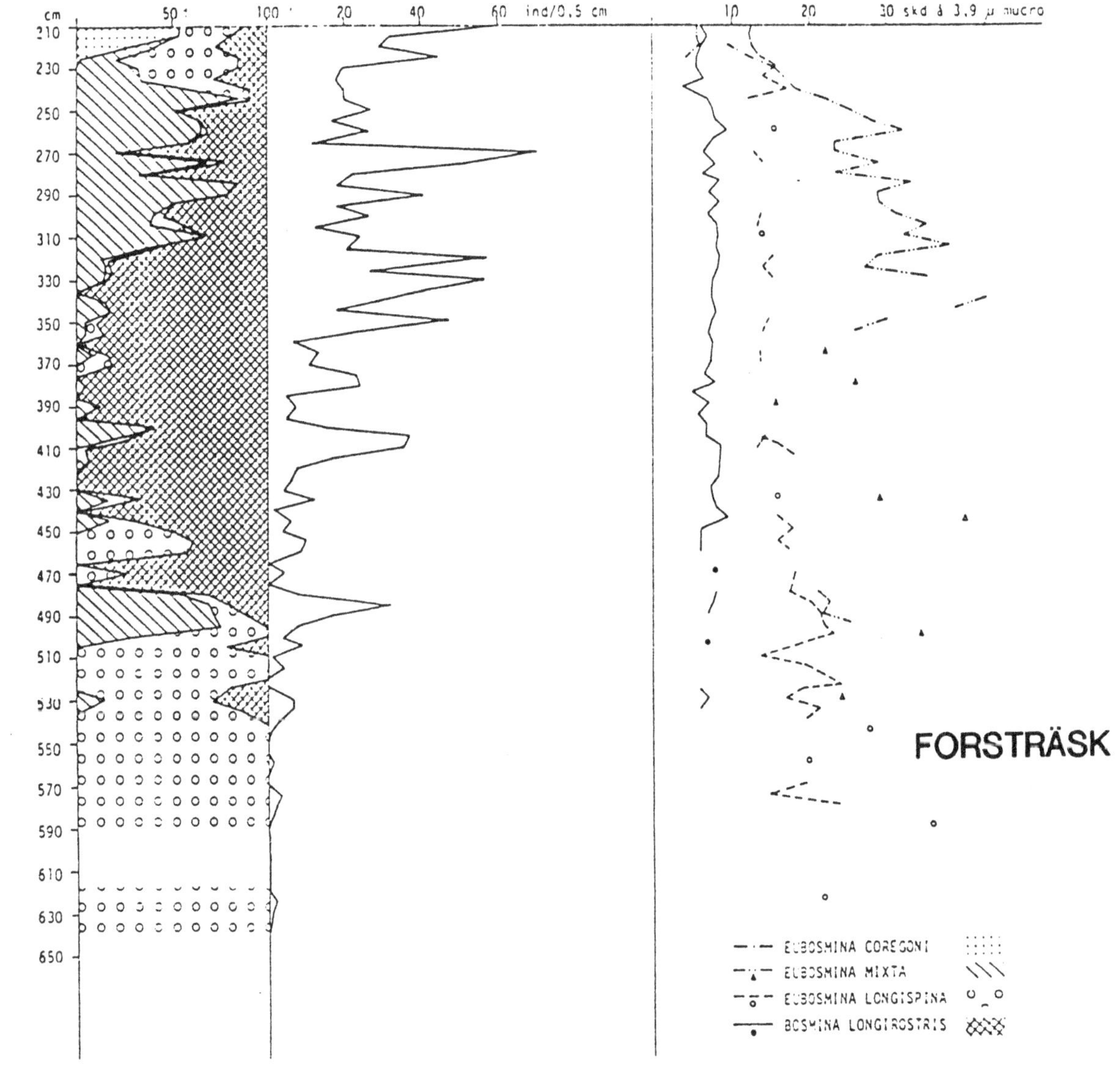

Fig. 13. Forsträsk, core from 2.1 m, *Bosmina* remains. From left: species composition, numbers/0.5 cm^3, average mucro lengths (scale units à 3.9 μ).

abruptly is replaced by *E. coregoni*. Finally in Lillträsk, *E. coregoni* appears without a preceeding *E. mixta*. Recent distribution of the species limits *E. mixta* to lakes below the HPC, while *E. coregoni* occurs only in the southeastern parts of the region, close to the Finnish border. This implies that *E. mixta* can be a relict from the Ancylus lake. Further evidence comes from the fact that *E. mixta* often is found associated with *Eudiaptomus gracilis*, which has been shown to be another possible Ancylus relict (Nauwerck, 1980). *E. coregoni*, on the other hand stands out as a late eastern immigrant. If we compare the time of earliest appearance of *E. coregoni* in different lakes with the actual geographical distribution of the species in Europe according to Lieder (1957, 1982), it seems apparent (Fig. 15) that the species still is spreading with a certain speed centrifugally from a center in north-eastern Europe.

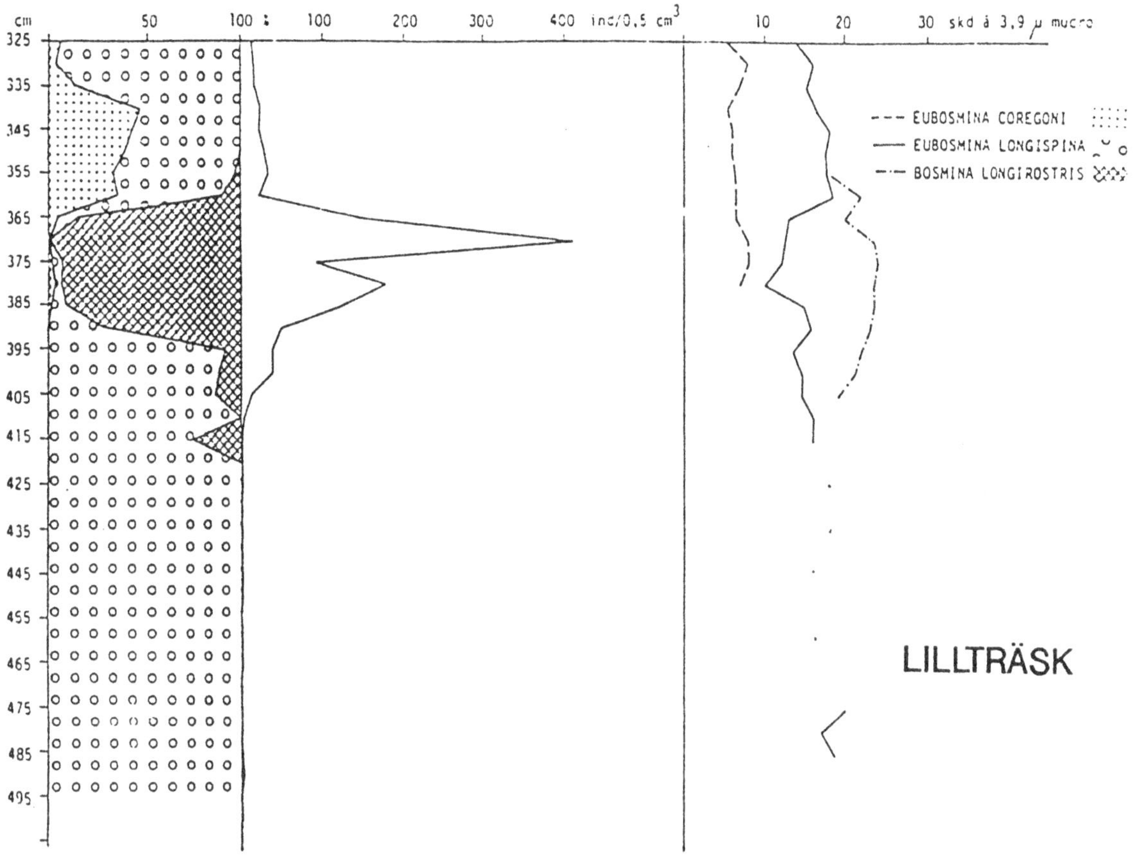

Fig. 14. Lillträsk, core from 3.25 m, *Bosmina* remains. From left: species composition, numbers/0.5 cm³, average mucro lengths (scale units à 3.9 μ).

Discussion

It seems that densities of bosminid remains in lake sediments are not closely related to lake productivity (Harmsworth & Whiteside, 1968). Nevertheless, obvious and sometimes dramatic changes of bosminid densities in our examples suggest connections with major changes of environmental conditions, such as changes of run-off from the watershed or changes of water level, events which in one way or the other also involve trophic changes.

More consistent than absolute densities as environmental indicators are changes in the relative abundance of species and in species replacements. The question is, how far do such changes depend on immigration and/or extinction of 'good species' and how far do they comprise morphological adaptation of 'relative species'.

According to our material, the 'good species' *Bosmina longirostris* occupies lakes that have reached a certain trophic state. Also *E. mixta* and *E. coregoni* appear as immigrants rather than as a result of morphological transgression from one ot another form. Their occupation of a lake may be favoured by suitable environmental conditions, but their arrival at the lake seems to be a matter of long-term zoogeographical spreading rather than of short-dated exploitation of a given opportunity.

On the other hand, within the *E. longispina* group as well as within the *E. mixta* group, one can recognize morphological successions from long-featured populations to short-featured popu-

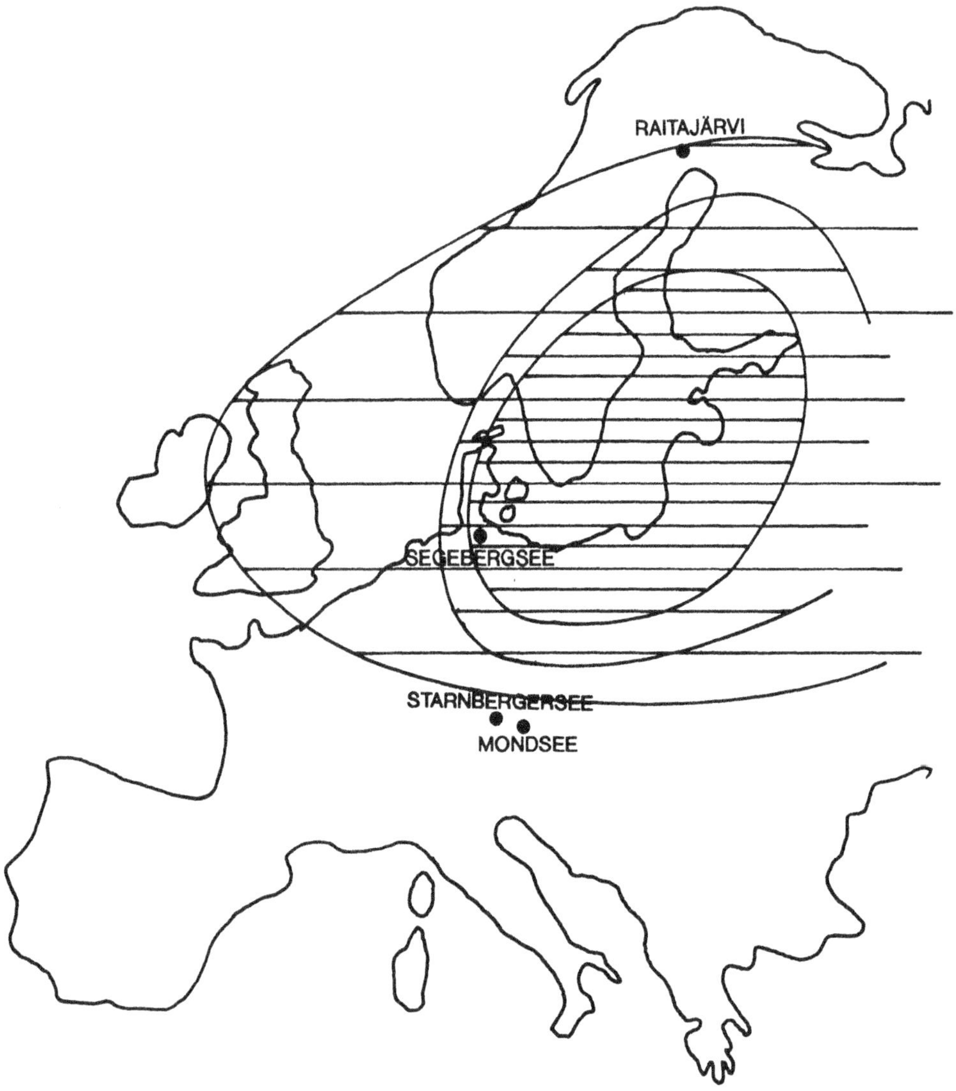

EARLIEST APPEARANCE IN LAKE SEDIMENTS:

Mondsee	~	5 cm (1985)
Starnbergersee	~	50 cm
Raitajärvi	~	100 cm
Segebergsee	~	300 cm (B.C.)

Fig. 13. Distribution of *Eubosmina coregoni* in Europe (modified from Lieder 1957). Earliest appearance in sediments in Mondsee, Raitajärvi (this study), Segebergsee, Starnbergersee (according to Hofmann 1977, 1987a).

lations or vice versa, which seem to be linked to environmental changes of the lakes. Our results support that changes are accomplished by 'clonal succession' (Black, 1980; Brock, 1980) or 'fluctuating regimes' (Kerfoot & Peterson, 1980).

Among the factors that may or may not favour one or the other *Bosmina* species, food supply can be expected to be a prominent one. Many authors have shown that *Bosmina longirostris* filters smaller particles, down to bacteria size, than does *Eubosmina coregoni* (Semenova, 1974; Demott, 1982; Müller, 1985), and that *E. longispina* prefers particles around 5 μm, which means mainly small algae (Johnsen & Börsheim, 1988). It has also been shown that *E. coregoni* successfully competes with *B. longirostris* under good food conditions (Matveev, 1985), and that *E. longispina* -related *E. fatalis* is more successful in high food concentrations than *B. longirostris*, which instead competes better at poor food conditions (Hanazato & Yasumo, 1987). Thus, a trophic gradient may be expected along the line *E. longispina – E. mixta – E. coregoni – B. longirostris*.

If we assume that large (usually long featured) forms prefer or are favoured by oligotrophic conditions (which means low densities of very small food particles and relatively high densities of small algae), and small (usually short featured or spineless) forms prefer or are favoured by eutrophic conditions (which means high densities of very small food particles and high densities of big, non-edible algae), the Swedish lakes at least would not contradict the assumption. Apparent contradictions in Mondsee can be explained by food quality changes: in spite of high phosphorus loads during the 1960s, allochthonous particles prevented the growth of nutritive algae, thus creating an oligotrophic situation for the bosminids. A similar situation arose during *Oscillatoria* blooms, when edible small algae were scarce. In both situations the mucro lengths of *E. longispina* increased. Decreasing mucro lengths after the decline of eutrophication went along with an increase of suitable food algae, particularly flagellates such as *Dinobryon*.

However, it is difficult to understand how food conditions should influence mucro length and how mucro length is related to feeding. Food supply may control species composition in the first hand, but predation pressure is more likely to influence the shape of individuals of a population. Since Dodson (1974) and others postulated size selection by predators as an important steering factor of population morphology, it has been shown by numerous authors that the shape of bosminids is squeezed between fish predation, which points at large specimens more than at small ones (Zaret & Kerfoot, 1975; Stenson, 1976; Johnson & Raddum, 1987; Sandlund *et al.*, 1987), and invertebrate predation, which strikes harder on small specimens than on big ones (Kerfoot, 1977; O'Brien, 1979; Kerfoot & Peterson, 1980; Wong, 1981; Johnson & Raddum, 1987), but that the latter also provokes defense mechanisms such as formation of long spines, etc. (Sprules *et al.*, 1984; Kerfoot, 1987).

Thus, in our examples, successive changes of species composition may be explained by eutrophication where small food particles became more numerous and small bosminids got a chance to compete with large *Eubosmina longispina*. Increasing predation pressure in eutrophied lakes can contribute to changes from large to small species and also from large to small specimens within the same species.

In Mondsee, *Bosmina longirostris* as a newcomer in the pelagical community has most likely been favoured by an increase of small food particles. *Eubosmina coregoni* may have been enabled by the same conditions to compete successfully with *E. longispina*, but it must be considered a late immigrant into the lake. Recent mucro length changes in the *E. longispina* group seem to be related to eutrophication. The relatively fast change of long featured *E. longispina ruehei* into short featured *E. longispina* proper in the 9–7 m level of the sediments more likely indicates a major change in marks the colonization of the earlier *Salvelinus*-dominated lake by planktivorous cyprinids, namely the arrival of the danubian bleach *Chalcalburnus chalcoides*.

Summary

Bosminid remains in sediment cores from lakes Mondsee and Höllerer See in Upper Austria, and from lakes Raitajärvi, Lillträsk, and Forsträsk in Northern Sweden were examined. In all lakes, long-featured *Eubosmina longispina* appears as the earliest postglacial colonist. Later, again in all lakes, successions can be observed from large long-featured to small, short-featured forms. In Mondsee, the original *E. longispina ruehei* is followed by *E. longispina*, which is being replaced by *E. coregoni coregoni*. In Höllerer See, *E. longispina ruehei* is replaced by *Bosmina longirostris*. In the Swedish lakes, all of them once parts of the Baltic, *B. longirostris* became dominant shortly after the cut-off of these lakes from the sea. Later *Eubosmina mixta* and *E. coregoni* were added to the earlier species. Transgressions between species seem not to have occurred. Immigration from elsewhere seems to be the primary reason for species succession. Suitable trophic conditions may allow newcomers to gain a foothold in the lakes. Temporal phase displacement of the arrival of *E. coregoni* in different lakes indicates ongoing spreading from a center in north-eastern Europe. *E. mixta* in northern Swedish lakes appears as a relict species of the Ancylus lake.

Obvious form variations, which may be connected to trophical changes of the lakes, appear only within the *E. longispina* group and within the *E. mixta* group. Such variations develop not only over centuries but also over a few years, which may contradict evolutionary conceptions. Essentially, the changes take place by an increment of certain size groups within the population and the fading away of others. Even if changed food conditions cannot be ruled out as a reason for morphological alterations of a population, predation seems to be the more important factor. The switch from *E. longispina ruehei* in old sediments of lake Mondsee into *E. longispina* proper may be explained by the immigration of planktivorous cyprinids into the lake at that time.

Acknowledgements

The author thanks I. Gradl for typing the manuscript, K. Maier for preparing the figures and Dr. C. Cotten for reviewing the English text.

References

Black, R. W., 1980. The nature and causes of cyclomorphosis in a species of the *Bosmina longirostris* complex. Ecology 61: 1122–1131.

Black, R. W. & N. G. Hairston Jr., 1983. Cyclomorphosis in *Eubosmina longispina* in a small north American pond. Hydrobiologia 102: 61–68.

Boucherle, M. M. & H. Züllig, 1983. Cladoceran remains as evidence of change in trophic state in three Swiss lakes. Hydrobiologia 103: 141–146.

Brock, D. A., 1980. Genotypic succession in the cyclomorphosis of *Bosmina longirostris* (Cladocera). Freshwat. Biol. 10: 239–250.

BMLF (Bundesministerium für Land- und Forstwirtschaft), 1989: Seenreinhaltung in Österreich.

Deevey, E. S., 1969. Cladoceran populations of Rogers Lake, Connecticut, during late- and postglacial time. Mitt. Int. Ver. Limnol. 17: 56–63.

Demott, W. R., 1982. Feeding selectivities and relative ingestion rates of *Daphnia* and *Bosmina*. Limnol. Oceanogr. 27: 518–527.

Dodson, S. I., 1974. Adaptive change in plankton morphology in response to size-selective predation: A new hypothesis of cyclomorphosis. Limnol. Oceanogr. 19: 721–729.

Dokulil, M. & A. Jagsch, 1987. Limnologische Untersuchung der Innviertler Seen. Report, Limn. Inst. Mondsee, 77 pp.

Einsele, W., 1963. Schwere Schädigungen der Fischerei und der biologischen Verhältnisse im Mondsee durch Einbringung von lehmig-tonigem Berg-Abraum. – Österr. Fischerei 17: 2–12.

Flössner, D., 1972. Kiemen- und Blattfüßer, Branchiopoda; Fischläuse, Branchiura. In Die Tierwelt Deutschlands, 60. Teil, VEB Gustav Fischer Verlag Jena.

Frey, D. G., 1964. Remains of animals in Quaternary lake and bog sediments and their interpretation. Arch. Hydrobiol. Beih. 2: 1–114.

Gerritsen, J., 1983. Cyclomorphosis of *Bosmina* (*Eubosmina*) *tubicen* in the Okefenokee Swamp. J. Plankton Res. 5: 713–722.

Goulden, C. E., 1964. The history of the cladoceran fauna of Esthwaite Water (England) and its limnological significance. Arch. Hydrobiol. 60: 1–52.

Hanazato, T. & M. Yasumo, 1987. Experimental studies on competition between *Bosmina longirostris* and *Bosmina fatalis*. Hydrobiologia 154: 189–199.

Harmsworth, R. V. & M. C. Whiteside, 1968. Relation of Cladoceran remains in lake sediments to primary productivity of lakes. Ecology 49: 998–1000.

Hofmann, W., 1977. *Bosmina* (*Eubosmina*) populations of the Großer Segeberger See during late glacial and postglacial times. Arch. Hydrobiol. 89: 349–359.

Hofmann, W., 1978. *Bosmina* populations of Großer Plöner See and Schöhsee Lakes during late glacial and postglacial times. Pol. Arch. Hydrobiol. 25: 167–176.

Hofmann, W., 1983. Stratigraphy of Cladocera and Chironomidae in a core from a shallow North German lake. Hydrobiologia 103: 235–239.

Hofmann, W., 1984a. Morphological variation in a late glacial population of *Bosmina longispina* Leydig (Crustacea, Cladocera) from the Lobsigensee (Swiss Plateau). Studies in the Late-Quaternary of Lobsigensee 9. Schweiz. Z. Hydrol. 46: 139–146.

Hofmann, W., 1984b. Postglacial morphological variation in *Bosmina longispina* (Crustacea Cladocera) from the Grosser Plöner See North Germany and its taxonomic implications. Z. Zool. Syst. Evolutionsforsch. 22: 294–301.

Hofmann, W., 1987a. The late Pleistocene/Holocene and recent *Bosmina* (*Eubosmina*) fauna (Crustacea: Cladocera) of the pre-alpine Starnberger See (FRG). J. Plankton Res. 9: 381–394.

Hofmann, W., 1987b. Stratigraphy of Cladocera (Crustacea) and Chironomidae (Insecta: Diptera) in Three Sediment Cores from the Central Baltic Sea as Related to Paleosalinity. Int. Revue ges. Hydrobiol. 72: 97–106.

Jagsch, A. & M. Dokulil, 1989. Mondsee. In Seenreinhaltung in Österreich, Schriftenreihe Wasserwirtschaft, BMLF Wien, 91–100.

Johnsen, G. H. & K. Y. Börsheim, 1988. Functional response and food selection of the water flea, *Bosmina longispina*. J. Plankton Res. 10: 319–325.

Johnsen, G. H. & G. G. Raddum, 1987. A morphological study of two populations of *Bosmina longispina* exposed to different predation. J. Plankton Res. 9: 297–304.

Kerfoot, W. C., 1977. Implications of copepod predation. Limnol Oceanogr. 22: 316–325.

Kerfoot, W. C., 1987. Translocation experiments: *Bosmina* responses to copepod predation. Ecology 68: 596–610.

Kerfoot, W. C. & C. Peterson, 1980. Predatory copepods and *Bosmina* replacement cycles and further influences of predation upon prey reproduction. Ecology 61: 417–431.

Klee, R. & R. Schmidt, 1987. Eutrophication of Mondsee (Upper Austria) as indicated by the Diatom Stratigraphy of a Sediment Core. Diatom Res. 2: 55–76.

Lieder, U., 1957. Beiträge zur Kenntnis des Genus *Bosmina* (Crustacea, Cladocera). Inauguraldiss. Humboldt-Universität Berlin, 247 pp.

Lieder, U., 1982. Zoogeography of the subgenera of *Bosmina* (Crustacea, Cladocera). Zool. Jahrb. Abt. Syst. Oekol. Geogr. Tiere 109: 514–519.

Lieder, U., 1983. Die Arten der Untergattung *Eubosmina* Seligo, 1900 (Crustacea: Cladocera, Bosminidae). Mitt. zool. Mus. Berlin 59: 195–292.

Lieder, U., 1986. *Bosmina* and *Daphnia*: taxonomically critical groups among the Cladocera (Crustacea Phyllopoda). Limnologica 17: 53–66.

Lilljeborg, W., 1901. Cladocera Sueciae. Reprint by Almquist & Wiksell, Uppsala 1982.

Matveev, V., 1985. Competition and population time lags in *Bosmina* (Cladocera, Crustacea). Int. Revue ges. Hydrobiol. 70: 491–508.

Müller, H., 1985. The niches of *Bosmina coregoni* and *Bosmina longirostris* in the ecosystem of Lake Constance. Verh. Int. Ver. Limnol. 17: 56–63.

Nauwerck, A., 1980. Die Verbreitung der Familie Diaptomidae Sars in Nordschweden. Arch. Hydrobiol. 89: 247–264.

Nauwerck, A., 1988. Veränderungen im Zooplankton des Mondsees 1943–1988. Ber. Nat.-Med. Ver. Salzburg 9: 101–133.

O'Brien, W. J., 1979. Arctic *Bosmina* morphology and copepod predation. Limnol. Oceanogr. 24: 564–568.

Patalas, J. & K. Patalas, 1966. The crustacean plankton communities in Polish lakes. Verh. Int. Ver. Limnol. 16: 204–215.

Pejler, B., 1965. Regional-ecological studies of Swedish freshwater zooplankton. Zool. Bidr. Upps. 36: 407–515.

Ravera, O. & S. Zarini, 1987. A century of variations in some zooplankton and zoobenthos organisms in Lake Comabbio (Northern Italy) 'read' from their remains in the lake sediment. Schweiz. Z. Hydrol. 49: 93–101.

Rühe, F. E., 1914. Beitrag zur Biologie der Bosminen des Salzkammergutes. Arch. Hydrobiol. 10: 77–90.

Sandlund, O. T., T. F. Naesje & G. Kjellberg, 1987. The size selection of *Bosmina longispina* and *Daphnia galeata* by co-occurring cisco *Coregonus albula*, whitefish *Coregonus lavaretus*, and smelt *Osmerus eperlanus*. Arch. Hydrobiol. 110: 357–364.

Schultze, E., 1985. Carotinoids from selected cores of the Trumer Lakes and the Mondsee (trophic development and human impact). Contributions to the Paleolimnology of the Trumer Lakes (Salzburg) and the lakes Mondsee, Attersee and Traunsee (Upper Austria). Limn. Inst. Österr. Akad. der Wissensch., D. Danielopol, R. Schmidt and E. Schultze Eds., 52–64.

Semenova, L. M., 1974. The feeding habits of *Bosmina coregoni* (Cladocera). Hydrobiol. J. 10: 28–34.

Sprules, W. G., J. C. H. Carter & C. W. Ramcharan, 1984. Phenotypic associations in the Bosminidae (Cladocera): Zoogeographic patterns. Limnol. Oceanogr. 29: 161–169.

Stenson, J. A. E., 1976. Significance of predator influence on composition of *Bosmina* spp. populations. Limnol. Oceanogr. 21: 814–822.

Wong, C. K., 1981. Cyclomorphosis in *Bosmina* and copepod predation. Can. J. Zool. 59: 2049–2052.

Zaret, R. E. & W. C. Kerfoot, 1975. Fish predation on *Bosmina longirostris*, body size selection vs. visibility selection. Ecology 56: 232–237.

Hydrobiologia **225**: 105–114, 1991.
V. Kořínek & D. G. Frey (eds), Biology of Cladocera.

Impact of prehistoric settlements on the Cladocera in the sediments of Lakes Suszek, Błędowo, and Skrzetuszewskie

K. Szeroczyńska
Institute of Geological Sciences of the Polish Academy of Sciences 02-089 Warszawa, Żwirki i Wigury 93, Poland

Key words: Cladocera, palaeolimnology, eutrophication, sediment, human settlement

Abstract

An analysis of Cladocera has been made on sediments from the three Polish lakes Suszek, Błędowo, and Skrzetuszewskie. The analysed sediments are gyttja from the Subboreal and Subatlantic periods, during which the lakes were subject variously to eutrophication. The eutrophication process has been analysed from the frequency of Cladocera, particularly the index species, and has been correlated with climatic and settlement changes. It was found that in Suszek lake, in which no significant human activity had taken place (palynologically documented), the eutrophication must have been slow and gradual. On the contrary, in the two remaining lakes, in which the human influence was more intense, the eutrophication process underwent intensive changes and coincided somewhat with settlement fluctuations.

Introduction

Complex investigations of Holocene deposits as well as of presentday lake sediments in Poland include the analysis of Cladocera. Cladocera, as is well known, are preserved in lake sediments as subfossil remains. A quantitative and qualitative study of cladoceran remains can give information about ecologic (Boucherle & Zülling, 1983; Frey, 1960; Günter, 1983; Smirnov, 1981) and climatic conditions (Alhonen, 1986; Frey, 1966; Hann & Karrow, 1984; Hofmann, 1983, 1987) as well as about lake characteristics (Havens & DeCosta, 1985; DeMott & Kerfoot, 1982; Uimonen-Simola & Tolonen, 1987).

The objective of the analysis of sediments from the 3 Polish lakes (Fig. 1) was to investigate the stages of eutrophication and their relation to human activities. Sediment cores were raised from the deepest part of each lake. Samples were taken every 5–20 cm, depending on the lithological changes. Analysis was made according to the method published by Frey (1986). The specific

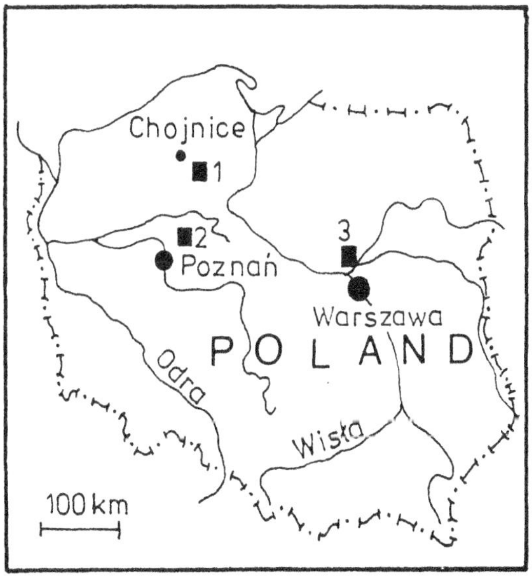

Fig. 1. The location of the study lakes. 1 – Suszek, 2 – Skrzetuszewskie, 3 – Błędowo.

BŁĘDOWO LAKE

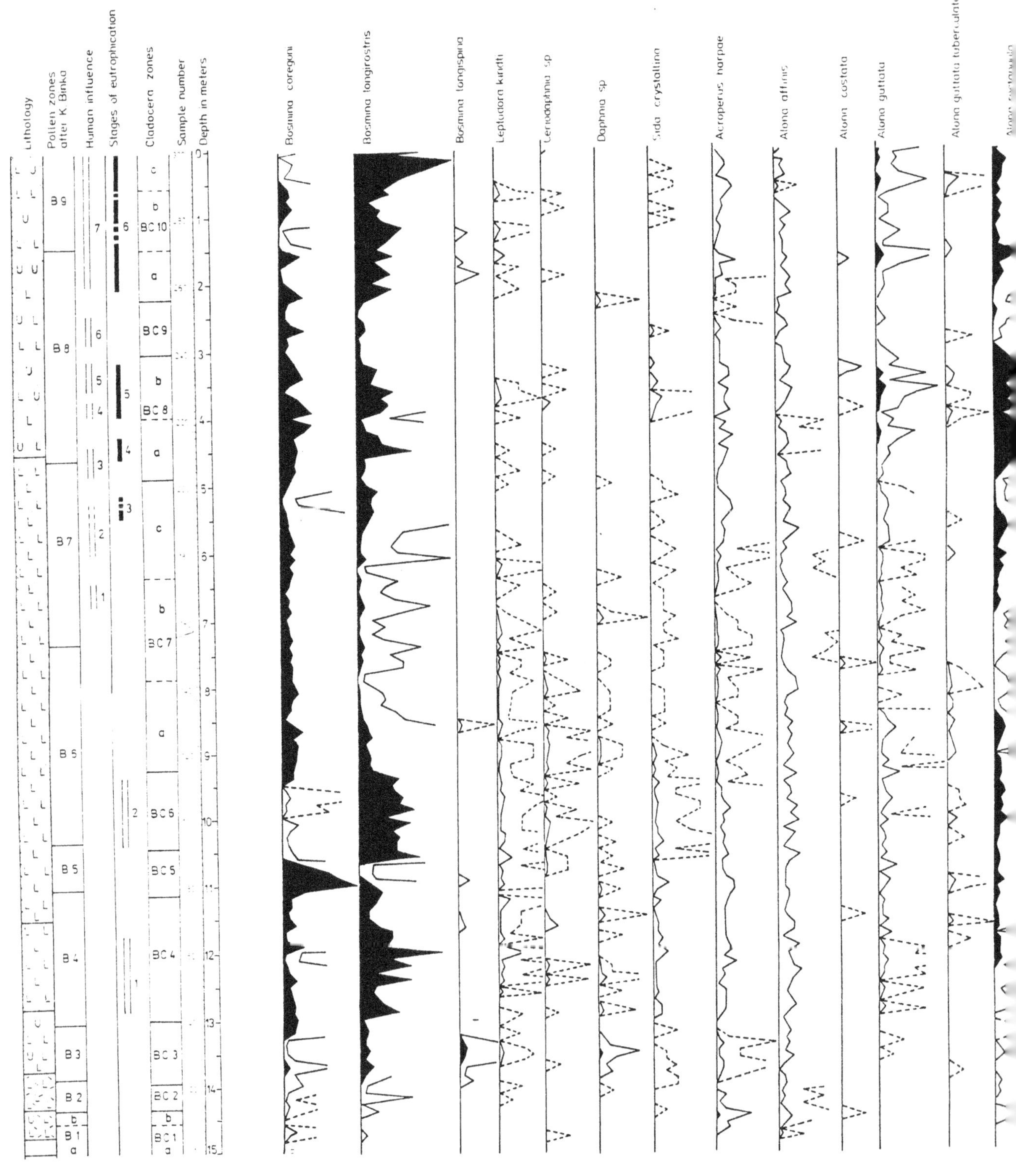

Fig. 2. Distribution of the Cladocera in the profile Błędowo III.

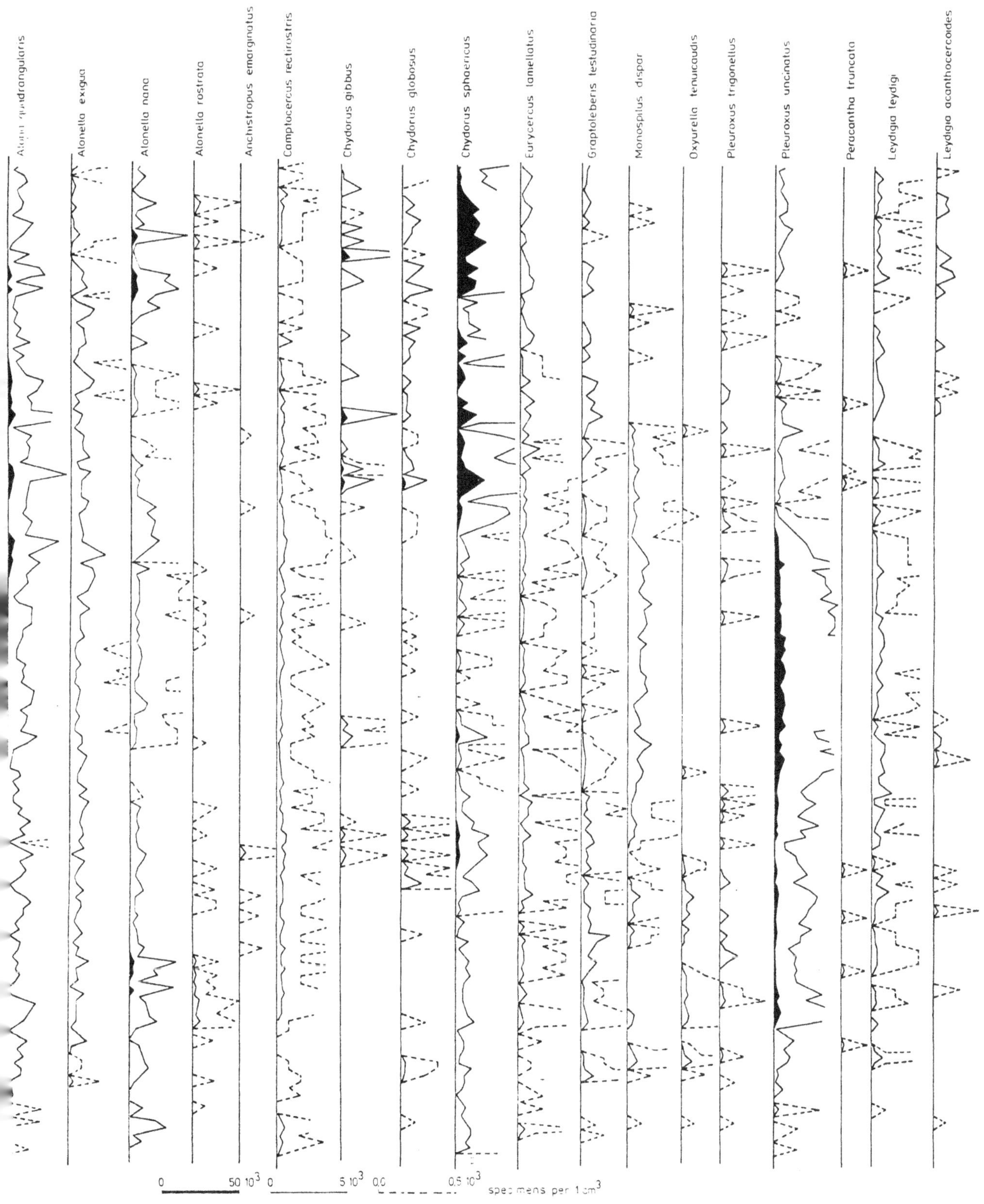

Alona quadrangularis
Alonella exigua
Alonella nana
Alonella rostrata
Anchistropus emarginatus
Camptocercus rectirostris
Chydorus gibbus
Chydorus globosus
Chydorus sphaericus
Eurycercus lamellatus
Graptoleberis testudinaria
Monospilus dispar
Oxyurella tenuicaudis
Pleuroxus trigonellus
Pleuroxus uncinatus
Peracantha truncata
Leydigia leydigi
Leydigia acanthocercoides
0 50 10^3
0 5 10^3
0.0 0.5 10^3
specimens per 1 cm^3

composition of Cladocera in the samples was estimated from an analysis of at least 2000 remains at each level in the cores. All remains of Cladocera that were been found – headshields, shells, postabdomens, postabdominal claws, and ephippia – were included in the calculations. Values given are the total number of specimens in 1 cm^3 of fresh deposits. The species composition and the frequency of species are illustrated in the diagrams (Fig. 2, 5, 7).

Area of study

Lake Błędowo

The lake is situated in Central Poland near Warsaw. It is the only one that presents the postglacial history of Cladocera development in Central Poland. This lake originated during Allerød time. Detailed results of complex investigations of this lake are in press (Binka *et al.*, 1991).

The Cladocera were analyzed in 170 samples at 10 cm intervals from the 15 m profile. Forty species of Cladocera belonging to five families – Bosminidae, Chydoridae, Daphniidae, Leptodoridae, and Sididae – were recorded, among which the Chydoridae comprised 32 species. The species composition of the Cladocera is dominated by the family Chydoridae, but attention should first be paid to species of the family Bosminidae, which are numerous and abundant in this lake. In Poland it was found for the first time that the sediment contains many specimens of *Bosmina longispina*, in contrast to lakes in other countries (Hofman, 1986; Boucherle & Züllig, 1983; Uimonen Simola & Tolonen, 1987).

The observed species composition and number of predominant plankton species of Cladocera (Fig. 2) enabled a climatic interpretation and the estimation of ecologic conditions of the lake during its development.

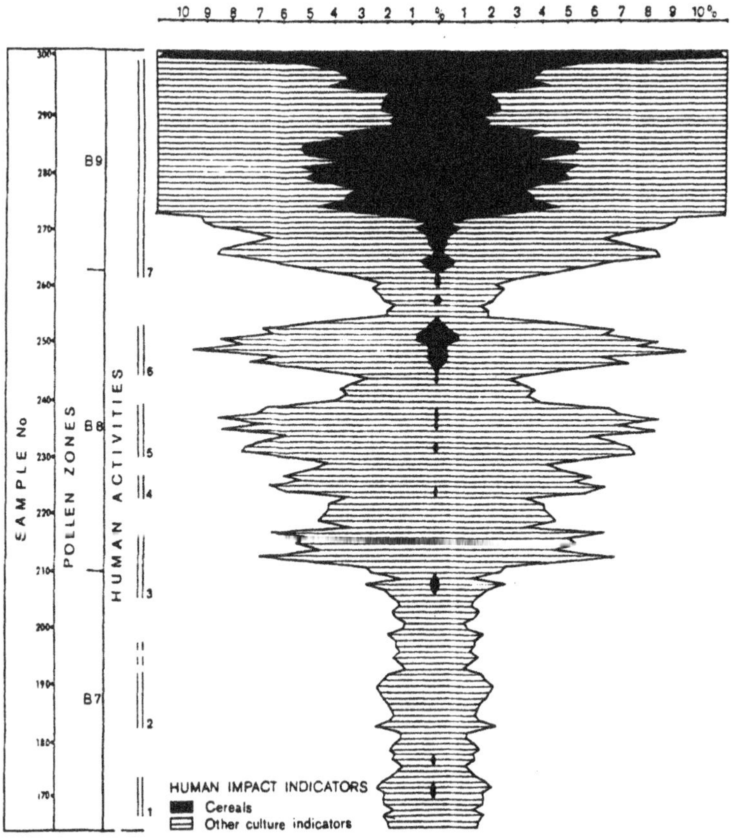

Fig. 3. Synthesized human impact pollen diagram from Błędowo Lake (after K. Binka *et al.*, 1989).

During the existence of the lake, 6 stages of eutrophication were noted as indicated in Fig. 2. The first two (numbers 1–2) resulted from more favorable climatic condition, which enabled development of the Cladocera. Stages 3 to 6 represent eutrophication probably connected with human activity. These last four periods of intensified trophy correspond to the periods of human settlement as indicated by pollen analysis (Binka *et al.*, 1989) (Fig. 3). The first two eutrophication periods, on the other hand, were presumably connected with the Neolithic Culture, which did not influence the eutrophication of the lake. Such a pattern is different from that observed in northern Poland, where the Neolithic settlement activity is reflected by strong eutrophication of the lakes (Pawlikowski *et al.*, 1982).

The influence of the Neolithic culture on Błędowo lake was less intensive, and due to its location, the lake was probably more isolated. The pollen of human indicators probably come from distant areas. All the stages shown in this diagram until the Mediaeval settlement (with exception of the 6th one), left distinct traces of their occurrence in the increasing trophic level of the lake.

A percentage distribution of pelagic-to-littoral species (Fig. 4) shows periods with a predominance of Bosminidae over the Chydoridae as well as the reverse. Certainly, the mass development of *Bosmina longirostris* often reflects intensive eutrophication rather than an extension of the pelagic zone. Periods with predominant Bosminidae /'*Eubosmina*'/ suggest that higher water levels in the Błędowo lake occurred during cladoceran zones 3, 5, and 9 in part, which were probably connected with climatic conditions.

Lake Skrzetuszewskie

This lake is being studied now, and hence results of many analyses are not yet published. For the Cladocera, only sediments connected with human

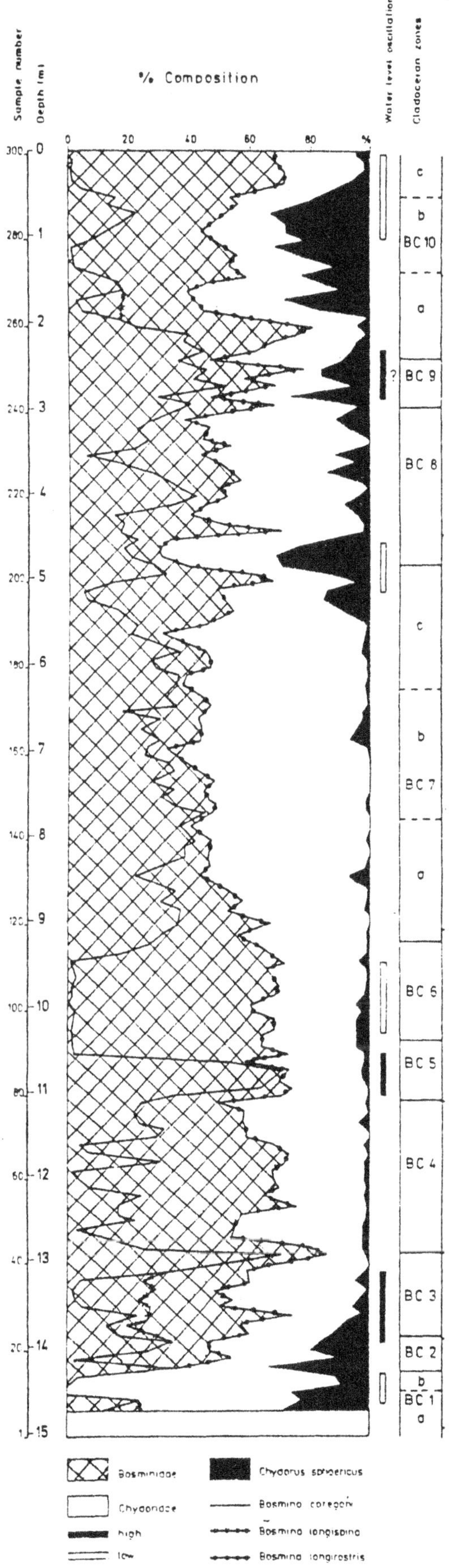

Fig. 4. Percentage distribution of species of pelagic and littoral Cladocera in the profile Błędowo III.

Fig. 5. Quantitative and qualitative distribution of the Cladocera in the profile Skrzetuszewskie/87.

activity (i.e. between 4 and 11 m) were studied. In the 42 samples, 37 species of Cladocera belonging to 5 families were recorded. The species composition (Fig. 5) is typical for eutrophic lakes, in which *Bosmina longirostris*, *Chydorus sphaericus*, *Alona rectangula*, *Alona guttata*, and *Pleuroxus* species predominated. Domination of *Bosmina longirostris* is very strong, its oscillation in abundance corresponding to the anthropogenic curves, meaning that its development is strongly connected with the intensification and subsequent reduction of human activities.

In the pollen diagram (Tobolski, 1987) the influence of human activities is clear from the depth of 9.7 m, i.e. from about 2800 years BP. This level represents also distinct boundary in the diagram of frequency of the Cladocera. Between 11 and 9.7 m, *Bosmina longirostris* comprises about 50% of population, while above 9.7 m it rises to 80–90%.

The palynologic analysis shows 5 stages of very active human influence (Tobolski, 1987). These stages correspond to periods of intensified eutrophication (numbers 2–6), based on Cladocera analysis, which correspond closely to the percentage curve of *Bosmina longirostris*. Eutrophication stage number 1 at a depth of about 11 m seems to be related to the Neolithic culture about 4000–3800 years BP. Palynologic analysis does not show significant changes in the composition of the flora.

Lake Suszek

Lake Suszek is located in the area of Bory Tuchowskie, eastward from Chojnice, near lake Wielkie Gacno, which represents a reference site for IGCP analyzed by Hjelmroos-Ericsson (1981).

The sediments of the lake are mainly calcareous gyttja and hence could not be dated satisfactorily by the C^{14} method. Their age however, was, estimated to be Early Sub-boreal from comparison

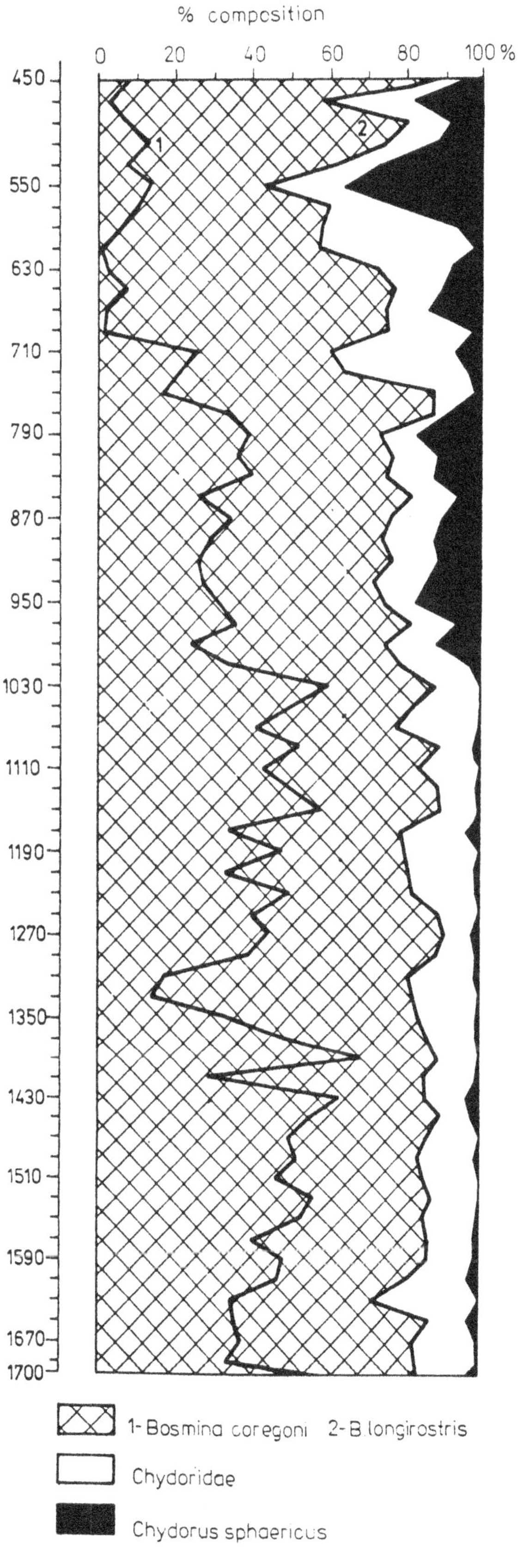

Fig. 6. Percentage distribution of species of Cladocera in the profile from Suszek lake.

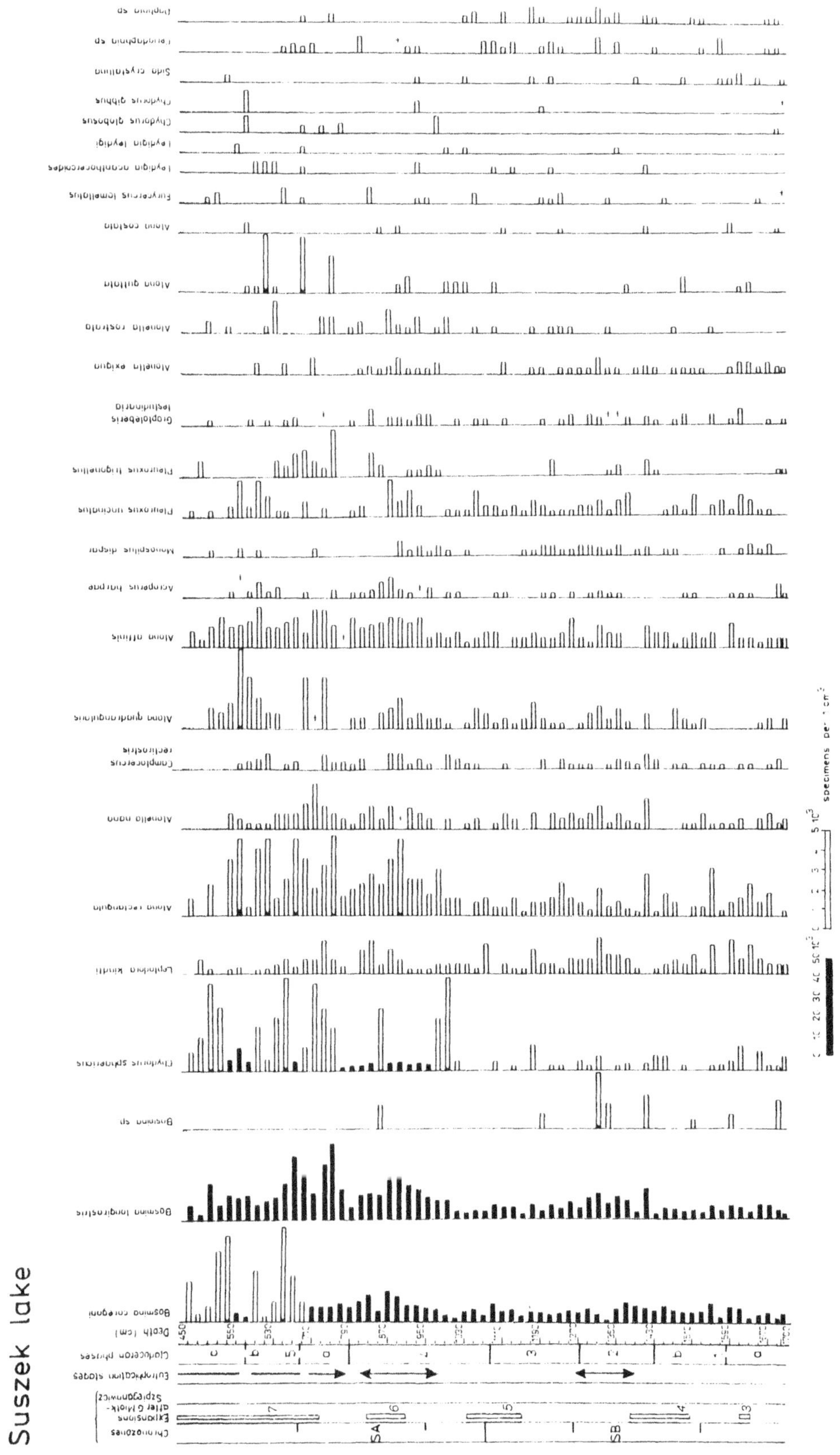

Fig. 7. Distribution of species of pelagic and littoral Cladocera in the profile from Suszek lake.

with the pollen curves of lakes Suszek and Wielkie Gacno. A cladoceran analysis was made for the sediments of the same profile (13 m) as for the pollen (Miotk-Szpiganowicz, 1989) and diatoms (Miotk *et al.*, 1986) at intervals of 20 cm. The occurrence of 31 Cladocera species was determined from a microscopic analysis of 70 samples.

A distinct dichotomy is present in the frequency diagram of individual Cladocera species (Fig. 6) at a depth of 10.30 m, which seems to reflect a climatic change between the Subboreal and Subatlantic chronozones.

Pollen analysis of the sediments revealed five

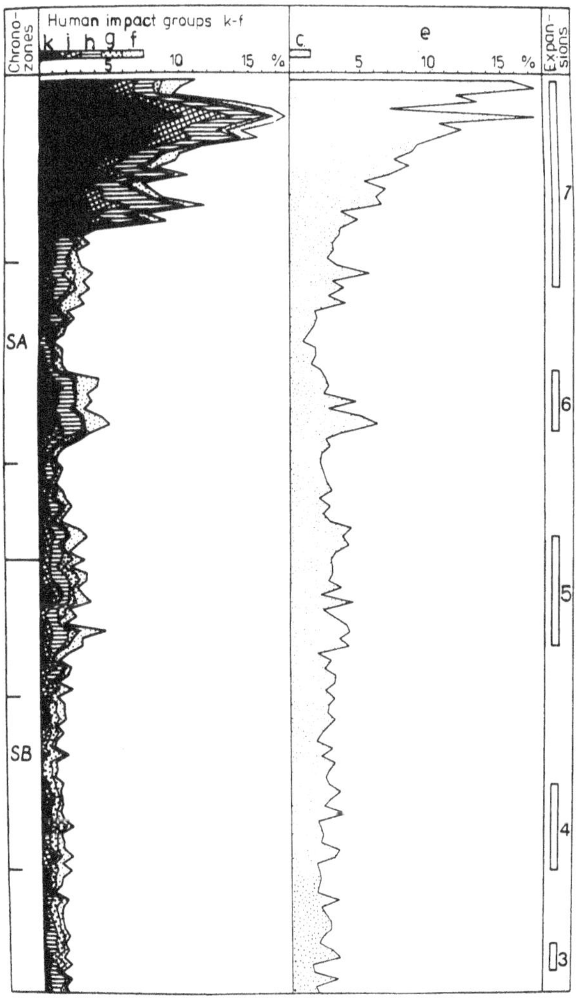

Fig. 8. Synthesized human impact pollen diagram from Suszek Lake (after Miotk-Szpiganowicz, 1989): *k* = cultivated land; *i* = fresh meadows; *h* = dry pastures; *g* = grazed woodlands; *f* = ruderal communities; *e* = general apophytes.

anthropogenic phases, that should be correlated with the eutrophication process. The Cladocera frequency pattern displays a distinct monotony, especially in the Subboreal chronozone. *Bosmina longirostris* is accompanied by *Bosmina coregoni* ('*Eubosmina*') (Fig. 7). The species that are connected with eutrophication seem not to be frequent, and they display low variation. From a detailed analysis of the frequency of eutrophic species, three stages of eutrophication were distinguished, but these stages do not fully correspond with the stages distinguished palynologically (Fig. 7, 8). Only the 6th and 7th stages of settlement are represented in the Cladocera frequencies. This correspondence seems to be connected with influence of Roman and Mediaeval culture. The indicator plants show clear dominance only during the above mentioned stage. The pollen of human indicators recorded during the earlier stages of settlement could have come from distant areas, and hence their influence on the lake's trophy was insignificant.

Conclusions

The results of this study clearly suggest a connection between the influence of human activity on the lake and the trophic status of the lake. The degree of development of the particular cultures is of great importance to the trophy of a lake. The character of pastoral-hunting, agricultural, or stock-raising influences the quality and quantity of eutrophic substances brought into the lake. The Błędowo lake may serve here as an example where the influence of pastoral hunting cultures were more prominent than the agricultural ones. The former have greatly increased the trophic status of this lake. Such effects are well illustrated by *Bosmina longirostris*, which is frequent in oligotrophic and eutrophic lakes, where it occurs with species of the genus *Bosmina*, but becomes more abundant during eutrophy. In lakes in which intensive eutrophication is going on it often occurs as the only species, and can make up to 90% of the whole plankton mass. It is possible that such a mass occurrence of *Bosmina longirostris* may be caused by the activity of fishes, which feed selectively on larger species of *Bosmina*. Such a supposition might be justified only if other small planktonic forms co-occurred. However, this is not the case, here, because during periods of very strong eutrophication there occur besides small *Bosmina longirostris* also abundant much larger individuals of such genera as *Pleuroxus* and *Leydigia*, which also can be regarded as indicators of eutrophication process.

In many cases a very detailed analysis of Cladocera from lacustrine sediments may supplement the results of other analyses, thus being very helpful in deciding about settlement phases in which the presence of the pollen of human indicators may derive from a very distant transport.

Acknowledgements

I am much indebted to Professor David G. Frey for critically reading the manuscript and for correcting the English text.

References

Alhonen, P., 1986. Holocene lacustrine microfossils and environmental changes. Bull. Geol. Soc. Finland, 58, Part 1: 57–69.

Binka, K., T. Madeyska, B. Marciniak, K. Szeroczyńska & K. Wieckowski, 1989. Błędowo Lake (Central Poland): history of vegetation and lake development during the last 12000 years. Bull. Pol. Ac. Sc. 36: 147–158.

Binka, K., A. Cieśla, B. Łacka, B. Marciniak, T. Madeyska, K. Szeroczyńska & K. Wieckowski, 1991. Lake evolution during the last 12000 years: Błędowo Lake (Central Poland). Studia Geologica Polonica 100: In press. Warszawa, 1991.

Boucherle, M. M. & H. Züllig, 1983. Cladoceran remains evidence of change in trophic state in three Swiss lakes. Hydrobiologia 103: 141–146.

DeMott, W. R. & W. C. Kerfoot, 1982. Competition among cladocerans: Nature of the interaction between *Bosmina* and *Daphnia*. Ecology 63: 1949–1966.

Frey, D. G., 1960. The ecological significance of Cladoceran remains in lake sediments. Ecology 41: 684–699.

Frey, D. G., 1966. Cladocera in space and time. Mar. Biol. Assoc. India, Proc. Intl. Symposium on Crustacea, Part I: 1–9.

Frey, D. G., 1986. Cladocera analysis. In: B. E. Berglund (ed). Handbook of Holocene Palaeoecology and Palaeohydrology. Wiley, Chichester: 667–692.

Günther, J., 1983. Development of Grossensee (Holstein, Germany): Variations in trophic status from the analysis of subfossil microfauna. Hydrobiologia 103: 231–234.

Hann, B. J. & P. F. Karrow, 1984. Pleistocene paleoecology of the Don and Scarborough Formations, Toronto, Canada, based on cladoceran microfossils at the Don Valley Brickyard. Boreas 13: 377–391.

Havens, K. & J. DeCosta, 1985. The effect of acidification in enclosures on the biomass and population size structure of *Bosmina longirostris*. Hydrobiologia 122: 153–158.

Hjelmroos-Ericsson, M., 1981. Holocene development of Lake Wielkie Gacno area, northwestern Poland. Thesis 10, Lunds Univ. LUNBDS/(NBGK-1010) /1–101/ (1981).

Hofmann, W., 1983. Stratigraphy of Cladocera and Chironomidae in a core from a shallow North German lake. Hydrobiologia 103: 235–239.

Hofmann, W., 1986. On the relationship between the *Bosmina* taxa *coregoni* and *thersites* (Cladocera), as indicated by subfossil remains. Hydrobiologia 143: 119–121.

Hofmann, W., 1987. Cladocera in space and time: Analysis of lake sediment. Hydrobiologia 145: 315–321.

Miotk-Szpiganowicz, G., 1989. The history of vegetation of Bory Tucholskie and the role of Man in the light of palynological investigations. Gdańsk, 1988. manuscript.

Miotk, G. & B. Bogaczewicz-Adamczak, 1986. Badania paleoekologiczne Borów Tucholskich. In press. Toruń 1986.

Pawlikowski, M., M. Ralska-Jasiewiczowa, W. Schönborrn, E. Stupnicka & K. Szeroczyńska, 1982. Woryty near Gietrzwałd, Olsztyn lake district, NE Poland vegetational history and lake development during the last 12000 years. Acta Palaeobot., 22: 85–116.

Smirnov, N. N., 1981. Historic biocenologic approach to animal remains in grounds of continental water bodies. Paleolimnology of lake Biwa and the Japanese Pleistocene. Vol. 9: 187–1900.

Tobolski, K., 1987. Badania palinologiczne osadów jeziora Skrzetuszewskiego (Reference site IGCP, 158B, p-15) w Polsce północono-zachodniej. Poznań 1987, manuscript.

Uimonen-Simola, P. & K. Tolonen, 1987. Effects of recent acidification on Cladocera in small clear-water lakes studies by means of sedimentary remains. Hydrobiologia 145: 343–351.

Hydrobiologia **225**: 115–120, 1991.
V. Kořínek & D. G. Frey (eds), Biology of Cladocera.

The influence of biomass and structure of the crustacean plankton on the water transparency in the Saidenbach storage reservoir

Wolfgang Horn
Sächsische Akademie der Wissenschaften zu Leipzig, Arbeitsgruppe 'Limnologie von Talsperren', D(O)-9343 Lengefeld, Germany

Key words: Crustacean plankton, *Daphnia*, biomass, size structure, Secchi depth

Abstract

Planktonic crustacean biomass as well as structure are important factors influencing water transparency. The significant dependence of the water quality (Secchi depth) on the concentration and the share of the *Daphnia* biovolume and not on the total Crustacea biovolume in the Saidenbach reservoir indicates that the density of the Crustacea is only a measure of the cleaning performance, if *Daphnia* dominates. Using the mean size, the influence of the crustacean structure on the Secchi depth can be recorded. If big size categories prevail (like *Daphnia*) the water transparency is high. The mainly occurrence of little species (*Mesocyclops*, *Bosmina*) results in lower Secchi depths. However, a well defined (significant) relationship is being prevented by the different feeding behaviour of the several species.

Introduction

The influence of the herbivorous filtering crustacean plankton on water quality and transparency is well known from field observations (spring or summer clearwater stage in ponds and lakes, water bodies with different fish stocks; Uhlmann, 1954, 1959; Lampert & Schober, 1978; Sommer *et al.*, 1986; and others), food web experiments (biomanipulation; Hrbáček *et al.*, 1961; Shapiro *et al.*, 1975; Henrikson *et al.*, 1980; Benndorf *et al.*, 1984) and laboratory tests (microcosm investigations, e.g. Uhlmann, 1954, 1959; filtering and selectivity experiments, e.g. Richman, 1966; Vanderploeg & Scavia, 1979). But with exception of the frequently stated clearwater phase, only few observations exist about the interactions and their effects between phyto-/zooplankton and the higher trophic levels in water bodies that are not at all or not much influenced by anthropogenic factors but instead only by natural variations in the plankton and higher trophic levels (Edmondson & Litt, 1982; Mothes & Krocker, 1986). Long term investigations are necessary to obtain more informations about the effects of such interactions over whole seasons or years. This study is an attempt to examine the influence of zooplankton concentration and structure on the water transparency using a ten-year data set.

Description of sites studied

The investigations were carried out in the meso-eutrophic Saidenbach drinking water reservoir (at maximum water level: volume 22.4 million m^3, surface area 146 ha, maximal depth 45 m, mean depth 15.3 m). During most of the year phosphorus is the limiting nutrient. The external load of soluble reactive phosphorus ranged from 0.4 to 1.0 g P m^{-2} yr^{-1} (1975–1988) and varied mainly with the changing yearly flushing rates. In the summer time the phytoplankton consists of 77% diatoms, 8.5% bluegreens, 3% colonial green algae, 3% nanoplankton, 3% *Cryptomonas* div. spc. and others (total biovolume average = 1.112 $mm^3 l^{-1}$, mean average 1975–1986).

Methods

For the recording of phytoplankton, separate samples were collected with a Ruttner sampler every 1–2 m in the epilimnion and every 2.5–5 m in the hypolimnion. Algae were counted under the Utermöhl microscope still in the same week. For sampling the zooplankton a cloth net was used (mesh size 175 μm, opening area 7 dm^2, filtering area 48 dm^2, towering speed about 10 $cm\ s^{-1}$, investigated layers: 40–25 m, 25–15 m, 15–10 m, 10–5 m, and 5–0 m). The samples were fixed with sugar-coated formalin (Haney & Hall, 1973) and counted in the same week under the Utermöhl microscope after division into two countable subsamples (for more details see Horn *et al.*, 1987). For calculating the biovolume, the length of about 10 to 200 animals (in dependence on abundance) of every category (species, males, females, juvenile stages) were measured under a microscope. The following equations were used:

Daphnia, Diaphanosoma: $y = 0.074\ L^{2.92}$

Copepoda: $y = 0.055\ L^{2.73}$

Bosmina: $y = 0.124\ L^{2.181}$

(L = length in mm, y = biovolume in mm^3)

Usually, phyto-, zooplankton, and Secchi depth were taken weekly (in the last 3 years every one or two weeks; 1982 was not considered, because the zooplankton was not regularly investigated) in the main part of the reservoir about 100 m in front of the dam. For all the parameters the summer average was calculated as a depth-and-time-integrated mean over the whole stagnation period. Significance of correlation was tested at a confidence level of $p < 0.05$ (two way analysis).

Results and discussion

Biomass, species composition, and size structure of the crustacean plankton community

The species composition, biomass, and size structure of the crustacean plankton community are depicted in the Figs. 1 and 2. Obviously there was no change in the species composition until 1986, but in the last two years a dramatic reduction of *Daphnia* occurred and many more small and/or nonfiltering organisms, like *Bosmina*, *Mesocyclops*, *Diaphanosoma*, and *Cyclops*, were present. Simultaneously a delayed population development of *Daphnia* can be registered. This tendency starts in 1984 with a decrease of the average crustacean body size (mean of all species) and a very sharp increase in the average concentration of animals. In contrast, the total biovolume concentration increased only slightly. Similar conditions (low average body size) were found in 1978, which was according to Schultz (1983) the only year in which the zooplankton was grazed effectively by yearlings (much perch and roach fry in 1977). Unfortunately, since 1985 no exact infor-

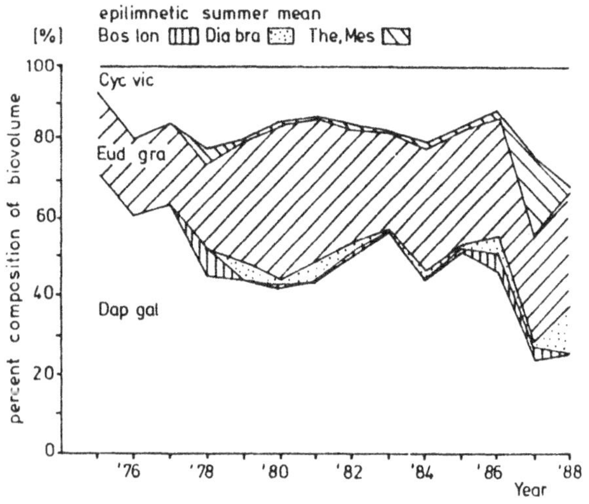

Fig. 1. The percentage of the different species or groups (without nauplius stages) on the total mean summer biovolume in the different years (in the first years *Thermocyclops/Mesocyclops* without copepodite stages). Abbreviations: Bos lon = *Bosmina longirostris*; Cyc vic = *Cyclops vicinus*; Dap gal = *Daphnia galeata*; Dia bra = *Diaphanosoma brachyurum*; Eud gra = *Eudiaptomus gracilis*; The, Mes = *Thermocyclops* sp./*Mesocyclops* sp.

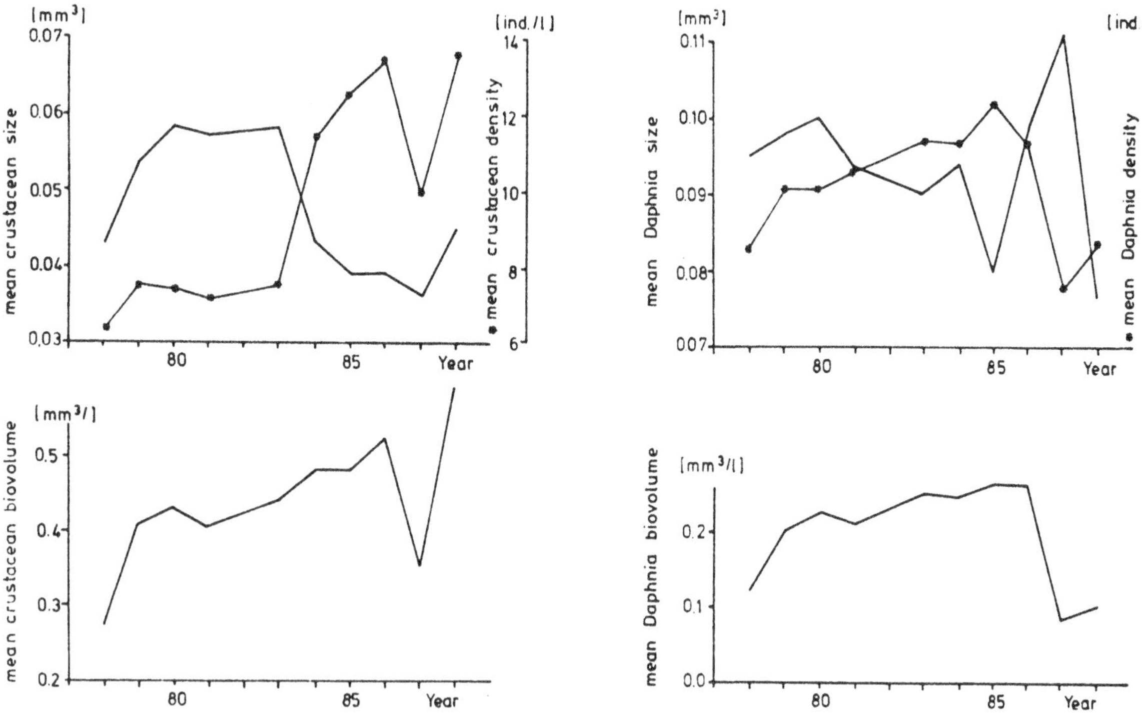

Fig. 2. Mean individual size, density, and biovolume concentration of the planktonic Crustacea without nauplius stages (left) and *Daphnia galeata* (right) in the whole waterbody in summer.

mation about the fish stock is available. Visual observations point to a mass development of zooplanktivorous fishes (fingerlings and a great number of sticklebacks, even in the pelagial).

Effects of planktonic crustacean biomass and structure changes on water transparency

It is known from different investigations that water transparency can be influenced by the crustacean plankton biomass. However, the biomass of Crustacea is only an incomplete parameter, because the feeding mechanisms, the selectivity, and the individual performance of the species or stages are very different. Very important are the herbivorous filter-feeding daphnids. They are most effective in retaining small particles like bacteria, ultra- and nanoplankton, which can't be grazed by *Eudiaptomus*, *Cyclops* (Vanderploeg & Scavia, 1979; Horn, 1985) and probably, *Mesocyclops*. Considering the special feeding behaviour of the different groups and the proportion of their biomass makes of the total, it is not surprising, that in the Saidenbach reservoir the Secchi depth is not controlled by the biovolume of the total crustacean plankton (Fig. 3). In contrast, a highly significant correlation between water transparency and *Daphnia* biomass (Fig. 4) respectively its relative proportion on the total crustacean plankton biovolume (Fig. 5) can be observed. The correspondence between Secchi depth and mean crustacean size in Fig. 6 confirms the above mentioned statement: Low water transparency can be found if smaller size categories dominate (*Bosmina*, *Diaphanosoma*, *Mesocyclops*, *Eudiaptomus*, and *Cyclops* copepodites). Greater Secchi transparency occurs during the dominance of larger species, especially *Daphnia galeata*. This species in the Saidenbach reservoir is one of the biggest animals and hence has a great influence on the mean crustacean size. However, this relationship is not statistically significant, which demonstrates that besides the size of the species, their feeding behaviour is important also. For example, in the Saidenbach reservoir, the big adults of *Cyclops* and *Eudiaptomus* are unable to retain small particles, which disturbs this relationship.

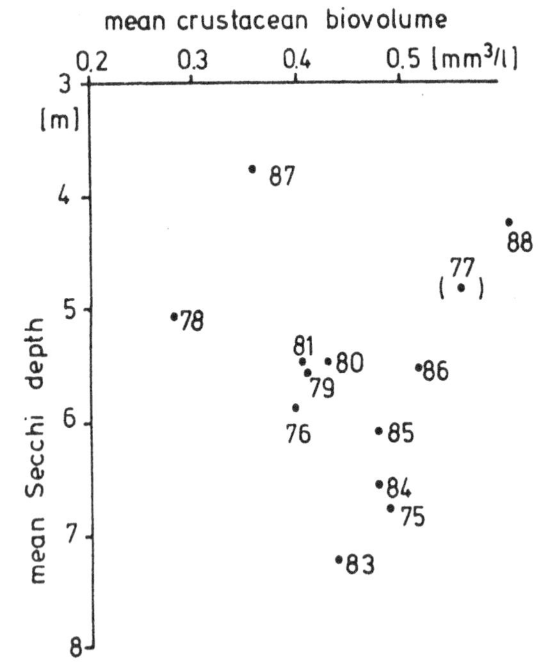

Fig. 3. The relationship between mean Secchi depth and the biovolume of the planktonic Crustacea (without nauplius stages) in the whole water body in summer (1977 = a year with high mineral turbidity was eliminated).

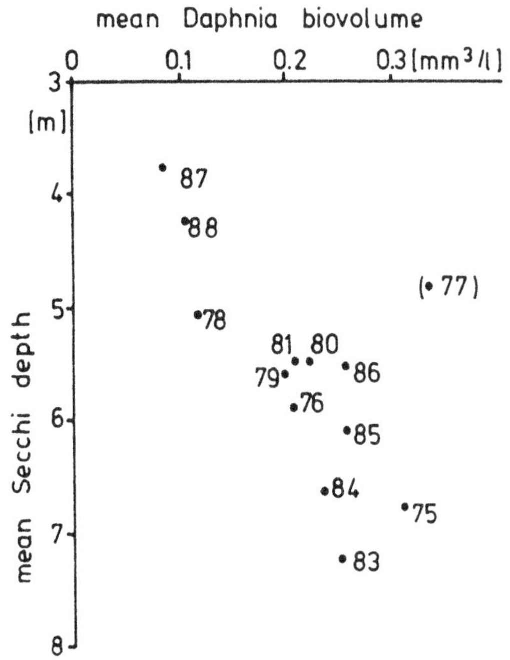

Fig. 4. The dependence of the mean Secchi depth on the mean *Daphnia galeata* biovolume in the whole water body in summer (1977 = a year with extremely high mineral turbidity was eliminated; $y = 3.19 + 11.87\times$, $r = 0.89$, $n = 12$).

The results obtained demonstrate clearly that biomass as well as structure influence water transparency. A high biomass of Crustacea alone is not sufficient if their species composition is unfavourable. The biomass concentration can be used as a measure only if the size structure is dominated

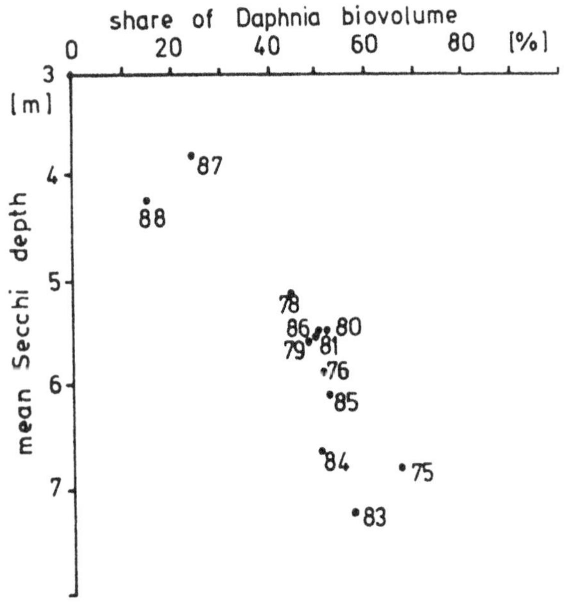

Fig. 5. The dependence of the mean Secchi depth on the relative proportion of *Daphnia* on the total crustacean-plankton biovolume in the whole water body in summer ($y = 2.69 + 0.062\times$, $r = 0.76$, $n = 12$).

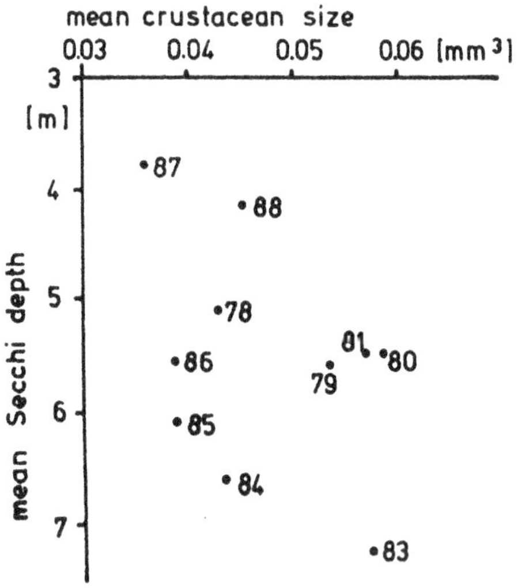

Fig. 6. The relationship between mean Secchi depth and mean size of planktonic Crustacea (without nauplius stages) in the whole water body in summer ($y = 3.21 + 49.34\times$, $r = 0.42$, $n = 10$).

by highly efficient filter feeders (e.g. *Daphnia* species; see Fig. 5).

The existence of the relationship between *Daphnia* biomass and water transparency can be also deduced from other field observations (such as the spring clearwater phase) and laboratory investigations (filtering and feeding experiments). The significant inverse relationships between the concentrations of nanoplankton (algae $<20\ \mu m$) and of *Asterionella* with respect to the biomass of *Daphnia* or the Crustacea (Horn & Horn, 1990) are further evidences. This means, that an increase in concentration of zooplankton is accompanied by a decrease of these algae. Among the Crustacea, *Daphnia* is the most important in that it exhibits the greatest correlation coefficients. Its grazing activity increases the Secchi depth. The shift to larger and colonial algae, promoted by the zooplankton (Porter, 1976), is an additional effect that improves water transparency.

The duration (or frequency) curves of the Secchi depth in the different years describing the probability (as percent of time) that the Secchi depth is smaller than the value of the ordinate, shows the development of the water quality during the whole investigation period in the summary. The Figure 7 demonstrates a clear decrease of the maximum Secchi depth over the investigated years (from 10–12 m to 6–8 m). The highest water transparencies are observed usually during the clearwater phase. Therefore the deterioration of the Secchi depth indicates the absence of this period in the last two years because of missing daphnids. Not only the highest values but also the minimum and the average values decrease. Another effect that must be explained is the higher visibility in the middle part (1983). In this year relatively many colonial bluegreens (*Aphanothece*, *Microcystis*) but scarcely any other forms occur. The colonies of big species do not retain so much light as the same biomass of smaller ones. On the other hand, there was in the last two years a sharp increase in the proportion of ultra- and nanoplankters as a result of the small *Daphnia* biomass and delayed population development. These smaller forms scatter and absorb more light,

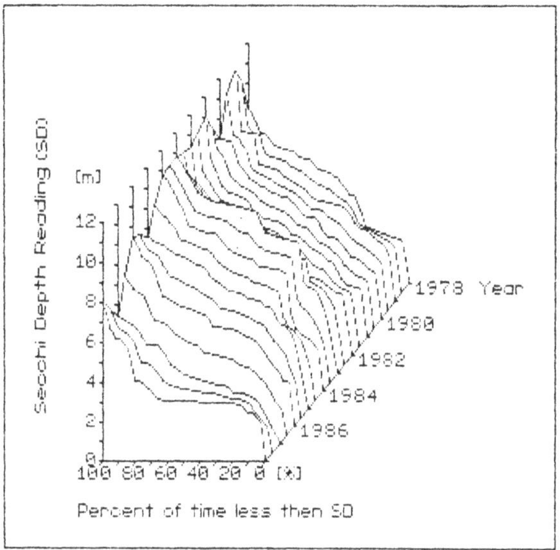

Fig. 7. Secchi depth (SD) duration curves from 1978–1988 (these curves describe the probability (as percent of time) that the SD is smaller than the value of the ordinate).

causing the transparency to be especially low despite the high biomass of bluegreens.

Conclusions

The influence of the herbivorous filtering crustacean plankton on the water transparency is well known from field observations (clearwater phase in ponds and lakes), biomanipulation experiments, and others. Among the crustacean plankton, the bigger daphnids are the most important. They have a high filtering efficiency covering a wide spectrum of particle size and a great individual filtration capacity. It enables them in dense animal populations to clean the water. This can be confirmed also in the Saidenbach storage reservoir by analysing and interpreting a long term record (through more than 10 years). For this investigation summer averages for the most important influencing factors were calculated. The statistical comparison of the essential functional parameters (water transparency, body size, and standing stock of zooplankton) of the different years show the greatest influence of the crustacean plankton on the water quality if *Daphnia* is the dominating form. The Saidenbach reservoir data show a significant relationship between the Secchi depth and the biovolume of *Daphnia galeata* and their mean proportion, but no correlation with the biovolume of the total crustacean plankton. Likewise the Secchi depth seems dependent on the mean size of the Crustacea, indicating the influence of *Daphnia* on water quality. Because of the occurrence of other species in the Saidenbach reservoir, at times with considerable concentrations of biomass, a significant relationship is obscured by the different feeding behaviour of the several species.

References

Benndorf, J., H. Kneschke, K. Kossatz & E. Penz, 1984. Manipulation of the pelagic food web by stocking with predacious fish. Int. Revue ges. Hydrobiol. 69: 407–428.

Edmondson, W. T. & A. H. Litt, 1982. *Daphnia* in Lake Washington. Limnol. Oceanogr. 27: 272–293.

Haney, J. F. & D. J. Hall, 1973. Sugar-coated *Daphnia*: A preservation technique for Cladocera. Limnol. Oceanogr. 18: 331–333.

Henrikson, L., H. G. Nyman, H. G. Oscarson & J. A. Stenson, 1980. Trophic changes, without changes in the external nutrient loading. Hydrobiologia 68: 257–263.

Horn, W., 1985. Investigations into the food selectivity of the planktic crustaceans *Daphnia hyalina*, *Eudiaptomus gracilis* and *Cyclops vicinus*. Int. Revue ges. Hydrobiol. 70: 603–612.

Horn, W. & H. Horn, 1990. Long-term relationships between phyto- and zooplankton in the meso-eutrophic reservoir Saidenbach. Arch. Hydrobiol. Beih. Ergebn. Limnol. 33: 749–762.

Horn, H., W. Horn & M. Kohlsdorf, 1987. Theoretical and practical investigations on the heterogeneous distribution of plankton in the main basin of the Saidenbach storage reservoir. Acta hydrochim. hydrobiol. 15: 327–350.

Hrbáček, J., M. Dvořakova, V. Kořínek & L. Procházkóva, 1961. Demonstration of the effect of the fish stock on the species composition of zooplankton and the intensity of metabolism of the whole plankton association. Verh. int. Ver. Limnol. 14: 192–195.

Lampert, W. & U. Schober, 1978. Das regelmäßige Auftreten von Frühjahrsalgenwachstum und 'Klarwasserstadium' im Bodensee als Folge von klimatischen Bedingungen und Wechselwirkungen zwischen Phyto- und Zooplankton. Arch. Hydrobiol. 82: 364–386.

Mothes, G. & M. Krocker, 1986. Beziehungen zwischen der Wasserqualität des Großen Müggelsees und der Zooplanktonentwicklung. Acta hydrochim. hydrobiol. 14: 263–282.

Porter, K. G., 1976. Enhancement of algal growth and productivity by grazing zooplankton. Science 192: 1332–1334.

Richman, S., 1966. The effect of phytoplankton concentration on the feeding rate of *Diaptomus oregonensis*. Verh. int. Ver. Limnol. 16: 392–398.

Schultz, H., 1983. Untersuchungen über die Größe und Struktur des Fischbestandes sowie den Nahrungskonsum ausgewählter Fischarten in zwei Stauseen unterschiedlichen Trophiegrades. Dissertation A, Dresden Univ. Technol., Fac. of Construction, Water Management and Forestry, 141 pp.

Sommer, U., Z. M. Gliwicz, W. Lampert & A. Duncan, 1986. The PEG-model of seasonal succession of planktonic events in fresh waters. Arch. Hydrobiol. 106: 433–471.

Shapiro, J., V. Lamarra & M. Lynch, 1975. Biomanipulation: An ecosystem approach to lake restoration. In P. L. Brezonik & J. L. Fox (eds), Water Quality Management through Biological Control, University of Florida, Gainesville. Proceedings: 85–96.

Uhlmann, D., 1954. Zur Kenntnis der natürlichen Nahrung von *Daphnia magna* und *Daphnia pulex*. Z. Fisch. N.F. 3: 449–478.

Uhlmann, D., 1959. Untersuchungen über die biologische Selbstreinigung häuslichen Abwassers in Teichen. Wiss. Zschr. Univ. Leipzig 8: 17–66.

Vanderploeg, H. A. & D. Scavia, 1979. Calculation and use of selectivity coefficients of feeding: Zooplankton grazing. Ecol. Modelling 7: 135–149.

Hydrobiologia **225**: 121–128, 1991.
V. Kořínek & D. G. Frey (eds), Biology of Cladocera.

Effect of food conditions on the bacterial feeding of *Daphnia galeata*

Jotaro Urabe[1] & Yasunori Watanabe[2]
[1]*Natural History Museum and Institute, CHIBA, Aoba-cho 955-2, Chiba, 280 Japan*; [2]*Department of Biology, Faculty of Science, Tokyo Metropolitan University, Minami-ohsawa 1-1, Hachiohji, 192-03 Japan*

Key words: Daphnia galeata, bacteria, *Chlamydomonas*, clearance rate, ingestion efficiency, feeding mechanism, zooplankton, adaptation

Abstract

Feeding selectivities of *Daphnia galeata* were measured using various density combinations of ^{14}C-labeled *Chlamydomonas* and ^{3}H-labeled natural bacteria. In all food combinations, *D. galeata* preferred algae. Selectivity coefficient (bacterial clearance rate : algal clearance rate) was 0.35 at the level of incipient food limitation (1.5×10^6 μm^3 ml^{-1}). However, it increased up to 0.9 with increasing total food concentration. A simple mechanical sieving model alone cannot explain such a change in the selectivity coefficient. A mechanism of 'clogging' suggestively explain the increase of bacterial ingestion efficiency by *D. galeata*, especially at high food concentrations. The change of selectivity coefficient related to food condition suggests that bacterial feeding by the *Daphnia* is incidental event associated with the algal feeding.

Introduction

The biomass of solitary bacterial cells smaller than 1 μm sometimes contributes a considerable proportion of the suspended organic matter in lakes and ponds (Sorokin & Paveljeva, 1978; Nagata, 1986) and daphniids may utilize these bacteria as a major food item (Børsheim & Olsen, 1984; Kankaala, 1988). This implies that nutritional flow from bacterial food source to daphniids may be an important trophic channel in aquatic food webs (Gophen *et al.*, 1974; Lampert, 1974; Gophen, 1977). However, to evaluate the importance of this linkage, it is necessary to clarify the mechanism of bacterial feeding by daphniids.

There are two models of the feeding mechanism of *Daphnia*. The difference between the two models are as follows: I) water entering the carapace cavity passes through thoracic limbs III and IV, which act as filters (Gophen & Geller, 1984; Brendelberger, 1985; DeMott, 1985; Hessen, 1985; Brendelberger *et al.*, 1986; Fryer, 1987), II) the water does not go through these limbs, which act as solid paddles (Gerritsen & Porter, 1982; Porter *et al.*, 1983; Ganf & Shiel, 1985; Gerritsen *et al.*, 1988). The 1st model implies that daphniids collect particles mostly by a sieving mechanism, and thus, bacteria larger than the intersetular distance are retained efficiently (Gophen & Geller, 1984). On the other hand, the 2nd model implies that bacteria are collected by direct interception or handling by these limbs and internal receptors, and that the efficiency of bacterial uptake is not necessarily determined by the intersetular distance *per se* (Ganf & Shiel, 1985; Gerritsen *et al.*, 1988). Porter *et al.* (1983) have also suggested that bacterial uptake is enhanced by the 'piggybacking' phenomenon, which may occur by association between bacterial and algal cells or by clogging of the feeding limbs with larger particles. Moreover, Brendelberger (1985) suggested that the process of transferring food from the thoracic limbs to the mouth may affect the efficiency of bacterial uptake.

The purpose of this study is to test whether daphniids collect bacteria only by mechanical sieving or by other mechanisms. We estimated simultaneously the bacterial and algal clearance rates by *Daphnia galeata* using the double labelling radioisotope technique (Gophen *et al.*, 1974; Lampert, 1974; Gophen, 1977; DeMott, 1982) to examine the feeding selectivity for the bacteria under various concentrations of the food components. The results would also provide information on the adaptive significance of bacterial utilization by daphniids.

Materials and methods

The experimental procedure is shown in Fig. 1. Planktonic bacteria used for the feeding experiments were isolated from a small experimental pond in the campus of Tokyo Metropolitan University by filtering the water through 1.0 μm Nuclepore filters. The bacteria were concentrated to make feeding suspensions of various bacterial densities as follows. The 1.0 μm filtrate was filtered through 0.2 μm Nuclepore filters under low vacuum (less than 10 cm Hg) to prevent the extrusion of particles larger than the filter pore size and the destruction of bacterial cells. Then, bacterial cells were rinsed out the filters with filtered (0.2 μm) pond water. The concentration of bacterial cells were determined by the AODC method (Hobbie *et al.*, 1977).

A green alga, *Chlamydomonas reinhardi* (IAM strain C-238), was cultured axenically in modified Bristol medium (Watanabe, 1960). Cells of 7-days-old cultures were centrifuged, washed twice with filtered (0.2 μm) pond water and used in the feeding experiments. The concentrations of algal cells were determined with a hemacytometer.

Daphniids (*D. galeata*, 1.2–1.4 mm carapace length) were isolated from the experimental pond 4 h before the experiments and kept in 1 l beakers containing filtered (1.0 μm) pond water with *C. reinhardi* (10^4 cells ml^{-1}) at room temperature (20 °C). These cladocerans were originally collected from the mesotrophic Ogochi Reservoir, located in the northwestern part of Tokyo, and cultivated in the experimental pond during a year.

Bacterial and/or algal clearance rates were estimated for 12 different food concentrations: 4 concentrations of *C. reinhardi* alone, one concentration of bacteria alone and 7 food concentrations with both alga and the bacteria (Table 1). Bacterial suspensions were made in duplicates. To one of each duplicate suspension, 70–160 μCi of [methyl-^{3}H] thymidine was added, so that the specific activity of labeled bacteria in the experimental feeding suspensions was above 5000 dpm ml^{-1}. Algal suspensions of various concentrations in modified Bristol medium were prepared in duplicates. Algal cells in one of each duplicate were labeled by 20–40 μCi NaH^{14}CO$_3$, so that the specific activity of labeled algae in the experimental feeding suspensions was about 30% lower than that of ^{3}H-labeled bacteria. Bacterial and algal suspensions without radioisotope labels were used for pre-experimental acclimatization. All suspensions were incubated under fluorescent lights at 20 °C.

After 10 h of incubation, non-labeled thymidine was added to the labeled bacterial suspensions at the final concentration of 1 mM in order to reduce labeled thymidine uptake by bacteria. The suspensions were filtered through 1.0 μm filters to exclude any particles larger than 1.0 μm and the filtrates were used as experimental feeding suspensions. Algal cells were centrifuged, washed twice with filtered (0.2 μm) pond water, and resuspended in the experimental feeding suspensions at appropriate densities. The bacterial and algal concentrations in feeding suspensions were adjusted with filtered (0.2 μm) pond water. The bacterial concentrations ranged between 0.25 and 2 times higher than that of the experimental pond. The sizes of bacterial cells did not differ significantly among feeding suspensions: more than 70% of bacterial cells were larger than 0.4 μm in diameter (Fig. 2), and the mean bac-

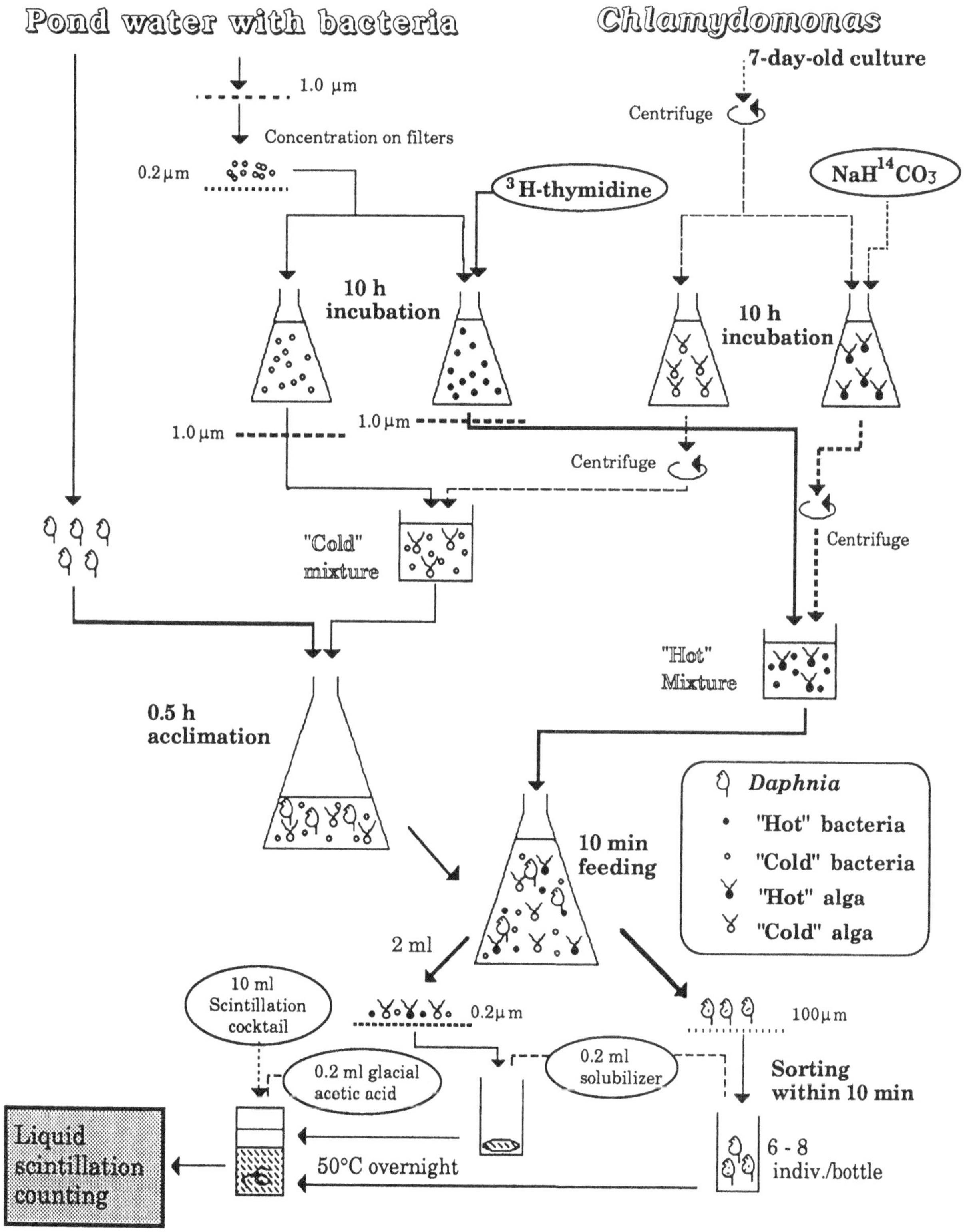

Fig. 1. Flow chart, showing methods used.

Table 1. Algal and bacterial densities in the feeding suspension, and rates of algal and bacterial clearance by *Daphnia galeata* in each experiment.

Exp. No.	Food concentration ($\times 10^6$ μm^3/ml)			Number of bottles	Algal clearance rate (ml/indiv./h) M ± 95% CI	Bacterial clearance rate (ml/indiv./h) M ± 95% CI
	Alga	Bacteria	Total			
1	0	0.394	0.394	8		0.375 ± 0.026
2	0.6	0.394	0.994	5	1.807 ± 0.227	0.526 ± 0.071
3	1.2	0.108	1.308	6	1.767 ± 0.126	0.635 ± 0.035
4	1.2	0.394	1.594	6	1.802 ± 0.155	0.652 ± 0.148
5	1.2	0.679	1.879	5	1.561 ± 0.061	0.760 ± 0.109
6	1.8	0.394	2.194	4	1.282 ± 0.148	0.474 ± 0.044
7	4.5	0.394	4.894	5	0.537 ± 0.129	0.307 ± 0.049
8	15.0	0.394	15.394	4	0.162 ± 0.032	0.140 ± 0.018
9	0.6	0	0.6	5	1.923 ± 0.284	
10	1.2	0	1.2	5	1.787 ± 0.161	
11	1.8	0	1.8	5	1.413 ± 0.199	
12	15.0	0	15.0	6	0.170 ± 0.043	

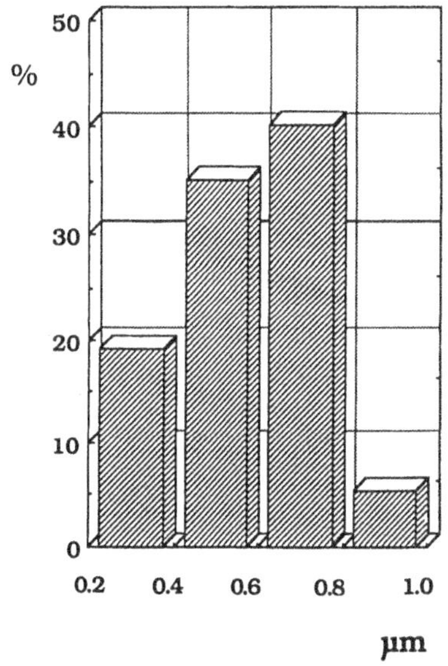

Fig. 2. Size-frequency distribution of the minimum diameter of bacteria used for feeding experiments.

terial biovolume was 0.083 μm^3 $cell^{-1}$, as determined by fluorescence micrographs of cells stained by acridine orange. The mean biovolume of *C. reinhardi* was 150 μm^3 $cell^{-1}$.

About 30 min before the feeding experiments, 30–40 individual daphniids were pipetted into a 500 ml flask containing 230 ml of non-radioactive suspension. After acclimatization, 230 ml of labeled food suspension with the same composition as the non-radioactive suspension was added gently to the flask to avoid any damage to animals (DeMott, 1982). Animals were allowed to feed for 10 min in all the feeding suspensions. During the incubation, two 2 ml aliquots of the feeding suspension were sampled and filtered through 0.2 μm Sartorius filters to determine the radioactivity of the food. The entire content of the flask was concentrated into a 30 ml bottle with a 3 × 3 cm net (96 μm mesh) covered window on the side wall. Immediately after that, CO_2 enriched water was poured into the bottle to stop the feeding. The anesthetized animals were collected and rinsed with 0.001 N HCl and thereafter with distilled water.

To prevent radioisotope leakage (Downing & Peters, 1980; DeMott, 1985; Nagata & Okamoto, 1988), the animals were transferred to scintillation vials within 10 min after each feeding trial. In each vial 6–8 animals were placed, and 0.2 ml Protsol (New England Nuclear) was added. Animals were dissolved at 50 °C during 24 h, then, the vials were filled with 10 ml of toluene-base scintillation cocktail, and 0.2 ml of glacial acetic acid was added to neutralize the solubilizer-Protsol (DeMott, 1982). The 0.2 μm filter with 2 ml of feeding suspension was treated in the same way. Radioactivity was determined by a liquid scintillation counter (Beckman LS3801). A preliminary experiment showed that direct uptake of [methyl-^{3}H]thymidine by *D. galeata* was negligible. Thus, ^{3}H-radioactivities of animals after feeding experiments were due to the bacterial ingestion.

The clearance rate (ml h^{-1}) of *D. galeata* was estimated using the equation presented by Peters (1984). The selectivity coefficient for the bacteria was calculated by the following equation:

$$SC_b = CR_b/CR_a ,$$

where SC_b is the selectivity coefficient, CR_b is the bacterial clearance rate and CR_a is the algal clearance rate. For this coefficient, a value of 1 indicates that the daphniids feed on bacteria as efficiently as on algae, and zero value indicates that the daphniids do not feed on bacteria at all. Differences in the selectivity coefficient were tested using ANOVA and the T′-method (Sokal & Rohlf, 1981). The selectivity coefficients were transformed to arcsine before the statistical tests.

Intersetular distances of thoracic limbs III and IV were measured by scanning electron microscopy on four *D. galeata* individuals collected from the same pond. The animals were anesthetized in carbonated water, fixed in 2% sugar formalin, rinsed in 2% glutaraldehyde, dehydrated in a graded ethanol series, and treated by critical-point drying. Then, they were dissected, coated with gold, and SEM photographs were taken (× 5000–10000). At least 75 intersetular distances were measured per daphniid.

Results

The cumulative frequency distribution of the intersetular distances of each four *Daphnia galeata* individuals is shown in Fig. 3. The intersetular distance differed among different position on the setae of thoracic limbs as noted by Gophen &

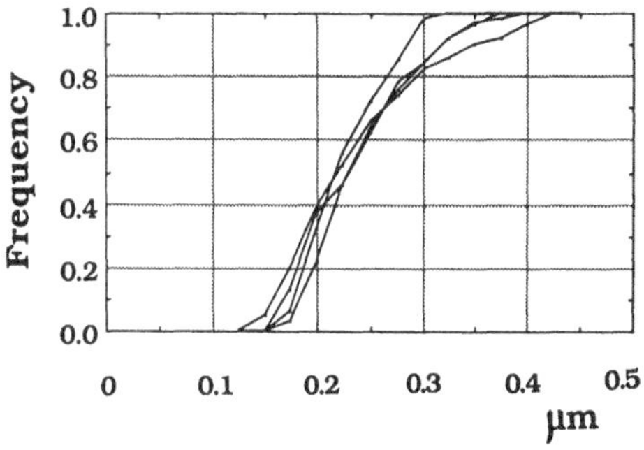

Fig. 3. Cumulative distributions of intersetular distance of the thoracic limbs III and IV in four specimens of *Daphnia galeata*.

Geller (1984), ranging from 0.15 to 0.45 μm with the mean value of 0.23–0.28 μm ($\pm$ SD 0.04–0.08). None of the intersetular frequency distributions differed significantly from the normal distribution (Kolmogrov-Smirnov test, $\alpha > 0.05$). Comparison between the intersetular distance and bacterial size predicted that *D. galeata* would be able to retain more than 70% of the bacteria if the intersetular distance is the only parameter controlling bacterial uptake efficiency.

The algal and bacterial clearance rates are presented in Table 1. Algal clearance rates in the mixed food suspensions did not differ significantly from that in the suspension of the algal alone with similar concentrations, and declined with increasing total density. The change of bacterial clearance rates with changing food concentrations was basically similar to that of the algal clearance rates. However, bacterial clearance rate in the suspension without alga (exp. 1) was significantly lower than that in the mixed food suspension containing the lowest algal concentration (exp. 1 vs. exp. 2: $t = 3.683$, $p < 0.001$).

The change of the ingestion rates related to food concentrations was closely similar when the mixture and single food sources were compared (Fig. 4). The ingestion rate increased with increasing food density and leveled off at 1.5 × 10^6 μm^3 ml^{-1} (the incipient limiting level: ILL). Such a functional response is quite similar to those for other daphniids (DeMott, 1982; Porter *et al.*, 1982).

Comparison between the algal and bacterial

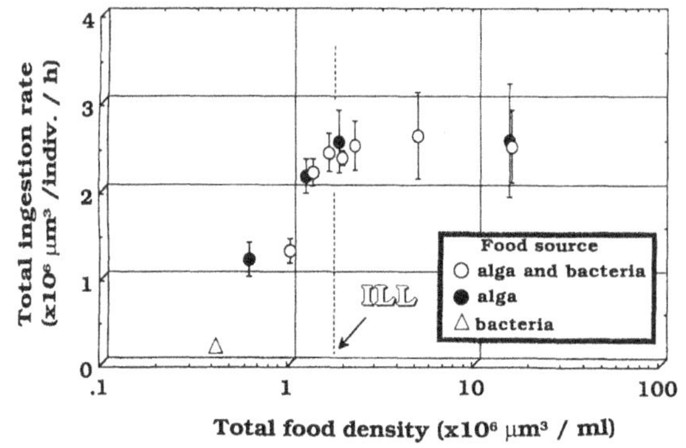

Fig. 4. Relationship between total food density and total ingestion rate for *Daphnia galeata*.

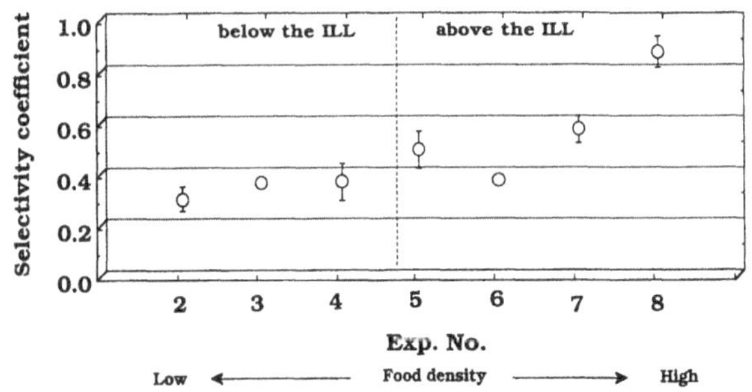

Fig. 5. mean and 95% CI of selectivity coefficients for bacteria obtained in the experiments using various mixed food conditions. Significant difference among experiments, as determined by ANOVA and the T′-method are 2 3 4 6 5 7 8.

clearance rates revealed that *D. galeata* ingested the alga more efficiently than the bacteria. However, the bacterial uptake efficiency differed with changing food density. *D. galeata* showed a selectivity coefficient much lower than that predicted from the intersetular distance when the food density was low (Fig. 5). The selectivity coefficient was constant at around 0.35 below the ILL, but showed a significant increase up to 0.9 with increasing food density above this level ($F_{4,19}$ = 46.048, $p < 0.001$).

Discussion

If the selectivity coefficient is constant irrespective of food conditions, it is most likely that daphniids collect the bacteria by simple mechanical sieving, since the intersetular distance of the limbs does not change. On the other hand, if the coefficient changes with changing food conditions, some process or mechanisms other than mechanical sieving would play an important role in determining the bacterial uptake efficiency. The results presented here clearly demonstrate that the bacterial uptake efficiency of *D. galeata* is strongly influenced by food conditions and, thus, cannot be fully explained only by mechanical sieving. Some other process or mechanism may play a substantial role in bacterial feeding by daphniids at high food concentrations.

When food concentration was lower than or slightly above the incipient food limiting level, the selectivity coefficient was stable under mixed food conditions. This stability seems never to be obtained by the handling process of the thoracic limbs proposed by Ganf & Shiel (1985). The results indicate that mechanical sieving is most likely to determine the bacterial uptake efficiency. However, although more than 70% of the bacteria supplied were larger than the maximum intersetular distance of thoracic limbs III and IV, the selectivity coefficient below the ILL was much lower than that predicted from the intersetular distance. This implies that a substantial amount of bacteria larger than the intersetular distance passes through the carapace cavity without being retained by the thoracic limbs, while *Chlamydomonas* is retained. Some studies also showed that bacteria and cyanobacteria are not necessarily retained as efficiently as algae by daphniids, even though the intersetular distances are fine enough to retain efficiently these particles by mechanical sieving (DeMott, 1985; Brendelberger, 1985). The intersetular distance may not be an absolute but rather a conservative indicator in determining the bacterial uptake efficiency by daphniids. Gerritsen *et al.* (1988) suggest that the surface area of the thoracic limbs available for particle capture, which is closely related to setule spacing, is an important parameter in determining the bacterial uptake efficiency.

The bacterial clearance rate in the suspension containing bacteria alone (exp. 1) was significantly lower than in the mixed suspensions of algae and bacteria below the ILL. Several authors have also noted that the bacterial clearance rate is enhanced by the presence of algae in the suspension (Porter *et al.*, 1983; Nagata & Okamoto, 1988). However, in this study, the low clearance rate in the suspension containing bacteria alone may be due to extremely low energy intake. A *Daphnia galeata* individual (1.3 mm carapace length) must assimilate 0.049 μg C h^{-1} to compensate for metabolic expenditure, as estimated from respiration measurements (Urabe & Watanabe, 1990). However, bacterial ingestion rate estimated from the bacterial clearance rate in the mixed suspension below ILL using the conversion factor, 105 fg C μm^{-3} (Nagata, 1986), was found to be less than 0.027 μg C h^{-1}. This implies that *D. galeata* was starved in the experiment using bacteria alone. Muck & Lampert (1980) found significant reduction in the clearance rate under extremely low food density due to starvation.

Enhancement of bacterial uptake by the presence of algae was clearly demonstrated at high food densities. Above the ILL, any change in the selectivity coefficient cannot be attributable to effects of starvation, since the total ingestion rate is high enough to compensate for metabolic expenditure. It may be considered that the increase of the selectivity coefficient is a consequence of a change in the effect of some process related to transfer of food in the food groove (Brendelberger, 1985): daphniids efficiently retain bacteria, whereas loss of bacteria in the process of transfer from the limbs to the mouth is high at low food concentration but becomes low with increasing food density. However, this idea cannot explain the significant correlation between the selectivity coefficient and intersetular distance of various species given in the literatures (DeMott, 1985; Brendelberger, 1985; Hessen, 1985), since the effect of this process is not related to the intersetular distance nor to the surface area of the feeding limbs.

Porter *et al.* (1983) suggested that the bacterial uptake efficiency might be enhanced by the 'piggy-backing' phenomenon. Change in the selectivity coefficient presented here supports this idea. In daphniids, it is known that the beat rate of feeding limbs decreases above ILL, and that rates of labral and postabdominal rejections of food in the thoracic chamber increase (Burns, 1968; Porter *et al.*, 1982). These patterns imply that excess food enters the thoracic chamber and encounters the limbs within a cycle of beating of the thoracic limbs above the ILL. Thus, phenomena such as clogging of the filter-mesh with foods are likely to occur frequently with increasing total food density. Furthermore, reduction in the beat rate of feeding limbs decreases the flow speed and increases the residence time of the particles in the vicinity of the thoracic limbs. As a result, the particles may easily be trapped by or stick the limbs. All of these phenomena would enhance the efficiency of bacterial uptake by daphniids above the ILL.

The present study suggests that the daphniid collects the bacteria basically by mechanical sieving. However, it also suggests that the mechanical sieving is not a sufficient explanation for the bacterial feeding, especially at the level of high food concentration. Not only the morphology but also the action and beat rate of the thoracic limbs may play an important role in determining the bacterial uptake efficiency. In copepods, it is known that the feeding appendage can act not only as a filter but also as a solid paddle, the function depending on its movement (Koehl, 1984; Price & Paffenhofer, 1986).

One further important point in the bacterial feeding of daphniids is to evaluate the adaptive significance of their feeding ability. According to foraging theory (Werner & Hall, 1974; Krebs, 1978), an animal should choose the most profitable food when various foods are supplied in abundance, whilst the optimal strategy would be to ingest even the less profitable foods under poor feeding conditions. Contrarily, *D. galeata* reduced the bacterial uptake efficiency with decreasing food density, although the algal clearance rate remained high. This suggests that *D. galeata* does not necessarily rely on planktonic bacteria even as a substitute food source, although change in the intersetular distance and surface area of thoracic limbs III and IV found in some populations (Kořínek & Machácek, 1980; Koza & Kořínek, 1985) may improve the ability to feed on bacteria. Thus, bacterial feeding by *D. galeata* may only be an incidental event associated with algal feeding.

Acknowledgements

The authors wish to thank Dr. M. Gophen for his invaluable comments and improvement of the manuscript.

References

Børsheim, K. Y. & Y. Olsen, 1984. Grazing activities by *Daphnia pulex* on natural populations of bacteria and algae. Verh. Int. Ver. Limnol. 22: 644–648.

Brendelberger, H., 1985. Filter mesh-size and retention efficiency for small particles: comparative studies with Cladocera. Arch. Hydrobiol., Beih. Ergebn. Limnol. 21: 135–146.

Brendelberger, H., M., Herbeck, H. Lang & W. Lampert, 1986. *Daphnia*'s filters are not solid walls. Arch. Hydrobiol. 107: 197–202.

Burns, C. W., 1968. Direct observation of mechanisms regulating feeding behavior of *Daphnia* in lake water. Int. Revue ges. Hydrobiol. 52: 83–100.

DeMott, W. R., 1982. Feeding selectivities and relative ingestion rates of *Daphnia* and *Bosmina*. Limnol. Oceanogr. 27: 518–527.

DeMott, W. R., 1985. Relations between filter mesh-size, feeding mode, and capture efficiency for cladocerans feeding on ultrafine particles. Arch. Hydrobiol., Beih. Ergebn. Limnol. 21: 125–134.

Downing, J. A. & R. H. Peters, 1980. The effect of body size and food concentration on the *in situ* filtering rate of *Sida crystallina*. Limnol. Oceanogr. 25: 883–895.

Fryer, G., 1987. The feeding mechanisms of Daphniidae (Crustacea: Cladocera): recent suggestions and neglected consideration. J. Plankton Res. 9: 419–432.

Ganf, G. G. & R. J. Shiel, 1985. Particle capture by *Daphnia carinata*. Aust. J. mar. Freshwat. Res. 36: 371–381.

Gerritsen, J. & K. G. Porter, 1982. Fluid mechanics, surface chemistry, and filter feeding by *Daphnia*. Science 216: 1225–1227.

Gerritsen, J., K. G. Porter & J. R. Strickler, 1988. Not by sieving alone: observation of suspension feeding in *Daphnia*. Bull. Mar. Sci. 43: 366–376.

Gophen, M., 1977. Feeding of *Daphnia* on *Chlamydomonas* and *Chlorobium*. Nature 265: 271–273.

Gophen, M., B. Z. Cavari & T. Berman, 1974. Zooplankton feeding on differentially labeled algae and bacterial. Nature 274: 393–394.

Gophen, M. & W. Geller, 1984. Filter mesh size and food particle uptake by *Daphnia*. Oecologia 64: 408–412.

Hessen, D. O., 1985. Filtering structures and particle size selection in coexisting Cladocera. Oecologia 66: 368–372.

Hobbie, J. E., R. J. Daley & S. Jasper, 1977. Use of Nuclepore filters for counting bacteria by fluorescence microscopy. Appl. Envir. Microbiol. 33: 1225–1228.

Kankaala, P., 1988. The relative importance of algae and bacteria as food for *Daphnia longispina* (Cladocera) in a polyhumic lake. Freshwat. Biol. 19: 285–296.

Koehl, M. A. R., 1984. Mechanisms of particle capture by copepods at low Reynolds number: possible modes of selective feeding. In D. G. Meyers & J. R. Strickler (eds.), Trophic interactions with aquatic ecosystems. Westview Press, Boulder: 135–166.

Kořínek, V. & J. Machácek, 1980. Filtering structures of Cladocera and their ecological significance. I. Daphnia pulicaria. Vestn. Cesk. Spol. Zool., 44: 213–218.

Koza, V. & V. Kořínek, 1985. Adaptability of the filtration screen in *Daphnia*: another answer to the selective pressure of the environment. Arch. Hydrobiol., Beih. Ergebn. Limnol. 21 :1 193–198.

Krebs, J. R., 1978. Optimal foraging: decision rules for predators. In J. R. Krebs & N. B. Davies, (eds.), Behavioural Ecology. Blackwell, Oxford: 23–63.

Lampert, W., 1974. A method for determining food selection by zooplankton. Limnol. Oceanogr. 19: 995–998.

Muck, P. & W. Lampert, 1980. Feeding of freshwater filter filter-feeders at very low food concentration; poor evidence for 'threshold feeding' and 'optimal foraging' in *Daphnia longispina* and *Eudiaptomus gracilis*. J. Plankton Res. 2: 367–379.

Nagata, T., 1986. Carbon and nitrogen content of natural planktonic bacteria. Appl. Envir. Microbiol. 52: 28–32.

Nagata, T. & K. Okamoto, 1988. Filtering rates on natural bacteria by *Daphnia longispina* and *Eudiaptomus japonics* in Lake Biwa. J. Plankton Res. 10: 835–850.

Peters, R. H., 1984. Methods for study of feeding, grazing and assimilation by zooplankton. In J. A. Downing & F. H. Rigler (eds.), A manual on methods for the assessment of secondary productivity in fresh waters, 2nd edition. Blackwell, Oxford: 336–412.

Porter, K. G., J. Gerritsen & J. D. Orcutt, Jr., 1982. The effect of food concentration on swimming patterns, feeding behavior, ingestion, assimilation, and respiration by *Daphnia*. Limnol. Oceanogr. 27: 935–949.

Porter, K. G., Y. S. Feig & E. F. Vetter, 1983. Morphology, flow regimes, and filtering rates of Daphnia, Ceriodaphnia, and Bosmina fed natural bacteria. Oecologia 58: 156–163.

Price, H. J. & G. A. Paffenhofer, 1986. Capture of small cells by the copepod *Eucalanus elongatus*. Limnol. Oceanogr. 31: 189–194.

Sokal, R. R. & F. J. Rohlf, 1981. Biometry, 2nd edition. Freeman New York. 859 pp.

Sorokin, Y. I. & E. B. Pavljeva, 1978. On structure and functioning of ecosystem in a salmon lake. Hydrobiologia 57: 25–48.

Urabe, J. & Y. Watanabe, 1990. Influence of food density on respiration rate of two crustacean plankters, *Daphnia galeata* and *Bosmina longirostris*. Oecologia 82: 362–368.

Watanabe, A., 1960. List of algal strains in collection at the Institute of Applied Microbiology, University of Tokyo. J. gen. appl. Microbiol. Tokyo 6: 283–292.

Werner, E. E. & D. J. Hall, 1974. Optimal Foraging and the size selection of prey by the bluegill sunfish (*Lepomis macrochirus*). Ecology 55: 1042–1052.

Hydrobiologia **225**: 129–154, 1991.
V. Kořínek & D. G. Frey (eds), Biology of Cladocera.

Mechanisms of food selection in *Daphnia*

Hans Julian Hartmann[1] & Dennis D. Kunkel[2]
[1]*School of Fisheries HF-15 and* [2]*Department of Neurological Surgery RI-20, University of Washington, Seattle, WA 98195, USA*

Key words: food selection, feeding model, phytoplankton, bacteria, SEM analysis

Abstract

A conceptual behavioural and mechanistic Holling-type model of food selection in *Daphnia pulicaria* is derived from SEM observations with animals feeding on mixtures of spherical-cylindrical diatoms, oblongate green algae, and filamentous cyanobacteria, as well as ultrafine particles. The algae used were *Stephanodiscus hantzschii* (≤ 6 μm length), *Monoraphidium setiforme* (≥ 20 μm), and *Oscillatoria aghardii* (strands, ≥ 80 μm). Cell (strand) selection can occur at any or all of three stages: (i) interception from the feeding currents, (ii) collection and channeling to the food groove, and (iii) compaction and transport to the mouth. During each stage, given equal initial cell densities, elongate cells are more likely to escape collection than spherical cells and are more likely to be rejected. In addition, filaments require increased handling time at stages (ii) and (iii) and promote entanglement with limb 5 and the postabdominal claw. Food is collected primarily with the aid of limbs 3 (and 4), but limbs 1 and 2 also intervene. Neither the leaky sieve hypothesis alone nor any other single-process hypothesis explains the observations on examined *in corpore* positions, morphology, and derived movements of the feeding limbs. Attachment and mucus appear to be important for the ingestion of bacteria and ultrafine particles.

The model is consistent with many experimental results of differential feeding by *Daphnia pulicaria* on mixtures of variously shaped algae and other observations on *Daphnia* feeding behaviour. The paradigm of invariate, nonselective feeding by *Daphnia* is rejected.

Introduction

Daphnia occupy an important position in many freshwater food webs. Their effectiveness as grazers on phytoplankton has been well documented (e.g. Edmondson & Litt, 1982; Lampert *et al.*, 1986). However, the mechanisms by which *Daphnia* interact with and collect food particles from picoplankton to mm-long filaments still remain in question.

Its feeding structures are complex, hidden within the carapace, and move rapidly, making direct visual observations on feeding extremely difficult. Real-time, *in vivo* observations require expensive and sophisticated high-speed cinematography equipment (e.g. Gerritsen *et al.*, 1988; Shiel *et al.*, 1988). Light-microscopy and comparative functional morphology of cladocera have helped to obtain information on general pathways of particles, particle movements within the food groove, and feeding currents, but lack sufficient resolution to explain the complex movements and function of all of *Daphnia*'s feeding limbs. Thus, controversies have arisen over several aspects of food collection.

A major point of discussion on *Daphnia's*

feeding mechanisms is whether or not particles are collected by sieving. Sieving is a widely accepted paradigm (Geller & Müller, 1981; Gophen & Geller, 1984; Brendelberger, 1985; Kořínek *et al.*, 1986). This derives from considerations, dating back to observations by Storch (1924) and Cannon (1933) that those limbs act as a mechanical suction pressure pump requiring passage of water across the setular meshes (e.g. reviews by Fryer, 1987; Lampert, 1987). Flow-model calculations emphasize that such a feeding mode is energy efficient (Brendelberger *et al.*, 1986; Geller & Knisely, 1988; W. Geller, comm. *in lit.*). But direct observations have shown that the models may not correctly represent the actual flow regime (Gerritsen *et al.*, 1988; Gophen *et al.*, pers. comm., unpublished results), and suggest that energy expenditures required for sieving may be higher than previously assumed (Gerritsen *et al.*, 1988; Philippova & Postnov, 1988).

In the absence of modifying factors, the 'leaky-sieve' hypothesis (Boyd, 1976) requires that *Daphnia* feed non-selectively on all particles exceeding the setular mesh size of limbs 3 and 4 up to an upper length determined by body size (Burns, 1968b). Indeed, many investigators discussing *Daphnia*'s grazing impact tacitly or explicitly assume this. Some observations done with spherical algae and particles appear to confirm it (Gophen & Geller, 1984; Brendelberger, 1985; DeMott, 1988, 1989).

Other factors may also intervene in small-particle collection, such as taste (Porter & Orcutt, 1980), surface chemistry (Gerritsen & Porter, 1982), actions of other trunk limbs (Ganf & Shiel, 1985a, b; Gerritsen *et al.*, 1988; Shiel *et al.*, 1988; Bernd, 1990) and gravitational and diffusion deposition (Rubenstein & Koehl, 1977).

The many indirect observations on selective feeding of *Daphnia* would seem to require the action of one or several of these mechanisms. Thus, *Daphnia* have been found to prefer smaller to larger spherical algae or plastic spheres (Meise *et al.*, 1985; Bernd, 1990), flagellates over coccales (Knisely & Geller, 1986), and small nearly spherical diatoms over elongate greens (Hartmann, 1985). They can capture and consume rotifers (Burns & Gilbert, 1986). Filamentous algae and cyanobacteria are either consumed without problems (Bogdan & McNaught, 1975; Holm *et al.*, 1983; Knisely & Geller, 1986; Fulton, 1988; DeMott, 1990), or are rejected or otherwise inhibit food collection (Burns, 1968a; Webster & Peters, 1978; Gliwicz & Siedlar, 1980; Porter & McDonough, 1984; Hartmann, 1985; Fulton, 1988; Hawkins & Lampert, 1989). Within one species, selectivity on the same alga may depend on animal size and age (Hartmann, 1985).

Evidently, experimental results are too diverse and seemingly contradictory to be explained by simple one-step models. They illustrate that neither simple mechanical models nor any other single mechanism, nor observations on the flow regime alone (Gophen, pers. comm.) suffice to explain *Daphnia*'s various feeding behaviours. The complex results may indicate that several selective modes operate together (e.g. DeMott, 1990), either in sequence or simultaneously, and that we have not yet been able to delineate their relative significance.

In this paper, a conceptual Holling-type behavioural-mechanistic *Daphnia* feeding model (i.e. Holling, 1966) is proposed. The model separates the feeding process into a series of steps (i.e. encounter, capture, transport, handling and rejection). Probabilities of particle retention at each step are proposed to relate to particle shape and specific movements of the feeding apparatus (e.g. thoracic appendages, mouthparts, post-abdominal claw), which in turn determine particle retention or loss, degree of particle compaction before ingestion, and handling time. The model is derived from an analysis of the positioning of food particles within *Daphnia*'s carapace ('*in corpore*') observed by SEM on undissected animals that had previously been exposed to mixtures of various food particles. The model explains how *Daphnia* selects among larger particles (those exceeding the setular mesh widths) based on their shape, from nearly spherical to elongate to filamentous, and proposes further evidence on mechanisms of bacterial feeding. These observations were made in parallel with selectivity experiments, and the model is consistent with evidence that selectivity varies with age, relative food proportions, and food concentration (Hartmann, 1985).

Material and methods

Algae and zooplankton

Experiments were done with three differently shaped algae, a small, spherical-to-cylindrical (length : width $\leq 2:1$) centric diatom (*Stephanodiscus hantzschii*), an elongate coccoid green alga (*Monoraphidium setiforme* [Nygaard, 1977]) and the filamentous cyanobacterium *Oscillatoria aghardii* (Table 1). *Stephanodiscus* and *Oscillatoria* were isolated from Lake Washington, and *Monoraphidium* originated from the University of Texas culture collection (Taub, 1985).

Algae were grown in axenic (*M. setiforme*) or unialgal (*S. hantzschii, O. aghardii*) semicontinuous cultures (2 l, 18 $\pm$ 2 °C, 81.4–103.0 μE m^{-2} s^{-1} of PAR, 12/12 h light/dark,) with Medium T80MV (derived from Taub & Dollar, 1964).

Daphnia pulicaria from Lake Washington were maintained in the laboratory under moderate light (18–22 °C, 1180–4300 lx; 12/12 to 16/8 h light/dark) in wide-mouth 4-liter glass jars with 0.45 μm filtered, sterile lake water (FLW) enriched with 1 ml l^{-1} of 1/100 strength Murphy's Vitamin mixture (Murphy, 1970) and 10^4–10^5 cells ml^{-1} of an algal mixture (*Stephanodiscus*, *Monoraphidium*, and *Chlamydomonas reinhardi*) for food. To increase clonal diversity, *Daphnia* cultures were occasionally replenished with field animals collected from Lake Washington. Cultures reached densities of 40 adults l^{-1} and 350 neonates l^{-1}, respectively. Details on media and algal and animal cultivation are given in Hartmann (1987). Algae and *Daphnia* parameters are summarized in Table 1.

Table 1. Characteristics of experimental algae and animals. Average linear dimensions (GALD), equivalent spherical diameter (E.S.D.), cell volumes, *Daphnia* carapace lengths and dry weights.

Species	Dimensions (GALD, μm)	E.S.D. (μm)	Volume (μm^3)
Monoraphidium setiforme	18–25 × 3.3	3.8–5.0	30–≥100
Stephanodiscus hantzschii	5–8 × 4.0	5.0–6.0	100–250
Oscillatoria aghardii (strands)	55–1,000 × 3.4		830–≥10,000
Daphnia pulicaria	Length (mm) range		Dry Weight (μg ind^{-1}) range
Large *Daphnia*	1.90–2.50		27–69
Medium *Daphnia*	1.30–1.90		12–26
Small *Daphnia*	0.95–1.30		5–11

Feeding and electron microscopy

Cultured *D. pulicaria* of all sizes were transferred to FLW by gently pouring laboratory cultures through appropriate ranges of nylon-mesh screen. The screens were placed in an overflow apparatus so that animals were always submerged in the water. The animals were left for 1-2 h to allow cleaning of the feeding structures and gut. Animals were then again strained by the overflow method, rinsed with FLW and transferred to 100-ml flasks with unialgal or mixed suspensions of *Monoraphidium* and *Stephanodiscus* (approximately 10^4 cells ml^{-1}, 50/50% by cell number) or *Monoraphidium* with 250 strands ml^{-1} *Oscillatoria*. After feeding for 60 min, the animals were anaesthetized with gently bubbling CO_2, transferred to size 00 Beem capsules fitted with 110 μm Nylon mesh bottoms and lids (Flood, 1973), rinsed with FLW and fixed in glutaraldehyde (Polysciences E.M. grade, 0.5% in FLW). After storing for 12–48 h at 4 °C in 3% glutaraldehyde, the animals were transferred to a phosphate buffer (1.9% Na H_2PO_4 in 0.1 N NaOH, pH 7.4), fixed for 1 h in 1% OsO_4/phosphate buffer, dried with ascending concentrations of ethanol/buffer (5, 15, 25, 35, 35% ethanol), ethanol/H_2O (60, 70, 85, 90, 95, 100% ethanol) freon TF/ethanol (50,

100% freon TF), and critically-point dried with Freon 13.

Individual animals were stub-mounted (double-stick tape and silver paste), dissected to expose desired portions of their feeding structures, gold-coated and examined with JEOL models JSM-U3 and JSM-35 Scanning Electron Microscopes.

Observations were judges valid for consideration only if they were replicated in different individuals. Observed replications ranged from 3 to ≥10. They were only used as a qualitative criterion for indicating the probability that one particular described behaviour occurred or not.

Results

Limb morphology and limb movements

The feeding morphology of *Daphnia* and other cladocerans has been described in detail by others (e.g. Cannon, 1933: Smirnov, 1971; Kořínek & Machácek, 1980; Watts & Petri, 1981; Glagolev, 1983; Ganf & Shiel, 1985a), usually from separately prepared limbs rather than from limbs properly in place on the animal. The descriptions here serve to emphasize relative limb positions and to deduce limb movements relative to each other (i.e. Tables 2, 3, 4).

Limbs 3 to 5

The surface formed by the gnathobasal setae of the 3rd and 4th thoracic limbs L3 and L4 dominate the interior of *D. pulicaria's* carapace chamber (Figs. 1, 3). The basal 30 to 50% of the setal length are stiff, while the remaining distal ends are flexible. They were observed curved either convex away from the median food groove (forming a tunnel-like space terminated posteriorly near the 5th thoracic limbs; Fig. 1), or concave towards the food groove, leaving a triangular space above it (Fig. 2). The concave position indicates that the limbs come to rest during the inward stroke (cf.

Table 2. Positions and derived movements of thoracic limbs and their setae inside the carapace of *Daphnia pulicaria*, from analysis of SEM photographs. S3, S4 = setae and mesh fields of limbs 3, 4.

Limb, Setae	Positions	Derived movements
L5	Posterior and inward of S3, S4, terminating in median food chamber	
L3 and L4	Dorsal (setae curved convex), Ventral (setae curved concave)	Dorsoventral and lateral, (setae changing curvature)
L2, Basal Fan	Anterior/dorsal: see Fig. 25A Posterior/ventral: see Fig. 25B	In 3 directions, predominating dorsoventral, and rotation
L2, Endite Setae		
Intermediate Seta	Dorsal: see Fig. 25A Ventral: see Fig. 25B	Dorsoventrally, sweeping over S3 ventral of L2's basal fan
Outer Setae	Covering anterior 80% of bases of S3, S4, in outer/median chamber	Anywhere within outer/central, *posterior* part of median chamber
L1, Endite Setae		
Outer Setae	Anterior: lateral of labrum Posterior: covering L2's gnathobase, and anterior 60% of S3's bases	Anywhere within outer/central, *anterior* part of median chamber
Anterior 2 Setae	Anteriormost between labrum and mandibles	Dorsoventrally and anterior-posterior, between mouth and labrum

Gerritsen *et al.*, 1988), while the convex position indicates a transition from backstroke to forestroke (Table 2). The tips of opposing setules commonly touch each other (Fig. 5). Together with their setae, they form a pleated mesh field of channels running perpendicular to the medial axis of the food groove (Figs. 4, 5, 6). Setules were rarely clean: they were frequently associated with bacteria (Figs. 4, 5).

Limb 3 of large specimens has 65 to 70 gnathobasal setae (Table 3). Small specimens have less, but no complete counts could be made from the SEM photographs. The setae were not uniformly spaced. Distal gaps occurred frequently, even in most carefully prepared animals (Figs. 6, 10). Setae of L3 are wider spaced (mean intersetal distance $ISD_3 = 7.78\ \mu m$) than those of L4 ($ISD_4 = 5.52\ \mu m$). Setal spacing decreases with animal size (Table 3). The setules of L3 (Table 4) are regularly spaced, with little difference between small and large individuals (mean intersetular distance $ISSD_3 = 0.64\ \mu m$). Mesh field widths between setules range from 0.31 to 0.48 μm (Table 4).

Limb 2

The complex 2nd thoracic limb L2 has two sets of morphologically different setae. Those inserted at the gnathobase form a fan (basal fan) spreading over part of L3's mesh field (Figs. 3, 10, 25). By

Table 3. Measurements from SEM preparations on setae of thoracic limbs 3 and 4 of *Daphnia pulicaria*. N = number of animals examined, n = total number of measurements taken, covering more than 2 setae. Means are calculated from N.

Parameter	Limb 3 mean ± s.d. (range)	*N*	*n*	Limb 4 mean ± s.d. (range)	*N*	*n*
1. Number of Setae	(65–70)	2	2			
2. Intersetal Distance (μm)						
large animals	7.78 ± 1.65 (5.88–10.34)	7	20	5.52 ± 1.18 (4.68–6.35)	2	15
medium animals	7.28	1	1			
small animals	3.71	1	1			
3. Setal Width (μm)	1.38 ± 0.55 (1.11–2.20)	2	4			
4. Intersetal Mesh (μm)	(3.25–5.25)	1	2			

Figures 1–24, explanations and abbreviations.
Scales are in μm. Arrows indicate direction of food movement, unless otherwise noted.

An = antennules,
Ba = bacteria,
Bo = bolus,
Bs_2 = bristle-like setules (limb 2),
Bu = bundle,
CS_2 = central seta (limb 2),
CSp = combed spine,
La = labrum,
L_5 = limb 5,
Lb_i = limb i gnathobase (i = 2...4),
Le_i = limb i endite (i = 1...2),
Ma = mandible,
Mo = *Monoraphidium setiforme*,
Mt = mandibular teeth,
Mu = mucus,
Mx = maxillule,
Os = *Oscillatoria aghardii*,
Pa = postabdominal claw,
Ro = rostrum,
Sb_i = limb i gnathobasal setae (i = 2...4),
Se_i = limb i endite setae (i = 1...2),
Sp = spine,
St = *Stephanodiscus hantzschii*,
Su_i = setules belonging to limb i setae (i = 1...4),
Us_2 = umbrella-like bristles (limb 2),
Va = valve edge.

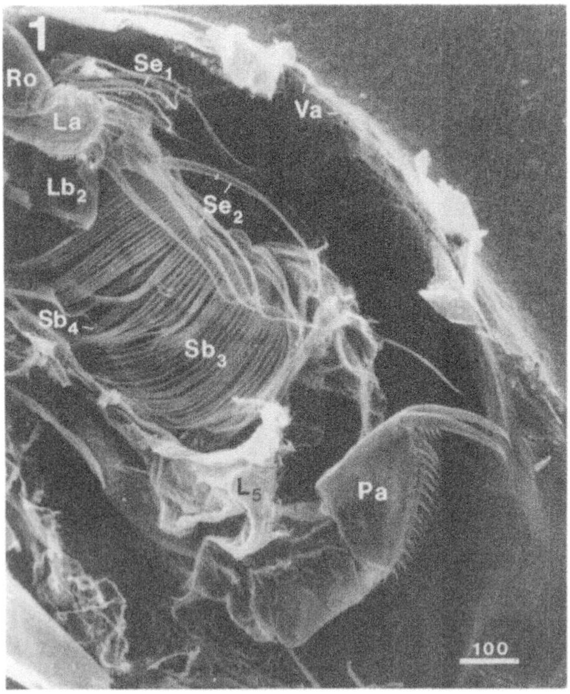

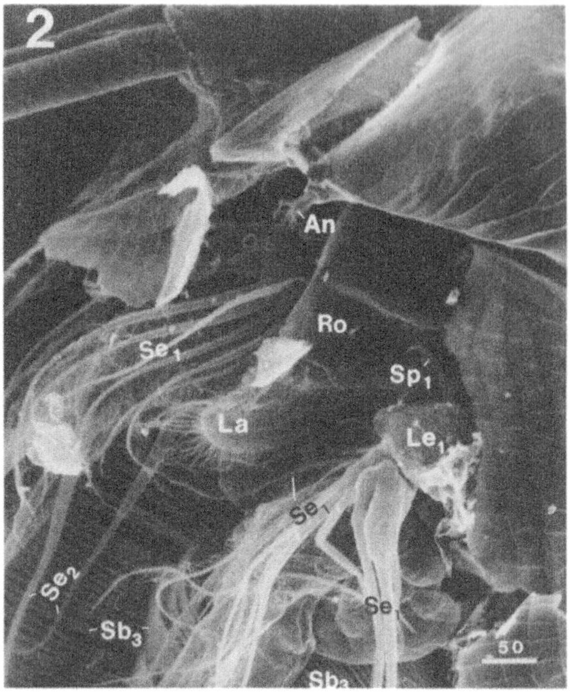

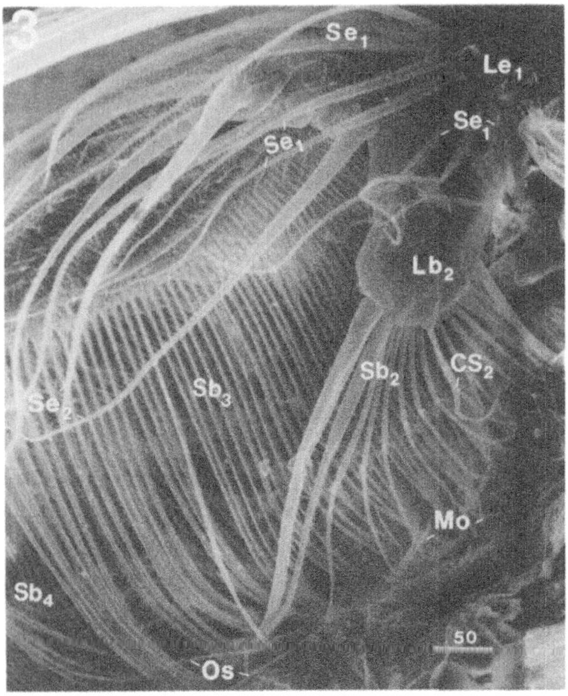

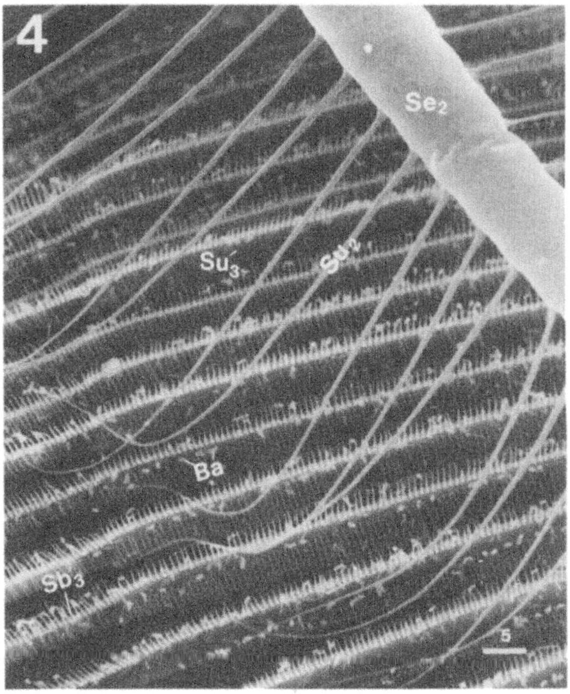

Fig. 1. Food-gathering chamber of a large female *Daphnia pulicaria*. Right half of the valve and the basal fan of the second limb removed. Gnathobasal setae of limb 3 are dominating the chamber and are curved convex.

Fig. 2. Details of anterior feeding structures of *Daphnia pulicaria*. Setae of limb 3 appear concave. Base of limb 1 with spines is exposed, its three outermost setae reach under the carapace, central setae reach toward sagittal plane, the two foremost setae reach under the labrum.

Fig. 3. Central food-gathering chamber of *Daphnia pulicaria*. Central group of setae of limb 1 cover most of the endite setae of limb 2. Intermediate endite seta of limb 2 covers basal portions of gnathobasal setae of limb 3. Gnathobasal fan of limb 2 is

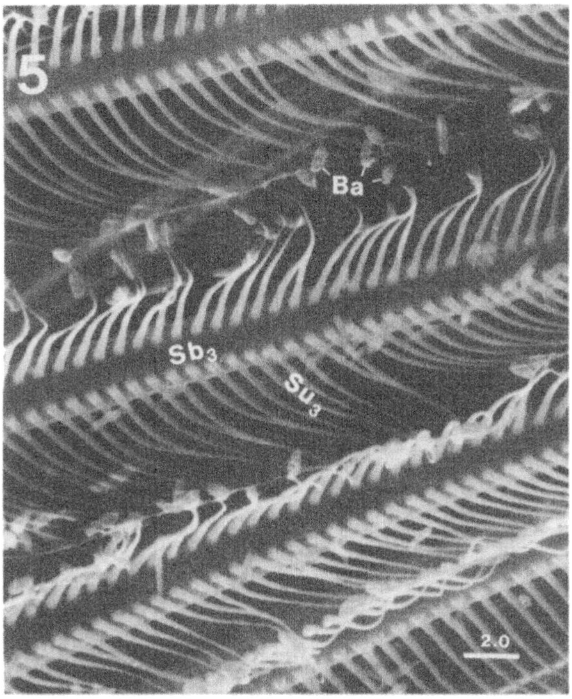

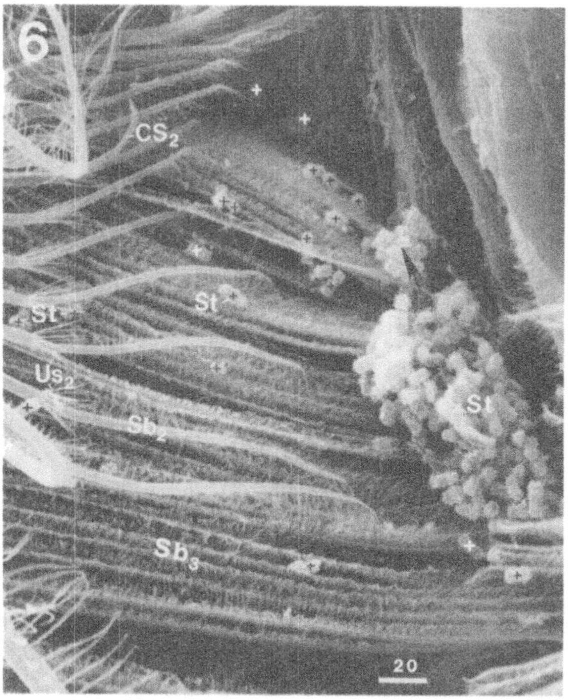

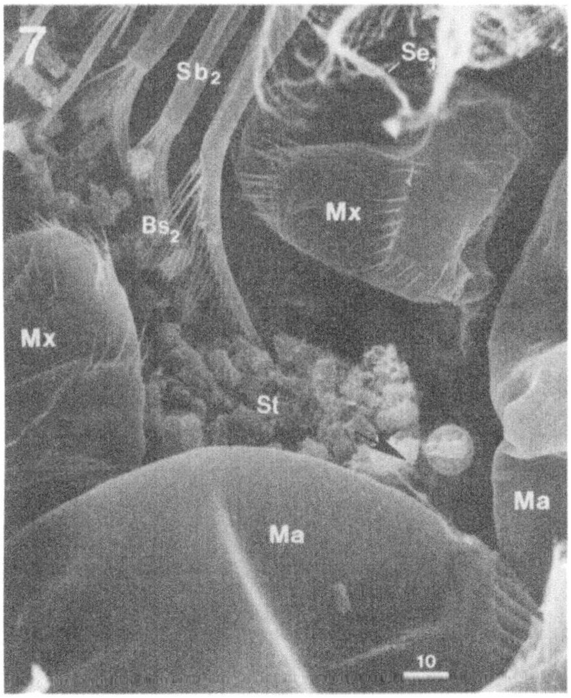

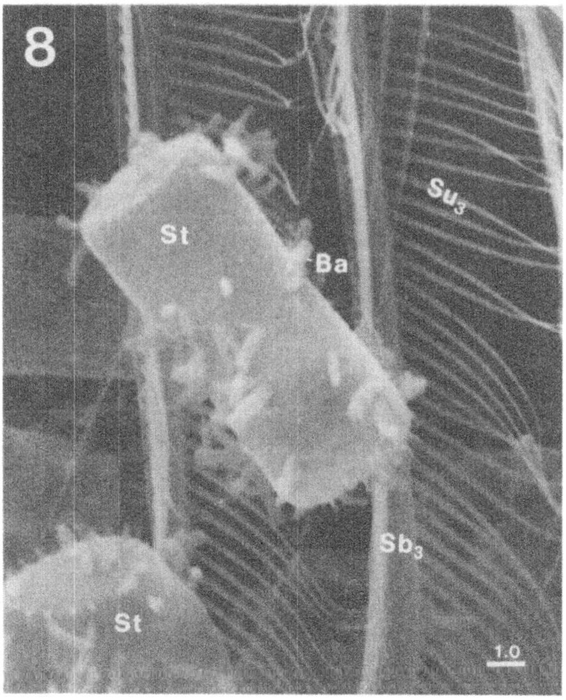

stopped in forward-dorsal position. Notice position of umbrella bristles and hook-shaped central seta of the fan (cf. Fig. 10). Gaps between the setae of limb 3 are partially penetrated by the tips of the fan-setae of limb 2. Some *Monoraphidium* cells and two *Oscillatoria* strands are loosely gathered in the food groove. Setae of limb 4 are visible in the lower left (background).

Fig. 4. Pleated mesh field of limb 3. A central section of the gnathobasal setae of limb 3 is being swept by setules of the long intermediate endite seta of limb 3. The tips of setules of limb 3 are not touching each other. Notice bacteria attached to setules of limb 3.

Fig. 5. Detail of channel-like pleating of setae of limb 3. Bacteria are attached preferentially to tips of setules.

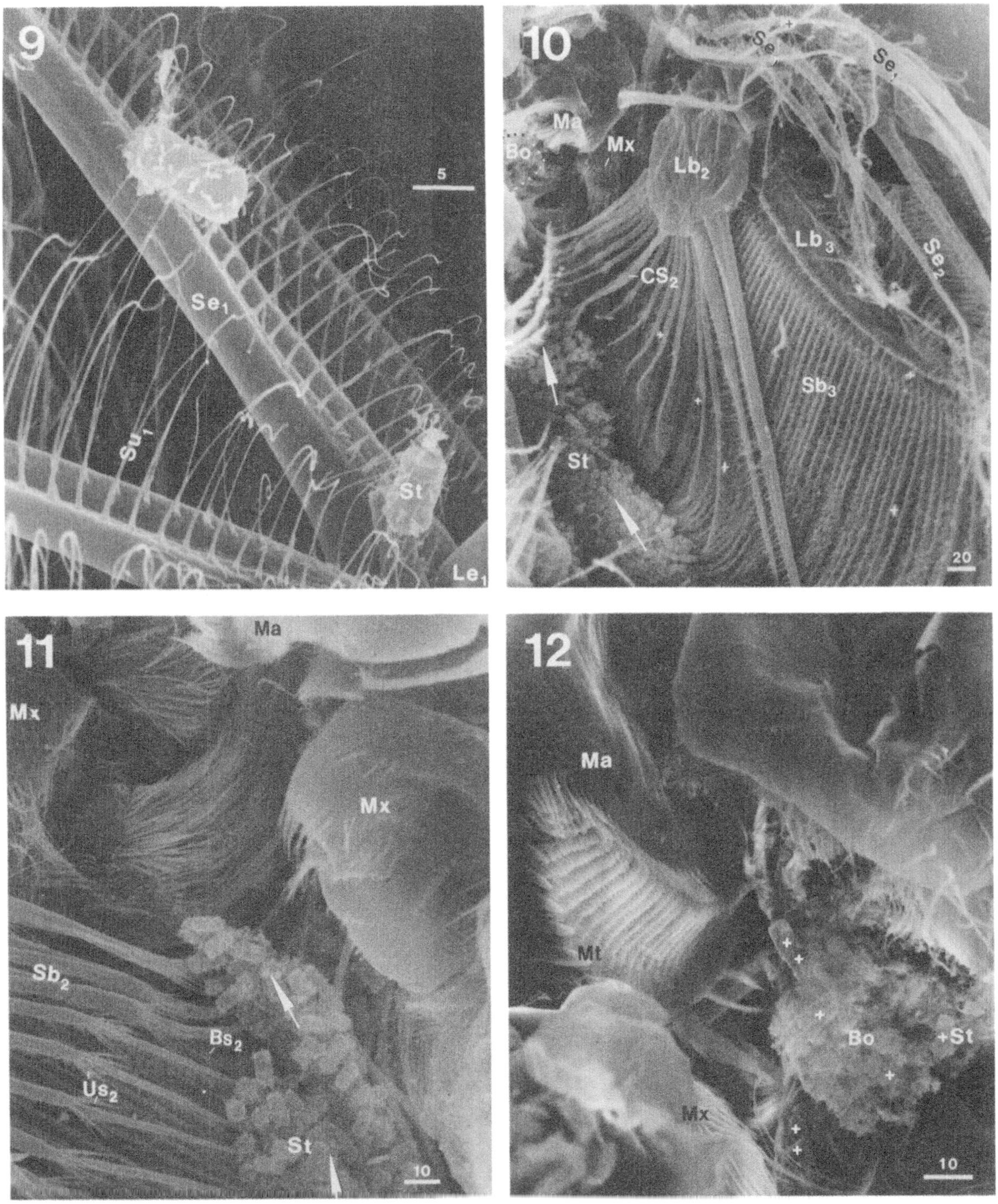

Fig. 6. Close-up of central part of feeding chamber, looking along the food groove towards mouth. Distal portions of limb 3-setae are being swept by gnathobasal fan-setae of limb 2. The fan is stopped in its most ventral position. On limb 2, umbrella bristle-setules are visible and forward/outward orientation of the distal setules is evident. *Stephanodiscus* cells are gathered in clumps near the food groove and are individually dispersed between the setae of limbs 3 and 2.

Fig. 7. Foremost part of food groove, near the mouth. The second limb's anterior three setae of gnathobasal fan showing bristle-like setules working on *Stephanodiscus* clumps. Maxillules and parts of mandibles are also visible.

Fig. 8. *Stephanodiscus* retention on limb 3. Cells are on setae of limb 3, showing attached bacteria.

Fig. 9. *Stephanodiscus* capture on limb 1. Cells entangled in setules of limb 1, near their bases.

Table 4. Measurements from SEM preparations on setules of thoracic limbs of *Daphnia pulicaria*. N = number of animals examined, n = total number of measurements taken, covering at least 2 setules. Means are calculated from N. ES = endite setae; BS_1 = posterior seta of basal fan.

Limb		Intersetular Distance (μm) mean ± s.d. (range)	N	n	Intersetular Mesh (μm) mean ± s.d. (range)	N	n
Limb 1		2.52 ± 0.37 (2.10–3.07)	3	5			
Limb 2	ES 1	5.52 ± 0.90 (4.50–6.58)	4	7			
	ES 2–5	4.46 ± 1.77 (3.21–5.72)	2	2			
	BS 1	4.68	1	1			
Limb 3		0.640 ± 0.076 (0.54–0.72)	5	13	0.391 ± 0.069 (0.31–0.48)	2	85

contrast, the distal feathered setae emerging from the anterior endite lobes reach far backward, running inside the carapace margin along the proximal ends of L3's setae and along the bases of L3 and L4 (Figs. 1, 3, 10).

As in other species of *Daphnia* (cf. Watts & Petri, 1981), an umbrella-like ring of bristles is inserted at the central setae of the basal fan (Figs. 6, 11). Distally, beyond the bristles, the setae are flexible but pointed, each with two rows of setules pointing forward (towards the mouth) and outward (towards L3; e.g. Fig. 6). The directional orientation differs from the tufty arrangements of other species reported by Watts & Petri (1981) and Ganf & Shiel (1985b). The two long and strong posterior setae of the basal fan lack bristles but each has rows of teeth or setules (Figs. 10, 14). There also is a single, separately inserted and rather flexible seta at the center of the fan (cf. Figs. 3, 10). The three anteriormost setae of the basal fan have long, stiff bristles on their curved tips, frequently touching food cells (Fig. 7).

Setules of L2 are variably spaced. Those of the long single intermediate seta (Fig. 4) are wider spaced (mean $ISSD2_{(ES1)}$ = 5.52 μm, Table 4) than the setules of the outer distal and two posterior basal setae (mean $ISSD2_{(ES2-5)}$ = 4.46 μm and $ISSD2_{(BS1)}$ = 4.68 μm, Table 4).

The observed positions of L2 relative to L3 are shown in Fig. 25. The basal fan (BF) moves along the sagittal plane (forward-backward), and laterally (inward-outward) while simultaneously rotating by 20–30° (cf. Fig. 25A–B). In their forward position, the 3 foremost bristled setae of the basal fan reach far forward underneath and in between the maxillules (Fig. 11). Overall, the setuled portion of the basal fan swept over maximally 35% of the mesh field of L3 (see Fig. 25B, dotted outlines). The single intermediate endite seta

Fig. 10. *Stephanodiscus* gathering and handling. Gnathobasal fan and endite setae of limb 2 are in most ventral position relative to limb 3. Setules of posterior two stiff setae of the fan are well visible; the central seta of the fan is only slightly curved (cf. Fig. 3). Notice gaps between distal sections of limb 3-setae. *Stephanodiscus* cells are gathered in central and forward section of the food groove and are passed forward: a bolus is being ingested (dashed-circled). Individual cells (+) are dispersed mostly between limb 2 and lower half of limb 3.

Fig. 11. Bolus formation of *Stephanodiscus*. Detailed view of anterior part of food groove, looking towards the mouth. Cells are handled (separated into a clump) by the three bristled setae of limb 2 in cooperation with the maxillule. Cells are also visible under (i.e. dorsal of) the maxillule.

Fig. 12. Ingestion of *Stephanodiscus*. A bolus of cells (+) mixed with mucus and bacteria is being ingested.

(IES) rotated and swept over another 15–20%. Thus, setae of L2 directly brushed over only 50–60% of the surface area of L3. We also noted that setae of the basal fan tended to penetrate the intersetal spaces of L3 (Figs. 3, 6, 21), reaching out toward L4 and creating large gaps in the mesh field.

In summary, the various positions of L2 indicate *Daphnia*'s ability to move this limb widely in three directions and changing its position with regard to L3 and L4, perhaps by pivotal rotation and twisting as well as by independent movements of its basal fan and of its distal endite lobes.

Limb 1

Four functional groups of endite setae can be identified on limb 1. (i) The outermost naked setae, which operate close to the carapace margin and can reach into the space between the limbs and the carapace (Fig. 2). The central feathered setae are divided into two groups: (ii) a central group of about 7 setae operating close to the labrum as well as around the outer feathered setae of L2, along L2's gnathobase and endite lobes (Figs. 2, 3, 21); and (iii), two anterior setae reaching around the labrum, their tips appearing between the maxillules (Figs. 2, 15, 16). The setules of L1'S feathered setae are more closely spaced than those of L2 (mean ISSD1 = − 2.52 μm), but still much more widely spaced than setules of limb 3 (Table 4). Finally (iv), spinelike setae emerge from the base of the limb (Fig. 2) similar to 'hooks' described in detail by Watts & Petri (1981) and Glagolev (1983).

The setae of L1 vary their positions both along the sagittal plane (anterior-posterior) and by rotating inward/outward (Table 2). The outermost endite lobe with its naked setae appears to move more independently than the other lobes (cf. Figs. 1, 2).

Overall, the central feathered setae of L1 appear to form, together with the distal feathered setae of L2, a complex and variable-mesh three-dimensional web moving within the median feeding chamber, inside the carapace edge (Figs. 2, 3, 10).

Cell positions and food handling modes

Stephanodiscus (Figs. 6–12)

The diatom cells were found with limbs L1, L2, and L3, in the food groove, and at the mouth. On L3, they were most commonly seen dispersed among the distal and forward 65% of L3's mesh field, i.e. mostly within reach of L2's basal fan (Fig. 6). In a large animal, cells on L3 covered 0.5 to 2.5 intersetal spaces and 10–25 intersetular spaces (Fig. 8). Large masses of cells gathered near and in the food groove near the tips of L3 (Figs. 6, 10). On limb 4, they were associated with the setal tips, as clumps around the food groove.

On limb 2, *Stephanodiscus* were usually associated with the basal fan, in various forms: individually around the umbrella bristles of the central setae (Figs. 6, 10) and as masses on the inside tips of the basal fan (Fig. 11). The fan occasionally functioned like a shovel or like a spread-out hand holding clumps (Fig. 21). Finally, in the food groove *Stephanodiscus* masses were being worked on by the anterior three bristled setal tips of L2, in cooperation with maxillules and other spined portions of the anterior food groove (Figs. 10, 11).

On limb 1, *Stephanodiscus* occurred on some outer feathered setae, close to their insertion, covering 2–5 setular spaces (Figs. 9, 10). *Stephanodiscus* were ingested as boluses, in association with mucus and with free and epiphytic bacteria (Figs. 10, 12, 24).

Monoraphidium (Figs. 13–16)

Monoraphidium and short *Oscillatoria* strands (<65 μm long) were found associated only with limbs 2, 3, and 4. On limb 3, they were oriented medially (parallel to axis of food groove) to latitudinally (parallel to axis of the meshed channels, Figs. 13, 14). Lengthwise, they covered a considerably greater handling surface than *Stephanodiscus*, i.e. 10–16 intersetal spaces or 1/3 to 1/2 setal lengths in small *D. pulicaria* (Fig. 13), and 6 to 11 spaces or 1/8 to 1/6 lengths, respectively, in medium-to large animals (Fig. 14). When oriented latitudinally, the umbrella-bristles of the basal fan setae of L2 apparently slide past them (Fig. 14). In the food groove, *Monoraphidium* cells were oriented randomly and were much less compacted than *Stephanodiscus* (Fig. 3, 21). Closer to the mouth, they remained only loosely assembled (Fig. 15). For ingestion, however, they were arranged parallel to the direction of food movement (Fig. 16). Unlike *Stephanodiscus*, *Monoraphidium* were not found with setae of L1 or distal setae of L2.

Oscillatoria (Figs. 17–20)
Short *Oscillatoria* strands ($<65\ \mu m$) were handled like *Monoraphidium* (Fig. 14). Long strands were found loosely bundled in the median chamber, i.e. ventral of the food groove and anterior of limb 5 (Fig. 19). When close to the food groove, they were arranged medially, like spaghetti (Fig. 17). Close to the mouth, those same strands were closely bundled for ingestion and associated with mucus (Fig. 18). Here, hairs and bristles, some associated with the maxillules, apparently push the strands along. Strands were found entangled between limb 5 and the post-abdominal claw, among the claw's teeth or wrapped around it (Figs. 19, 20).

Table 5. In situ distribution of bacteria on setules along various lengths of different setae of limb 3, in two different specimens of SEM-prepared *Daphnia pulicaria*. Chi-square test for index of dispersion $I = s^2/x = 1$; x = mean.

	Position on Setule			c^2 ¥
	Tip	Center	Base	
Specimen 1	7	1	0	
	6	2	7	
	5	2	5	
	13	4	4	
	13	2	9	
	9	5	3	
	1	2	6	
Sums	54	18	34	18.42***
Specimen 2	17	6	1	
	5	6	5	
	12	7	2	
	2	2	0	
Sums	36	21	8	18.12***

¥ $c^2 = (s^2\ df)/x$; $df = 2$.
*** $p < 0.005$.

Mixed food and bacteria (Figs. 21–24)
Handling of cell mixtures depended on cell-type combinations. *Stephanodiscus* and *Monoraphidium* were stored within the food groove (Fig. 21). They were never seen being ingested simultaneously (no mixed-cell bundles or boluses near the mouth). Conversely, *Monoraphidium* mixed easily with *Oscillatoria* strands. Both were found traveling intermixed and in longitudinal orientation along the anterior food groove toward the mouth (Fig. 22), and were ingested simultaneously. Apparently the spaghetti-arrangement of *Oscillatoria* precluded a separate bundling of *Monoraphidium*.

Free rod-shaped bacteria were typically seen on or close to the setules of L3 (Figs. 4, 5). They were preferentially attached to the setule tips (Table 5), but did not attach to setules of limbs 1 or 2. Bacteria were ingested with copious quantities of mucus (Fig. 23).

Attached bacteria commonly were associated with *Stephanodiscus* cells (Figs. 8, 9). Those bacteria may have originated from the non-axenic *Stephanodiscus* cultures. Indeed, axenic *S. hantzschii* did not grow well in Medium T82MV and became more quickly invaded by bacteria than other axenic cultures. *Stephanodiscus hantzschii* may also have preferentially attracted bacteria for attachment within their phycosphere. Those attached bacteria were readily ingested (Fig. 24) along with *Stephanodiscus*. Bacteria were occasionally attached to *Oscillatoria*, near the mouth (Fig. 23), but were never seen on *Monoraphidium*.

In order to summarize and quantify the association of the different cells with the individual limbs, frequencies of occurrence of cells (strands) were scored in at least 10 different animals on 6 functional limb components (L1: anterior two endite setae (FLC1) and outer endite setae (FLC2); L2: basal fan (FLC3) and endite setae (FLC4); gnathobasal setae of L3 (FLC5) and of L4 (FLC6)). Scores were from absent (score 0) to abundant (score 3). For individual cells or strands, the sums of scores of all FLC's (SSC)

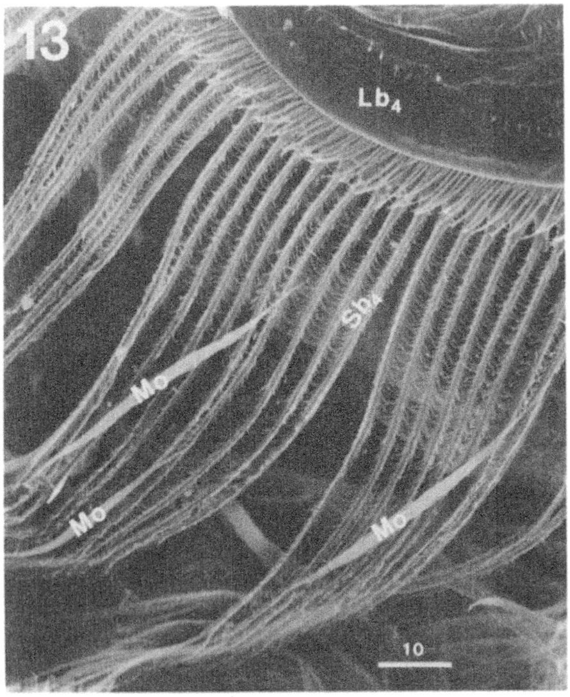

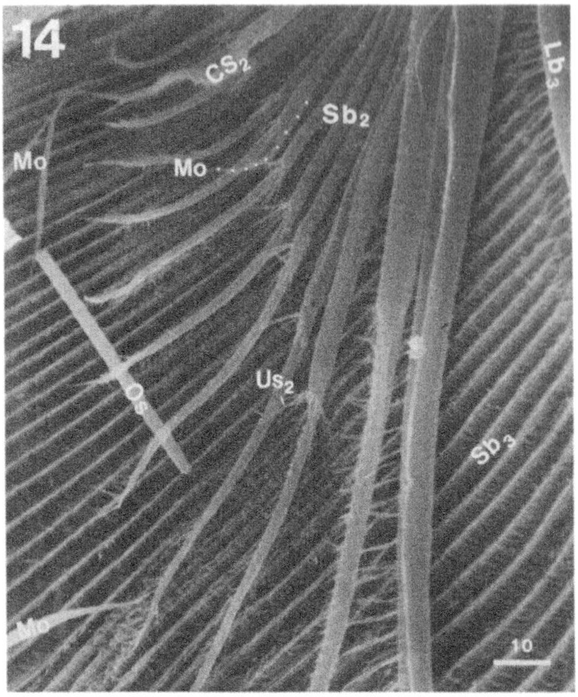

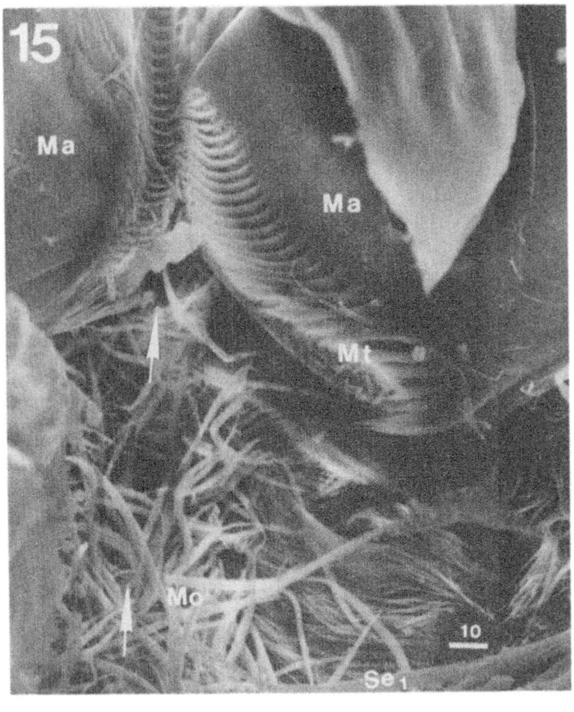

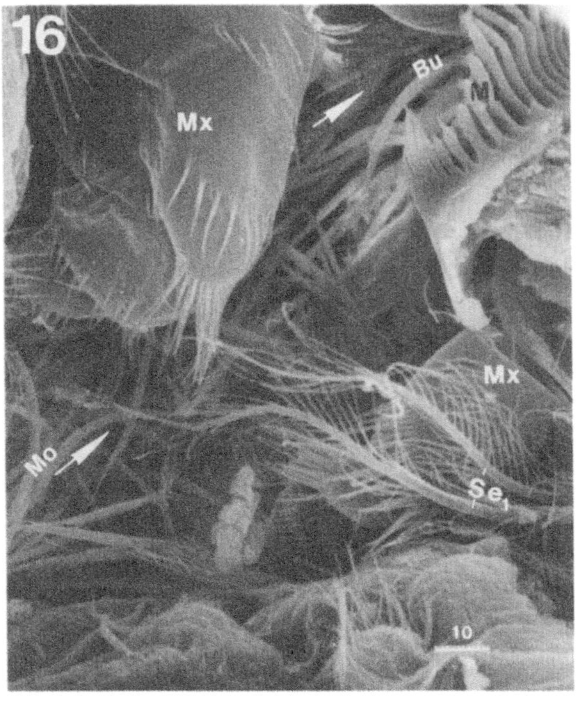

Fig. 13. *Monoraphidium* retention. Large cells are located on setae of limb 4, in a small *Daphnia pulicaria*. Gaps between the setae are artificial, due to dissection.

Fig. 14. Gathering of *Monoraphidium* cells and short *Oscillatoria* strands. Limb 2 is in its most ventral position. Cells (strands) are either positioned parallel or perpendicular to axis of food groove. The umbrella-brushes slip past one perpendicularly oriented *Monoraphidium* cell (dots mark the cell position).

Fig. 15. Handling and transport of *Monoraphidium*. The cells are only loosely and randomly compacted prior to reaching mandibles. Notice the foremost seta of limb 1 at bottom of the picture.

Fig. 16. Preparation of *Monoraphidium* cells for ingestion. Cells are randomly oriented posteriorly (bottom left), but bundled lengthwise anteriorly (top right). The two foremost setae of limb 1 are positioned above the right maxillule.

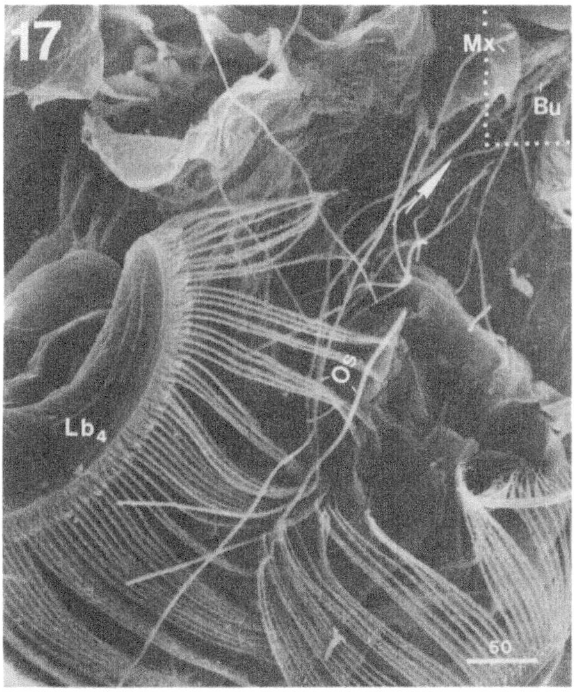

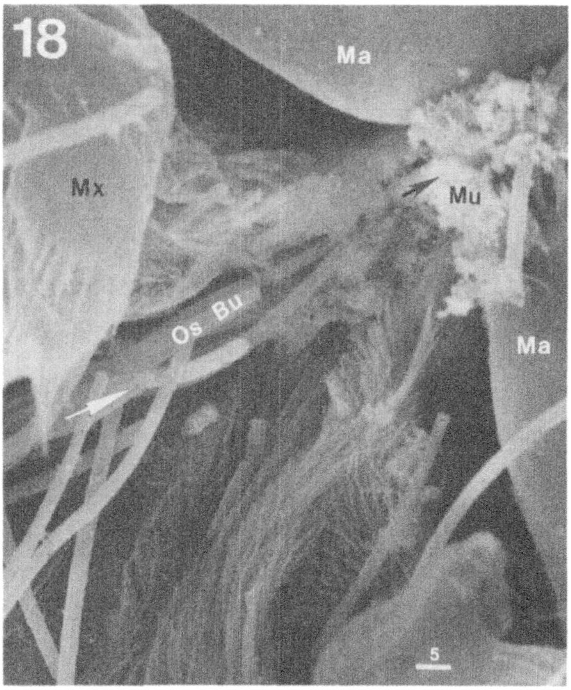

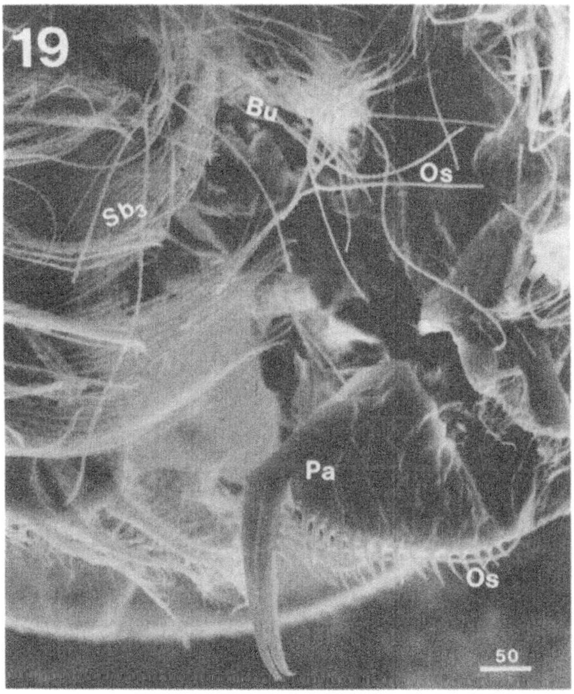

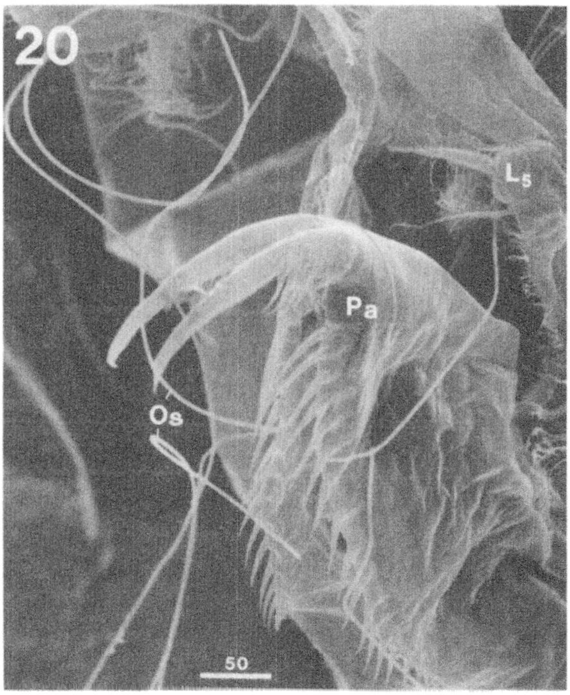

Fig. 17. *Oscillatoria* strand handling and ingestion. The strands are lined up spaghetti-like along the setae of limb 4 (center-bottom), then bundled in the anterior food groove (top right) and ingested (topmost right dashed corner, see Fig. 18). Limbs 1 to 3 have been removed, for visibility.

Fig. 18. Ingestion of *Oscillatoria* strands: same animal as in Fig. 17, showing close-up of mandibular region.

Fig. 19. *Oscillatoria* strand collection and entanglement. A loose strand-bundle is located in the posterior part of the food groove/feeding chamber. One strand is tangled among the postabdominal teeth.

Fig. 20. Entanglement with *Oscillatoria*. Strands are being tangled around the postabdominal claw and its teeth, and between limb 5 and the claw.

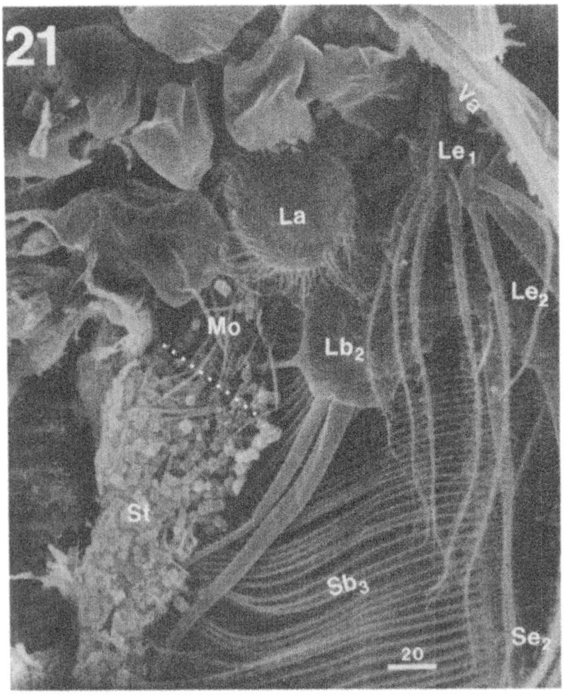

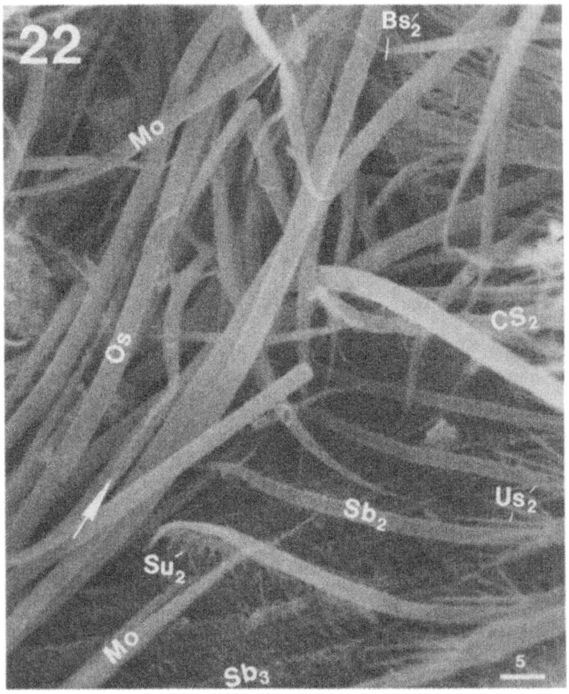

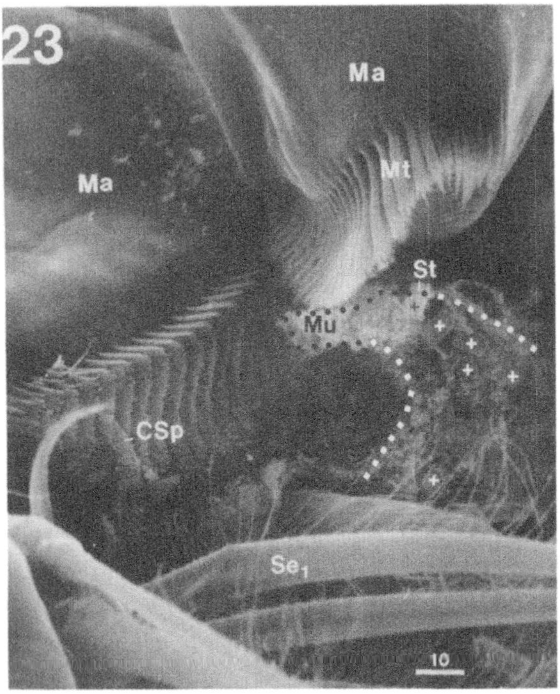

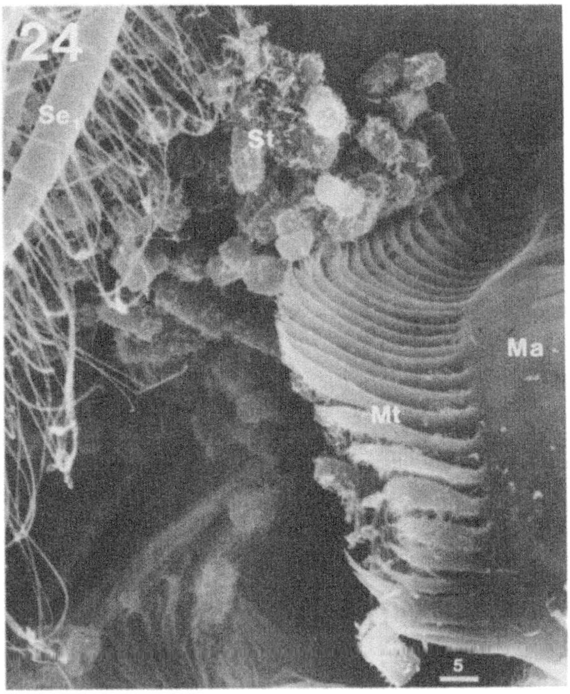

Fig. 21. Cell sorting by *Daphnia pulicaria*. *Stephanodiscus* are in the central part and most *Monoraphidium* in the anterior part of the food groove (separated by a dashed line). Limb 2 is in forward-dorsal position relative to limb 3, 'scooping' cells. Notice gaps between setae of limb 3 and penetration of gaps by posterior fan-setae of limb 2. The two foremost setae of limb 1 are reaching forward below the labrum (their distal tips are hidden from view).

Fig. 22. Mixed transport of *Monoraphidium* cells and *Oscillatoria* strands in the anterior part of the food groove. Most *Monoraphidium* cells are aligned lengthwise with the strands. Gnathobasal setae of limb 2 are enmeshed with cells and strands.

Fig. 23. Ingestion of bacteria. The food cluster (outlined by black dots) consists of bacteria, mucus and a few *Stephanodiscus* cells (+). Notice setae of limb 1 and combed spine (probably belonging to limb 1) brushing over teeth of left mandible.

Fig. 24. Ingestion of attached bacteria. A clump of *Stephanodiscus* cells with attached bacteria is ingested with the aid of the mandibles. Notice lack of mucus, as opposed to Fig. 23.

were highest for *Stephanodiscus* (SSC = 8), intermediate for *Monoraphidium* (SSC = 5), and lowest for *Oscillatoria* (SSC = 2). For grouped cells (clumps, bundles, boluses), the rank sequence was the same, but the distances between ranks were less (SSC = 11, 9, and 8.5, for *Stephanodiscus*, *Monoraphidium*, and *Oscillatoria*, respectively).

Discussion

The SEM observations clearly illustrate that the different algae have different degrees of association and distribution within *Daphnia pulicaria*'s feeding chamber and food groove. They must be captured, handled, and ingested in different ways. Despite equal initial cell densities, *Stephanodiscus* was preferentially retained in the feeding chamber (highest SSC score), followed by the elongate *Monoraphidium* cells and the *Oscillatoria* filaments. The experimental approach chosen here – feeding in appropriate food environments, then anaesthetizing and SEM preparation – has been criticized as being meaningless for drawing conclusions on the limbs' particle collection functions (Fryer, 1987). However, it must be noted that our method differed from Ganf & Shiel's (1985b): animals were gradually anaesthetized without transferring them to a new environment, such as soda water, and preparations were more gentle than usual standard methods. Furthermore, the distributions of particles on the limbs could not have arisen from haphazard attachment: particles were nearly all found between setae or limbs rather than resting on top of them, or else distributed along the food groove up to the mouth in ways they could not have been deposited there by chance. Almost no observations could be made without partial dissection, i.e. removal of the labrum or a part of the carapace. When cells were only loosely associated with limbs, e.g. Figs. 8 and 13, they served only for size comparisons. Finally, replicate observations removed the probability that our findings were artifacts. We consider that our approach has minimized Fryer's objections to the highest possible degree.

Limb functions

Limb 1

Although limb 1 lacks a developed gnathobase, it appears to have more diverse functions than previously proposed (i.e. Cannon, 1933; Fryer, 1968; Watts & Petri, 1981; Ganf & Shiel, 1985b). The basal spines of limb 1 are supposed to function as scrapers/cleaners/ejectors, working around the labrum and inside the carapace. The positions of those spines in *Daphnia pulicaria* suggest similar tasks (e.g. Figs. 2, 23). One particular task was directly observed: one combed spine appeared on the mandibular teeth (Fig. 23). Obviously, the spine either cleans the mandibles and/or pushes food (boluses, bundles, strands) into the mouth. It probably belonged to limb 1, since its distal combs resemble those described by Watts & Petri (1981).

The feathered setae of limb 1 do not act as a unit, as proposed by some observers (Smirnov, 1971), but have divided tasks: the two anterior ones have a relatively restricted action range: they reach under the labrum, and their tips handle food boluses and cell bundles (Figs. 15, 16). Only the remaining feathered setae have a wide range of action, laterally from the labrum back to the central median chamber. They are not in touch with food gathered in the median food groove. Smirnov (1971) suggested that they 'evidently take part in filtration', and Ganf & Shiel (1985b) propose a collection function for *D. carinata*. Since the only cells associated with those setae were *Stephanodiscus*, small spheroid cells may indeed occasionally be retained by the setae. Longer cells and strands, however, would not be retained, and should upon contact be redirected or rejected.

Finally, the outermost, distal, spinelike setae of limb 1 were seen to operate only along the outer edges of the feeding chamber. They are inserted a very distinct lobes, suggesting independent movement from the other setae. Overall, their function remains unclear. Ganf & Shiel (1985b) assign them a tactile and chemosensory function, without substantiation. The lack of setules suggest that they could easily slip underneath the valves, behind limbs 2 and 3, perhaps for cleaning.

Limb 2

The complexities of limb 2, its range of positions along the food groove and central feeding chamber, and its many associations with all cell types suggest that it plays a central role in food handling and selection. Previous accounts on the role of the foremost 3 setae of the gnathobasal fan have been deduced from comparative morphology, i.e. pushing and brushing (Watts & Petri, 1981; Ganf & Shiel, 1985b). Our observations suggest more precise functions: they help form boluses only of *Stephanodiscus* cells, but not of the elongate *Monoraphidium* cells or of *Oscillatoria* strands. Posterior to those 3 setae, *Stephanodiscus* cells are packed solid in the food groove (Fig. 10), underneath them the mass is breaking up (Fig. 11), and anterior to them, separate boluses appear (Fig. 10, dot-enclosed area). Conversely, *Monoraphidium* are not separated into boluses, and cells are being closely bundled only once situated in front of the mandibles (Figs. 15, 16), perhaps with the aid of the maxillules. *Oscillatoria* strands are being lined up along the entire food groove, and the three anterior bristled setae of L2's basal fan might help push them forward and bundle them, perhaps aided by mucus (Figs. 17, 18, 20).

The widening distal spacing of the fan's central setae, their restricted (distal) flexibility, and the forward/outward orientation of their setules, already noted by Cannon (1933) and Storch (1924) do not suggest a role in direct interception of particles (cf. Ganf & Shiel, 1985b). Most authors assign them a brushing function, i.e. 'brushing off food particles retained by the third limb' (Watts & Petri, 1981), without substantiation. Indeed, the forward/outward orientation of the setules would favor brushing only during a forward stroke relative to limb 3 (cf. Fig. 25). Brushing alone by those setae, however, cannot explain all observations: (i) brushing cannot be totally efficient, since elongate cells oriented parallel to the axis of a seta may be missed during the inward stroke (Fig. 14) despite the presence of the umbrella bristles; (ii) the umbrella bristles do not prevent cells from getting too close to the gnathobase, as suggested by Watts & Petri (1981); and (iii) brushing alone could not explain how cells can gather inside the fan (Fig. 21). Thus, as the fan rotates mouthward with respect to the gnathobasal setae of limb 3 (Fig. 25), it may assist cells traveling along L3's channeled mesh field toward the food groove, but simultaneously shovel up cells previously gathered in the food groove. This very complex task may require independent rotation of each fan-setule and also involve action of the separately inserted central seta.

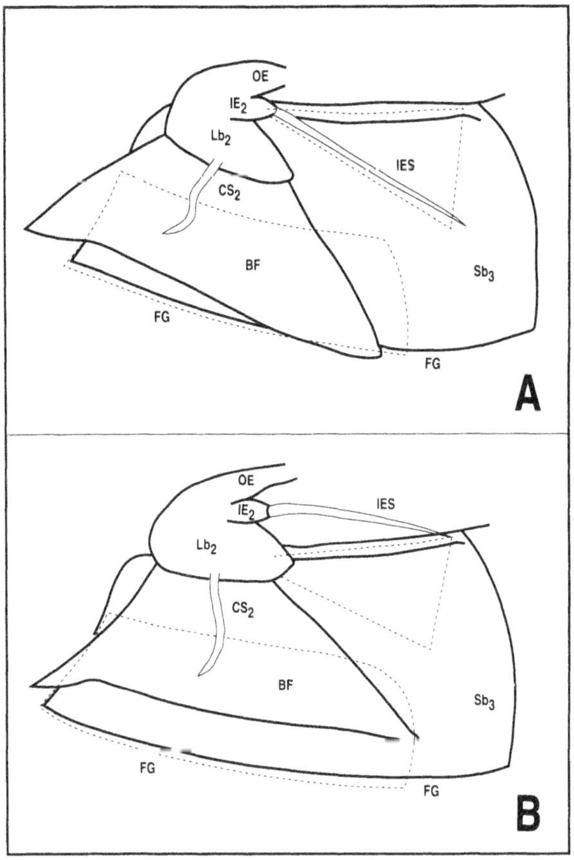

Fig. 25. Positions of limb 2 relative to the gnathobasal setae of limb 3 in *Daphnia pulicaria*. View along the sagittal plane. Stippled regions are areas swept by brushing of setules of limb 2. A. Forward/dorsal position. B. Posterior/ventral position. Individual setae are not shown except for CS_2 and IES. BF = gnathobasal fan of limb 2; FG = food groove; IE_2 = intermediate endite lobe; IES = seta of intermediate endite; OE = outer endite lobes; other abbreviations are given in the list for Figs. 1–24.

Our SEM photographs reveal for the first time that the basal fan setae of L2 are able to penetrate behind the setae of limb 3, effectively breaking gaps into the mesh field. Thus, they assist in dislodging or transporting cells that are near the distal ends of the basal setules of limb 4, assuring their collection in the food groove. However, we were not able to observe cells gathered by limb 4 under those conditions.

The endite setae of limb 2 are all feathered, as opposed to to *D. magna* (Watts & Petri, 1981), *D. longispina* (Smirnov, 1971), *D. obtusa* and *D. pulex* (Koza & Kořínek, 1985). Nevertheless, their tasks are differentiated. Only the innermost, median seta was seen in contact with the setae of limb 3, sweeping along the bases of those setae (Fig. 25). Its insertion on a separate lobe suggests that it has a wide and independent range of action. Since its wide setular spacing nearly corresponds to intersetal spacing of limb 3, it may help individual cells to move toward the food groove.

The remaining distal endite setae of limb 2 operate along the gnathobase of limbs 3 and 4 and in the central median food chamber. We did not observe them reaching behind limb 3, as suggested by Ganf & Shiel (1985b). Particle retention by those setae was proposed by Eriksson (1934), implied by Smirnov (1971), and mentioned by Gerritsen *et al.* (1988). Burns & Gilbert (1986) observed that they prevented rotifers from reaching the food groove. We hypothesize that they form in *Daphnia pulicaria* a dynamic, three-dimensional coarse-mesh field together with the feathered seta of limb 1, that may retain, redirect, or reject particles, depending on particle size and shape, and on flow conditions. We only observed *Stephanodiscus* cells associated with them, never *Monoraphidium* or *Oscillatoria*. *Stephanodiscus* cells are retained perhaps because they easily get caught in the setules, while larger or more elongate cells will get redirected or rejected upon setule contact. Recently, Herold (pers. comm.) reported that the relative setal length of all limbs of *D. longispina*, not just of limbs 3 and 4 (cf. Koza & Kořínek, 1985; Kořínek *et al.*, 1986) adjusted to different feeding conditions, indicating that limbs 1 and 2 could also play a role in particle collection by *Daphnia*.

Limbs 3 and 4

The setular meshes of *Daphnia pulicaria*'s gnathobasal setae of limbs 3 and 4 varied within ranges reported by others (Kořínek & Machácek, 1980; Gophen & Geller, 1984; Brendelberger & Geller, 1985; Kořínek *et al.*, 1986). Intersetal distances, however, were more variable than reported elsewhere from dissected limbs. These discrepancies emphasize the great degree of adaptability of setal length and total mesh field area found for *Daphnia pulicaria* as compared to other daphniids (Kořínek *et al.*, 1986). But the high variability also arose from the gaps forming between the setae (cf. Figs. 3, 6, 10, 21; compare, however, with artificial gaps caused by dissection: Figs. 1, 13, 17). The gaps are due to several phenomena: (i) Movement of the postabdominal claw into and out of the food groove during a rejection stroke (McMahon & Rigler, 1963; Watts & Petri, 1981; Porter *et al.*, 1982; Burns & Gilbert, 1986). We could not confirm this. (ii) Flexibility of setal tips: individual setae may get hung up on food masses in the food groove, especially when they move backwards against the direction of food transport. (iii) Action of one or several setae of the gnathobasal fan of limb 2 (Figs. 3, 6, 21).

Limb 3 was the only limb associated with all types of cells both single and grouped, and also with bacteria. Most of the single cells were found between the basal fan of limb 2 and the inner surface of the setae of limb 3. Very few were outside this region (i.e. Fig. 10). If the cells would have been deposited there as a result of anaesthetization, they would have been distributed much more regularly over the entire mesh surface. This clearly was not the case. The tips of the setae of limbs 3 and 4 most likely served to certain cell masses and strands close to the food groove. Our observations did not indicate very active mouthward 'sweeping' by the setae as proposed by Storch (1924) and Fryer (1987). Such forward sweeping can more readily be assigned to the basal fan of limb 2.

Feeding model

The predation model of Holling (1966) separates the feeding processes into four steps: probability directly without settling on the meshes (i.e. Hessen & Nordby, 1988). In this case, *Stephanodiscus* cells would be retained preferentially by the animal, since they are easily compacted into the food groove, while elongate cells and strands which are not oriented parallel with the food groove would likely be lost. Indeed, we rarely observed *Monoraphidium* and *Oscillatoria* in the posterior end of the food groove (e.g. Figs. 3, 21), while bulks of *Stephanodiscus* cells were always present here (Fig. 10). As a net result, spheroid cells like *Stephanodiscus* have a higher probability of retention during abstraction from the feeding currents than elongate cells and strands.

Particle transport to the mouth, handling (ingestion; Fig. 26C)

The remaining particle selection will then take place during their transport to the mouth. This occurs during several operations. First, cells earlier retained on limbs 3 and 4 must reach the food groove. We were not able to observe this process on limb 4. For limb 3, several processes contribute to transport. First, tangential flow, when it prevails, assists travel of all cells previously retained by sieving. Furthermore, the ventral-dorsal component of the sweeping action of the gnathobasal fan of limb 2 and its median feathered seta will preferentially gather *Stephanodiscus*, because elongate cells and strands will be missed by the sweeping action unless they are favorably oriented (e.g. Fig. 14). Second, once in the food groove, spheroid cells are better compacted than elongate cells, and therefore more spheroid cells can be transported per operation. Therefore, more *Stephanodiscus* cells should reach the mouth than *Monoraphidium* or *Oscillatoria* strands due to better *a priori* packing of cells in the food groove. Third, elongate cells and strands must be aligned in bundles before ingestion; such an operation especially for strands, appears complex. Indeed, *Monoraphidium* cells are aligned once close to the mouth, perhaps with the aid of the maxillules, while the long *Oscillatoria* strands must be aligned all along the food groove, certainly requiring complex cooperation of feeding limbs. Handling time during transport to the mouth thus increases with increasing particle length, and could be a major factor for *D. pulicaria*'s preferential feeding on *Stephanodiscus* (Fig. 26C).

The differential handling of cells during transport to the mouth may have other, indirect consequences. It may be responsible for the apparent cell sorting observed (Fig. 21). Furthermore, handling will induce loss of cells, either directly, as a result of the manipulations themselves, or indirectly, when the handling apparatus becomes clogged and cells are rejected with the post-abdominal claw. In this case, if cells are unsorted, not just the target cells are rejected, but also cells already prepared for ingestion. Cell sorting would be a mechanism to reduce risk of claw-rejection of desirable cells. Finally, differences in handling time for different cell types will decrease with increasing *Daphnia* size, since larger animals can easier accommodate elongate cells and strands and may even have acquired 'experience' (e.g. Price & Paffenhöfer, 1984).

Model summary and implications

The proposed food handling and selection processes are summarized in a series of four mechanical-behavioural hypotheses:

1. Feeding currents (Fig. 26A): outside the animal's carapace, only long (e.g. *Oscillatoria*) strands are prevented from entering the feeding chamber.
2. During food collection (Fig. 26B), small round cells are more likely to reach the food groove due to: (i) probability of retention among the outer feathered setae of limbs 1 and 2, and rejection of elongated cells and filaments; and (ii) their better compaction when directly reaching the food groove.
outer feathered setae of limbs 1 and 2, and rejection of elongated cells and filaments; and (ii) their better compaction when directly reaching the food groove.
3. During food transport to the mouth (Fig. 26C), more small round cells reach the mouth because: (i) spheroid cells are swept toward the food groove by the basal fan of limb 2 without loss, while collection of elongate cells

depends on their orientation with respect to the basal setae of limb 3; (ii) spheroid cells are densely compacted in the groove and can be separated into boluses; (iii) spheroid cells require less handling time, since they do not need to be reoriented and bundled for ingestion as do elongate cells; strands must be aligned along the entire food groove, requiring even more handling time; (iv) elongate cells and strands are directly lost during handling, a loss which exceeds indirect loss of spheroid cells from inadvertent rejection with those undesirable cells.

4. Rejection of strands from the feeding chamber or the food groove contains a risk of entanglement, thus increasing energy costs and decreasing feeding efficiency. The risk is significantly higher for juveniles than for adults.

The model lacks details on important functions for which we did not have replicate observations: capture and transport of cells along the gnathobasal setae of limb 4, the roles of endopodite setae of limbs 3 and 4 (e.g. Cannon, 1933), and of limb 5 (e.g. Gerritsen *et al.*, 1988). Nevertheless, the model helps to explain several selective behaviours of *Daphnia* observed in experiments:

1. Small *Daphnia* would select small rounded cells over elongate cells more significantly than large *Daphnia* because of higher probability of rejection of the latter in the feeding current, greater loss from the food groove, and increased handling time. Indeed, in double-label feeding experiments, juvenile *Daphnia pulicaria* fed more selectively on *Stephanodiscus* over *Monoraphidium* than did adults (Hartmann, 1985).
2. At very high food concentrations, the advantages for selecting the more spherical cells would decrease, because the animal must spend more time handling and transporting elongate cells, increasing the probability of inadvertent loss of already gathered spherical cells due to claw rejection. Scavia *et al.* (1984) observed increased loss of *Chlamydomonas* cells by *Daphnia* with increasing food concentration, confirming the indicated potential for cell loss at high food concentrations. In our double-label feeding experiments with *D. pulicaria* (Hartmann, 1985), we showed a general decrease in preference for *Stephanodiscus* at high food concentrations ($> 10^3$ to 5×10^4 cells ml^{-1}) in all *Daphnia* sizes. Those results were later confirmed in detail (Hartmann, 1987). Indeed, at very high food concentrations, the selection may well get reversed. Ganf & Shiel (1985a) reported that *D. carinata* preferred the elongate *Ankistrodesmus falcatus* over the spherical *Dictyosphaerium pulchellum* of comparable cell volume at a very high cell concentration (2×10^4 cells ml^{-1}). Inadvertent loss of already gathered spheres due to excessive *Ankistrodesmus* handling by *Daphnia carinata* may explain this.
3. The model also explains how *Daphnia* would prefer smaller over larger spherical algae, as found by Meise *et al.* (1985). Some of the smaller algae may be more easily intercepted by the feathered setae of limbs 1 and 2, and the small algae will be better compacted in the food groove than the larger ones.
4. The model furthermore predicts that *Daphnia* would have no problem ingesting moderately-sized filaments, perhaps in length not exceeding their carapace. Feeding efficiencies, however, would be reduced due to excessive handling time and poor interception of strands. Thus, small *Daphnia* would be especially vulnerable to entanglement with long strands. Food gathering rates measured by Hartmann (1987) of large *D. pulicaria* on moderate concentrations of *Oscillatoria* strands alone were low (mean = 0.10 ml an^{-1} hr^{-1} for a 2.4-mm animal) compared to feeding on *Monoraphidium* and *Stephanodiscus* (0.94 and 1.22 ml an^{-1} hr^{-1}, respectively), while rates for small *Daphnia* feeding on *Oscillatoria* alone were unusually high (Table 6). Furthermore, Hartmann (1985) showed that small *Daphnia* preferentially gathered *Oscillatoria* strands from *Monoraphidium*/*Oscillatoria* mixtures, while large *Daphnia* rejected those same strands in favor of *Monoraphidium*.

Our SEM observations now support in part the earlier interpretation of these results that the high food gathering rates and selective gathering of *Oscillatoria* strands by small *Daphnia* were due to entanglement, rather than actual ingestion. There remains, however, the possibility that *Oscillatoria* inhibited feeding due to toxicity; large *Daphnia* may have fed more *Oscillatoria* during prefeeding, therefore becoming more inhibited than small animals (W. DeMott, comm. *in lit.*). *D. pulicaria* were also reported to feed at much lower rates on the filamentous cyanobacterium *Cylindrospermopsis raciborskii* than on *Scenedesmus acutus* (Hawkins & Lampert, 1989). The comparatively high feeding rates on *Oscillatoria* strands reported by DeMott (1990), i.e. 2.5 ml an^{-1} hr^{-1} for adult *D. pulicaria*, may be due to the very low filament concentrations used in his experiments. Infante & Litt (1985) reported that *D. pulicaria* ingested strands of the diatom *Melosira italica*, but had difficulty handling them. Those strands are ingested in spaghetti fashion, as we show for *Oscillatoria* (Infante & Edmondson, 1985). Other *Daphnia* species have been shown to feed on single filaments of algae, sometimes with reduced efficiencies (Arnold, 1971; Geller, 1975; Porter & Orcutt, 1980; Holm *et al.*, 1983; Knisely & Geller, 1986; Fulton, 1988).

Table 6. Mean clearance rates of three sizes of *Daphnia pulicaria* feeding on *Stephanodiscus hantzschii* cells, *Monoraphidium setiforme* cells, and *Oscillatoria aghardii* strands alone, at low concentrations (below the respective incipient limiting food concentrations). The values are taken from data on regressions of clearance rates versus animal lengths in Hartmann (1987).

Daphnia size (mm)	Clearance Rate (ml an^{-1} h^{-1})		
	Stephanodiscus	*Monoraphidium*	*Oscillatoria*
2.40	1.22	0.94	0.10
1.40	0.23	0.12	0.25
0.90	0.06	0.02	0.56

Loss of cells during transport and handling, and differences in handling time have been acknowledged as potential factors, but have not been seriously investigated. Thus, Porter *et al.* (1983) speculated that residence time of particles within the filtering chamber may contribute to particle selection. Brendelberger (1985) admits that feeding rates could, outside of sieving, also be influenced by the efficiencies of transfer of retained food to the mouth. Gliwicz (1980) showed rather convincingly that large beads are more abundant in the food groove than in the gut, suggesting selection during food transport and handling. Recent results by Bern (1990) suggest that large spherical particles (plastic beads) are rejected near the mouth region, immediately before ingestion. Ringelberg (1988) demonstrated that processing of food limited the rate of *Chlorella* ingestion by *Daphnia magna*. Our observations suggest that food handling is an important precursor of cell ingestion, and perhaps more readily determines food selection and energy expenditure for *Daphnia* suspension feeding than the mesh size distribution and setulation of the thoracic limbs. The model rejects simple leaky-sieve feeding (i.e. no selection possible on large particles), and proposes that ingestion efficiency depends largely, but not exclusively, on cell handling and transport processes within the feeding chamber.

Feeding on bacteria and ultrafine particles

Bacteria were preferentially attached to the tips of setules (Table 5), or were associated with algae (Fig. 8) or with mucus in food boluses (Fig. 23). From those observations, we hypothesize different respective pathways of ingestion of free and attached ultrafine particles by *Daphnia*.

First, free bacteria can attach to tips of setules of limbs 3 and 4 as the feeding currents flow past them. Attachment may be due to surface charging (i.e. Gerritsen & Porter, 1982) or stickiness, and is a form of passive adherence. Indeed, when investigating bacterial feeding of Crustacea, Marvalin & Lazarek (1988) found that passively adhering bacteria constituted 9% of total activity of *Daphnia longispina* feeding on radiolabeled natural bacteria. Since they did not find bacteria adhering to *Daphnia*'s outside carapace, adhesion on the setules of feeding limbs could have been likely.

Transport of bacteria to the food groove would require their eventual detachment from the setules. Namely, there are ways of maintaining bacteria in the food stream. First, they may travel by passive 'setule-hopping' from tip to tip. Then, piggybacking on sticky larger cells would almost certainly assure ingestion, as we demonstrated for bacteria associated with *Stephanodiscus*. Finally, closer to the mouth, abundant mucus would be able to bind fine particles. Indeed, Hessen & Nordby (1988) detected acid mucopolysaccharides near the labral and mandibular glands of *Holopedium gibberum*. Other authors proposed packaging of food by sticky substances (Cannon, 1933; Watts & Young, 1980), but did not explicitly observe it. If the meshes of limbs 3 and 4 would be the only or major factors in food selection, then ingestion rates for bacteria and ultrafine particles by *Daphnia* should be much lower than commonly measured (e.g. Peterson *et al.*, 1978; Porter *et al.*, 1982; DeMott, 1982, 1985; Gophen & Geller, 1984; Hadas *et al.*, 1982; Kankaala, 1988; Marvalin & Lazarek, 1988). Again, as for algae, this suggests that more attention should be paid to selection or retention of particles during the transport phase rather than just during particle capture.

Secondly, bacteria that enter the feeding chamber already attached to other particles are easily transported to the mouth, as we could see for *Stephanodiscus* (Fig. 24). While free bacteria may only form a small percentage of energy intake in *Daphnia* feeding in nature (Kankaala, 1988), attached bacteria could be an important source of energy for daphniids feeding in detritus-rich environments (Schoenberg & McCubbin, 1985). Attached bacterial biomass has been found to exceed free bacterial biomass in eutrophic systems during periods of peak zooplankton production: Marvalin *et al.* (1989) reported that in eutrophic Lake Aydat, attached bacteria formed 50–60% of epilimnetic bacterial biomass in summer during peak cladoceran production and biomass. Algae were large, gelatinous, and colonial (Marvalin, *op. cit.*), yet *Daphnia longispina* was the dominant crustacean (Lair, 1990). Thus, in nature, attached bacteria may be more important for daphniids than free bacteria, since free bacteria tend to dominate at times when easily handled and nutritious phytoplankton food is abundant (Marvalin *et al.*, 1989).

In summary, our observations and results strongly suggest that selective feeding of *Daphnia* on larger particles is the result of a series of complex behavioural-mechanistic processes. Other factors, such as surface charge, mucus, and stickiness, appear to play a role in feeding on smaller, bacteria-sized particles. Processing of food by *Daphnia* and other Cladocerans beyond simple filtering or initial particle capture deserves further study, before general food selection models can be quantified.

Acknowledgements

Contribution Nr 831 from the School of Fisheries, WH-10, University of Washington. This project was partially supported from the U.S. Department of Interior and Washington State Water Research Center Grants B-089-WASH and B-0124-WASH to F.B. Taub and H.J. Hartmann. The Friday Harbor Laboratories of the University of Washington provided SEM facilities and laboratory space at a critical phase of the project. Manuscript preparation was in part supported by the Université Blaise Pascal, Clermont-Ferrand, France.

Thanks to F.B. Taub, W.T. Edmondson, M. Gophen, R. Shiel, and W. Geller for valuable discussions and information, and to W.R. DeMott and D.G. Frey for a critical review of an earlier manuscript. A. Litt helped with field collections. T. MacFarlin assisted with specimen preparation, and P. Schwartz facilitated darkroom use.

References

Arnold, D. E., 1971. Ingestion, assimilation, survival, and reproduction by *Daphnia pulex* fed seven species of blue-green algae. Limnol. Oceanogr. 16: 906–920.

Bern, L., 1990. Postcapture particle size selection by *Daphnia cucullata* (Cladocera). Limnol. Oceanogr. 35: 923–926.

Bogdan, K. G. & D. C. McNaught, 1975. Selective feeding by *Diaptomus* and *Daphnia*. Verh. int. Ver. Limnol. 19: 2935–2942.

Boyd, C. M., 1976. Selection of particle sizes by filter feeding copepods: a plea for reason. Limnol. Oceanogr. 21: 175–180.

Brendelberger, H., 1985. Filter mesh-size and retention efficiency for small particles: comparative studies with Cladocera. Arch. Hydrobiol. Beih. Ergebn. Limnol. 21: 135–146.

Brendelberger, H. & W. Geller, 1985. Variability of filter structures in eight *Daphnia* species: mesh sizes and filtering areas. J. Plankton Res. 7: 473–486.

Brendelberger, H., M. Herbeck, H. Land & W. Lampert, 1986. *Daphnia*'s filters are not solid walls. Arch. Hydrobiol. 107: 197–202.

Burns, C. W., 1968a. Direct observations of mechanisms regulating feeding behavior of *Daphnia* in lakewater. Int. Revue ges. Hydrobiol. 53: 83–100.

Burns, C. W., 1968b. The relationship between body size of filter-feeding Cladocera and the maximum size of particles ingested. Limnol. Oceanogr. 13: 675–678.

Burns, C. W. & J. B. Gilbert, 1986. Direct observations of the mechanisms of interference between *Daphnia* and *Keratella cochlearis*. Limnol. Oceanogr. 31: 859–866.

Cannon, H. G., 1933. On the feeding mechanism of Branchipoda. Phil. Trans. r. Soc., Lond., Ser. B. 222: 267–352.

DeMott, W. R., 1982. Feeding selectivities and relative ingestion rates of *Daphnia* and *Bosmina*. Limnol. Oceanogr. 27: 518–527.

DeMott, W. R., 1985. Relations between filter mesh-size, feeding mode, and capture efficiency for cladocerans feeding on ultrafine particles. Arch. Hydrobiol. Beih. Ergebn. Limnol. 21: 125–134.

DeMott, W. R., 1988. Discrimination between algae and artificial particles by freshwater and marine copepods. Limnol. Oceanogr. 33: 397–408.

DeMott, W. R., 1989. Optimal foraging theory as a predictor of chemically mediated food selection by suspension feeding copepods. Limnol. Oceanogr. 34: 140–154.

DeMott, W. R., 1990. Retention efficiency, perceptual bias, and active choice as mechanisms of food selection by suspension feeding zooplankton. In R. N. Hughes (ed.), Behavioural Mechanisms of Food Selection. NATO ASI Series A: Life Sciences. Springer Verlag, Heidelberg; New York: 569–594.

Edmondson, W. T. & A. H. Litt, 1982. *Daphnia* in Lake Washington. Limnol. Oceanogr. 27: 272–293.

Eriksson, S., 1934. Studien über die Fangapparate der Branchiopoden. Zool. Bidr. Upps. 15: 21–287.

Flood, P., 1973. A simple technique for the prevention of loss or damage to planktonic specimens during preparation for transmission and scanning electron microscopy. Sarsia 54: 67–74.

Fryer, G., 1968. Evolution and adaptive radiation in the Chydoridae (Crustacea: Cladocera): a study in comparative functional morphology and ecology. Phil. Trans. r. Soc., Lond., Ser. B. 254: 221–385.

Fryer, G., 1987. The feeding mechanisms of the Daphniidae (Crustacea: Cladocera): recent suggestions and neglected considerations. J. Plankton Res. 9: 419–432.

Fulton, R. S., 1988. Grazing on filamentous algae by herbivorous zooplankton. Freshwat. Biol. 20: 263–271.

Ganf, G. G. & R. J. Shiel, 1985a. Feeding behaviour and limb morphology of two cladocerans with small intersetular distances. Aust. J. mar. Freshwat. Res. 34: 49–84.

Ganf, G. G. & R. J. Shiel, 1985b. Particle capture by *Daphnia carinata*. Aust. J. mar. Freshwat. Res. 34: 371–381.

Geller, W., 1975. Die Nahrungsaufnahme von *Daphnia pulex* in Abhängigkeit von der Futterkonzentration, der Temperatur, der Körpergrösse und dem Hungerzustand der Tiere. Arch. Hydrobiol. 48: 47–107.

Geller, W. & C. Knisely, 1988. Drag forces and energetic costs in *Daphnia* filter-feeding. (Abstract). Verh. int. Ver. Limnol. 23: 2061.

Geller, W. & M. Müller, 1981. The filtration apparatus of Cladocera: filter mesh-sizes and their implications on food selectivity. Oecologia 49: 316–321.

Gerritsen, J. & K. G. Porter, 1982. The role of surface chemistry in filter feeding by zooplankton. Science 216: 1225–1227.

Gerritsen, J., K. G. Porter & J. R. Strickler, 1988. Not by sieving alone: observations of suspension feeding in *Daphnia*. Bull. mar. Sci. 43: 366–367.

Glagolev, S. M., 1983. Morphology of thoracic limbs of some species of the genus *Daphnia* and its importance for the systematics of the genus [in russian]. In N. N. Smirnov (ed.), Biocenosy mezotrofnogo ozera Glubokogo. Moscow: 61–93.

Gliwicz, Z. M., 1980. Filtering rates, food size selection, and feeding rates in Cladocerans – another aspect of interspecific competition in filter-feeding zooplankton. In W. C. Kerfoot (ed.), Evolution and Ecology of Zooplankton Communities. The University Press of New England, Hanover (N.H.); Lond.: 282–291.

Gliwicz, Z. M. & E. Siedlar, 1980. Food size limitation and algae interfering with food collection in *Daphnia*. Arch. Hydrobiol. 88: 155–177.

Gophen, M. & W. Geller, 1984. Filter mesh size and food particle uptake by *Daphnia*. Oecologia 64: 408–412.

Hadas, B. Z., Y. K. Cavari, Y. Kott & U. Bachrach, 1982.

Preferential feeding behavior of *Daphnia magna*. Hydrobiologia 89: 49–52.

Hartmann, H. J., 1985. Feeding of *Daphnia pulicaria* and *Diaptomus ashlandi* on mixtures of unicellular and filamentous algae. Verh. int. Ver. Limnol. 22: 3178–3183.

Hartmann, H. J., 1987. Effects of differential grazing and phosphorus recycling on phytoplankton community dynamics. Ph.D. Thesis, Univ. of Washington, Seattle, WA, USA, 326 pp.

Hawkins, P. & W. Lampert, 1989. The effect of *Daphnia* body size on filtering rate inhibition in the presence of a filamentous cyanobacterium. Limnol. Oceanogr. 34: 1084–1089.

Hessen, D. O. & O. Nordby, 1988. Limb morphology and the process of particle capture in the cladoceran *Holopedium gibberum* Zaddach. Verh. int. Ver. Limnol. 23: 2038–2044.

Holling, C. S., 1966. The functional response of invertebrate predators to prey density. Mem. entom. Soc. Can. 48: 1–86.

Holm, N. P., G. G. Ganf & J. Shapiro, 1983. Feeding and assimilation rates of *Daphnia pulex* fed *Aphanizomenon flos-aquae*. Limnol. Oceanogr. 28: 677–687.

Infante, A. & W. T. Edmondson, 1985. Edible phytoplankton and herbivorous zooplankton in Lake Washington. Arch. Hydrobiol. Beih. Ergebn. Limnol. 21: 161–171.

Infante, A. & A. H. Litt, 1985. Differences between two species of *Daphnia* in the use of 10 species of algae in Lake Washington. Limnol. Oceanogr. 30: 1053–1059.

Kankaala, P., 1988. The relative importance of algae and bacteria as food for *Daphnia longispina* (Cladocera) in a polyhumic lake. Freshwat. Biol. 19: 285–296.

Knisely, K. & W. Geller, 1986. Selective feeding of four zooplankton species on natural lake phytoplankton. Oecologia 69: 86–94.

Kořínek, V. & J. Machácek, 1980. Filtering structures of Cladocera and their ecological significance. I. *Daphnia pulicaria*. Věst. čs. Společ. zool. 44: 213–218.

Kořínek, V., B. Krepelová-Machácková & J. Machácek, 1986. Filtering structures of Cladocera and their ecological significance. II. Relation between the concentration of seston and size of filtering combs in some species of the genera *Daphnia* and *Ceriodaphnia*. Věst. čs. Společ. zool. 50: 244–258.

Koza, V. & V. Kořínek, 1985. Adaptability of the filtration screen in *Daphnia*: another answer to the selective pressure of the environment. Arch. Hydrobiol. Beih. Ergebn. Limnol. 21: 193–198.

Lair, N., 1990. Effects of invertebrate predation on the seasonal succession of a zooplankton community: a two year study in Lake Aydat, France. Hydrobiologia 198: 1–12.

Lampert, W., 1987. Feeding and nutrition in *Daphnia*. Mem. Istit. ital. Idrobiol. 45: 143–192.

Lampert, W., W. Fleckner, H. Rai & B. E. Taylor, 1986. Phytoplankton control by grazing zooplankton: a study on the spring clear-water phase. Limnol. Oceanogr. 31: 478–490.

Marvalin, O. & S. Lazarek, 1988. Improved method for determining bacterial filtration in zooplankton. Appl. envir. Microbiol. 54: 2149–2151.

Marvalin, O., L. Aleya, H. J. Hartmann & S. Lazarek, 1989. Coupling of the seasonal patterns of bacterioplankton and phytoplankton in a eutrophic lake. Can. J. Microbiol. 35: 706–712.

McMahon, J. W. & F. H. Rigler, 1963. Mechanisms regulating the feeding rate of *Daphnia magna* Strauss. Can. J. Zool. 41: 321–332.

Meise, C. J., W. R. Munns Jr. & N. G. Hairston Jr., 1985. An analysis of the feeding behavior of *Daphnia pulex*. Limnol. Oceanogr. 30: 862–870.

Murphy, J. S., 1970. A general method for the monoxenic cultivation of the Daphnidae. Biol. Bull. 139: 321–332.

Nygaard, G., 1977. New or interesting plankton algae. The Royal danish Academy of Sciences and Letters, Botany, Zoology and General Biology. Publn 21: 52–107.

Peterson, B. J., J. E. Hobbie & J. F. Haney, 1978. *Daphnia* grazing on natural bacteria. Limnol. Oceanogr. 23: 1039–1044.

Philippova, T. G. & A. L. Postnov, 1988. The effect of food quantity on feeding and metabolic expenditure in Cladocera. Int. Revue ges. Hydrobiol. 73: 601–615.

Porter, K. G. & R. McDonough, 1984. The energetic cost of response to blue-green algal filaments by cladocerans. Limnol. Oceanogr. 29: 365–369.

Porter, K. G. & J. D. Orcutt Jr., 1980. Nutritional adequacy, manageability, and toxicity as factors that determine the food quality of green and blue-green algae for *Daphnia*. In W. C. Kerfoot (ed.), Evolution and Ecology of Zooplankton Communities. The University Press of New England, Hanover (N.H.); Lond.: 268–281.

Porter, K. G., Y. S. Feig & E. F. Vetter, 1983. Morphology, flow regimes, and filtering rates of *Daphnia*, *Ceriodaphnia*, and *Bosmina* fed natural bacteria. Oecologia 58: 156–163.

Porter, K. G., J. Gerritsen & J. D. Orcutt Jr., 1982. The effect of food concentration on swimming patterns, feeding behavior, ingestion, assimilation, and respiration by *Daphnia*. Limnol. Oceanogr. 27: 935–949.

Price, H. J., 1988. Feeding mechanisms in marine and freshwater zooplankton. Bull. mar. Sci. 43: 327–343.

Price, H. J. & G.-A. Paffenhöfer, 1984. Effects of feeding experience in the copepod *Eucalanus pileatus*: a cinematographic study. Mar. Biol. 84: 35–40.

Ringelberg, J., 1988. Clearance and ingestion in *Daphnia* during the first half hour of feeding after starvation. Verh. int. Ver. Limnol. 23: 2063–2066.

Rubenstein, D. I. & M. A. R. Koehl, 1977. The mechanisms of filter feeding: some theoretical considerations. Am. Nat. 111: 981–984.

Scavia, D., G. L. Fahnenstiel, J. A. Davis & R. G. Kreis Jr., 1984. Small-scale nutrient patchiness: some consequences and a new encounter mechanism. Limnol. Oceanogr. 29: 785–793.

Schoenberg, S. A. & A. E. MacCubbin, 1985. Relative

feeding rates on free and particle-bound bacteria by freshwater macrozooplankton. Limnol. Oceanogr. 30: 1084–1090.

Shiel, R. J., G. G. Ganf & J. Gormley, 1988. Particle capture by *Daphnia*: evidence from high-speed microcinematography. (Abstract). Verh. int. Ver. Limnol. 23: 2060.

Smirnov, N. N., 1971. Morpho-functional grounds of mode of life of Cladocera. V. Morphology and adaptive modifications of trunk limbs of Anomopoda. Hydrobiologia 37: 317–345.

Storch, O., 1924. Morphologie und Physiologie des Fangapparates der Daphniden. Ergebn. Fortschr. Zool. 6: 125–234.

Taub, F. B., 1985. Toward interlaboratory (round-robin) testing of a standardized aquatic microcosm. In J. Cairns, Jr, (ed.), Multispecies Toxicity Testing. Pergamon Press, Oxford, U.K.: 165–186.

Taub, F. B. & A. N. Dollar, 1964. A *Chlorella-Daphnia* food-chain study: the design of a compatible chemically defined culture medium. Limnol. Oceanogr. 9: 61–74.

Watts, E. & M. Petri, 1981. A scanning electron microscope study of the thoracic appendages of *Daphnia magna* Strauss. J. nat. Hist., Lond. 15: 464–473.

Watts, E. & S. Young, 1980. Components of *Daphnia* feeding behavior. J. Plankton Res. 2: 203–212.

Webster, K. E. & R. H. Peters, 1978. Some size-dependent inhibitions of larger cladoceran filterers in filamentous suspensions. Limnol. Oceanogr. 23: 1238–1245.

Hydrobiologia **225**: 155–167, 1991.
V. Kořínek & D. G. Frey (eds), Biology of Cladocera.

Feeding behaviour and morphology of filtering combs of *Daphnia galeata*

Evžen Stuchlík
Department of Hydrobiology, Charles University, Viničná, CS-128 44 Prague 2, Czechoslovakia

Key words: feeding behavior, filtering combs, *Daphnia*

Abstract

The population of *Daphnia galeata* Sars from the fish pond Velký Pálenec (Blatná, Czechoslovakia) living in high food conditions (7 mgC l^{-1}) was characterized by a small size of the filtering comb on the thoracic limb 3, measured as seta length, length of the base of the comb and number of setae (population 1). One month cultivation of this population in low food conditions (1.5 mgC l^{-1}) in the laboratory resulted in twofold increase in size of the filtering comb (population 2). Filtering and ingestion rates of both populations were measured at eight concentrations of food (approximately 0.025–3.2 mgC l^{-1}) using ^{14}C labeled *Scenedesmus acutus*. The results show that size of the filtering combs influences considerably feeding behavior of *Daphnia*. The comparison of animals with the same body length suggests that the population with a large comb feeds at concentration of food below 0.4 mgC l^{-1} more intensively and reaches the maximum of the filtering rate at a lower concentration than the population with a small comb. The situation is opposite at concentration above 0.4 mgC l^{-1}. The higher values of theoretical flow in population with a small projection of filtering area suggest that this population has to compensate disadvantage of a small comb with the higher appendages beat frequency.

Introduction

Egloff & Palmer (1971) first showed the possible positive correlation between the area of the filtering combs of limbs 3 and 4 and the filtering rate in *Daphnia*. They suggested a very complex approach to investigation of morphology and feeding behavior that had unfortunately only a few followers. Nevertheless very important discoveries were made in the field of morphology of the filtering combs of the thoracic limbs concerning a large scale change of these structures under variable food conditions (Kořínek *et al.*, 1981, 1986; Koza & Kořínek, 1985). The aim of this study was to find appropriate consequences of the change of morphology of the filtering comb on feeding behavior of *Daphnia galeata*.

Materials and methods

The experiments presented in this paper were conducted at the Hydrobiological station of Charles University near the town Blatná in south-western Bohemia from August to September of 1986.

Experimental animals

Two populations of *Daphnia galeata* Sars were used in experiments. The first population (population 1) was collected from the fish pond Velký Pálenec (area 31 ha, mean depth 1.4 m) August 30, 1986, 24 hours before the experiment. Mean concentration of seston varied about 7 mgC l^{-1}

in this pond. Approximately 300 animals were immediately separated from other zooplankton with a wide pipette and placed in water from the locality filtered through a 39 μm mesh, at a temperature 20 °C and natural illumination. The second population (population 2) coming from the same lake was collected August 17, 1986, kept under laboratory conditions at a temperature of 20 °C and natural illumination and fed with *Scenedesmus acutus* at a concentration of about 1.5 mgC l^{-1} till September 21, 1986 when the experiment was carried out.

Feeding suspension

The axenic culture of *Scenedesmus acutus* Mayen used for preparation of both active and non-active feeding suspensions was received from the collection of algae of the Institute of Botany, Czechoslovak Academy of Sciences, Třeboň.

Non-active suspension was produced on Bold basal medium – Bristol (Nichols, 1975) in a semicontinual sterile cultivator (a glass tube 60 cm long and 6 cm in diameter with two parallel fluorescent tubes, aeration and peristaltic pump for input of fresh medium) and harvested daily. To be received more easily, the required concentration of food was prepared according to the calibration graph between extinction (5 cm cuvette, 750 nm) and total particulate volume (TPV, $mm^3\ l^{-1}$) of cells in the peak of size distribution, and between extinction and organic carbon (C, mg l^{-1}). In the log-log plot the relation was linear in the range of concentrations 1–100 mgC l^{-1}. The same size distribution is essential for validity of this relation. I took care the feeding suspension was composed only from single cells; in other case, sonication was used to destroy two or four cell cenobia. Besides, the feeding suspension used in the experiment was checked directly for content of organic carbon and size distribution.

A Coulter counter – model ZB extended with an 18 channel analyzer of our own construction, microcomputer and dot-matrix printer was used to analyze cell-size distribution and TPV of non-active algal culture (Fig. 1). Samples were gently prefixed with the Lugol solution (100 μl to 100 ml) at first and then a solution containing NaCl and formalin was added. Final concentrations of NaCl and formalin in samples were 2% and 0.9%, respectively. The electrolyte with formalin was filtered carefully through nitro-cellulose membrane filters Synpor® (Barvy a laky, Praha), finishing this procedure with a 0.3 μm pore sized filter (Synpor No 7).

Organic carbon was estimated by multiplication of the value of dichromate reducing capacity of organic material by factor 3 according to Mackereth *et al.* (1978).

Active suspension of *Scenedesmus acutus* was prepared from cultures growing on Bristol in 100 ml Erlenmayer flasks under controlled axenic conditions. Flasks were placed in a shaker at 20 °C, under continuous illumination of 1000 Lux. About one week before the experiment an appropriate volume of the culture was resuspended in a flask filled with the minimum maintenance medium (MMM) prepared by

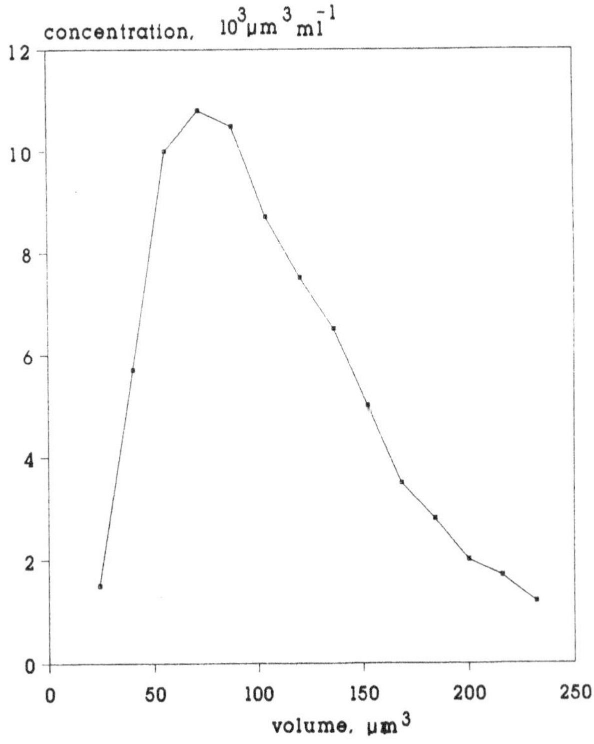

Fig. 1. Arithmetical size spectrum of the culture of *Scenedesmus acutus* (single cells) used in experiments.

Thompson *et al.* (1982). This medium is free of inorganic carbon and pH 8.6 is maintained by TRIS. After some days 5.5–11 kBq ^{14}C-labelled $NaCO_3$ was added to 50 ml of the culture. Immediately before the beginning of the experiment, the labelled culture was checked under microscope (if necessary, sonication was carried out), centrifuged three times (15 min, 2000 RPM) and redispersed again in the fresh MMM. After this procedure all activity was in cells only. Extinction was measured and concentration of the basic suspension was calculated from the calibration graph; required concentrations were prepared then by diluting the basic suspension. Mean activity in experiments ranged, depending on food concentration, from 16.7 to 200 Bq ml^{-1}.

Diluting water for the experiment was taken from the fish pond Smyslov, near the Hydrobiological station, which was in a clear water phase during the experimental period. Water was filtered through a Whatman GF/C filter and a series of Synpor filters, diluted ten times with distilled water to decrease the content of dissolved organic carbon and finally filtered through Synpor No 7. After several hours of aeration in a thermostat the temperature of the water was 20 °C, pH 7.1, calcium about 7 mg l^{-1}, DOC 0.9 mg l^{-1} and the water was saturated with oxygen to 100%.

Experimental design

Eight hours before the planned start of each experiment, the logarithmic scale of eight concentrations of the feeding suspension (approximately 3.2, 1.6, 0.8, 0.4, 0.2, 0.1, 0.05 and 0.025 mgC l^{-1}) was prepared using non-active suspension and diluting water. About 150 individuals of *Daphnia* were transported with a pipette to 120 ml narrow necked stoppered glass bottles containing feeding suspensions through several 'washing baths' with diluting water. *Daphnia* were sorted to have all size groups proportionally represented in each concentration. There were 15–20 animals in a bottle to keep the possible effect of crowding on the filtering rate constant (Helgen, 1987). Bottles were not closed with stoppers but their narrow necks protected *Daphnia* from catching on the surface of the water. Then all bottles were put into the thermostat (20 °C, 1000 Lux) for *Daphnia* adaptation to a new concentration of food. Time of adaptation was set according to data of Muck & Lampert (1984) who found a dramatic increase in the filtering rate in animals transferred from high to low concentrations of food, with the maximum two hours after the change. During the adaptation the content of the bottles was gently mixed with a pipette several times to protect sedimentation of cells. After 4–6 hours the bottles were brought to the laboratory where the same experimental conditions were arranged (20 °C, 1000 Lux). Approximately 30 minutes before the experiment all bottles were mixed again, almost the whole volume of feeding suspension were gently sucked off from each bottle (with a 200 ml capacity syringe with a rubber tube, covered with a 100 μm mesh on the mouth) and refilled with a fresh feeding suspension. This step was essential to return oxygen saturation and concentration of food to original values.

The proper measuring of feeding activity took place at midnight to avoid the influence of diurnal variations of feeding behavior (Starkweather, 1983) though in experiments with *Daphnia* from shallow lakes no significant diurnal rhythms were found (Stuchlík, 1982). The exposure started with replacement of an appropriate volume of the feeding non-active suspension with the ^{14}C-labelled one. The time of exposure was 6 minutes for concentrations 3.2–0.4 mgC l^{-1}, 20–30 minutes for 0.2–0.1 mgC l^{-1} and 40–60 minutes for concentrations 0.05–0.025 mgC l^{-1}. As it follows from published data (Geller, 1975; Muck & Lampert, 1984) and from conducted methodical experiment (Fig. 2), these times of exposure did not probably exceed the gut passage time. During the exposure the content of the bottles was mixed gently several times to protect sedimentation of food. After a fixed time period the content of each bottle was filtered through a sieve with 99 μm mesh. *Daphnia* remaining on the sieve were narcotized in soda water, gently but thoroughly

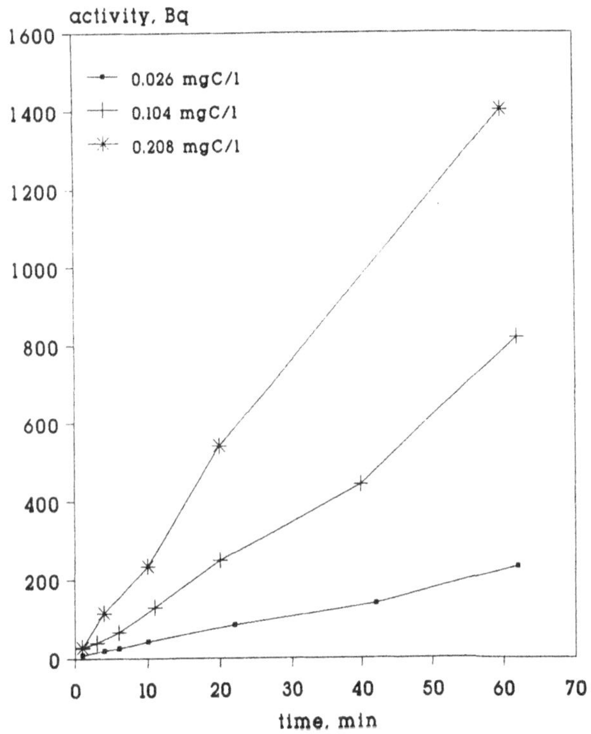

Fig. 2. Dependence of activity of adult *Daphnia pulicaria* on time of exposure at various concentrations of food.

rinsed with soda water again and placed with a pipette into Petri dishes on small GF/C filters with a diameter of 8 mm which layed on a wet filter paper. Closed dishes were stored in a refrigerator. 250 μl of the feeding suspension was pipetted into the vial and mixed with 1 ml of NCS tissue solubilizer (Amersham). Alternatively 1 ml of the feeding suspension was filtered through a GF/C filter to assure the measurable part of activity was not released from the labelled cells.

When the experiment was finished but not later than in two following days, body length of animals without a spine was measured under a microscope, directly on the wet small GF/C filters that are sufficiently transparent. The whole filter with *Daphnia* was then put into the vials. An appropriate volume of the tissue solubilizer (0.4 ml as a rule) was added immediately. 3–5 filters with a drop of water but without *Daphnia* were used as a blank.

All samples with the solubilizer were heated to 50 °C overnight. After completing 10 ml of the scintillation cocktail SLT 61 (Spolana Neratovice), the activity of samples was measured by the liquid scintillation counter Tri-Carb 300C (Packard Canberra).

The proposal of the method for storing *Daphnia* in a refrigerator within a period before measuring is based on the supposition that labelled cells in the gut or in the food grove of *Daphnia* will lose their activity more slowly than after preservation with formalin, ethanol or Lugol solution. Several experiments were conducted to find time dependent losses of activity. The results showed that the percentage of losses is linear in time and can be described with the following relation:

$$\% \text{ of losses} = 12.44 + 10.36 \log t,$$

where t = time in hours. Correction factors calculated according to this relation ranged from 1.35 to 1.44 for the presented experiments. In spite of the fact that the losses were lower than in formalin or ethanol (Holtby & Knoechel, 1981; Mourelatos, 1990), the method does not satisfy requirements regarding the losses of activity and can be recommended only if animals are measured within one hour. In this case losses do not exceed 15%. Manipulation with animals laying on small GF/C filters is extremely fast and easy. In any case the filters protect against next contingent losses of activity in particles that can be released from gut and food grove. Handling animals on filters could be combined with rapid freezing on dry ice (Sierzen & Waltras, 1987) or with freezing and lyophilization (Berberovic & Pinto-Coelho, 1989), thus providing a good tool for individual

Filtering rate (FR) was calculated according to the following relation:

$$FR = \frac{(A_D - B_D) \cdot 60}{(A_{FS} - B_{FS}) \cdot t} \text{ (ml ind}^{-1}\text{ h}^{-1}\text{)},$$

where A_D and A_{FS} are activities in *Daphnia* and feeding suspension (ml^{-1}), resp., B_D and B_{FS} are appropriate blanks and t is time of exposure in minutes. Ingestion rate is then:

$$IR = FR \cdot c \text{ (mgC ind}^{-1}\text{ h}^{-1}\text{)},$$

where c is concentration of food in mgC l^{-1}.

Morphology of the filtering screen

Total body length, length of the base of the comb (comb length), seta length and number of setae of the filtering comb of the thoracic limb 3 within the whole range of size of both populations were measured under a microscope. Usually about 12 animals were used. Parameters listed above were found to be the most important for the change of projection of filtering area (Kořínek *et al.*, 1986). Relative (size specific) comb length and seta length recognized as closed to constant for a population (Pop, 1991) were then calculated.

Results

Experimentally induced change of the filtering screen

Transferring *Daphnia galeata* from the fish pond, with high concentration of food and with heavy fish predation to a laboratory and, 35 days of cultivation under conditions of relatively low concentration of food and without predation, changed the original population completely. Compared populations 1 and 2 differ significantly both in range of size and length-weight relationship (Novotná, 1987) and especially in morphology of the filtering comb of thoracic appendages (Table 1). Length of the base of the comb and length of seta of the comb on limb 3 depend linearly on body length with the slope close to 1. The mean number of setae in the comb, just like relative comb and seta lengths, are independent of body length and thus are advantageous criteria for investigations of morphology of the filtering screen (Kořínek *et al.*, 1986; Pop, 1991). Mean values of these parameters within the population have very narrow confidence intervals (Table 1) and they are usually expressed in percentage of body length (Fig. 3). The results presented here show that population 2, coming from a lower concentration of food, exceeds population 1 in all parameters under investigation.

Feeding behavior of population 1

Unfortunately I could not secure all size groups of this population to be significantly distributed in the experiment and thus mean values only for animals in the range of size 1.15–1.40 mm were finally calculated (Table 2). If concentration of food decreased, the mean filtering rate increased reaching the maximum at 0.14 mgC l^{-1}. The next

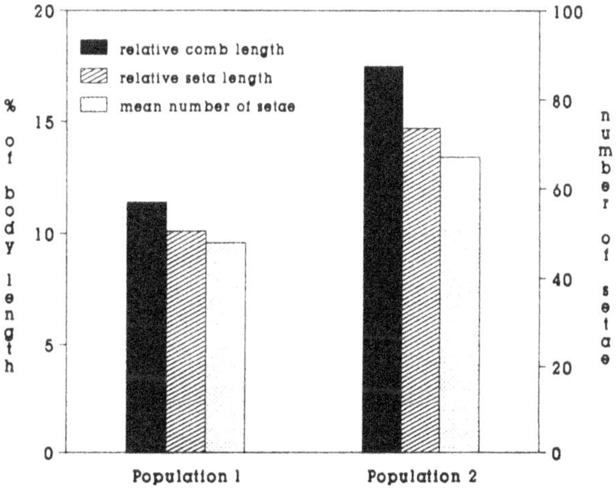

Fig. 3. Comparison of size of the filtering comb between populations 1 and 2 of *Daphnia galeata*.

Table 1. Coefficients of power regression of the comb length (CL) and seta lengths (SL), both in μm, on body length (L) in mm and mean size specific comb length (CL/L) and seta length (SL/L) and number of setae (NS) including 95% confidence limit of limb 3 in both populations.

	CL			SL			CL/L	± cl	SL/L	± cl	NS	± cl
	a	b	r	a	b	r						
Population 1	112.17	0.82	0.98	99.48	0.89	0.98	114.50	5.02	101.80	3.55	47.60	1.15
Population 2	181.27	0.85	0.99	149.91	0.91	1.00	175.00	6.88	146.90	3.68	66.70	2.44

Table 2. Experimental conditions and feeding behaviour of population 1. Animals in the range of size 1.15–1.40 mm.

Concentration, mgC l^{-1}	4.16	2.08	0.37	0.53	0.27	0.14	0.07	0.038
Mean L, mm	1.23	1.24	1.24	1.24	1.25	1.26	1.24	1.26
(± 95% cl)	(0.04)	(0.07)	(0.04)	(0.04)	(0.05)	(0.04)	(0.04)	(0.05)
Mean FR, ml ind^{-1} h^{-1}	0.07	0.11	0.20	0.21	0.18	0.39	0.20	0.14
(± 95% cl)	(0.01)	(0.02)	(0.03)	(0.03)	(0.01)	(0.12)	(0.04)	(0.04)
Number of animals	5	6	7	8	11	6	9	9

Table 3. Experimental conditions and coefficients of power regression of filtering rate on body length of population 2.

Concentration, mgC l^{-1}	0.03	0.05	0.10	0.19	0.37	0.72	1.44	2.84
Number of animals	6	15	16	13	18	18	17	17
a	0.05	0.04	0.06	0.15	0.17	0.21	0.23	0.30
b	2.23	3.19	2.82	2.28	3.07	3.14	2.76	2.81
r	0.95	0.92	0.76	0.78	0.98	0.92	0.91	0.93

reduction of food concentration was accompanied with a drop of the filtering rate. Ingestion rate grew together with concentration of food, identically to the known model. It seems to reach saturation at concentration about 0.7 mgC l^{-1} but next two points defy this idea.

Feeding behavior of population 2

This experiment provided me with more detailed information about feeding behavior of *Daphnia galeata*. Filtering rate is a power function of the body length with a slope ranging from 2.23 to 3.19 (Fig. 4, Table 3). There is no relation of value of the slope to the concentration of food while the value of the intercept increases with the concentration of food. Filtering and ingestion rates for animals with a size of 0.8, 1.0, 1.5 and 2.0 mm were calculated then and plotted against the concentration of food (Fig. 5). Filtering rate increases in relation to the drop of the concentration of food but does not reach the plateau. Ingestion rate, like in the experiment with population 1, fit the saturation model rather well excluding last two points. The model I used had to suit the following assumptions: 1. saturation course is used for ingestion rate, 2. the drop after reaching the maximum occurs if it is used for the filtering rate, 3. it must go through the origin in both the cases. Then

$$IR = \frac{k_1 \cdot c^2}{k_2 + c^2} \text{ and } FR = \frac{k_1 \cdot c}{k_2 + c^2},$$

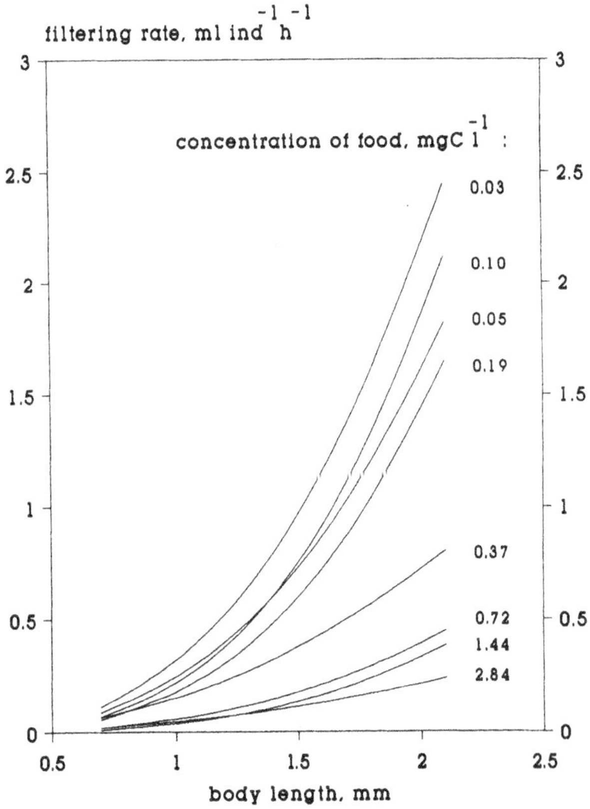

Fig. 4. Power curve fit of filtering rate on body length at eight concentrations of food in population 2.

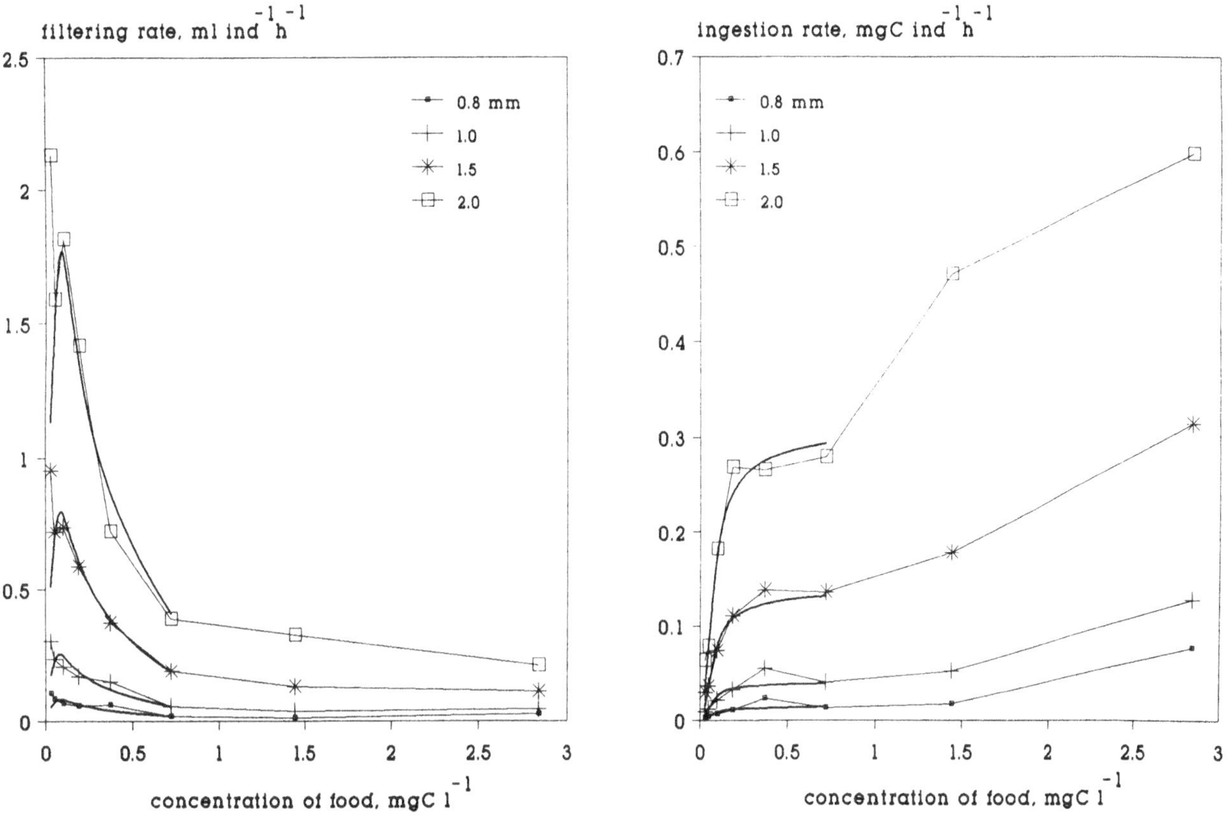

Fig. 5. Plots of filtering and ingestion rates on concentration of food within animals of population 2 with varied body length. Calculation of the points based on parameters of power regression from Table 3 (thin lines). Coefficients of the curves provided by the model (thick lines) are in Table 4.

where c is concentration of food and k_1, k_2 are coefficients. To estimate values of the coefficients k_1, k_2, I plotted 1/IR against $1/c^2$ and calculated intercept and slope of the linear regression, where a = $1/k_1$ and b = k_2/k_1. The last two points lay out of this model and the first one had to be excluded because it affected unrealistically the fitting of the model after the linearization (Fig. 5, Table 4).

Table 4. Coefficients (a, b, r) of linear regression of 1/IR on $1/c^2$ and coefficients (k_1, k_2) of relation IR = $(k_1 \cdot c^2)/(k_2 + c^2)$, where a = $1/k_1$ and b = k_2/k_1, IR is ingestion rate, in mgC ind^{-1} h^{-1}, c is concentration of food, in mgC l^{-1}. Calculation based on concentrations 0.05–0.72 mgC l^{-1}.

Size, mm	a	b	r	k_1	k_2
0.8	71.118	0.463	0.966	0.014	0.007
1.0	24.863	0.151	0.976	0.040	0.006
1.5	7.467	0.051	0.998	0.134	0.007
2.0	3.359	0.023	0.998	0.298	0.007

Comparison of both populations

It is not easy to decide whether the feeding behavior is to be compared for animals with the same body size, or body mass or finally with the same filtering combs size. As regards population 1, I have only the size group 1.15–1.40 mm at my disposal. The mean body length is about 1.25 mm (Table 2), dry weight 8.85 μg (Novotná, 1987) and the average length of the seta is 121 μm. The members of population 2 with equal length of seta have, however, body length of only 0.8 mm. Finally I computed filtering and ingestion rates and morphological parameters for population 2 as a mean of values calculated in the range of size 1.15–1.40 mm, using appropriate regression equations (mean body length 1.28 mm and dry weight 8.47 μm – Novotná, 1987). The fact that dry weights of both populations are close allows me to suppose that the filtering and ingestion rates presented for animals with the same body length are in a similar ratio as their size-specific values (Fig. 6). The same equation and method were applied to fit data with the model as described above. Filtering and ingestion rates of the population with a large filtering comb (population 2) significantly exceed values of population 1 below concentration of food 0.4 mgC l^{-1}. At a concentration where the filtering rate of population 1 is decreasing, that of population 2 still increases and does not reach the true plateau. Starting from the data provided by the model, the maximum of the filtering rate of population 2 is almost two times higher and incipient limiting level nearly at half concentration if compared with population 1. This means the ingestion rate at incipient limiting level is practically the same in both populations (Fig. 7). Different results were recorded at the concentration above 0.4 mgC l^{-1}. Population 1 tends to higher filtering and ingestion rates but differences are not so manifest (Fig. 6).

Flow regimes

Considering that the true projection of filtering area (PFA) was not measured for populations under study, I used a converting equation for the relation between seta length and projection of filtering area which is available for the limb 3 of *Daphnia pulicaria* (Pop, 1991). On the supposition that projection of filtering area of limb 4 is 55.4% of limb 3 (calculation based on data of Křepelová (1980)), total projection of filtering area can be estimated from seta length of the filtering comb of limb 3 (SL):

$$PFA = 5.840\ SL^{1.996}.$$

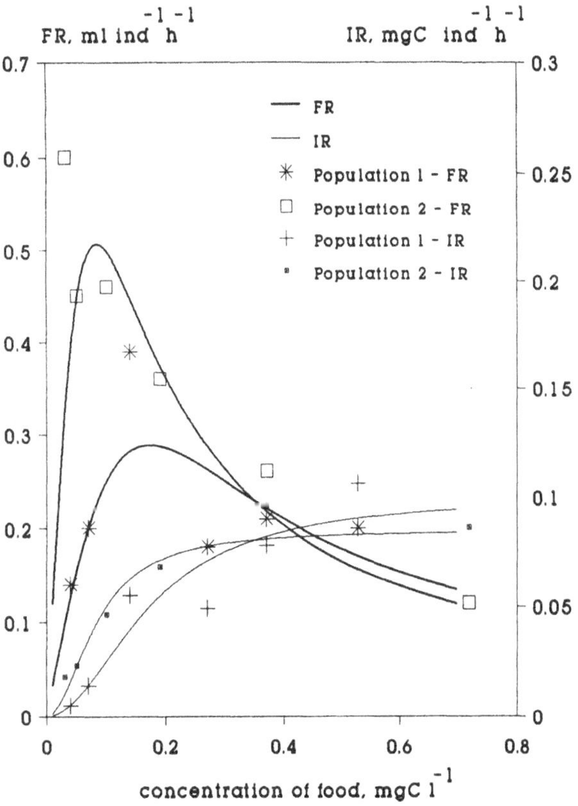

Fig. 6. Comparison of feeding behavior of populations 1 and 2 of *Daphnia galeata* with the same body length.

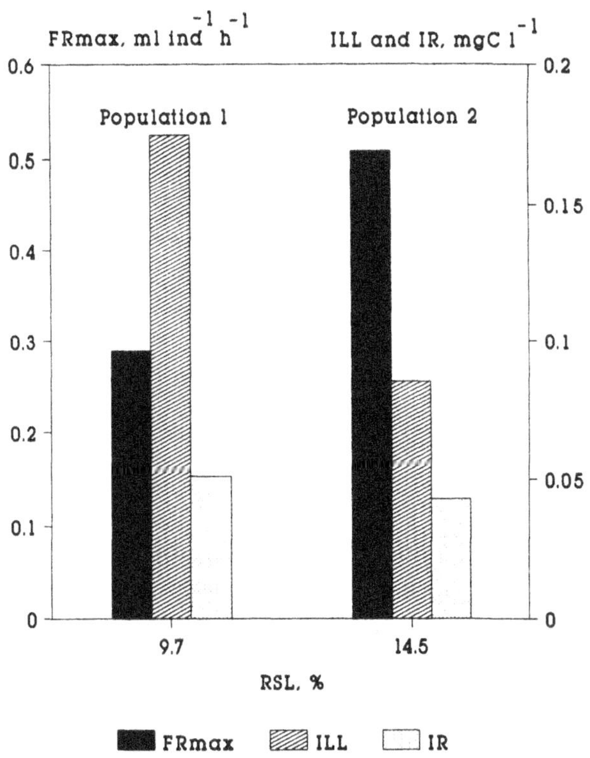

Fig. 7. Basic parameters of feeding response curves and relative seta length summarized for populations 1 and 2 of *Daphnia galeata*.

The ratio of the filtering rate to projection of filtering area sets the value of a theoretical flow. Plots of maximal filtering rate, projection of filtering area and flow vs. body length, maximal filtering rate vs. projection of filtering area and flow vs. concentration of food are presented in Fig. 8. For the values of coefficients of the power regression see Table 5. Excluding the dependence of projection of filtering area on body length, the slopes of the relations are somewhat higher than could be expected from theoretical considerations. There are manifest differences between populations 1 and 2 within all relations shown. In plots of flow on body length and on concentration of food, the values of a theoretical flow calculated for population 1 exceed those for population 2 with one exception at the lowest concentration. This means that population 1, which is handicapped by smaller projection of filtering area, has to compensate for it, probably by a higher beat frequency of limbs.

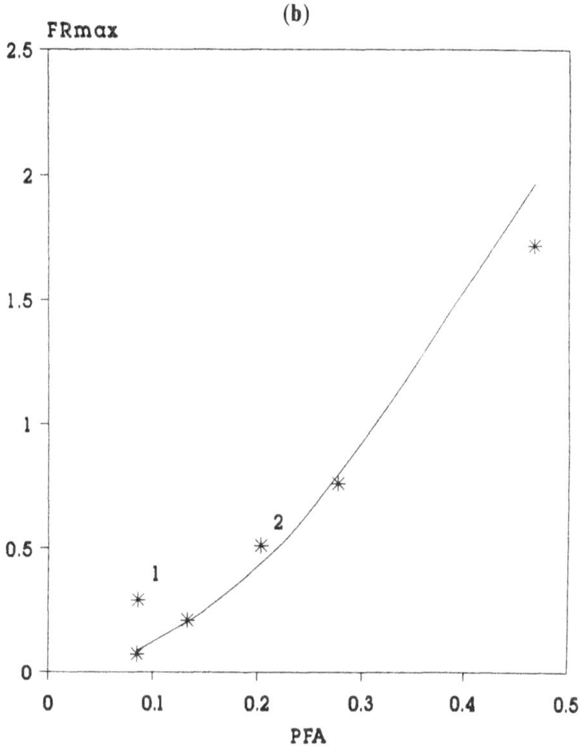

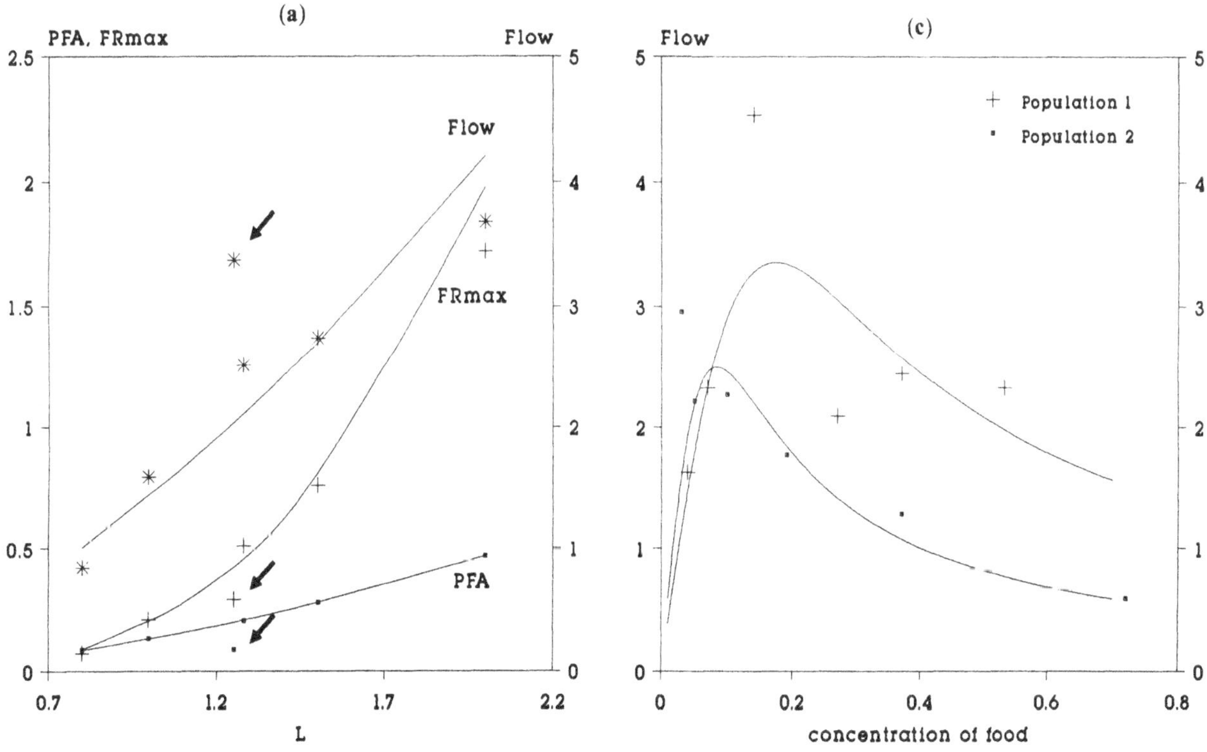

Fig. 8. Relations of projection of filtering area (PFA), maximal filtering rate (FR_{max}) and flow to body length (L) – a; relation of maximal filtering rate to projection of filtering area – b and relation of flow to concentration of food – c for both populations of *Daphnia galeata*. PFA is in mm^2, FR_{max} is in $ml\ ind^{-1}\ h^{-1}$, flow is in $ml\ mm^{-2}\ h^{-1}$, concentration of food is in $mgC\ l^{-1}$ and L is in mm. Arrows indicate points for population 1, figures 1 and 2 are points for animals from populations 1 and 2 with the same body length. For parameters of power regression curves, see Table 5.

Table 5. Coefficients of power regression of dependence of maximum filtering rate (FR_{max}), in ml $ind^{-1} h^{-1}$, on body length (L), in mm, and projection of filtering area (PFA), in mm^2, and projection of filtering area and flow, in ml $mm^{-2} h^{-1}$, on body length of population 2.

Dependent variable	Independent variable	a	b	r
FR_{max}	L	0.18	3.42	0.99
PFA	L	0.13	1.85	1.00
Flow	L	1.42	1.56	0.96
FR_{max}	PFA	8.03	1.85	0.99

Discussion

Today there is not doubt that cladocerans can change the morphology of their filtering screen (especially the area of the combs of limbs 3 and 4) depending on the change of concentration of food (Kořínek *et al.*, 1981, 1986). This ability is specific to some species and reflects their ecological strategy and is the most developed in the species *Daphnia pulicaria*. Its ratio of minimal to maximal projection of filtering area is more than two (Koza & Kořínek, 1985). An experimentally induced change and recorded differences within populations 1 and 2 show this ability is practically the same in *Daphnia galeata*. Projection of filtering area calculated for population 1 is the lowest which has ever been found for this species while projection of filtering area calculated for population 2 is equal to the maximal values published for the population from Lake Constance (Brendelberger & Geller, 1986) and from the drinking water reservoir Želivka, in Czechoslovakia (Kořínek *et al.*, 1986). In the past it was supposed that the maximal change of projection of filtering area within a natural populations need several weeks to manifest itself (Kořínek *et al.*, 1986); a more recent study showed however that the reaction of an individual animal is immediate (Pop, 1991). Nevertheless the data presented here for *Daphnia galeata* are the first data recording such a dramatic increase of projection of filtering area, the comb and the seta length and number of setae. The observations published formerly present rather the decrease of projection of filtering area as a result of a growth of food concentration (Kořínek *et al.*, 1986).

The original rectilinear model of the dependency of filtering and ingestion rates on concentration of food in daphnids according to which each filtering rate below incipient limiting level is equal to maximal filtering rate (McMahon & Rigler, 1965) became invalid when data appeared that filtering rate can decrease again after reaching its maximum (Horton *et al.*, 1979; Muck & Lampert, 1980; Nikolaev & Postnov, 1981; Filippova & Postnov, 1988), as described in calanoids (Frost, 1975). Muck & Lampert (1980), who worked with daphnids from Lake Constance (probably a mixture of *Daphnia galeata* and *D. hyalina*), explained the drop of filtering rate at very low concentrations of food as exhaustion of animals in the process of adaptation and supported this hypothesis experimentally (Muck & Lampert, 1984; Lampert & Muck, 1985). In my experiments I tried to minimize the possible effect of exhaustion by an appropriate time of adaptation. In experiments by Nikolaev & Postnov (1981) and Filippova & Postnov (1988) with *Daphnia longispina*, *Ceriodaphnia quadrangula*, *Moina macrocopa*, *Daphnia magna* and *Simocephalus vetulus* where the cell count method and 5–10 hour exposures were used, the mechanism of exhaustion is improbable. This drop in filtering rate was also recorded by Ganf & Shiel (1985) in *Daphnia carinata* and *Ceriodaphnia quadrangula* (Coulter counter method and 2–4 hour exposures), by Holm *et al.* (1983) in *Daphnia pulex* (cell count method) and by several authors who used the radio-isotopic method: Horton *et al.* (1979) in *Daphnia pulex* (only one hour adaptation) and Porter *et al.* (1982) in *Daphnia magna* where moreover the decrease of beating rate of thoracic appendages and mandibles was recorded, too. The first data regarding retardation of thoracic appendages and mandibles beat frequency at very low concentrations of food were published by Burns (1968). This mechanism seems to be involved in the drop of filtering rate mentioned, though the beating of thoracic limbs does not have to be connected with filtering of food (M. Gophen, pers. com.). In my opinion, whether the drop of filtering rate was found or not depended 1. on the range of concentrations in the experiment and 2. on the feeding history of the examined population which also conspicuously influences morphology of the filtering screen (Fig. 6). Population 2 probably did not reach this point and its filtering rate curve is similar to the response found in starved *Daphnia magna* (Ringelberg & Royackers, 1985). Geller (1975) in experiments with *Daphnia pulex* also found filtering rate at very low concentrations to be higher for starved than for prefed animals though typical response curves were not reached in either cases.

The curves of ingestion rate seem to reach saturation at concentration 0.7 mgC l^{-1}. The high values of ingestion rates at higher concentrations indicate possible influence of so called excessive feeding (Ivanova, 1970). Animals probably can not decrease filtering rate because it may be necessary for respiration (Gutelmacher *et al.*, 1988), and postabdomen rejection rate, although it grows with concentration of food, is not able to prevent from gut filling. Thus I suppose short retention time and partial digestion of food in the gut are mechanisms that regulate real ingestion rate. Ingestion without saturation is also typical for hungry animals (Geller, 1975; Ringelberg & Royackers, 1985). Paffenhöfer & Orcutt (1986) received similar results in an experiment with juveniles of *Penilia avirostris*; they explain this phenomenon as an absence of food satiation.

There are many equations which were used to describe the relation between filtering and ingestion rates and concentration of food (Mullin *et al.*, 1975; Nikolaev & Postnov, 1981; Porter *et al.*, 1982; Peters, 1984; Lampert, 1987; Gutelmacher *et al.*, 1988). I also tried to find the 'true' model to fit my data. The generalization was finally done after excluding the two highest concentrations and the lowest concentration (in both populations) provided the best fitting of data. The most questionable point of the presented models was excluding of the lowest concentration which was necessary because of the method used for computation of the coefficients. The low number of points did not allow distinguishing what is the trend of filtering rate of population 2 – whether it is still growing or falling already. In any case the model supposes the final drop of filtering rate at the end.

The fact that the population with a smaller projection of filtering area gathers food with a lower efficiency below food concentration about 0.4 mgC l^{-1}, and is limited by deficiency of food earlier than the population with a larger projection of filtering area, in contrast to the situation above this concentration, corresponds to 'oligotrophic' and 'eutrophic' types of feeding behavior recognized by Kasprzak *et al.* (1986). The data presented demonstrate that *Daphnia galeata* is able to change from the first type to the second one within one month.

Regarding the positive correlation between filtering rate and projection of filtering area my findings are rather close to the older data (Egloff & Palmer, 1971; Arruda, 1983; Ganf & Shiel, 1985) though there are some slight differences in the values of slopes. The fact that *Daphnia* with larger area of the filtering comb also has a higher filtering rate was recorded for the first time by Egloff & Palmer (1971). Unfortunately they did not measure filtering rate regularly and used only published data. The following authors were more skeptical in evaluating the influence of area of the filtering comb on filtering rate. They ascribed the main effect rather to some other factors such as gut fullness and/or capture efficiency (Arruda, 1983; Porter *et al.*, 1983 and Ganf & Shiel, 1985). On the other hand, if any concentration below incipient limiting level is used, we can hardly expect to measure the true maximal filtering rate. Moreover, it is really difficult to carry out measurements of filtering rate where only one simple factor varies and the other remain constant. This may results in poor comparability of data.

Acknowledgements

I thank Zuzana Stuchlíková for analysis of organic carbon, Jaroslav Pazourek for helpful suggestions regarding mathematical model and Vladimír Kořínek for critical comments on the manuscript.

References

Arruda, J. A., 1983. Daphnid filtering comb area and the control of filtering rate. J. Freshwat. Ecol. 2: 219–224.

Berberovic, R. & R. Pinto-Coelho, 1989. Dry first, measure later: a new procedure to preserve a measure zooplankton for ecophysiological studies. J. Plankton Res. 11: 1109–1116.

Brendelberger, H. & W. Geller, 1985. Variability of filter structures in eight Daphnia species: mesh size and filtering areas. J. Plankton Res. 7: 473–487.

Burns, C. V., 1968. Direct observations of mechanisms regulating feeding behavior of Daphnia, in lake water. Int. Revue ges. Hydrobiol. 53: 83–100.

Egloff, D. A. & D. S. Palmer, 1971. Size relations of the filtering areas of two Daphnia species. Limnol. Oceanogr. 16: 900–905.

Filippova, T. G. & A. L. Postnov, 1988. Relationship between metabolism and feeding in Cladocera (Crustacea). Zh. Obshch. Biol. 49: 540–551.

Frost, B. W., 1975. A threshold feeding behavior in Calanus pacificus. Limnol. Oceanogr. 20: 263–266.

Ganf, G. G. & R. J. Shiel, 1985. Feeding behaviour and limb morphology of two cladocerans with small intersetular distances. Aust. J. mar. Freshwat. Res. 36: 69–86.

Geller, W., 1975. Die Nahrungsaufnahme von Daphnia pulex in Abhängigkeit von der Futterkonzentration, der Temperatur, der Körpergrösse und dem Hungerzustand der Tiere. Arch. Hydrobiol. Suppl. 48: 47–107.

Gutelmakher, B. L., A. P. Sadchikov & T. G. Filippova, 1988. Pitanie zooplanktona. VINITI, Moskva, 156 pp.

Helgen, J. C., 1987. Feeding rate inhibition in crowded Daphnia pulex. Hydrobiologia 154: 113–119.

Holm, N. P., G. G. Ganf & J. Shapiro, 1983. Feeding and assimilation rates of Daphnia pulex fed Aphanizomenon flos-aquae. Limnol. Oceanogr. 28: 677–687.

Holtby, L. B. & R. Knoechel, 1981. Zooplankton filtering rates: Error due to loss of radioisotopic label in chemically preserved samples. Limnol. Oceanogr. 26: 774–780.

Horton, P. A., M. Rowan, K. E. Webster & R. H. Peters, 1979. Browsing and grazing by cladoceran filter feeders. Can. J. Zool. 57: 206–212.

Ivanova, M. B., 1970. Vliyanie koncentracii pishchi na skorost' filtracii u Cladocera. Zh. obshch biol. 31: 721–731.

Kasprzak, P., V. Vyhnálek & M. Straškraba, 1986. Feeding and food selection in Daphnia pulicaria. Limnologica (Berlin) 17: 309–323.

Kořínek, V. & J. Macháček, 1980. Filtering structures of Cladocera and their ecological significance. I. Daphnia pulicaria. Věst. čs. Společ. zool. 44: 213–218.

Kořínek, V., B. Křepelová & J. Macháček, 1981. Filtering structures of Cladocera and their ecological significance. II. Species of the genera Daphnia and Ceriodaphnia. (Abstract). Verh. Int. Verein. Limnol. 21: 1667–1567.

Kořínek, V., B. Křepelová-Macháčková & J. Macháček, 1986. Filtering structures of Cladocera and their ecological significance. II. Relation between the concentration of the seston and the size of filtering combs in some species of the genera Daphnia and Ceriodaphnia. Věst. čs. Společ. zool. 50: 244–258.

Koza, V. & V. Kořínek, 1985. Adaptability of the filtration screen in Daphnia: Another answer to the selective pressure of the environment. Arch. Hydrobiol. Beih. Ergebn. Limnol. 21: 193–198.

Křepelová, B., 1980. Ekologický význam rozdílů ve stavbě filtračního aparátu druhů rodu Daphnia. M.Sc. Thesis. Charles University, Prague, 99 pp. [Ecological significance of differences in morphology of the filtering screen in species of the genus Daphnia].

Lampert, W., 1987. Feeding and nutrition in Daphnia. In R. H. Peters & R. de Bernardi (eds.), 'Daphnia' Mem. Ist. Ital. Idrobiol. 45: 143–192.

Lampert, W. & P. Muck, 1985. Multiple aspects of food limitation in zooplankton communities: the Daphnia – Eudiaptomus example. Arch. Hydrobiol. Beih. Ergebn. Limnol. 21: 311–322.

Mackereth, F. J. H., J. Heron & J. F. Talling, 1978. Water analysis: Some revised methods for limnologists. FBA Scientific publication No 36. 120 pp.

McMahon, J. W. & F. H. Rigler, 1965. Feeding rates of Daphnia magna Straus in different foods with radioactive phosphorus. Limnol. Oceanogr. 10: 105–113.

Mourelatos, S., 1990. A new technique for long preservation of ^{14}C-labelled cladocerans. Hydrobiologia 190: 147–154.

Muck, P. & W. Lampert, 1980. Feeding of freshwater filter-feeders at very low food concentrations: poor evidence for 'threshold feeding' and 'optimal foraging' in Daphnia longispina and Eudiaptomus gracilis. J. Plankton Res. 2: 367–379.

Muck, P. & W. Lampert, 1984. An experimental study on the importance of food conditions for the relative abundance of calanoid copepods and cladocerans. 1. Comparative feeding studies with Eudiaptomus gracilis and Daphnia longispina. Arch. Hydrobiol. Suppl. 66: 157–179.

Mullin, M. M., E. Fuglister Stewart & F. J. Fuglister, 1975. Ingestion by planktonic grazers as a function of concentration of food. Limnol. Oceanogr. 20: 259–262.

Nichols, H. W., 1975. Growth media-freshwater. In J. R. Stein (ed.), Handbook of phycological methods. Culture methods and growth measurements. Cambridge University press, Cambridge: 7–24.

Nikolaev, T. M. & A. L. Postnov, 1981. Izuchenie racionov i skorostey filtracii u trekh vidov Cladocera (Crustacea). Dokl. Akad. Nauk SSSR 257: 509–512.

Novotná, Z., 1987. Vliv koncentrace potravy na parametry biomasy u několika druhů rodu Daphnia. M.Sc. Thesis. Charles University, Prague, 69 pp. [Effect of concentration of food on biomass parameters within a several species of the genus Daphnia].

Paffenhofer, G. A. & J. D. Orcutt, Jr., 1986. Feeding, growth and food conversion of the marine cladoceran Penilia avirostris. J. Plankton Res. 8: 741–754.

Peters, R. H., 1984. Methods for the study of feeding, grazing and assimilation by zooplankton. In J. A. Downing & F. H. Rigler (eds.), A manual on methods for the assessment of secondary productivity in fresh waters. IBP handbook 17. Blackwell Scientific Publications, Oxford.: 336–412.

Pop, M., 1991. Mechanisms of the filtering area adaptation in Daphnia. In: Kořínek, V. & D. G. Frey (eds), Biology of Cladocera. Developments in Hydrobiology 71. Kluwer Academic Publishers, Dordrecht: 169–176. Reprinted from Hydrobiologia 225.

Porter, K. G., J. Gerritsen & J. D. Orcutt, Jr., 1982. The effect of food concentration on swimming patterns, feeding behavior, ingestion, assimilation, and respiration by Daphnia. Limnol. Oceanogr. 27: 935–949.

Porter, K. G., Y. S. Feig & E. F. Vetter, 1983. Morphology, flow regimes, and filtering rates of Daphnia, Ceriodaphnia, and Bosmina fed natural bacteria. Oecologia (Berlin.) 58: 156–163.

Ringelberg, J. & K. Royackers, 1985. Food uptake in hungry cladocerans. Arch. Hydrobiol. Beih. Ergebn. Limnol. 21: 199–207.

Sierszen, M. E. & C. J. Watras, 1987. Rapid-freeze preservation minimizes radioisotope leakage from zooplankton in feeding experiments. J. Plankton Res. 9: 945–953.

Starkweather, P. L., 1983. Daily patterns of feeding behavior in Daphnia and related microcrustacea: implications for cladoceran autecology and the zooplankton community. Hydrobiologia 100: 203–221.

Stuchlík, E., 1982. Filtrační rychlosti rybničního zooplanktonu. M.Sc. Thesis. Charles University, Prague, 79 pp. [Filtering rate of zooplankton in fish ponds].

Thompson, J. M., A. J. D. Ferguson & C. S. Reynolds, 1982. Natural filtration rates of zooplankton in a closed system: the derivation of a community grazing index. J. Plankton Res. 4: 545–560.

Hydrobiologia **225**: 169–176, 1991.
V. Kořínek & D. G. Frey (eds), Biology of Cladocera.

Mechanisms of the filtering area adaptation in *Daphnia*

Michal Pop
Department of Hydrobiology, Charles University, Viničná 7, CS-12844 Prague 2, Czechoslovakia

Key words: *Daphnia*, filtration, morphology

Abstract

The changes of selected parameters of the filtering comb of the third thoracic limb were studied in a natural population of *Daphnia pulicaria* Forbes, as well as in experimental enclosures and in lab cultures, including individual life history. Two hypotheses were tested: 1. either these changes are related to the succession of clones coexisting within one population, or 2. the size of the filtering area changes gradually as an individual adaptation during the moulting. No evidence supporting the 'clonal hypothesis' was found. On the contrary, the adaptability of the filtering comb is the same in a natural population as it is in a clone and in individuals.

Introduction

The phenomena of changes in the size of the filtering comb in daphnids in relation to the food available has been examined for many years. Kořínek & Macháček (1980) measured several morphological parameters of filtering combs in *Daphnia pulicaria* and found significant differences between populations from various habitats. Kořínek *et al.* (1986) measured and compared many parameters of filtering combs on the 3rd and 4th thoracic limbs in several species of *Daphnia* and *Ceriodaphnia* from many habitats and indicated that the 3rd pair of daphnid's limbs has the biggest ability to change. They also studied changes of filtering combs parameters in the pond-dwelling population of the species *Daphnia pulicaria* during the spring-autumn season and recognized large changes of the projected filtering area (PFA) within one population in consequence of the concentration of phytoplankton in water. Koza & Kořínek (1985) compared the ability to change the size of the filtering area within clones in several species of the Daphnia pulex group and suggested that it is largest in *D. pulicaria*. They explain differences in this ability in accordance with diverse ecological strategies of particular species.

This phenomena seems to be obligatory for pelagic species of daphnids regardless of the mechanism of the way of particle capture (Rubenstein & Koehl, 1977; Gerritsen & Porter, 1982; Ganf & Shiel, 1985; Fryer, 1987; and others). The aim of the present paper is to elucidate the mechanism by which this change occurs.

Material and methods

Females of *Daphnia pulicaria* were used in experiments, which were performed on four levels.

1. The *natural population* in the carp pond Smyslov near Blatná, South Bohemia (Fott, 1975), was followed over the whole season in 1986. Samples were taken from April 1986 to March 1987 at intervals of about two weeks (in

winter irregularly). A Schindler-Patalas sampler was used for zooplankton sampling.

Besides filtering areas, a few other parameters of the population were studied, e.g. population density or production of eggs, which are not mentioned in this paper. The amount of food available was determined as a chlorophyll-*a* content (Lorenzen, 1976).

2. A part of the natural population was collected and cultivated as a *subpopulation* in six plastic barrels under artificially increased food conditions (Lampert, 1975; Vijverberg, 1989).

Six 50 l plastic barrels were used for cultivation of natural zooplankton, collected by a Schindler-Patalas sampler. The density of the *D. pulicaria* population in the barrels varied between 50 and 500 ind l^{-1}, being affected by the different amount of food available and by sampling. A dense mixture of algae (with dominance of *Scenedesmus quadricauda*) was added to pond water with common pond phytoplankton (Fott, 1987) in order to obtain a medium with a substantially higher amount of food in the barrels in comparison to the original habitat of the observed population. This amount varied, being affected by the filtering activity of zooplankton during the experiment, but did not drop under 10 μg of chlorophyll-*a* l^{-1}. Medium was changed daily and sampling of zooplankton was performed at intervals of about 6 days.

The experimental barrels were divided into two groups; B1–B3 with a very high amount of food (over 100 μg of chlorophyll-*a* l^{-1}) and B4–B6 with a medium amount of food (20–50 μg of chlorophyll-*a* l^{-1}). Results from the B1 and B4 barrels are presented in this paper.

The barrels were placed in the pond (close to the surface) in order to maintain physical conditions comparable with the original population.

3. *Individual cultivations* were performed in order to obtain data on the development of the filtering comb during the life of a particular individual. Ten parthenogenetic females from a clone held in a laboratory were cultivated semi-continuously under different food conditions in 30 ml tubes (Vijverberg, 1989). The amount of food was controlled by enriching of the medium with various volume of a nonsterile culture of *Scenedesmus obliquus*. Cultivations were observed several times every day, dead animals being removed. Newborn neonates were replaced and cultivated separately. By this method I was able to monitor direct lines of parthenogenetic females.

4. Variability of filtering comb parameters within a *clone* was tested in the clone, held in the laboratory.

Two parameters of the filtering comb, seta length in the central portion of the filtering comb (SL) and number of setae (NS), were measured and counted respectively by using a projection microscope (Visopan-Reichert) after the dissection of an animal. These parameters were chosen due to relatively easy gathering of great numbers of pinpoint data (necessary for the method of cluster analysis, that I was expecting to use); 200 animals were dissected for the measurement from several samples of the natural population and from the barrel experiments. Smaller numbers were used from the rest of the samples studied.

A very useful and simple parameter is the relative seta length (RSL), that is the mean seta length expressed in percent of body length. This parameter is constant in all size groups and has very low variability. It is possible to use if for a simple expression of the size of the filtering area, characteristic for a population in specific food conditions (see Fig. 1).

Data on the parameters of filtering combs during the life history of individual animals were obtained by measuring the exuviae after each moulting.

Besides this I computed original data gathered by Křepelová in 1980.

Results

The natural population

The usual situation in the natural population in rather steady food conditions is shown in Fig. 1.

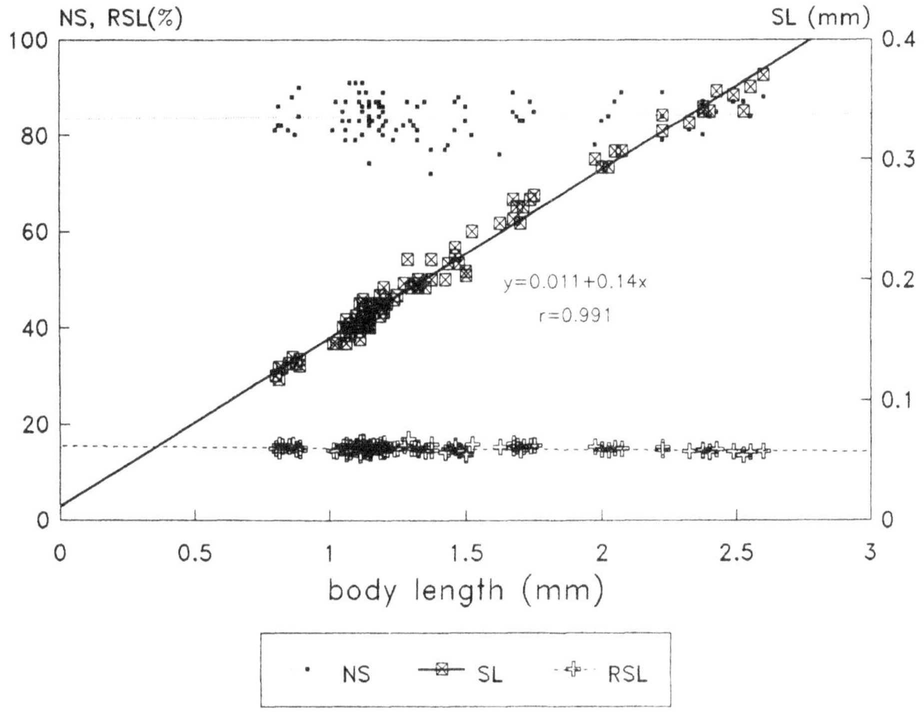

Fig. 1. Number (NS), length (SL) and relative length of setae (RSL) in Smyslov pond population; (n = 100).

Both the number of setae in a filtering comb (NS) and the relative seta length (RSL) are independent of the body length, while the mean length of seta (SL) grows proportionally to the body length with very high correlation ($r > 0.97$, in most cases $r > 0.99$).

RSL changes within the population during the season in consequence of changes in the seston concentration, and probably in food quality, too. The impact of temperature on the filtering screen is not clear yet (T. Herold, personal communication, 1989).

The barrel experiments

After an increase in seston content, as for instance in the experimental barrels, we can detect a very fast decrease in the relative seta length (Fig. 2). The seston content in the B-4 barrel was approximately ten times higher and in the B-1 barrel from 20 to 30 times higher than that in the original habitat of the animals used in this experiment. This decrease was later balanced and the RSL returned back to higher values.

Individual cultivations

Figure 3a reflects the development of RSL after very fast increase of food concentration (overfeeding) in an adult daphnia from medium food conditions. The change in the growth trend in the seta length appeared immediately after the first moulting after overfeeding. The growth in the body length was not followed by corresponding growth in the seta length, which stopped elongation. Relative seta length decreased from about 14% to 11.7% during this experiment, which lasted for 15 days.

Figure 3b describes changes in RSL in two animals from the same clutch. While the first daphnia (L9) was kept in original medium, the second one (L8) was removed and placed into rich food conditions.

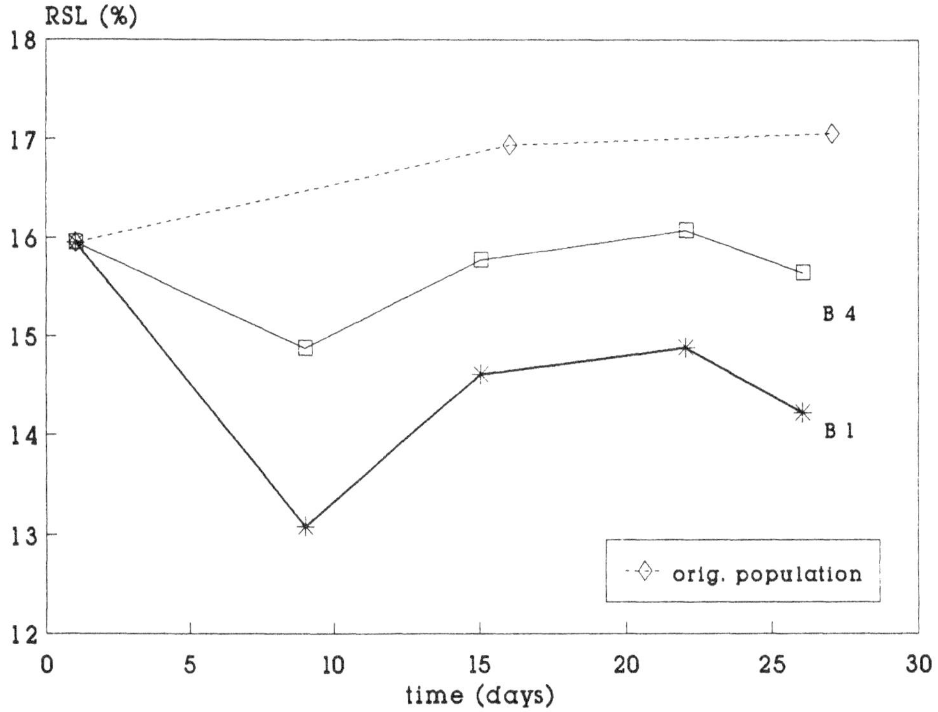

Fig. 2. Changes of RSL in the barrel experiments (barrels B1 and B4, $n = 200$) in comparison to the original population.

The number of setae (NS)

Concerning the mean number of setae in the comb, the situation is rather different. In constant conditions, the number of setae is independent of the body length and usually has a normal distribution (see Fig. 1). Another situation appears in the case of changes in seston concentration, as it was in the barrel experiment. In such a case it is useful to study numbers of setae in the comb separately in adults and juveniles. While adults follow the trend from the original population, the distribution in juveniles shifts to lower or higher numbers of setae. This shift is later gradually balanced by the growth and replacement of individuals in a population (Fig. 4).

This observation suggests, that the number of setae is constant during the whole life of an individual. This suggestion was later supported by the individual cultivation experiments. In these experiments, changes in the number of setae in the comb during the life of an animal were observed very rarely, and I think that it was caused more likely by some accidental injury than by a regular way of development of the filtering comb during growth.

Furthermore, the number of setae in the comb is closely similar in both limbs of the pair, the mean difference being 3 setae. Since the variability of NS in the clone was not less than that in the natural population, no impact of heredity on this parameter was identified.

NS in neonates

When counting the number of setae in neonates, I mostly was not able to count this number exactly. I found that the setal comb of the 2nd and older instars is terminated on the proximal side by a distinctly different, thicker seta. But this 'terminal seta' was found only rarely in the comb of neonata. This observation suggests that the neonata keeps the ability to change this parameter (NS), which is later unchangeable for the rest of the animal's life.

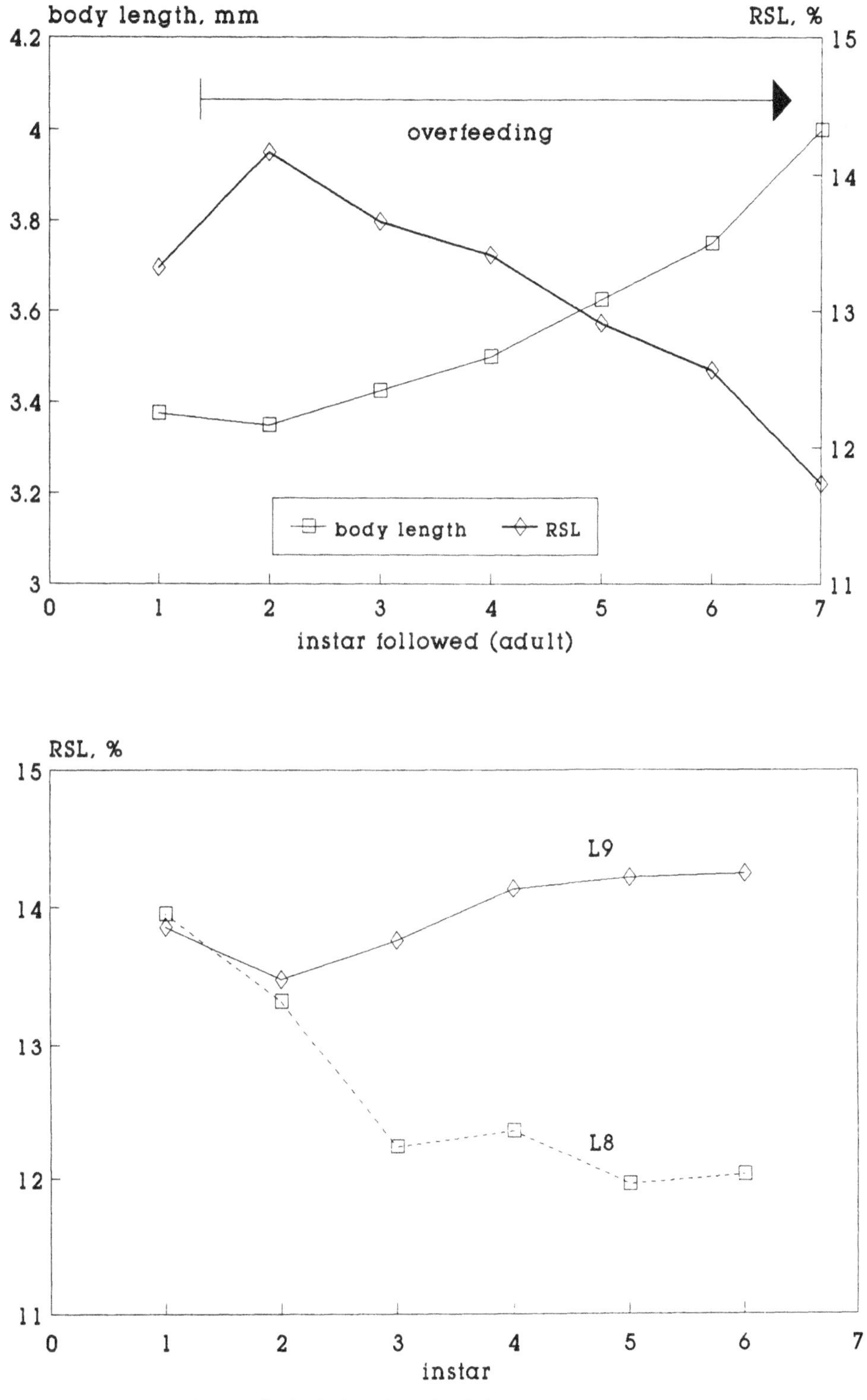

Fig. 3. Individual cultivations: a) changes in body length and relative seta length (RSL) in an adult daphnia after strong overfeeding. b) changes in RSL in two animals from one clutch being cultivated in different food conditions.

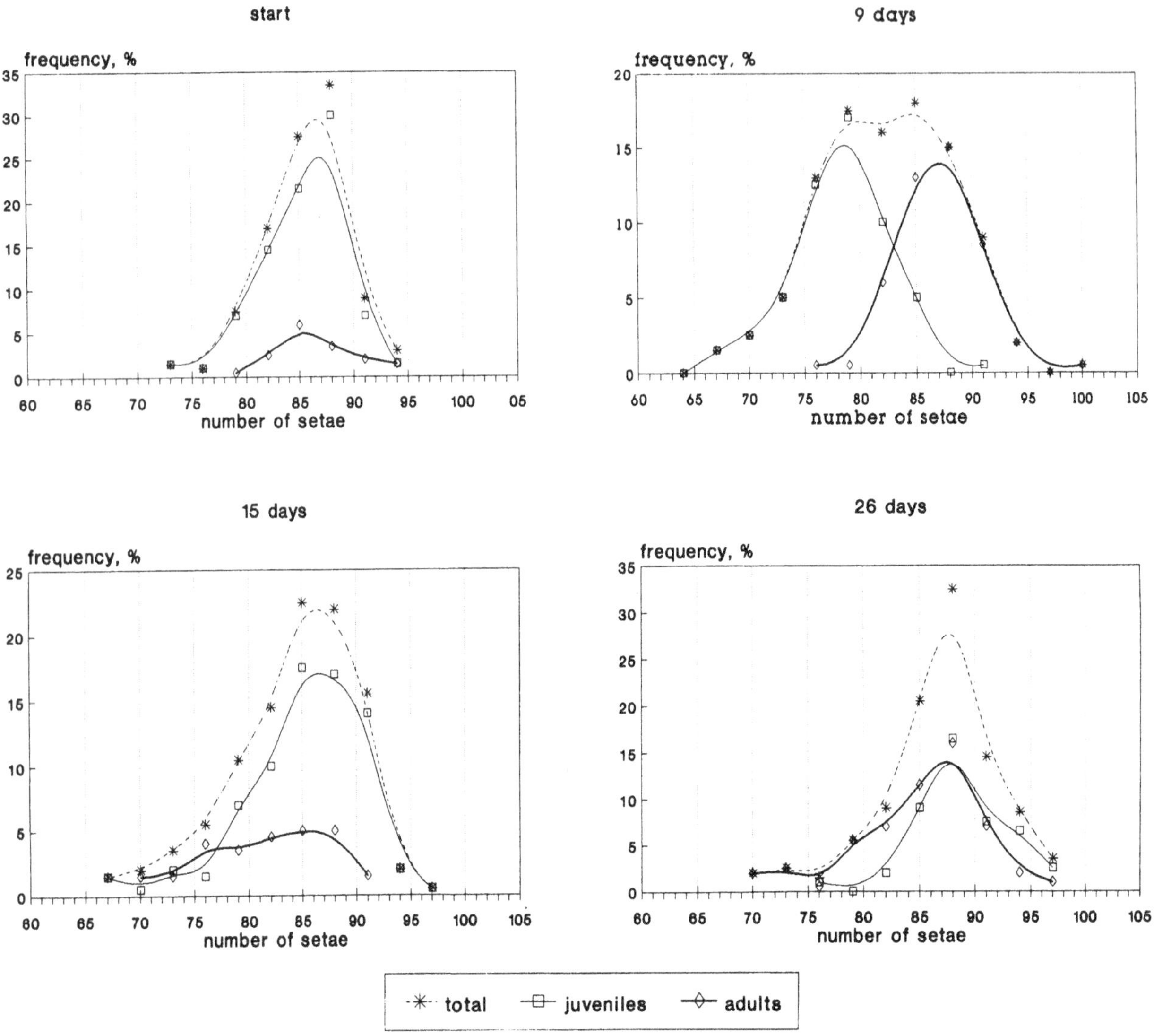

Fig. 4. Development of the distribution of numbers of setae in the subpopulation in the B1 barrel during 26 days.

Křepelová data

When computing original data, gathered by Křepelová in 1980 (Kořínek *et al.*, 1986), I found a close relation between seta length and the projection of the filtering area (1). This calculation involves several populations from Europe and Canada, from different food and temperature conditions. A very high level of correlation enables us to use this equation for calculation of the size of projected filtering area (PFA) from the length of setae without using a planimeter (Egloff & Palmer, 1971; Crittenden, 1981), and, moreover, there is evidence that the setae growth proportionally to the elongation of the base of the comb.

$$y = 1.879 \cdot x^{1.996};\ r = 1 \qquad (1)$$

y – projected filtering area (PFA) of one comb (in mm^2)
x – seta length (SL) (in mm)

Discussion

1. There are two steps in the changing of the size of filtering areas in a population:

1st – practically immediate change of the RSL in every individual of the population affected by a change in the seston concentration. The lag time is equal the moulting period between two stages.

2nd – this fast changes is followed by a slower change in the mean number of setae within the comb, which resulted from the replacement of individuals in a population.

This suggests that only the seta length determines the size of PFA (of course proportionally to the length of the base of the comb), while the number of setae has likely the function of holding the mechanical solidity of the filtering structure of the comb, as well as taking part in the optimization of the filter mesh-size of the filtering comb.

2. No separate groups of animals were found, neither in the RSL, nor in the number of setae, appearance of which would support the previously mentioned, 'clonal hypothesis', which was rejected.

Acknowledgements

I would like to thank Dr. V. Kořínek for original ideas and helpful support during my work, as well as for valuable comments on the manuscript; Dr. E. Stuchlík for critical comments on an earlier version of this paper.

References

Arruda, J. A., 1983. Daphnid filtering comb area and the control of filtering rate. J. Freshwat. Ecol. 2: 219–224.

Bleiwas, A. H. & P. M. Stokes, 1985. Collection of large and small food particles by Bosmina. Limnol. Oceanogr. 30: 1090–1092.

Brendelberger, H., 1985. Filter mesh-size and retention efficiency for small particles: comparative studies with Cladocera. Arch. Hydrobiol. Beih. Ergebn. Limnol. 21: 135–146.

Brendelberger, H. & W. Geller, 1985. Variability of filter structures in eight Daphnia species: mesh size and filtering areas. J. Plankton Res. 7: 473–486.

Brendelberger, H., M. Herbeck, H. Lang & W. Lampert, 1986. Daphnia's filters are not solid walls. Arch. Hydrobiol. 107: 197–202.

Burns, C. V., 1968. Direct observations of mechanisms regulating feeding behavior of Daphnia, in lakewater. Int. Revue ges. Hydrobiol. 53: 83–100.

Burns, C. W., 1968. The relationship between body size of filter-feeding Cladocera and the maximum size of particle ingested. Limnol. Oceanogr. 13: 675–678.

Cannon, H. G., 1933. On the feeding mechanisms of Branchiopoda. Phil. Trans. r. Soc. B, 222: 267–352.

DeMott, W. R., 1985. Relation between filter mesh-size, feeding mode, and capture efficiency for cladocerans feeding on ultrafine particles. Arch. Hydrobiol. Beih. Ergebn. Limnol. 21: 125–134.

Egloff, D. A. & D. S. Palmer, 1971. Size relations of the filtering areas of two Daphnia species. Limnol. Oceanogr. 16: 900–905.

Fott, J., 1975. Seasonal succession of phytoplankton in the fish pond Smyslov near Blatná, Czechoslovakia. Arch. Hydrobiol. Suppl. (Algological Studies 12) 46: 259–279.

Fott, J., 1987. Phytoplankton of carp ponds. Arch. Hydrobiol. Beih. Ergebn. Limnol. 27: 251–256.

Fryer, G., 1987. The feeding mechanisms of the Daphniidae (Crustacea: Cladocera): recent suggestions and neglected considerations. J. Plankton. Res. 9: 419–432.

Ganf, G. G. & R. J. Shiel, 1985. Particle capture by Daphnia carinata. Aust. J. Mar. Freshwat. Res. 36: 371–381.

Ganf, G. G. & R. J. Shiel, 1985. Feeding behaviour and limb morphology of two cladocerans with small intersetular distances. Aust. J. mar. Freshwat. Res. 36: 69–86.

Geller, W. & H. Müller, 1981. The filtration apparatus of Cladocera: Filter mesh-sizes and their implications on food selectivity. Oecologia (Berlin) 49: 316–321.

Gerritsen, J. & K. G. Porter, 1982. The role of surface chemistry in filter feeding by zooplankton. Science 216: 1225–1227.

Gophen, M. & W. Geller, 1984. Filter mesh size and food particle uptake by Daphnia. Oecologia (Berlin) 64: 408–412.

Haney, J. F., 1985. Regulation of cladoceran filtering rates in nature by body size, food concentration, and diel feeding patterns. J. Plankton Res. 30: 397–411.

Hayward, R. S. & D. N. Gallup, 1976. Feeding, filtering and assimilation in Daphnia schoedleri Sars as affected by environmental conditions. Arch. Hydrobiol. 77:

Hebert, P. D. N. & T. J. Crease, 1980. Clonal coexistence in Daphnia pulex (Leydig): another planktonic paradox. Science 207: 1363–1365.

Hebert, P. D. MN., 1980. The genetics of Cladocera. in: Kerfoot, W. C. (ed.), Evolution and Ecology of Zooplankton communities, Hanover, N.H.: 329–336.

Helgen, J. C., 1987. Feeding rate inhibition in crowded Daphnia pulex. Hydrobiologia 154: 113–119.

Hessen, D. O., 1985. Filtering structures and particle size selection in coexisting Cladocera. Oecologia 66: 368–372.

Horn, W., 1985. Results regarding the food of the planktic crustaceans Daphnia hyalina and Eudiaptomus gracilis. Int. Revue ges. Hydrobiol. 70: 703–709.

Horton, P. A., M. Rowan, K. E. Webster & R. H. Peters, 1979. Browsing and grazing by cladoceran filter feeders. Can. J. Zool. 57: 206–212.

Janicki, A. & J. DeCosta, 1984. The filtering rates of four herbivorous crustaceans on nannoplankton and net plankton fractions in an acid lake in west Virginia. Int. Revue ges. Hydrobiol. 69: 643–652.

Kořínek, V. & J. Macháček, 1980. Filtering structures of Cladocera and their ecological significance. I. Daphnia pulicaria. Vest. ceskoslov. Spolec. zool. 44: 213–218.

Kořínek, V., B. Křepelová-Macháčková & J. Macháček, 1986. Filtering structures of Cladocera and their ecological significance. II. Relation between the concentration of the seston and the size of filtering combs in some species of the genera Daphnia and Ceriodaphnia. Vest. ceskoslov. Spolec. zool. 50: 244–258.

Koza, V. & V. Kořínek, 1985. Adaptability of the filtration screen in Daphnia: Another answer to the selective pressure of the environment. Arch. Hydrobiol. Beih. Ergebn. Limnol. 21: 193–198.

Lampert, W., 1975. A laboratory system for the cultivation of large numbers of Daphniae under controlled conditions. Arch. Hydrobiol./Suppl. 48 1: 138–140.

Lorenzen, C. J., 1967. Determination of chlorophyll and pheo-pigments: spectrophotometric equations. Limnol. Oceanogr. 12: 343–346.

McMahon, J. W. & F. H. Rigler, 1963. Mechanisms regulating the feeding rate of Daphnia magna Straus. Can. J. Zool. 41: 321–332.

Porter, K. G., Y. S. Feig & E. F. Vetter, 1983. Morphology, flow regimes and filtering rates of Daphnia, Ceriodaphnia and Bosmina fed natural bacteria. Oecologia 58: 165–163.

Richman, S. & S. I. Dodson, 1983. The effect of food quality on feeding and respiration by Daphnia and Diaptomus. Limnol. Oceanogr. 28: 948–956.

Rubenstein, D. I. & M. A. R. Koehl, 1977. The mechanism of filter feeding: some theoretical consideration. Am. Nat. 111: 981–994.

Vijverberg, J., 1989. Culture techniques for studies on the growth, development and reproduction of copepods and cladocerans under laboratory and *in situ* conditions: a review. Freshwat. Biol. 21: 317–373.

Watts, E. & M. Petri, 1981. A scanning electron-microscope study of the thoracic appendages of Daphnia Magna Straus. J. Nat. Hist. 15: 463–473.

Hydrobiologia **225**: 177–183, 1991.
V. Kořínek & D. G. Frey (eds), Biology of Cladocera.

Effect of previous feeding conditions on survival in *Daphnia pulicaria* (Crustacea: Cladocera)

Zuzana Stuchlíková
Department of Hydrobiology, Charles University, Viničná 7, CS-12844 Prague 2, Czechoslovakia

Key words: Daphnia, starvation, survival

Abstract

Survival of natural *Daphnia* populations without previous acclimatization in filtered water under laboratory conditions (starvation) was used as a parameter describing the actual competitive ability of populations in periods of food limitation. Two populations (1 and 2) of *Daphnia pulicaria* Forbes coming from two enclosures with a low (< 1 mgC l^{-1}) and a high (about 3 mgC l^{-1}) seston level were examined in their resistance to starvation. Juveniles of both populations survived better than adult females in the conditions without seston as well as in the control experiment, despite their much lower biomass. Animals from population 1 had lower mortality than those from population 2 in all variants of the experiment. The results confirm the hypothesis that the ability of daphnids to starve depends also on adaptation to previous feeding conditions.

Introduction

Starvation ability (resistance), or more exactly survival in very low food concentrations, seems to be an important part of the ecological strategy of daphnids. Since advantageous life strategy, as well as effective gathering of food and effective metabolism, are the decisive factors in competition, the success of species (populations, clones) in competition and their consequent (co)existence in the ecosystem can be determined under critical feeding conditions (e.g. during clear water periods) by their ability to withstand starvation (Tessier *et al.*, 1983).

When starving, animals lose their energy reserves and biomass and they die gradually (Lemcke & Lampert, 1977). As proved experimentally, starvation resistance depends not only on individual biomass (Threlkeld, 1976) but also on energy reserves of individuals; the influence of the latter is probably more important (Tessier *et al.*, 1983). I aimed to find whether and how previous feeding conditions influence survival of daphnids in starvation, i.e. to assess the effect of a long-term adaptation to life in low or high concentrations of food on their starvation time. Further, I wanted to use starvation ability as a parameter describing the actual competitive ability of populations under food limitation.

Materials and methods

Two natural populations of *Daphnia pulicaria* Forbes from enclosures (limnocorrals) in the carp pond Velký Pálenec near Blatná in SW Bohemia were examined under laboratory conditions in their starvation ability. The experiment was conducted at the Hydrobiological station of Charles University in late summer of 1985 (August 23 –

September 8). For detailed information on the field enclosures project, see Černý & Bytel (1991).

The two populations under study developed from the original pond population of *Daphnia pulicaria* by their short-term isolation in different fish predation and food conditions (Table 1). Population 1 came from enclosure 1 which was kept fishless from the end of July (the enclosures were installed in the pond in mid-June). Immediately after removing planktivorous coarse fish, *Leucaspius delineatus*, from the enclosure, *Daphnia pulicaria* and *D. magna* began to dominate over *D. galeata* in this enclosure. Intensive filtering activity of the daphnids resulted in a decrease in concentration of seston; the clear water (Kořínek *et al.*, 1987) period lasted up to the beginning of my experiment. Population 2 was collected from enclosure 3 with low predation (coarse fish) and a higher seston level. Throughout the season, *Daphnia pulicaria* coexisted there with *D. galeata* (the latter species dominated up to the beginning of September).

In the experiment, both adult females (without embryos and ephipia) and early instars of juveniles were used – always a random selection from a representative quantitative sample collected from the enclosure at night with a Schindler-Patalas sampler (28.1 l). Parameters of initial body mass of examined animals are given in Table 1. Quantitative samples of zooplankton from both enclosures collected on August 18 were stored at natural density in a polyethylene barrel, then individuals were collected and kept 1–2 days before the start of the experiment in water from original localities at 20 °C.

In the *starvation* experiment, I studied survival of daphnids in filtered water from enclosure 1. Water was gradually filtered through a sieve with mesh size 39 μm, a glass fibre filter, a series of nitro-cellulose membranc filters Synpor® (Barvy

Table 1. Origin and some characteristics of the examined populations of *Daphnia pulicaria* and experimental design. Fish predation, species composition of daphnids and seston concentration in enclosures 1 and 3 before the beginning of the experiment are shown (1). Individual body mass as dry weight (DW), organic carbon content and percentage carbon content at the beginning of the experiment is recorded for adults and juveniles of each population (2). 3 – design of the experiment: numbers of animals (n) and parameters of water used in the starvation and control experiments. DP – *Daphnia pulicaria*, DG – *D. galeata*, DM – *D. magna*; POC – particulate organic carbon, TPV – total particulate volume, DOC – dissolved organic carbon.

	Population 1		Population 2	
	Adults	Juveniles	Adults	Juveniles
1 origin	enclosure 1		enclosure 3	
fish predation	no predation		low predation	
daphnids' composition	DP, DG, DM		DG, DP	
POC, mg l^{-1}	0.5–0.7		2–3.5	
chlorophyll-*a*, μg l^{-1}	1–2		9–30 (mostly around 20)	
TPV, mm^3 l	5–9		16 (August 18)	
2 DW, μg	45.0	6.8	36.9	4.3
org. C, μg	18.7	2.4	13.3	–
org. C as % of DW	42	34	36	–
3 n (starvation/control)	14/15	20/19	15/14	18/20
starvation:	filtered water from enclosure 1			
DOC, mg l^{-1}	7.5–9.4			
control:	water from enclosure 1		water from enclosure 3	
POC, mg l^{-1}	0.6–2.7		3.2–5.2	
chlorophyll-*a*, μg l^{-1}	2–11		30–48	
TPV, mm^3 l	6–9		18–22	

a laky, Praha) with a decreasing porosity (Synpor No 2 and 4) and a bacteriological glass candle to remove bacterial size particles. The whole procedure was finished by filtration through a filter Synpor No 7 equivalent to porosity 0.3 μm and by aeration to 100% saturation. Dissolved organic carbon (DOC) was measured daily.

In the *control* experiment, animals were cultivated in water from original habitats (from enclosure 1 for population 1, from enclosure 3 for population 2) filtered through a sieve with mesh size 39 μm and aerated to 100% saturation before use. Seston concentration in this water was checked regularly (once per 2 days) – see Table 1.

14–15 adults and 18–20 juveniles of each population were placed into 250 ml polyethylene bottles. Bottles were fixed at a rotary desk (4 RPM) to prevent sedimentation of particles, and immersed in a thermostated water bath (20 °C), under illumination of 1000 Lux and light : dark regime 12 : 12. Animals in all variants of the experiment were changed every 1–2 days to fresh water and clean bottles; numbers of dead and new-born animals were recorded. Concentration of oxygen in bottles (measured by an electrode, Orion) did not decrease below 50% in intervals between changes of water, concentration of food, however, decreased considerably.

Seston was determined as chlorophyll-*a* using hot extraction in an acetone : methanol mixture (Pechar, 1987), as total particulate volume (TPV) measured on Coulter counter – model ZB extended with an 18 channel analyzer in the range of size 12 288–0.375 μm^3, and as reducing capacity by a dichromate oxidation micromethod (Mackereth *et al.*, 1978). Particulate organic carbon (POC) was estimated by multiplication of reducing capacity of organic material by the factor 3. The same method was used for determination of dissolved organic carbon (DOC) and organic carbon in daphnids.

Results

The results of my experiments are presented in Figs 1 and 2. Survival curves for the two populations are summarized in Fig. 1 separately in starvation and in the control experiments. Survival curves both in starvation and in the control experiments are combined in Fig. 2, and compared for populations 1 and 2. Table 2 gives calculations of starvation time in the populations under study in comparison with the theoretical value based on the Tessier's model (Tessier *et al.*, 1983).

Regarding the *control* experiment (Fig. 1), high

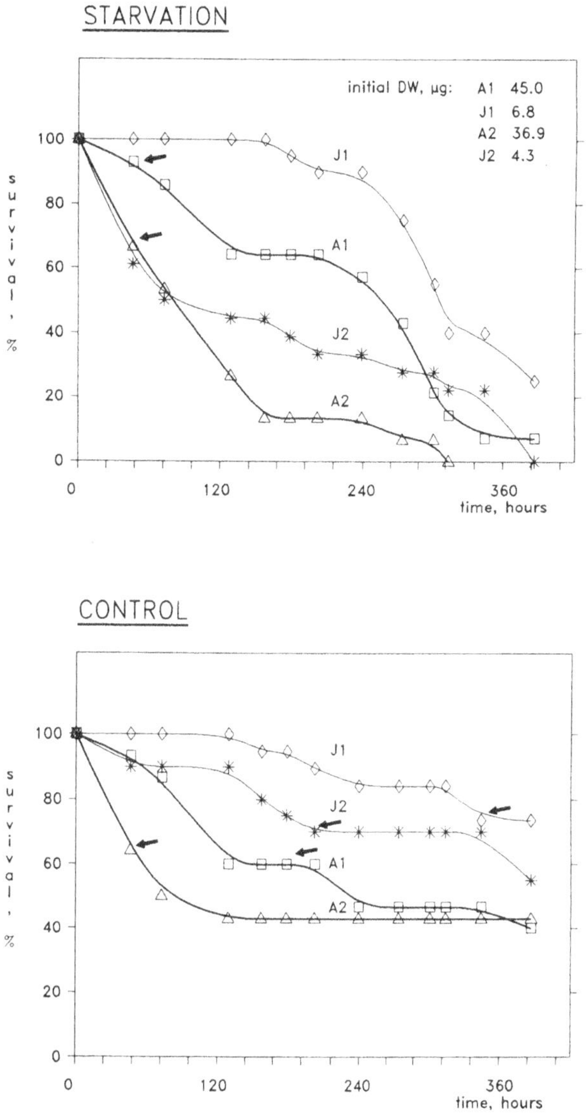

Fig. 1. Survival of *Daphnia pulicaria* populations 1 and 2 in the starvation experiment and in the control. Individual body mass as dry weight (DW) of adults (A) and juveniles (J) at the beginning of the experiments is denoted. Arrows indicate the time when the first neonates were found.

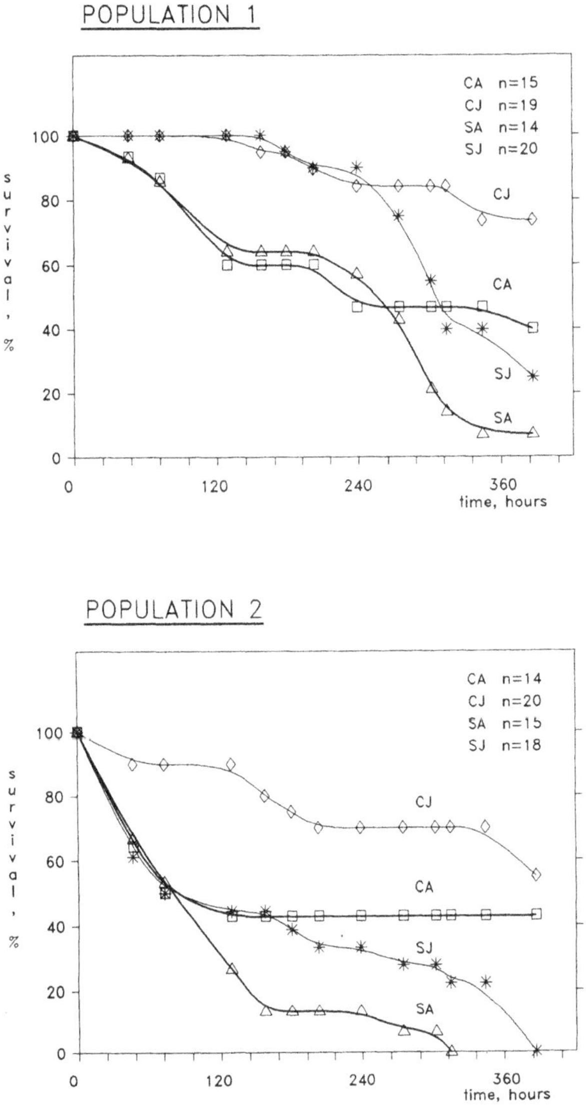

Fig. 2. Survival of *Daphnia pulicaria* in the starvation (S) and control (C) experiments – comparison between populations 1 and 2. The same data as in Fig. 1 are shown. A – adults, J – juveniles; for both populations numbers of animals in the experiments (n) are indicated.

mortality of adults was surprising at the very beginning of the experiment (during the first 129 hours). Population 2 was affected more than population 1. The initial mortality caused a decrease in numbers of adults to 60% (in population 1) and 43% (in population 2); at the end of the experiment (after 387 hours), about 40% of adults survived in both populations, 73% of juveniles of population 1 and 55% of juveniles of population 2.

Different previous feeding histories of examined populations and different concentrations of food in the experiment led to different total development times and different production of newborns. In juveniles of population 1, the first neonates appeared after 344 hours (development time approximately 14 days), while in juveniles of population 2 after 203 hours (development time 8.5 days). Adult females of population 1 (without eggs at the beginning of the experiment) started to produce neonates after 179 hours only. The production of neonates was very low throughout the experiment, increasing from 0.1 to 1.3 neonates per female. In adults of population 2, 16 neonates were found after 47 hours (1.8 neonates per female); later, neonates appeared only after 157 hours and their number increased from 1.3 to 10 neonates per female.

A conspicuous reduction in numbers of animals was recorded from the beginning of each variant of the *starvation* experiment (Fig. 1); only juveniles of population 1 were not affected. Population 1 withstood starvation better than population 2, and juveniles of both populations were more successful than adults as in the control experiment. Adult females of population 1 produced neonates throughout the experiment but they were few in numbers (altogether 8, of which 5 were found after 239 hours of the experiment). In adults of population 2, the only 16 neonates appeared after 47 hours as in the control; further on only females with empty brood chambers were observed. Juveniles of neither population reached maturity despite their regular moulting (exuviae were found in experimental bottles).

To evaluate survival in the starvation experiment (Table 2), it was necessary to consider the high initial mortality in the control. In juveniles of population 1, survival curves in both starvation and the control experiments were almost identical up to the 250th hour (10.5–11 days); in adults of population 1, the situation was similar (Fig. 2). It means that the negative effect of starvation was noticeable only from the moment when the number of animals in the control remained almost constant while survival curves of starving individuals dropped. Survival curves of adults of

Table 2. Comparison between various calculations of starvation time in the starvation experiment and starvation time predicted by the model ($T = 66.0\ W^{0.165}$, where T is starvation time in hours and W is body mass in μg: Tessier *et al.*, 1983). LT50 – time when 50% of animals survived in the experiment, $\bar{x}$ – arithmetical weighted mean of starvation time, $\hat{x}$ – time of maximal mortality. LT50 and $\hat{x}$ are corrected for mortality in the control experiment. A1, A2, J1, J2 – adults and juveniles of populations 1 and 2.

	Starvation time (hours)			
	LT50	$\bar{x}$	$\hat{x}$	Model
A1	301	>220	302	124
J1	321	>>301	302	91
A2	132	120	129	120
J2	132	173	47	84

population 2 diverged after 73 hours of the experiment. From the 129th hour, the number of animals did not change in the control but it decreased with the same rate in the starvation experiment. In juveniles, there was a great difference between survival curves in starvation and in the control experiments from the beginning: animals survived well in the control while the number of starving individuals decreased.

In cases when survival curves in the starvation experiment first followed survival curves in the control, the moment when the two curves diverged was taken as an actual beginning of the experiment. In starving juveniles of population 1, 85% of individuals still survived after 239 hours and maximal mortality was recorded after 302 hours of the experiment. Adults of population 1 had two maxima of mortality. The first one (after 129 hours) corresponds with the situation in the control and its cause is far from clear; however, the second maximum (after 302 hours) is without any doubt the result of starvation. The number of juveniles of population 2 sharply decreased at the beginning of the starvation experiment as well as close to the end of the experiment. Considering that only a small initial drop was recorded in the control, I supposed starvation to affect mortality immediately (47 hours) after the start of the experiment. The initial mortality of adults of population 2 was probably the consequence of some other factors and only the next drop in number of animals (after 129 hours) was the result of starvation.

Various attempts at expressing starvation time with respect to the initial mortality which could have been caused by other factors rather than by starvation, and values of starvation time provided by the model (Tessier *et al.*, 1983) are compared in Table 2. The relatively high initial mortality in starvation as well as in the control, the fact that survival curves differ from a normal distribution, and the presence of several maxima of mortality make the calculations difficult. The corrected values of 50% mortality time (LT50) coincide closely with times of the maximal decrease in numbers of animals ($\hat{x}$). The value for juveniles of population 2 is loaded with the largest error. A good agreement between the model and the experiment was found only for adults of population 2 while mean starvation times ($\bar{x}$) in the other variants of the experiment were longer (more than three times) than those predicted by the model. Juveniles of population 2 reached similar mean starvation time as adults of the same population; time of their maximal mortality ($\hat{x}$), however, was much shorter. Juveniles of population 1 have an even longer mean starvation time than do adults. Similar results are provided also by Fig. 1 (starvation).

Discussion

The interpretation of experimental results is hampered by the high and unexpected mortality in the control. Tessier *et al.* (1983) showed in starvation experiments with *Daphnia galeata* and *D. magna* that at high food levels more than 95% of animals survived after the first 25 days. However, the investigated populations were acclimatized for two generations before these experiments. No previous acclimatization was performed in my experiments aimed to test the starvation (and competitive) ability of natural populations. An enhanced sensitivity of late summer natural populations of daphnids to any change of conditions, including a transfer to the laboratory, may have also been a source of mortality. There are usually problems with reproduction of populations collected in this period of a year in a culture. On the other hand, development times in well fed population 2 in the control experiment are in a good agreement with data of Hrbáčková & Hrbáček (1978).

Relatively long starvation times seem to suggest that true 'hungry' conditions were not ensured in the experiment. Non-sterile water with a high content of dissolved organic compounds (7.5–9.4 mgC l^{-1}) is an environment suitable for the growth of bacteria and heterotrophic flagellates (Salonen & Hammar, 1986) which are a potential source of food. Besides, organic compounds can be utilized directly (Porter *et al.*, 1982). In the starvation experiment, guts of the daphnids were full of live, dark, relatively large, rod-shaped bacteria. I have no data to decide whether these bacteria developed in water during intervals between changes of water, and whether they were gathered by daphnids. I suppose rather that bacteria lived permanently in a gut assisting to utilize the dissolved organic compounds. However, starvation times in adults of population 2, close to the values predicted by the model, weaken this assumption. Consequently, longer starvation times in population 1 indicate more likely that a markedly lower investment into the growth and reproduction (as a base of the life strategy developed during a long-time exposure to low food) influenced positively the resistance of the population to starvation. Neonates of *Daphnia pulicaria* growing in low food conditions need more instars to reach maturity, and their relative length increments are smaller than in neonates from high food conditions (Hrbáčková-Esslová, 1963). The mechanism seems to be essential for saving energy in a period of food limitation and for prolongation of starvation time. Three times longer starvation times in juveniles in my experiment can hardly be explained only on the basis of 20% differences in dry weight and/or in organic carbon content. Better ability of juveniles to withstand starvation originates probably in a sufficient amount of maternal lipid (Tessier *et al.*, 1983) and in the fact that they do not spend energy for reproduction.

A several week isolation of a population of *Daphnia pulicaria* in enclosures with different fish predation resulted in quite different food concentrations, owing to the filtering activity of daphnids (Černý & Bytel, 1991). Adaptation to the new feeding conditions is probably responsible for longer starvation time of the population from low food level, in comparison with the population from high food level. It seems that adaptation to low food plays a more important role in starvation resistance than individual biomass (Threlkeld, 1976) and/or the amount of lipid reserves (Tessier *et al.*, 1983). Unfortunately it is impossible to decide whether differences in starvation resistance between populations 1 and 2, and between adults and juveniles, resulted from differences in energy reserves; however, Novotná (1987) found that populations from low food conditions have a lower lipid index.

The use of starvation ability as a parameter to estimate which population will be competitively superior in nature under critical feeding conditions is somewhat problematic. The results show that too many factors influence survival of daphnids in starvation and their persistence in competition. Moreover, a transfer of natural *Daphnia* to the laboratory without previous acclimatization results in a high initial mortality in the experiment.

Acknowledgements

I am very grateful to Evžen Stuchlík for helpful suggestions and assistance in the laboratory, and to Dr. Vladimír Kořínek for comments on the manuscript.

References

Černý, M. & J. Bytel, 1991. Density and size distribution of *Daphnia* populations at different fish predation levels. In: Kořínek, V. & D. G. Frey (eds), Biology of Cladocera. Developments in Hydrobiology 71. Kluwer Academic Publishers, Dordrecht: 199–208. Reprinted from Hydrobiologia 225.

Hrbáčková-Esslová, M., 1963. The development of three

species of Daphnia in the surface water of the Slapy reservoir. Int. Revue ges. Hydrobiol. 48: 325–333.

Hrbáčková, M. & J. Hrbáček, 1978. The growth rate of Daphnia pulex and Daphnia pulicaria (Crustacea: Cladocera) at different food levels. Věst. čs. Společ. zool. 42: 115–127.

Kořínek, V., J. Fott, J. Fuksa, J. Lellák & M. Pražáková, 1987. Carp ponds in Central Europe. In R.G. Michael (ed.), Managed Aquatic Ecosystems. Elsevier Science Publishers B.V., Amsterdam: 29–62.

Lemcke, H. W. & W. Lampert, 1977. Veranderungen im gewicht und der chemischen zusammensetzung von Daphnia pulex im hunger. Arch. Hydrobiol. Suppl. 48: 108–137.

Mackereth, F. J. H., J. Heron & J. F. Talling, 1978. Water analysis: Some revised methods for limnologists. FBA Scientific publication No. 36, 120 pp.

Novotná, Z., 1987. Vliv koncentrace potravy na parametry biomasy u několika druhů rodu Daphnia. M.Sc. Thesis, Charles University, Prague, 69 pp. [Effect of concentration of food on biomass parameters within several species of the genus Daphnia.]

Pechar, L., 1987. Use of an acetone : methanol mixture for the extraction and spectrophotometric determination of chlorophyll-*a* in phytoplankton. Arch. Hydrobiol. Suppl. 78: 99–117.

Porter, K. G., J. Gerritsen & J. D. Orcutt, Jr., 1982. The effect of food concentration on swimming patterns, feeding behavior, ingestion, assimilation, and respiration by Daphnia. Limnol. Oceanogr. 27: 935–949.

Salonen, K. & T. Hammar, 1986. On the importance of dissolved organic matter in the nutrition of zooplankton in some lake waters. Oecologia (Berlin) 68: 246–253.

Tessier, A. J., L. L. Henry, C. E. Goulden & M. W. Durand, 1983. Starvation in Daphnia: Energy reserves and reproductive allocation. Limnol. Oceanogr. 28: 667–676.

Threlkeld, S. T., 1976. Starvation and the size structure of zooplankton communities. Freshwat. Biol. 6: 486–496.

Hydrobiologia **225**: 185–191, 1991.
V. Kořínek & D. G. Frey (eds), Biology of Cladocera.

Reproductive effort in *Daphnia galeata*: comparison of reservoir and pond populations

Zdeněk Brandl & Monika Wittingerová
Hydrobiological Institute, Czechoslovak Academy of Sciences, Na sádkách 7, CS – 370 05 České Budějovice, Czechoslovakia

Key words: Reproduction, *Daphnia galeata*, ponds and reservoirs, clones and populations

Abstract

Populations of *Daphnia galeata* from the deep stratified moderately eutrophic reservoir and the shallow highly eutrophic carp ponds differed in the reproductive effort of the first laboratory generation cultivated at a rich algal diet. The reservoir animals were smaller at the first adult instar and exhibited longer time between hatching of one brood and forming the eggs of the next one.

Introduction

Daphnia galeata Sars is a common cladoceran species in both the shallow eutrophic carp ponds and deep stratified reservoirs of southern Bohemia. Nevertheless, the pond and reservoir populations apparently differ in some properties. The aim of this study was to compare some characteristics of the population growth and reproduction of *Daphnia galeata* from these two different habitats under the same laboratory conditions at a high algal food level.

Material

The reservoir population of *Daphnia galeata* was obtained from the Římov Reservoir, a moderately eutrophic stratified reservoir (maximum depth 44 m, surface area 2.1 km^2, mean retention time 95 days; for other details see Brandl *et al.*, 1989). The reservoir was built for the retention of drinking water and is inhabited by dense populations of perch, roach, and bream (Kubečka, 1989). Concentration of the chlorophyll-*a* in the 'edible' size fraction of algae (less than 0.04 mm) varied during the vegetation season of the years 1987 and 1988 between 2 and 20 mg m^{-3} in the upper 5 m layer (Komárková, 1989).

The nearest large system of carp ponds, i.e. the group of ponds between České Budějovice and Hluboká n. Vlt., was chosen as source of a pond material, as well as the source of the phytoplankton-rich pond water. The pond system consists of numerous carp ponds with the size from 0.1 to almost 1 km^2, the mean depth rarely exceeding 1 m, some of them very close to one another and all situated in the watersheds of two neighbouring brooks. They are highly eutrophic and used for fish rearing, mostly common carp, which is periodically present (usually for about 18–20 months from the spring of one year till the autumn of the next year).

Daphnia galeata was collected from the pond Naděje nr. Hluboká (0.31 km^2), whereas algal food for the experiments was obtained from the ponds Starý Haklovský, Starý Vrbenský, and Nový Vrbenský. These ponds had fish stocks high enough to eliminate larger cladocerans and chlorophyll-*a* concentration usually between 10 and 300 mg m^{-3}. Since extensive limnological research was carried out in the other pond area, namely in the ponds nr. Blatná, southwestern Bohemia (see e.g. Fott *et al.*, 1974, 1980 and Kořínek *et al.*, 1987), we used another pond population of *Daphnia galeata* from the pond Starý nr. Blatná for some comparative experiments, too.

Methods

Water from the Římov Reservoir filtered through 0.04 mm mesh was used for the stock cultures of *Daphnia* and all the experiments in which laboratory culture of *Scenedesmus dimorphus* was used as a food. For the experiments in which *Daphnia* were feeding on the natural algal assemblage, fresh pond water was collected every second day and filtered through 0.04 mm mesh to contain small-cell algae only. These were represented by numerous species of green algae and cryptomonads, giving the chlorophyll-*a* concentration of 0.1–0.3 mg l^{-1}. (The dominants: *Asterionella formosa*, *Cryptomonas curvata*, *C. reflexa*, *Scenedesmus acuminatus*, *S. dimorphus*, *S. quadricauda*, *Monoraphidium* sp., *Nitzschia* sp., *Pediastrum* sp., *Tetrastrum* sp.).

The laboratory clones of *Daphnia galeata* were derived as parthenogenetic descendants of individual females randomly selected from the field population. They were kept in two replicates at 15 °C and fed with *Scenedesmus dimorphus*. Similarly, all experiments, except for some of those designed to measure the embryonic development rate, were carried out at 15 °C in a 12 h : 12 h light : darkness regime.

We have made several types of experiments differing in purpose and design.

The ability to reproduce at a high level of natural food was tested as follows: 50–60 females ('grandmothers') were randomly chosen from the field population, brought to the laboratory and allowed to reproduce in water from their habitat. The neonates produced within 24 h were separated and cultivated for 10 days, i.e. the period slightly shorter than the length of time of the post-embryonic development at 15 °C. During this period, water was replaced every 48 h. The mixture 1 : 1 of water from the original habitat (either the reservoir or the pond) and from another phytoplankton-rich pond was used when the first exchange of water came about. Later only the water from another pond was supplied. After 10 days, 30 females ('mothers') were randomly selected from the cohort, for each replicate, placed into 3 l of pond water and allowed to reproduce for another 10 days. Water from these batch cultures was siphoned off and replaced by fresh pond water every 48 h. After 10 days, the cultures were preserved and the number of neonates produced per female and day was calculated (Fig. 1).

Experiments similar to those mentioned above were also made with clonal material after some period of its laboratory cultivation. In these experiments, groups consisting of 5 pond or reservoir clones were compare, each clone represented by two or three replicates. *Scenedesmus dimorphus* in concentrations exceeding 100 mg m^{-3} of chlorophyll-*a* was supplied as food during every 48 h exchange of water.

Long-term batch cultures similar in starting and feeding design to those derived from the field populations were established but then kept for the period of several weeks (up to 10). A definite part (up to 3/4) of each culture was periodically removed after a thorough mixing and censused. The remaining stock was allowed to grow further.

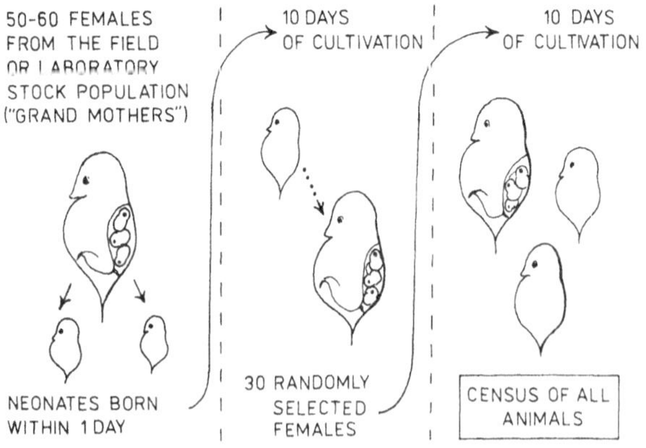

Fig. 1. The basic design of experimental procedure.

All the cultures were identically treated. Their reproduction was evaluated in approximately 10-day intervals as an exponential growth rate calculated as
$r = (\ln N_t - \ln N_o) \cdot t^{-1}$, where N_o and N_t are the respective numbers at the beginning and at the end of a t-day interval.

The duration of the embryonic and postembryonic development was measured in two ways.

1. We measured the cumulative length of time from the hatching of an animal to the hatching of its first brood, at a constant temperature and at a food level of more than 100 mg m^{-3} of chlorophyll-*a* (*Scenedesmus*).
2. Both embryonic and postembryonic development in the same food and temperature conditions as above were monitored in detail. The length of the individual embryonic stages was measured in 1-h intervals, as well as the length of the time between the hatching of one brood and the deposition of new eggs into the brood pouch after molting. The growth of subadult instars was measured daily.

Results

Reproductive effort of the pond and the reservoir populations

The first daughter generation of the pond stock of *Daphnia galeata* reared under laboratory conditions at a high food level produced a significantly higher number of offspring during the ten-day period than did the reservoir animals in two series of experiments started in March and April. Similarly, the laboratory descendants of the April stock of the reservoir population exhibited a lower production of neonates than the pond population after 7 weeks, although the difference was not so great (Table 1).

However, the reservoir population sampled in May supplied the experimental material, which practically did not differ from the contemporary pond population.

The long-term experiment starting in April and lasting till July revealed that despite the changing food conditions (high at the beginning, temporarily low, and then again with a high level of chlorophyll), the growth rate of the two populations differed during the first seven weeks and became almost identical later (Fig. 2). In a similar experiment starting at the end of October, the growth rates of both populations differed for about one month.

Table 1. Number of offspring produced by one female per day in 10-day experiments.

Origin of mothers of experimental females	Pond	Reservoir	t-test
Natural populations:			
March	3.14 ± 0.49	0.54 ± 0.15	*
April	4.26 ± 0.40	0.60 ± 0.15	*
May	4.62 ± 0.14	4.10 ± 0.06	NS
Laboratory population:			
April stock after 7 weeks of cultivation	4.21 ± 0.01	1.82 ± 0.23	*

* different at P 0.05
NS no statistically significant difference

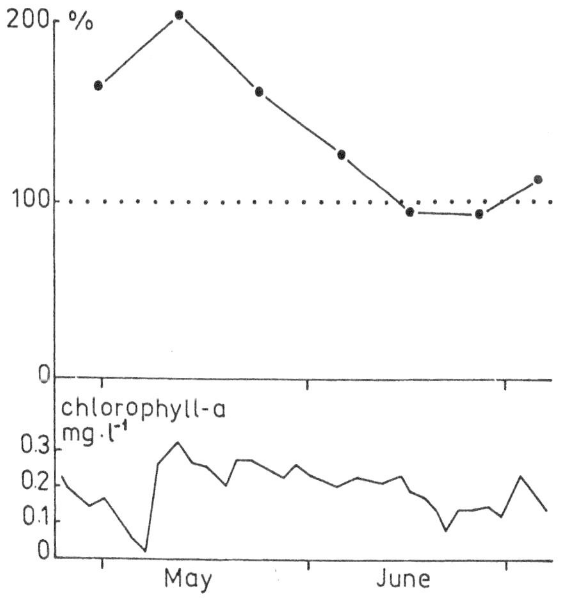

Fig. 2. Population growth rate r of the pond population in % of r of the reservoir population, 10-week batch experiment with the periodical reduction of the population size and feeding by the natural pond phytoplankton.

Reproductive effort of the laboratory clones

Laboratory clones isolated from both the pond and reservoir populations and cultivated for various periods lasting from a few weeks to about one year were used for similar experiments, in which the number of the offspring produced during the period from the 10th through the 20th day of life and being alive at the end of this time was determined. We compared five clones of the reservoir population with two similar groups of pond clones, one groups from the pond Naděje nr. České Budějovice and the other from the Blatná Pond District. The analysis of variance of the hierarchical experimental design (populations × clones × replicates, Table 2) has shown that most of the variation is due to the differences existing among clones within a population. Although the 'population' means (i.e. means for group of five clones from each population) differ only slightly (Table 2), the values within these groups of clones exhibit relatively high variation coefficients (up to 41%). When the same clones were examined in the repeated experiments after three or six months, their mutual relation (a higher or lower number of offspring produced by a definite clone) was similar even when the absolute number of produced neonates changed.

Table 2. Analysis of variance of the number of offspring produced by the laboratory clones of *Daphnia galeata* isolated six months earlier from two pond and one reservoir populations.

Source of variation	df	msq	F	
Among populations	2	0.585	0.26	NS
Among clones	12	2.262	22.06	**
Replicates	15	0.103		
Total	29			

Mean number of neonates per one female per day produced by the population from:

	Mean ± 1 SD
Římov Reservoir	2.85 ± 1.18
Pond Naděje	3.21 ± 1.21
Pond Starý	3.31 ± 0.57

Table 3. The duration of the cumulative length of the embryonic and postembryonic development of *Daphnia galeata* at 15 °C, in days (Means of the clonal means ± 1 SD, total $n = 525$).

Římov Reservoir	12.46 ± 0.39
Pond Naděje	12.03 ± 0.91
Pond Starý	12.04 ± 0.25

Length of the embryonic and postembryonic development

Since the hatching of neonates is easy to register, we have determined the cumulative length of the embryonic and postembryonic development (i.e. the time from the hatching of an animal to the hatching of its first brood) on the extensive material of about 525 animals. The results (Table 3) were very similar to those in the previous paragraph concerning reproduction. There was only little difference between the pond and the reservoir populations in the mean duration of the cumulative embryonic and postembryonic development, with a lower development rate in the reservoir population. However, very great variation was found again among the means of individual clones, much exceeding the difference between the populations. In contrast to the reproductive rate, there was also a great residual variation represented by the variability of individual animals.

Other characteristics of the populations

We have examined some other characteristics on a limited extent of the experimental material (groups of about 30 individually reared animals). Two pond and two reservoir clones reared for more than six months under laboratory conditions were used.

We did not find any significant difference between the pond and the reservoir populations in some characters including the duration of the embryonic development, the duration of the postembryonic development, the size of a neonate, the number of instars from hatching to the first brood

Table 4. Analysis of variance of the length at the stage of primipara of *Daphnia* individually reared under identical conditions:

Source of variation	df	msq	F	
populations	1	0.2612	35.3	*
clones	2	0.0466	6.3	*
mothers of examined animals	8	0.0094	1.3	NS
residual	19	0.0074		

(to the stage of a primipara), and the number of eggs in the first brood.

A significant difference was found in two characteristics: the size at the age of primipara and the time lapsed between the hatching of one brood and the deposition of new eggs into the brood chamber. Both the total length of the primipara and the width of its carapace was bigger in the pond animals (Table 4). The period between hatching and new egg deposition was quite different in the pond and the reservoir animals, averaging less than one hour in the pond populations and more than four hours in the reservoir ones.

Discussion

The spring increase of the population density of *Daphnia galeata* in the Římov Reservoir is delayed after the growth of algal populations as shown by Seďa (1989). Therefore, the *Daphnia* population is unable to manage an efficient top-down control of phytoplankton, and the phenomenon known as the 'clear-water phase' (Sommer *et al.*, 1986) develops only when a stabilized thermal stratification separates the euphotic layer from the hypolimnetic nutrient input and causes the nutrient limitation of algal growth (Komárková, 1989). Although predation on *Daphnia* may be an important key factor determining the density and structure of *Daphnia* population in the Římov Reservoir (Kubečka, 1989), the question we asked was whether the population of *Daphnia galeata* inhabiting this reservoir is able to react to the plentiful food to the same extent as the pond populations do. The egg production in *Daphnia* is known to depend partly on the quantity of available food (e.g. Hrbáčková-Esslová, 1963; Horn & Horn, 1987 or – for the Římov Reservoir – Seďa, 1989). To avoid the direct influence of the previous field conditions on the egg production of experimental females, we used the animals born just after the transfer to the laboratory. The number of progeny produced in the early stage of life of a cladoceran depends not only on the number of eggs per ovigerous female but also on the length of the subadult period, the length of embryonic development of each brood and – last but not least – the reproductive ratio, i.e. the percentage of ovigerous females in the total number of adult females.

Hrbáčková & Hrbáček (1979) reported that, after the food enrichment, the postembryonic development was shorter in the laboratory clones derived from the populations inhabiting more eutrophic water bodies.

The results of our experiments revealed that the animals obtained as the first laboratory generation of either the pond or the reservoir stock differed greatly in their population response to the rich algal food in the very first period of their mature life. The difference between these two habitat types was great in March and April but disappeared in May, when the field 'grandmother' population in the reservoir experienced the best food conditions of the year. The difference also disappeared after a long laboratory cultivation if the culture was established as a 'population culture' originating from the mixture of neonates obtained from the field sample of females. However, when isolated clones were cultivated, they differed from each other in the number of progeny produced during the second 10-day period of their life, even within a group of clones originating from the same population.

The changing ability of the reservoir stock of *D. galeata* to increase its reproductive effort after transfer into the non-limiting food conditions both during the season (March, April vs. May) and after a few weeks of laboratory cultivation contrasts with great and relatively constant interclonal differences. This suggests that the field population consists of numerous clones of different reproductive potential and that mutual proportion of these clones changes during the seasons. Hebert (1974) was the first to shown that *Daphnia* populations consist of a number of ecologically differentiated clones. These phenomena have received more attention recently: Pace *et al.* (1984) found clones or groups of clones in a population of *Daphnia parvula* differing in life history characters. Korpelainen (1986a) studied interclonal variation in *Daphnia magna*. Weider (1987) examined the variation of life history characters among nine clones of *Daphnia pulex* and found significant interclonal differences in the age at first reproduction, the number of eggs, and intrinsic rate of natural increase *r*. The temporal changes of clonal frequencies during the seasons are reported by Pace *et al.* (1984), Weider (1985), Korpelainen (1986b) and Carvalho & Crisp (1987).

Besides the brood size, other life history traits determine the results of a population reproductive effort. Among them, length of both embryonic and postembryonic development are highly determined by the environmental factors. Resulting variation makes it uneasy to disclose the inter-population variability on field collected material. Great differences among clones isolated from the same population and cultivated under identical laboratory conditions emphasize the need to determine the clonal structure of a population.

The size at the age of primipara is another life history trait reflecting the extent of resource allocation into growth and into reproduction. Hrbáček & Hrbáčková-Esslová (1960) when studying the growth rate of *Daphnia* under the identical laboratory conditions observed that strains from waterbodies with high stocks of planktivorous fish had smaller size at the instar of primipara. The difference we have found between the laboratory clones of reservoir and pond origin corresponds with the similar difference between the respective field populations and may have resulted from different field predation pressure in each of the respective habitats.

Conclusion

We examined the life history response of two habitat types of *Daphnia galeata* to a rich algal diet, using the generations in the laboratory produced by the field population sample. The population from the deep stratified moderately eutrophic reservoir with dense planktivorous fish differed greatly in its reproductive effort during the early stage of mature life (measured as the production of progeny during the second 10 day period of life, at 15 °C) from populations deriving from shallow and highly eutrophic carp ponds. The former were also smaller at the first adult instar and had longer time between hatching of one brood and the deposition of next clutch into a brood pouch. The difference in reproductive effort disappeared later in cultures derived from the population sample. Individual clones cultivated under identical conditions for half a year or more retained great interclonal variability in their reproductive effort. The results suggest that there is a changing clonal structure of the population both in the field and in the laboratory culture. The results also greatly emphasized the need of exact determination of clonal structure of the populations studied.

References

Brandl, Z., B. Desortová, J. Hrbáček, J. Komárková, V. Vyhnálek, J. Seďa & M. Straškraba, 1989. Seasonal changes of zooplankton and phytoplankton and their mutual relation in some Czechoslovak reservoirs. Arch. Hydrobiol. Beih. Ergebn. Limnol. 33: 597–604.

Carvalho, G. R. & D. J. Crisp, 1987. The clonal ecology of *Daphnia magna* (Crustacea: Cladocera). I. Temporal changes in the clonal structure of a natural population. J. anim. Ecol. 56: 453–468.

Fott, J., V. Kořínek, M. Pražáková, B. Vondruš & K. Forejt, 1974. Seasonal development of phytoplankton in fish ponds. Int. Revue ges. Hydrobiol. 59: 629–641.

Fott, J., L. Pechar & M. Pražáková, 1980. Fish as a factor controlling water quality in ponds. In J. Barica & L. R. Mur (eds.), Hypertrophic ecosystems. Developments in Hydrobiology, 2: 255–261. Dr. W. Junk Publ., Hague.

Hebert, P. D. N., 1974. Ecological differences between genotypes in a natural population of *Daphnia magna*. Heredity 33: 327–337.

Horn, W. & H. Horn, 1987. Field studies concerning the

fecundity of *Daphnia galeata*, its causes and consequences. Int. Revue ges. Hydrobiol. 72: 559–573.

Hrbáček, J. & M. Hrbáčková-Esslová, 1960. Fish stock as a protective agent in the occurrence of slow-developing dwarf species and strains of the genus *Daphnia*. Int. Revue ges. Hydrobiol. 45: 355–358.

Hrbáčková-Esslová, M., 1963. The development of three species of *Daphnia* in the surface water of the Slapy reservoir. Int. Revue ges. Hydrobiol. 48: 325–333.

Hrbáčková, M. & J. Hrbáček, 1979. Rate of the post-embryonic development in several populations of the group of the species *Daphnia hyalina* Leydig at various concentrations of food. Věstník Čs. spol. zoologické 43: 253–259.

Korpelainen, H., 1986a. Intraspecific variation in lifespan and fecundity under controlled conditions in the cyclic parthenogen, *Daphnia magna*. Arch. Hydrobiol. 105: 321–327.

Korpelainen, H., 1986b. Temporal changes in the genetic structure of *Daphnia magna* populations. Heredity 57: 5–14.

Komárková, J., 1989. Changes of phytoplankton assemblage during the spring period in the moderately eutrophic Římov reservoir (Czechoslovakia). Arch. Hydrobiol. Beih. Ergebn. Limnol. 33: 419–433.

Kořínek, V., J. Fott, J. Fuksa, J. Lellák & M. Pražáková, 1987. Carp ponds of Central Europe. In R. G. Michael (ed.), Managed aquatic ecosystems. 29: 29–62.

Kubečka, J., 1989. Development of the ichthyofauna of the Římov Reservoir and its management. Arch. Hydrobiol. Beih. Ergebn. Limnol. 33: 611–613.

Pace, M. L., K. Porter & Y. S. Feig, 1984. Life history variation within a parthenogenetic population of *Daphnia parvula* (Crustacea: Cladocera). Oecologia (Berlin) 63: 43–51.

Seďa, J., 1989. Main factors affecting spring development of herbivorous Cladocera in the Římov Reservoir (Czechoslov.). Arch. Hydrobiol. Beih. Ergebn. Limnol. 33: 619–630.

Sommer, U., Z. M. Gliwicz, W. Lampert & A. Duncan, 1986. The PEG-model of seasonal succession of planktonic events in fresh waters. Arch. Hydrobiol. 106: 433–471.

Weider, L. J., 1985. Spatial and temporal heterogeneity in a natural *Daphnia* population. J. Plankton. Res. 7: 101–123.

Weider, L. J., 1987. Life history variation among low-arctic clones of obligately parthenogenetic *Daphnia pulex*: a diploid-polyploid complex. Oecologia (Berlin) 73: 251–256.

Hydrobiologia **225**: 193–197, 1991.
V. Kořínek & D. G. Frey (eds), Biology of Cladocera.

Indirect effect of planktivorous fish on the growth and reproduction of *Daphnia galeata*

Jiři Macháček
Czechoslovak Academy of Sciences, Hydrobiological Institute Na sádkách 7, CS-370 05 České Budějovice, Czechoslovakia

Key words: *Rutilus rutilus*, *Daphnia galeata*, predator-induction, growth, reproduction

Abstract

Both the size of *Daphnia galeata* primiparae and the average egg volume were smaller in individuals reared from neonata in water inhabited for the preceding 24 hours by Y-O-Y roach when compared with control animals reared in the same water with the exclusion of fish. The smaller size of experimental primiparae resulted from smaller increments in some instars and from the earlier maturation of about two-thirds of those that matured in the 4th instar, while the rest of the experimental animals and all the controls matured in the 5th instar. There were larger clutches and smaller eggs in experimentals when compared with the similar sized controls, but the difference hasn't proved to be statistically significant.

Introduction

In the paper by Hrbáček & Hrbáčková-Esslová (1960) the populations of some *Daphnia* species from localities with higher fish predation were found to be smaller on the average than those from localities where fish populations were rare or absent. Smaller size was due to smaller relative increments, resulting in smaller primiparae, which produced smaller neonates.

The aim of this study was to find out whether some chemical agent produced by the planktivorous fish and released into water can be the proximate cause of this kind of phenomenon.

Material and methods

In a series of experiments, individuals of *Daphnia galeata* were reared from the neonata to the primipara stage in aged tap water previously inhabited for 24 hours by roach (*Rutilus rutilus*, about 6 cm body length) at a density of 1 spec. per 5 l. The same water without fish served as a control medium. A laboratory culture of *Scenedesmus dimorphus* was used as food. Its concentration was measured spectrophotometrically and expressed as concentration of chlorophyll *a*. The preparation of both control and experimental media is shown in Fig. 1.

The *Daphnia galeata* clone used had been isolated from the Římov Reservoir on 26th April, 1988, and kept under laboratory conditions for about one year. Neonates were measured several hours after their release from the brood pouch and placed individually into the 150 ml flasks, with individuals from the same clutch evenly distributed between the control and the experiment. The animals were measured daily and then transferred into a freshly prepared medium. The characteristics investigated were the body and carapace lengths (as no allometric growth was

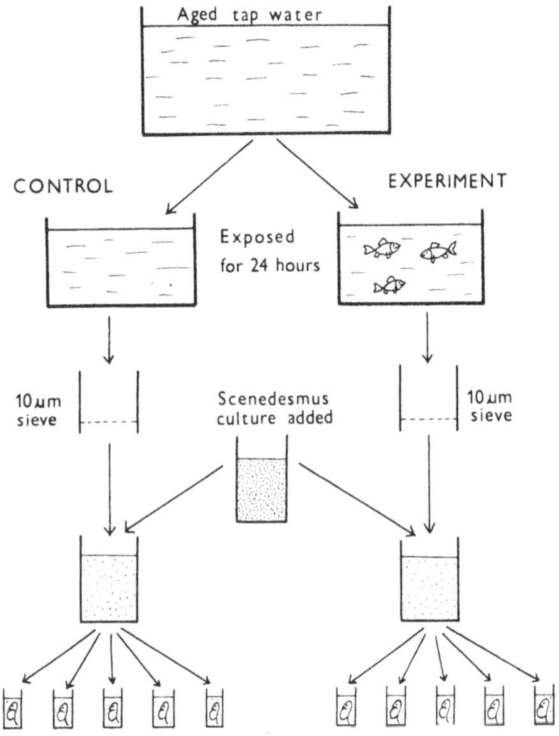

Fig. 1. The scheme of daily preparation of both control and experimental media for *D. galeata* rearing.

recorded, only the body length is presented below), the size and age of primiparae, clutch size, and the size of eggs or neonates.

Most of the experiments were carried out with the algal concentration adjusted to the level of about 100 $\mu g\, l^{-1}$ chlorophyll *a*. This initial concentration dropped within the period between the medium exchange (24 hours) due to sedimentation to about the half of the initial level. In spite of this, the food concentration was considered to be high enough so that potential differences between the control and experiment due to the different algal growth could not substantially influence the results. Nevertheless, the experiments with very high and low initial food concentrations (150 and 10 $\mu g\, l^{-1}$ chlorophyll *a*, respectively) were carried out to determine the influence of food concentration alone and in concert with the 'fish factor'. To estimate the potential influence of different bacterial growth in the control and experiment, another trial was carried out in which bacteria were removed from the water, and baker's yeast were used as food.

The statistical significance of the differences between the control and experiment was tested by using the t-test.

Results

Results of four experiments are given in Fig. 2. There is a clear tendency to earlier maturation and

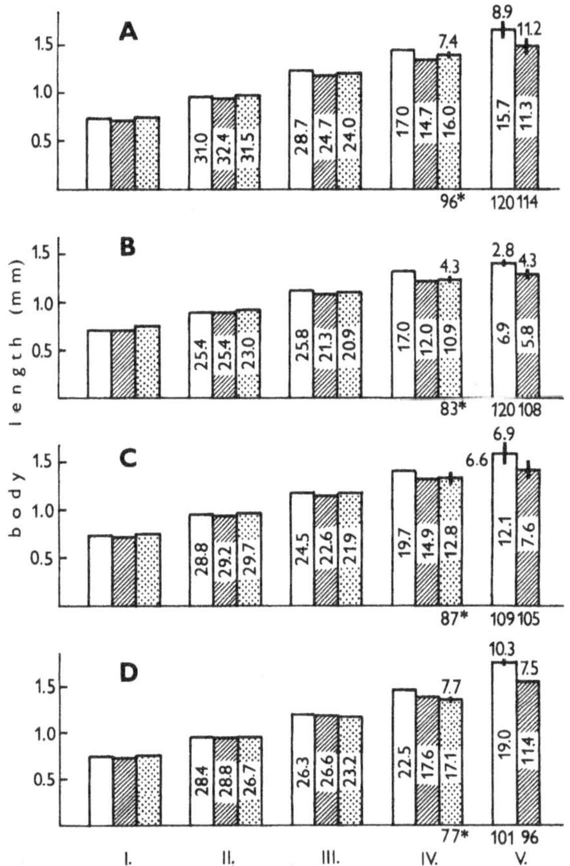

Fig. 2. Postembryonic development of *D. galeata* in four experiments with different food conditions: A, C, D – *Scenedesmus dimorphus* in respective concentrations daily adjusted to 100, 10, and 150 $\mu g\, l^{-1}$ chlorophyll *a*, B – bacteria-free water, yeast concentration daily adjusted to 1.8 $\mu g\, C\, l^{-1}$. Empty bars – control animals, stippled and hatched bars – experimental animals maturing in the 4th and 5th instar, respectively. Relative increment (%) per instar, average clutch size (eggs per female), and the length of postembryonic development (hours) up to the first ovulation are indicated inside, above, and below the bars, respectively. 95% C.I. of the primiparae body length is indicated, and the values significantly different from the control are marked with asterisks. (*P* 0.95). The number of individuals in every trial was on the average about 10 in both control and experimental groups.

smaller size of primiparae in experimental animals compared with the control ones. On the average about 65% of experimental animals matured in the 4th instar. The remainder of experimental animals plus all the control animals matured in the 5th instar. Relative increments were smallest in the experimental animals that matured in the 4th instar, somewhat higher in the experimental animals that matured in the 5th instar, and highest in the control animals.

Clutch size is proportional to the size of the female. There is, however, remarkable, though not statistically significant (at $P = 0.05$) due to large variation, shift of the regression line towards larger clutches in experimental animals, i.e. there are larger clutches in experimental animals in comparison with the similar sized control ones (Fig. 3). The same holds, when clutch size is replaced by the total clutch volume, obtained as the multiple of the clutch size and average egg volume. The plot of the clutch size against the total clutch volume (Fig. 4) suggests that the same volume of the germinal matter is divided into smaller eggs in experimental animals. The shift of the regression line, however, hasn't proved to be significant at $P = 0.05$.

Experiment B with bacteria-free water and yeast as food is somewhat outstanding. There were apparently some nutrients lacking, because relatively lower increments and smaller clutches were recorded in general, and the neonates, when released from the brood pouch, were not fully developed and mostly non viable. Enhanced reproductive investment in experimental animals, however, can be seen even in this case.

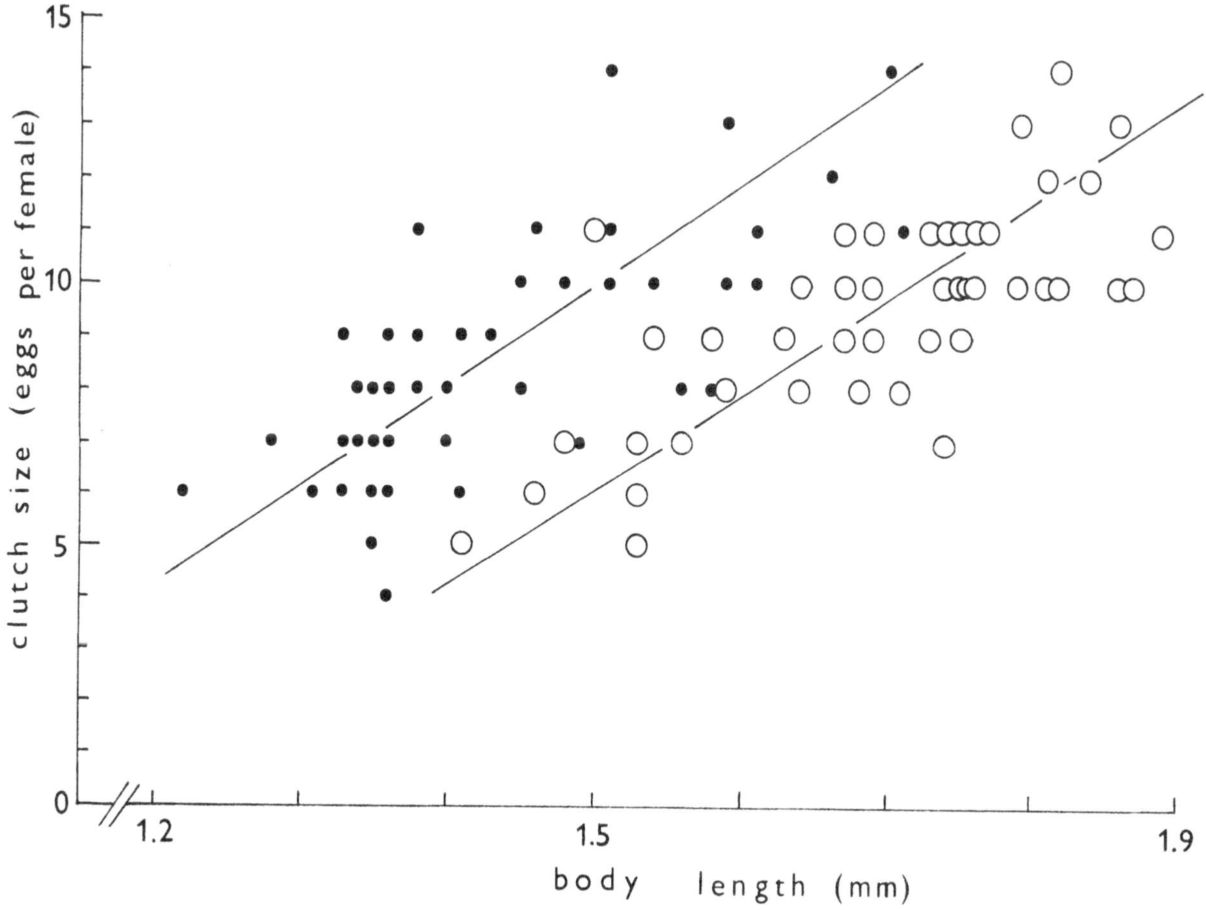

Fig. 3. Relationship between the body length and the clutch size in *D. galeata* primiparae. Circles – the control animals, dots – the experimental ones. Corresponding regression equations and coefficients are: $Y = 0.0184\ X - 21.55$, $r^2 = 0.53$, and $Y = 0.0199\ X - 19.8$, $r^2 = 0.51$, respectively.

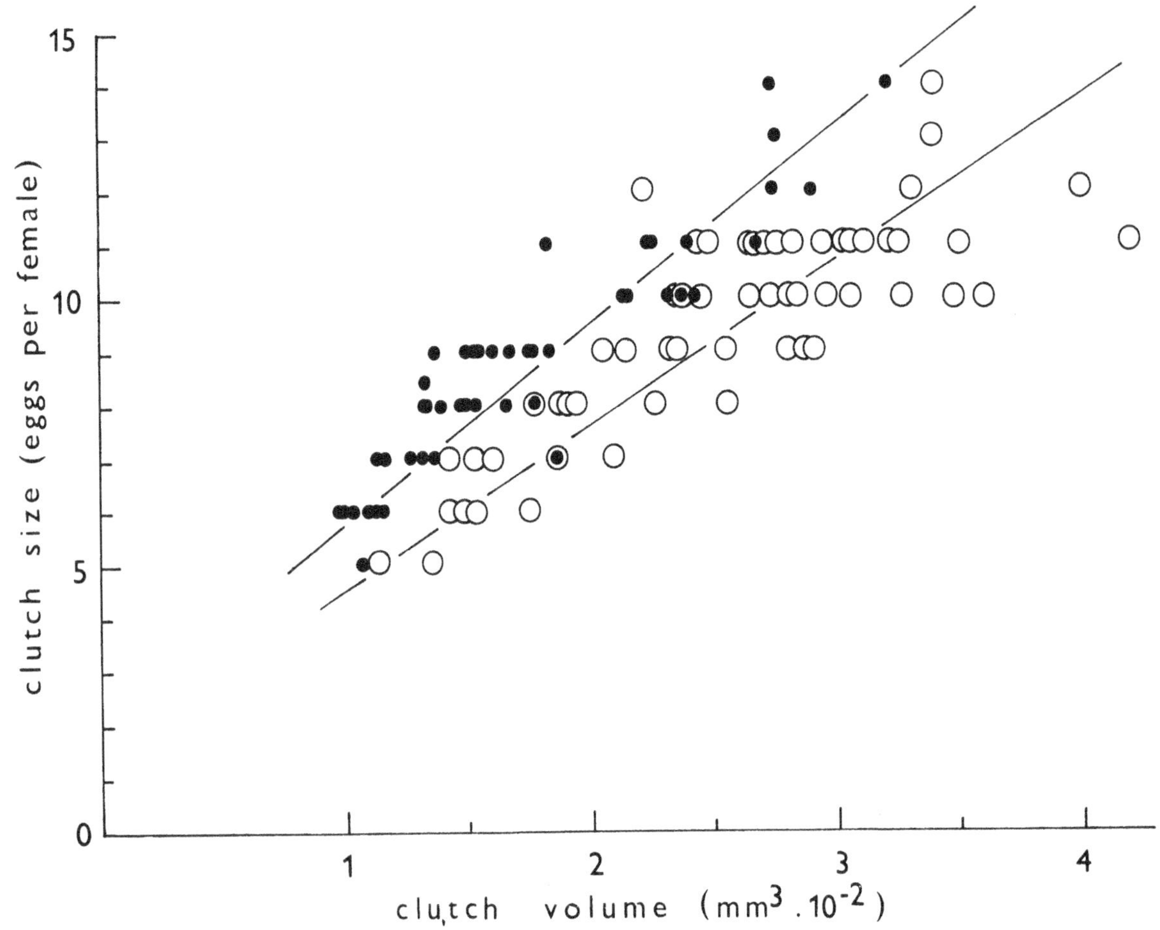

Fig. 4. Relationship between the total clutch volume and clutch size in *D. galeata* primiparae. Circles – the control animals, dots – the experimental ones. Corresponding regression equations and coefficients are: $Y = 0.0031$ X + 1.46, $r^2 = 0.68$, and equations $Y = 0.0039 + 2.09$, $r^2 = 0.79$, respectively.

The comparison of control groups in experiments C and D provides information on the influence of food concentration on the parameters under study. Smaller increments, slightly retarded maturation, and smaller egg clutches resulted from a lower food concentration.

Discussion

Much attention has recently been paid to the role of a predator-produced chemical stimulus as a proximate cue triggering the defensive responses of various *Daphnia* species and clones. This kind of interaction is under study in both invertebrate predators (Krueger & Dodson, 1981; Grant & Bayly, 1981; Havel, 1985; Hebert & Grewe, 1985; Dodson, 1988a, b; Vuorinen *et al.*, 1989; Walls & Ketola, 1989) and planktivorous fish, for which somewhat fewer data are available (Dodson, 1988a, b, 1989). The results of these studies suggest that the response of daphniids is quite widespread and is also consistent with current theories of adaptive responses for the various sizes of *Daphnia* to tactile and visual predators. Hence, the smaller size of primiparae, neonates, and eggs presented here provides quite an expected reaction to visually oriented planktivorous fish. Similar results were reported by Dodson (1988a, 1989). In his experiments various *Daphnia* species and clones exhibited, among other things, the reduced body size and increased egg clutches when exposed for one generation to *Lepomis macrochirus* enclosed in a mesh bag. The time course of growth and maturation of juveniles was not investigated in detail in his experiments, so it is not clear whether an earlier maturation occurred in animals from the fish-treated water, as was the case in my experiments.

Larger egg clutches in experimental animals, reported by Dodson (1989), are not apparent from the comparison of average values, presented here in Fig. 2. However, a comparison of equally sized animals, illustrated in Fig. 3, demonstrates the increase quite well. It seems, anyway, that reproduction at the population level is promoted by the 'fish factor', due either to the earlier maturation of juveniles or to larger egg clutches or both. The reduced size of experimental animals suggests that this occurs at the expense of the energy for growth.

Both the growth and reproduction of daphniids are influenced by the food conditions, as proved by many authors (Hrbáčková-Esslová, 1963; Weglenska, 1971; Hrbáčková, 1974). Reduced growth in experimental animals presented here cannot, however, be interpreted as a result of food limitation, because simultaneously enhanced reproduction contradicts this.

The results of experiments with a low food concentration and with the yeast show that even under conditions of nutrient limitation a similar kind of response is obtained, unlike the results of Dodson & Havel (1988), who found that the response of *Daphnia pulex* to *Notonecta undulata* disappeared at low food levels.

Conclusion

The results presented suggest that there is some water soluble chemical factor produced by the roach that probably triggers the change of energy allocation pattern in the postembryonic development of *Daphnia galeata*. The part of energy allocated to growth is reduced in favour of energy allocated to reproduction.

As indicated by the relatively rare and very recent data in the literature the investigation of the subject is at its beginning and further studies are needed.

References

Dodson, S. I., 1988a. Cyclomorphosis in *Daphnia galeata mendotae* Birge and *D. retrocurva* Forbes as a predator-induced response. Freshwat. Biol. 19: 109–114.

Dodson, S. I., 1988b. The ecological role of chemical stimuli for the zooplankton: predator-avoidance behavior in *Daphnia*. Limnol. Oceanogr. 33: 1431–1439.

Dodson, S. I., 1989. The ecological role of chemical stimuli for the zooplankton: predator-induced morphology in *Daphnia*. Oecologia 78: 361–367.

Dodson, S. I. & J. I. Havel, 1988. Indirect prey effects: some morphological and life history responses of *Daphnia pulex* exposed to *Notonecta undulata*. Limnol. Oceanogr. 33: 1274–1285.

Grant, J. W. G. & I. A. E. Bayly, 1981. Predator induction of crests in morphs of the *Daphnia carinata* King complex. Limnol. Oceanogr. 26: 201–218.

Havel, J., 1985. Cyclomorphosis of *Daphnia pulex* spined morphs. Limnol. Oceanogr. 30: 853–861.

Hebert, P. D. N. & P. M. Grewe, 1985. *Chaoborus*-induced shifts in the morphology of *Daphnia ambigua*. Limnol. Oceanogr. 30: 1291–1297.

Hrbáček, J. & M. Hrbáčková-Esslová, 1960. Fish stock as a protective agent in the occurrence of slow-developing dwarf species and strains of the genus *Daphnia*. Int. Rev. ges. Hydrobiol. 45: 355–358.

Hrbáčková, M., 1974. The size of primiparae and neonates of *Daphnia hyalina* Leydig (Crustacea: Cladocera) under natural and enriched food conditions. Věst. čs. spol. zool. 38: 98–105.

Hrbáčková-Esslová, M., 1963. The development of three species of *Daphnia* in the surface water of the Slapy Reservoir. Int. Revue ges. Hydrobiol. 48: 325–333.

Krueger, D. A. & S. I. Dodson, 1981. Embryological induction and predation ecology in *Daphnia pulex*. Limnol. Oceanogr. 23: 219–223.

Vuorinen, I., M. Ketola & M. Walls, 1989. Defensive spine formation in *Daphnia pulex* Leydig and induction by *Chaoborus crystallinus* De Geer. Limnol. Oceanogr. 34: 245–248.

Walls, M. & M. Ketola, 1989. Effects of predator-induced spines on individual fitness in *Daphnia pulex*. Limnol. Oceanogr. 34: 390–396.

Weglenska, T., 1971. The influence of various concentrations of natural food on the development, fecundity and production of planktonic crustacean filtrators. Ekol. Pol. 19: 427–471.

Hydrobiologia **225**: 199–208, 1991.
V. Kořínek & D. G. Frey (eds), Biology of Cladocera.

Density and size distribution of *Daphnia* populations at different fish predation levels

Martin Černý & Jiří Bytel
Department of Hydrobiology, Charles University, Viničná 7, CS-12844 Prague 2, Czechoslovakia

Key words: fish predation, *Daphnia*, size distribution

Abstract

Density and size structure changes of natural daphniids populations were studied in enclosures with a different level of fish predation. *Daphnia pulicaria* was totaly removed in high predation variants, and the differences of the mean body length both in adults and juveniles are apparent between low predation enclosure and enclosure without fish. *Daphnia galeata* was replaced by *D. magna* and *D. pulicaria* in the enclosure without fish. The decline of densities and the substantial (30–50%) and fast (during 12 days) shift of mean body length both in adults and juveniles of *D. galeata* was induced by the fish (carp fry) introduction to the high predation enclosures.

Introduction

Fish predation has an important effect on zooplankton structure (Hrbáček, 1957) and through it on the whole plankton (Hrbáček *et al.*, 1961). Cladocera, and especially the genus *Daphnia*, whose members belong to the bigger zooplankters, are considerably exposed to the fish predation. Moreover, as dominating group of zooplankton in many freshwater waterbodies, daphniids can significantly regulate the amount of phytoplankton and through competition affect other zooplankters including themselves. A condition in which *Daphnia pulicaria* outcompetes *D. galeata* is common in Czech carp ponds. This phenomenon occurs when the fish stock is lower, but, on the contrary, the more robust *D. pulicaria* is removed under high fish stock while the smaller *D. galeata* persists. We can assume there are two main biological factors that affect the structure and life history traits of *Daphnia*'s populations, at least in carp ponds: the size selective predation by fish and the competition among certain species. The consequence of invertebrate predation for *Daphnia*'s populations in carp ponds is not yet clear.

We attempted to evaluate the effect of size selective predation on the *Daphnia* populations under different levels of predation pressure and to examine the mechanisms playing a role in this interaction. Our investigation included an experiment, which was carried out in the summer season of 1985. This experiment with enclosures investigated the interactions between fish, daphniids, and phosphate leading to the blue-green algal blooms. This study presents a small part of results of this experiment.

Material and methods

Experimental design

The experiment was carried out at the Hydrobiological field station near the town Blatna (South Bohemia, about 100 km SW from Prague) over a 119 day period from 14.6.1985 to 11.10.1985.

Four circular enclosures (volume 110 m^3, diameter 10 m, depth 1.4 m) were installed in the carp pond Velky Palenec. This pond is typical of eutrophic waterbodies used for carp aquaculture in Central Europe (Kořínek *et al.*, 1987), with an area of 0.3 km^2, max. depth 3.5 m, mean depth 1.4 m. The pond fish stock consisted mainly of 2-year old carp up to several hundreds per hectare. There were also numerous populations of perch and coarse fish, such as *Leucaspius delineatus* and *Gobio gobio*. The enclosures were constructed of polyethylene sheet anchored well in bottom sediments and floating on the surface by means of a collar filled by styrofoam. Walls were fastened to poles to ensure a rigid circular shape.

Within four enclosures an original pond plankton and few hundred planktivorous fish *Leucaspius delineatus* survived up to the fourth decade of experiment. Then we removed nearly all fish from the first and the fourth enclosure, and at the same time we introduced approximately six hundred carp (*Cyprinus carpio*) offspring (5.5 ind m^{-3}) into both the second and the fourth enclosure. Different levels of fish predation pressure on the zooplankton were established in each enclosure: enclosure 1 – without predation (no fish stock), enclosure 2 – high predation (carp offspring and coarse fish), enclosure 3 – low predation (coarse fish), enclosure 4 – high predation (carp offspring).

Phosphate treatment (4 g m^{-3} addition) of the enclosures 3 and 5 did not affect the nutrient concentrations (Fott & Macháčková, 1989).

Sampling

Temperature and transparency as Secchi depth were measured daily; dissolved oxygen, pH, and major nutrients (N, P) were measured irregularly, approximately twice a month.

Phytoplankton was sampled with a tube sampler at fourteen-day interval, more frequently during August and September. We used chlorophyll-*a* concentration as a measure for phytoplankton biomass. The other part of sample was preserved with Lugol's solution for the species determination and volume analyses.

Zooplankton was sampled with Schindler-Patalas sampler (volume 28.1 l) during the night to minimize error from light-induced crowding of zooplankton. This sample was partioned into two portions. Both portions were screened into four fractions by sieves of different meshsize: 0.04–0.515 mm (mainly rotifera, nauplii, young copepodids, *Bosmina*, and small specimens of daphniids), 0.515–0.85 mm (larger copepodids and daphniids), 0.85–1.00 mm (the largest copepodids of *Diaptomus* and *Cyclops*, daphniids) and > 1.00 mm (only daphniids, mainly adults of *Daphnia pulicaria* and *D. magna*). The fractions of one portion were preserved with 4% formalin for zooplankton counting and size analysis. The zooplankton was counted under microscope in Sedwick-Rafter chambers until at least 300 specimens from each of the dominant groups (e.g. nauplii, rotifers, certain *Daphnia* species) had been examined. The fractions of the other portion were dried for biomass assessment. Simultaneously with sampling several tows by the plankton net (meshsize 0.04 mm) were taken for additional analyses (e.g. clutch, morphology). Samples were taken at fourteen-days intervals, more frequently only in August.

Daphnia size structure analysis

Size structures of populations of *Daphnia galeata* and *D. pulicaria* were determined. Body length from top of head to the base of shell-spine was measured. At least 100 females (males were excluded) were measured in each sample. Only on a few occasions when the density was under about ten animal per liter was the number smaller.

Results

Daphnia abundances and patterns of plankton succession in enclosures

Abiotic conditions

The temperature regime (Fig. 1) was the same in all enclosures, with the differences either among enclosures or between surface and bottom temperature not exceeding 1 °C. The oxygen concentration did not fall below 4.5 mg l^{-1} with the exception of enclosure 1, where the oxygen concentration varied between 1 and 5 mg l^{-1} from the fourth decade. The major nutrient (phosphorus, nitrogen) concentrations did not exceed the common values found in carp ponds (Kořínek *et al.*, 1987). Only in enclosure 1 was there an unusually high concentration of ammonia (1.0–2.26 mg l^{-1}) in the second period when fish were removed.

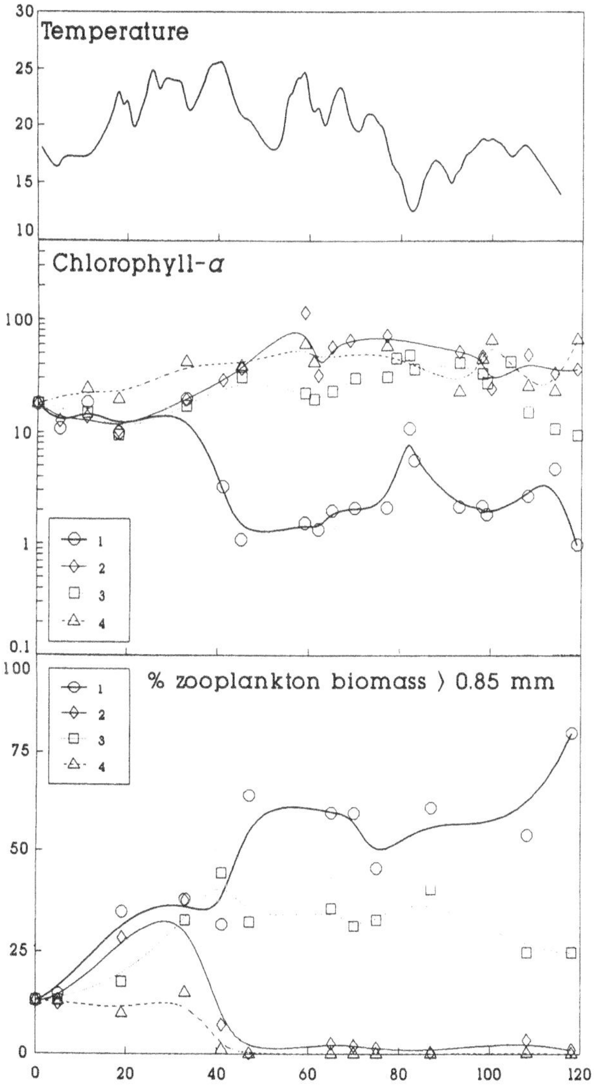

Fig. 1. Temperature (°C), phytoplankton biomass as concentration (mg m^{-3}) of chlorophyll-*a*, percentage of biomass of zooplankton (almost only daphniids) caught on sieve with 0.85 mm mesh size. 1 – without fish, 2 and 4 – high predation, 3 – low predation.

Starting point

All enclosures contained an original pond plankton at the beginning of the experiment. *Daphnia galeata* populations were large (Fig. 2), those of *Daphnia pulicaria* scanty (Fig. 2). The density of *Bosmina longirostris* was about 10 ind l^{-1}. Among rotifers *Polyarthra* ssp. dominated, with a density of about 60 ind l^{-1}. Cyclopoid copepodids were about 10 ind l^{-1}, calanoid copepodids 2 ind l^{-1},

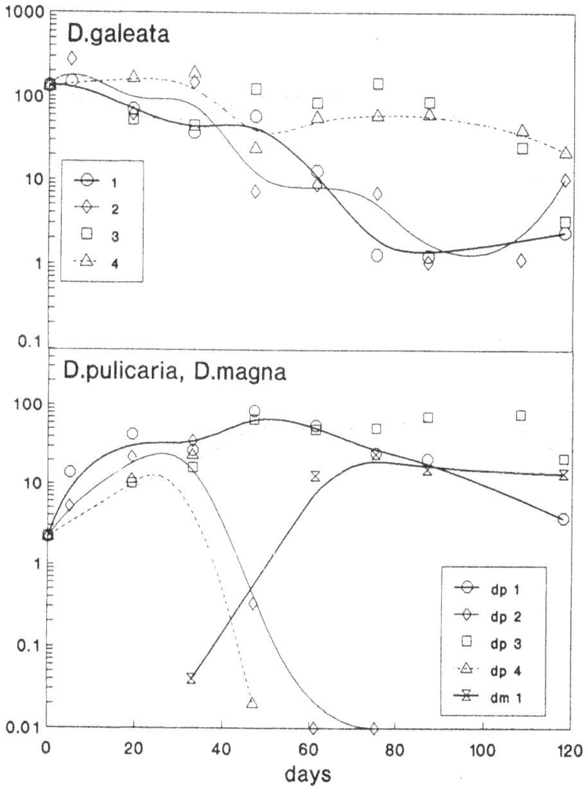

Fig. 2. Densities (ind l^{-1}) of *Daphnia pulicaria*, *Daphnia galeata* and *Daphnia magna* in enclosures. 1 – without fish, 2 and 4 – high predation, 3 – low predation, dp – *Daphnia pulicaria*, dm – *Daphnia magna*.

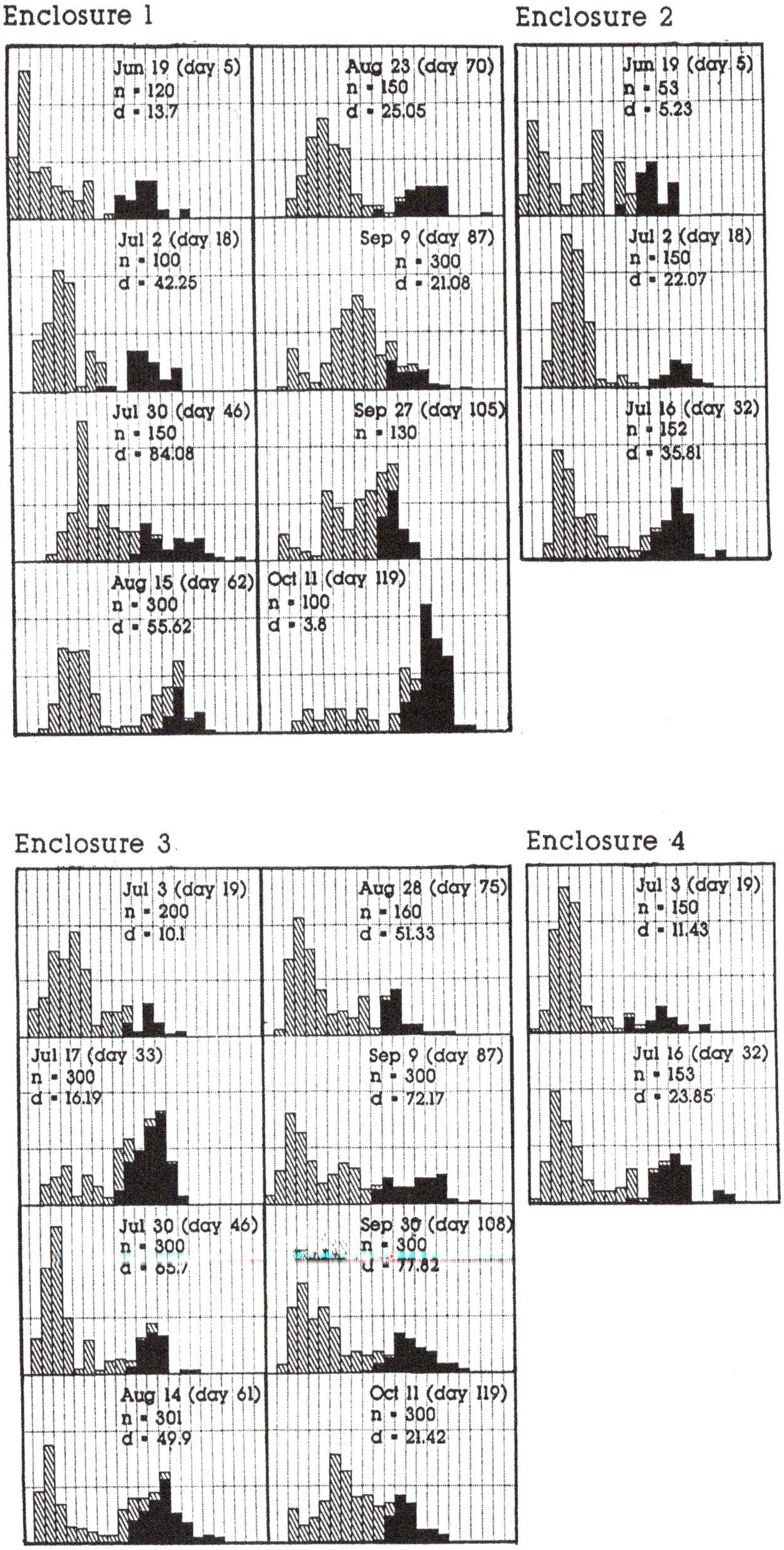

Fig. 3A. Size distribution of *Daphnia pulicaria*. Stripped bars – juveniles, black bars – adults. Horizontal grid lines by 10%. Vertical grid lines by 0.1 mm, size axis begins at 0.5 mm. n = number of examined animals, d = density (ind l^{-1}). 1 – without fish, 2 and 4 – high predation, 3 – low predation.

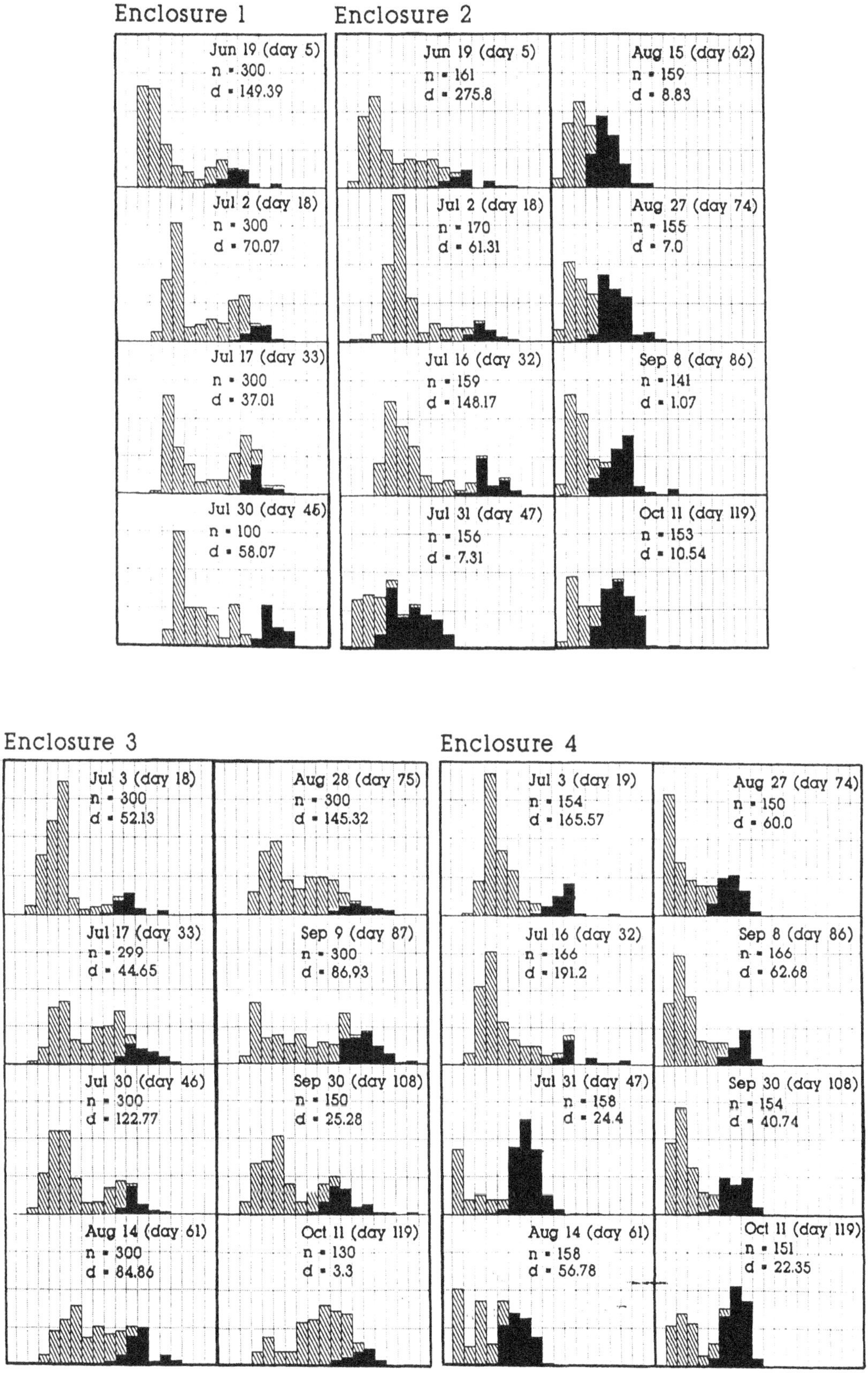

Fig. 3B. Size distribution of *Daphnia galeata*. Comments are the same as at Fig. 3A, only size axis begins at 0.3 mm.

mediately after the carp were introduced into enclosures 2 and 4 (Fig. 2). The size-distribution in the enclosure without predation varied approximately within the same range as in the enclosure with low predation (Fig. 3A), but there were some differences. The trend to attain larger body size is evident in conditions without predation. The range of mean body length of adult females (Fig. 4A) and the size of primipara (first reproducing instar) (Fig. 3A) extended to higher values in enclosure 1 without predation than in enclosure 3 with low predation.

The slender *Daphnia galeata* persisted in all enclosures, but its density fell to 1 ind l^{-1} in enclosure 1 without fish and in enclosure 2 with the highest fish stock (Fig. 2). The introduction of fish into enclosures 2 and 4 had a dramatic effect on size distribution within the *Daphnia galeata* populations. The downward shift covered the whole range of size distribution (Fig. 3B). The mean body length of both adults and juveniles decreased by about 50% in enclosure 2 and 30–40% in enclosure 4 (Fig. 4B). The size of primipara dropped from 1.3 mm to 0.6 and to 0.8 mm in enclosures 2 and 4, respectively. The shift of adult size was accompanied by corresponding shift in the offspring, the minimal neonate body length falling from 0.5 mm to 0.3 mm in enclosure 2 and to 0.4 mm in enclosure 4. This shift occurred very rapidly. Fish were being introduced from the 35th to the 40th day of the experiment, and the change in the size distribution was already considerable on the 47th day. The higher percentage of adults is apparent in the high predation enclosures.

A comparison of the maximal size of *Daphnia*

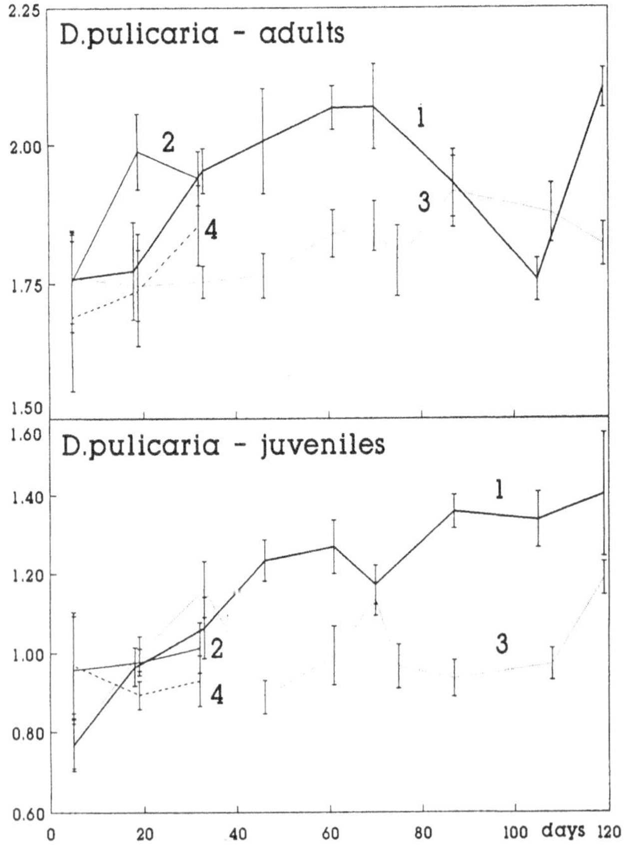

Fig. 4A. Mean body length (mm) of adults and juveniles of *Daphnia pulicaria*. Error bars indicate 99% confidence interval. 1 – without fish, 2 and 4 – high predation, 3 – low predation.

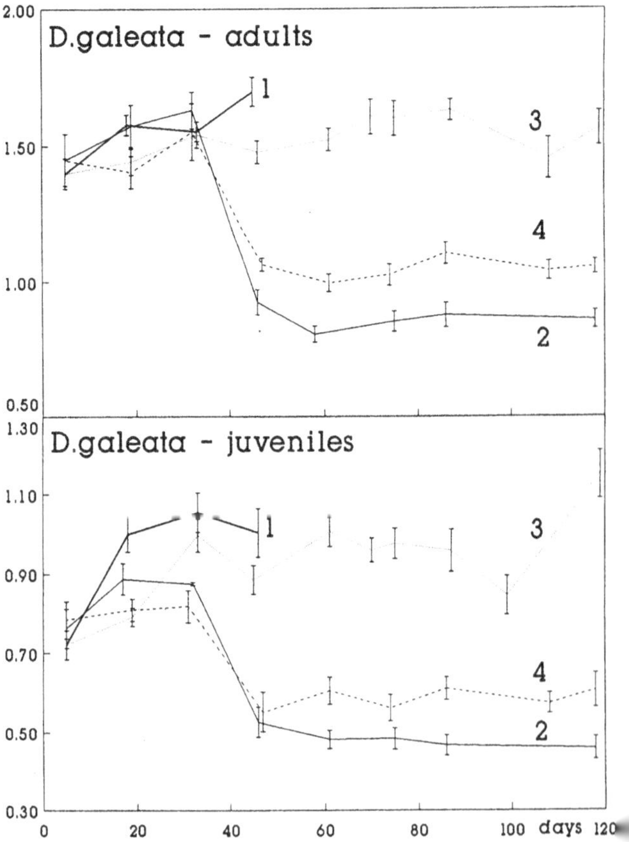

Fig. 4B. Mean body length (mm) of adults and juveniles of *Daphnia galeata*. Comments are the same as at Fig. 4A.

galeata in high predation enclosures and the minimal size of first adult instar of *Daphnia pulicaria* in enclosures with low predation reveals that the survival of *Daphnia pulicaria* populations under size selective predation was limited by its inability to decrease the size of primipara below 1.3–1.4 mm.

Discussion

The changes of the whole plankton community structure in enclosures are in a good agreement with changes observed in carp ponds with different fish management (Kořínek *et al.*, 1987), and correspond well with numerous investigations started by Hrbáček (1957).

Our results clearly demonstrate that predation by carp offspring was strongly size selective and hence negatively affected populations of the largest species of *Daphnia*. We assumed that the predator was visually orientated, because the bigger but transparent *Leptodora kindtii* specimens persisted in enclosures with high predation pressure whereas the smaller *Daphnia pulicaria* did not. The boom of small herbivores in the high predation enclosures and the dominance of large *Daphnia* species in conditions without fish generally fit in with the size-efficiency hypothesis (Brooks & Dodson, 1965).

However, this hypothesis doesn't cover the phenomenon of body-size shift within population of certain species as it occurred in *Daphnia galeata* populations in this experiment. It is clear that the downward change in body-size distribution is not due to simply removing the largest size-classes by fish. The decreasing size at maturity and consequently the shift in the whole range of body-size distribution gives evidence that this change constitutes an adaptive mechanism for avoiding size selective predation. A similar strategy was observed in natural populations of *Daphnia hyalina* in Lago Maggiore (de Bernardi *et al.*, 1975), and in enclosure experiments with *Daphnia galeata* (Kokvsik & Langeland, 1988) and *Ceriodaphnia* and *Diaphanosoma* (Vanni, 1987) populations.

There are two possible mechanisms leading to this size-shift within a population. The first is a clone replacement and the second a morphological adaptation of an existing population. The first is a genetically based hypothesis. It is supported by data of Hrbáček and Hrbáčková-Esslová (1960), who observed the smaller primipara size in *Daphnia* clones coming from conditions with high fish stock. Although we have no data about clonal structure of *Daphnia* populations, this hypothesis seems irrelevant to our results. It is known from many observations (e.g. Hrbáčková-Esslová, 1963), that the generation time from neonata to primipara at 20 °C is about 5 days and that the time for embryonic development is about 2 days under good feeding conditions. Therefore the population needs at least 7 days for the establishment of only the first generation of a new selected clone, whereas in the experiment the size-shift was strongly pronounced during 7–12 days after fish had been introduced into the population. Moreover, no single specimen from a potential dwarf clone was recorded before fish introduction.

The 'phenotypic' hypothesis seems to be more probable in our experiment. Morphological adaptations to predation are known from experiments with well defined clones of *Daphnia* and both vertebrate and invertebrate predators (Hebert & Grewe, 1985; Havel & Dodson, 1987; Dodson & Havel, 1988; Dodson, 1988). The decrease in body-size of adults was observed in the presence of fish (Dodson, 1988) and *Notonecta* (Dodson & Havel, 1988), even when the predator was separated from its prey by a mesh but mixing of water on the two sides was possible. On the average, the decrease in size was 17,7% in the presence of *Notonecta* and at good feeding conditions during the preadult development (4–5 days). Such results suggest that there is some chemical signal inducing the morphological adaptations.

Acknowledgements

We thank J. Fott, a supervisor of the experiment, and Z. Stuchlíková, E. Stuchlík for field and/or lab assistance. We are grateful to V. Kořínek for critical review and comments on this manuscript.

References

Brooks, J. L. & S. J. Dodson, 1965. Predation, body size and competition of plankton. Science 150: 28–35.

de Bernardi, R., G. Giussani & M. Manca, 1987. Cladocera: Predators and prey. Hydrobiologia 145: 225–243.

Dodson, S. I., 1988. Cyclomorphosis in *Daphnia galeata mendotae* Birge and *D. retrocurva* Forbes as a predator-induced response. Freshwat. Biol. 19: 109–114.

Dodson, I. S. & J. E. Havel, 1988. Indirect prey effects: Some morphological and life history responses of *Daphnia pulex* exposed to *Notonecta undulata*. Limnol. Oceanogr. 336: 1274–1285.

Fott, J. & J. Macháčková, 1989. Vliv rybi obsadky na vyskyt ruznych analytickych frakci sloucenin fosforu a dusiku v pokusnych ohradach. Zbornik referatov vedeckej konferencie Vyznam malych polnohospodarskych nadrzi pre rybarstvo a ochranu vodneho prostredia krajiny, Nitra: 106–111. [in Czech].

Havel, J. E. & I. S. Dodson, 1987. Reproductive cost of *Chaoborus*-induced polymorphism in *Daphnia pulex*. Hydrobiologia 150: 277–281.

Hebert, P. D. N. & P. M. Grewe, 1985. *Chaoborus*-induced shifts in the morphology of *Daphnia ambigua*. Limnol. Oceanogr. 306: 1291–1297.

Hrbáček, J., 1957. Density of the fish population as a factor influencing the distribution and speciation of the species in the genus *Daphnia*. XVth Inter. Congr. Zool., Sect. 10, Pap. 27, London, 2 pp.

Hrbáček, J., M. Dvořáková, V. Kořínek & L. Procházková, 1961. Demonstration of the effect of the fish stock on the species composition of zooplankton and intensity of metabolism of whole plankton association. Verh. Int. Ver. Limnol. 14: 192–195.

Hrbáček, J. & M. Hrbáčková-Esslová, 1960. Fish stock as a protective agent in the occurrence of slow-developing dwarf species and strains of the genus *Daphnia*. Int. Revue ges. Hydrobiol. 453: 355–378.

Hrbáčková-Esslová, M., 1963. The development of three species of *Daphnia* in the surface water of the Slapy reservoir. Int. Revue ges. Hydrobiol. 482: 325–333.

Kokvsik, J. I. & A. Langeland, 1987. Effects of size selective predation by whitefish *Coregonus lavaretus* L. on *Daphnia galeata* Sars and *Cyclops scutifer* Sars in limnocorral experiments. Pol. Arch. Hydrobiol. 341: 67–80.

Kořínek, V., J. Fott, J. Fuksa, J. Lellák & M. Pražáková, 1987. Carp ponds in Central Europe. In R. G. Michael (ed.), Managed Aquatic Ecosystems. Elsevier Science Publishers B.V., Amsterdam: 29–62.

Vanni, M. J., 1987. Effects of food availability and fish predation on a zooplankton community. Ecol. Monographs 571: 61–88.

Hydrobiologia **225**: 209–216, 1991.
V. Kořínek & D. G. Frey (eds), Biology of Cladocera.

Impact of fishery management on Cladoceran populations

Miroslava Pražáková
Department of Hydrobiology, Charles University, Viničná 7, CS-12844 Prague 2, Czechoslovakia

Key words: Fish predation, Daphnia, fish ponds

Abstract

Zooplankton of the fish pond reveals a two-year periodicity, which is induced by the two-year cycle of fishery management. In the first year of each cycle, the biomass of large cladocerans prevailed – mainly adult specimens of the species Daphnia pulicaria Forbes, which were absent in the second year. The species Daphnia galeata Sars was present in both years, however, it is more numerous in the second year. The fraction of small zooplankton (rotifers, nauplii, Bosmina, newborn Daphnia) was abundant in the second years, but scarce in the first years, respectively.

Introduction

The zooplankton and namely Cladocera of carp ponds in south-west Bohemia have been studied for many years (Kořínek *et al.*, 1987). Since the end of the sixties some ponds have been managed in two-year cycles (Fott *et al.*, 1980). Simultaneously we observed distinct two-year periodicity of the zooplankton, phytoplankton and transparency. Therefore we started the study of the phenomenon in three following cycles (six years) and subsequent one more year in the pond.

The investigated pond was Velký Pálenec (A = 31 ha, z = 1.4 m), more details on the morphometry in Hrbáček (1966). The main fish species raised is common carp (Cyprinus carpio) which is usually stocked as yearlings reaching the marketable size in two years. Whitefish (Coregonus lavaretus maraena) is usually introduced as fry together with carp (Table 1).

The results from the first three years of the study were already published by Fott *et al.* (1980). They found that in the first year of each cycle, transparency and light penetration were high, chlorophyll-*a* content and the production of phytoplankton were low. Large specimens the genus Daphnia were abundant and the small zooplankton were scarce. In the second year of each cycle the situation was opposite.

We continued the investigation for four more years. In this paper I deal with the impact of the biannual management on the zooplankton biomass and species composition in the whole seven year period under study.

Methods

Since 1977 till 1983 we have sampled the pond in about 2-week intervals, three times in May and as well as in the period from August to September. On a sampling date two composite representative samples of 50 l were taken by a Friedinger bottle of two litre capacity on several stations and layers with respect to their volume (Hrbáček, 1966). One sample of zooplankton was filtered through the metallic netting of 820, 350 and 40 μm mesh size, preserved with formalin and later counted for

Table 1. Numbers and weight of fish in the pond V. Pálenec.

Year	Fish stocked (spring)	Per hectare		Fish out (autumn)	Per hectare	
		Kilograms	Numbers		Kilograms	Numbers
1977	Carp	55	1550			
	Whitefish	fry	16000			
1978				Carp	1880	1410
				Whitefish	60	160
1979	Carp	17	870			
	Whitefish	fry	16000			
	Carp	56	806 – (restocked in autumn)			
1980				Carp	1085	990
				Whitefish	145	160
1981	Carp	194	720			
	Tench	3	100			
	Whitefish	fry	16000			
1982				Carp	1355	720
				Tench	39	?
				Whitefish	94	312
1983	Carp	174	870			
	Whitefish	fry	10000			

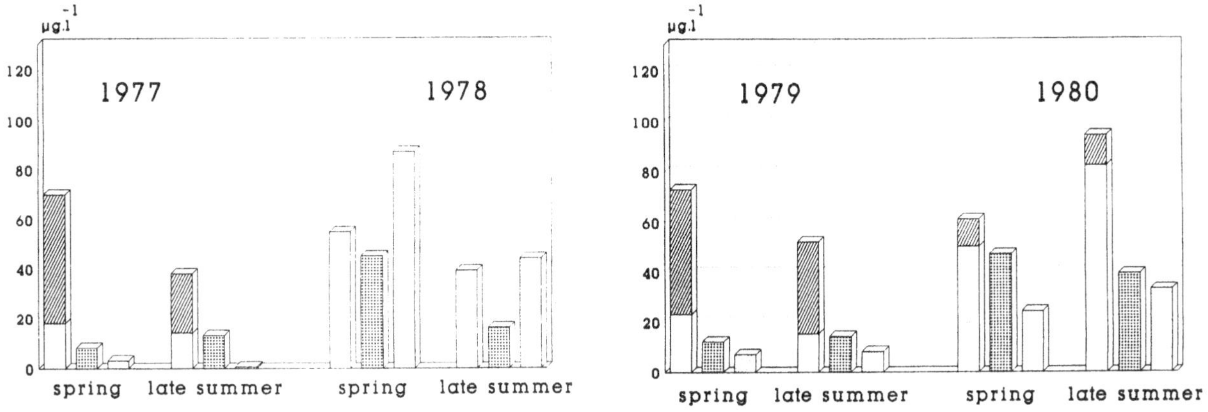

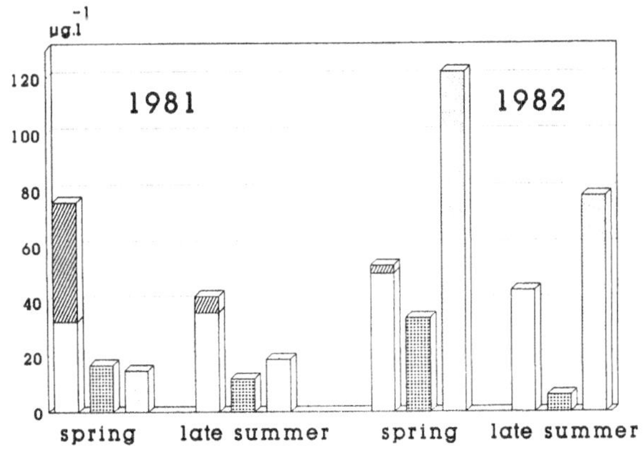

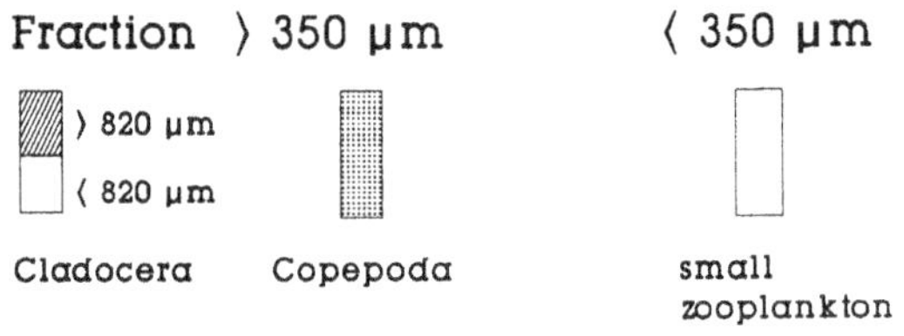

Fig. 1. The mean biomass of the zooplankton (in protein nitrogen).

Table 2A. Numbers of zooplankton in the pond Velký Pálenec – in the first year (ind.l^{-1}).

	1977					
	8.5.	23.5.	5.6.	13.8.	5.9.	25.9.
Bosmina longirostris	+	+	–	+	21	38
Ceriodaphnia affinis	–	–	–	16	44	12
Daphnia galeata	42	37	28	7	6	+
Daphnia pulicaria	14	37	52	52	42	16
Diaphanosoma brachyurum	–	–	–	+	+	+
f. Chydoridae	+	+	+	8	18	+
f. Cyclopidae ad. + cop. st.	28	16	10	24	20	40
f. Diaptomidae ad. + cop. st.	+	7	26	22	35	28
nauplia f. Cyclop. + Diapt.	57	68	128	108	29	82
Rotatoria	31	27	35	206	121	238
	1979					
	10.5.	24.5.	8.6.	2.8.	20.8.	6.9.
Bosmina longirostris	+	64	4	+	–	–
Ceriodaphnia affinis	–	–	+	+	+	5
Daphnia galeata	44	71	27	+	+	+
Daphnia pulicaria	6	18	43	70	31	43
Diaphanosoma brachyurum	–	–	–	+	+	8
f. Chydoridae	+	+	+	+	+	+
f. Cyclopidae ad. + cop. st.	24	19	22	7	18	16
f. Diaptomidae ad. + cop. st.	+	+	24	18	26	23
nauplia f. Cyclop. + Diapt.	32	114	96	17	38	54
Rotatoria	21	32	66	54	30	35
	1981					
	4.5.	18.5.	1.6.	5.8.	19.8.	3.9.
Bosmina longirostris	+	+	+	+	–	–
Daphnia galeata	96	42	62	209	47	7
Daphnia pulicaria	50	25	23	12	5	+
Diaphanosoma brachyurum	–	–	–	8	6	+
f. Chydoridae	+	+	16	+	–	–
f. Cyclopidae ad. + cop. st.	102	27	25	64	137	203
f. Diaptomidae ad. + cop. st.	+	–	15	+	+	+
nauplia f. Cyclop. + Diapt.	29	43	81	41	264	154
Rotatoria	328	82	169	187	169	521

numbers of zooplankton. The other parallel was used for biomass determination. The sample was again divided into four fractions using sieves of 820, 350 and 40 μm mesh size and the method of Straškraba (1964) was applied to separate cladocerans from copepods. The biomass was analyzed for protein nitrogen according to Blažka (1965). Detailed description of methods is in Fott *et al.* (1974).

In some years we sampled the pond from March to October, but for comparison of the first (odd) and the second (even) year we used only three dates from spring and late summer.

Results

During our investigation the zooplankton in Velký Pálenec reveals a two-year periodicity.

In the first year of each cycle, the biomass of

Table 2B. Numbers of zooplankton in the pond Velký Pálenec – in the second year (ind.l^{-1}).

	1978					
	2.5.	13.5	29.5.	14.8.	1.9.	19.9.
Bosmina longirostris	437	1180	573	16	+	28
Ceriodaphnia affinis	–	–	+	+	7	10
Daphnia galeata	37	84	102	85	94	101
Diaphanosoma brachyurum	–	–	–	–	–	–
f. Cyclopidae ad. + cop. st.	228	98	57	331	117	110
f. Diaptomidae ad. + cop. st.	+	+	+	–	–	–
nauplia f. Cyclop. + Diapt.	566	139	136	407	382	182
Rotatoria	712	80	245	644	551	1245
	1980					
	7.5.	22.5.	5.6.	11.8.	21.8.	3.9.
Bosmina longirostris	–	+	–	+	–	–
Ceriodaphnia affinis	–	–	–	–	+	–
Daphnia galeata	41	253	117	202	151	152
Diaphanosoma brachyurum	–	–	–	12	24	6
f. Cyclopidae ad. + cop. st.	185	204	39	82	95	208
f. Diaptomidae ad. + cop. st.	+	+	5	11	+	+
nauplia f. Cyclop. + Diapt.	405	193	44	165	232	325
Rotatoria	27	37	142	1148	447	1270
	1982					
	28.4.	12.5.	27.5.	9.8.	23.8.	6.9.
Bosmina longirostris	243	1260	3391	199	368	6
Ceriodaphnia affinis	–	–	–	–	+	8
Daphnia galeata	11	29	108	96	131	115
f. Chydoridae	–	+	+	–	+	+
f. Cyclopidae ad. + cop. st.	191	265	33	504	240	144
f. Diaptomidae ad. + cop. st.	–	–	–	–	–	–
nauplia f. Cyclopidae	414	444	147	292	555	129
Rotatoria	287	204	411	1701	2896	2406

large cladocerans prevailed – mainly adult specimens of the species Daphnia pulicaria Forbes and the smaller species Daphnia galeata Sars (Table 2). Cladocerans larger than 820 μm participated in 40–60% on the whole biomass (except late summer 1981). The fraction of the small zooplankton was scarce (Fig. 1).

In the second year of each cycle, Daphnia galeata Sars was more numerous than in the first year. Daphnia pulicaria Forbes was absent, also copepodites and adult stages of the species Eudiaptomus vulgaris Schmeil and Eudiaptomus gracilis G.O. Sars were present only in low numbers. The fraction of the small zooplankton (rotifers, naupliar stages, newborns of the genera Daphnia and Bosmina) was abundant (Table 2B). Especially Bosmina longirostris (O.F. Müller) was very abundant in May 1978 and 1982, but in 1980 it was absent. In this year also the biomass of small zooplankton was lower than in other even years (Fig. 1).

The mean biomass of the total zooplankton was nearly two-fold higher in the even years than in the odd ones.

In Table 3 the differences are given between the first and the second year. Many differences are significant at the 5% level of significance.

Species Daphnia pulicaria and Daphnia galea-

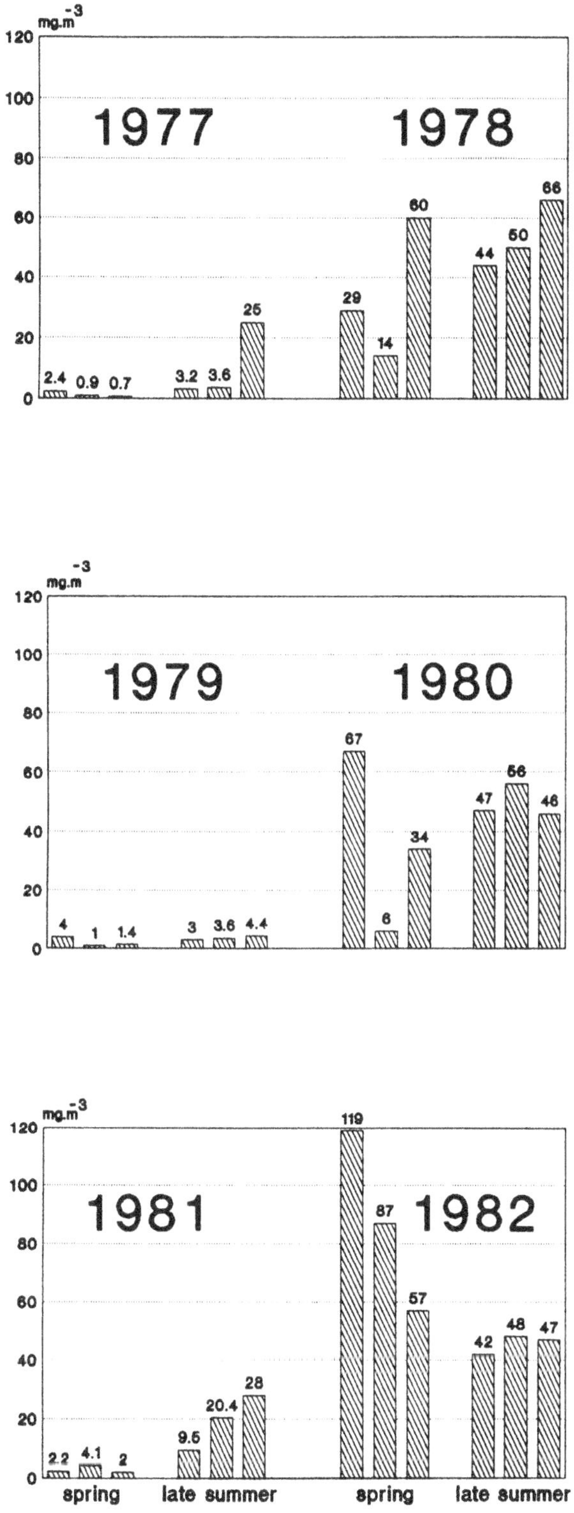

Fig. 2. Chlorophyll-*a* in phytoplankton.

ta were present simultaneously in zooplankton in the first year of each cycle. Data presented in the Fig. 3 show that in the years 1977 and 1979, Daphnia pulicaria prevailed from beginning of June, but in the years 1981 and 1983 the species Daphnia galeata prevailed throughout the whole season.

Conclusions

As we have discussed in the previous paper (Fott *et al.*, 1980), the main features of the plankton of the pond under study seem to be determined by fish. During the two-year period of each management cycle, the biomass of carp increases considerably, while the numbers do not change or decrease. In the first year, the pond is relatively understocked and is overstocked in the second year. Therefore the zooplankton is exposed to less intensive predation in odd years and species Daphnia pulicaria is dominant. This species is very sensitive to fish predation and very efficient grazer with ability to gather food particles from thin suspension (Hrbáček & Hrbáčková, 1960; Hrbáčková & Hrbáček, 1978). When Daphnia pulicaria is present, the nannoplankton are scarce (Fig. 2). In the even years the nannoplankton are very abundant and the density of large cladocerans is limited by fish predation.

The observed regular pattern of the pond plankton is to be retained only as long as the two-year management is applied. In the years 1979–80, the stock of carp was lower than in the years 1977–78 and 1981–1982 (Table 1). Consequently, the year 1979 was a typical 'understocked' one and 1980 was intermediate. Bosmina longirostris was absent and the fraction of small zooplankton was low.

Carp predation influences also the proportionality of two species of thc genus Daphnia. In 1977 and 1979 Daphnia pulicaria prevailed over Daphnia galeata, but in 1981 and 1983 the dominance switched. We hypothesize that carp with higher individual weight predated on large specimens of Daphnia pulicaria more intensively (Table 1).

Acknowledgements

I wish to express my thanks to J. Fott who was a supervisor of this project. V. Kořínek and J. Fott kindly read the manuscript and offered suggestions.

References

Blažka, P., 1966. Bestimmung der Proteine in Material aus Binnengewässern. Limnologica 4: 387–396.

Fott, J., V. Kořínek, M. Pražáková, B. Vondruš & K. Forejt, 1974. Seasonal development of phytoplankton in fish ponds. Int. Revue ges. Hydrobiol. 59: 629–641.

Fott, J., L. Pechar & M. Pražáková, 1980. Fish as a factor controlling water quality in ponds. In: J. Barica and L. R. Mur (Editors), Hypertrophic Ecosystems. Dev. Hydrobiol. 2: 255–261.

Hrbáček, J., 1966. A morphometrical study of some backwaters and fish ponds in relation to the representative plankton samples. Hydrobiol. Studies 1: 221–265, Prague.

Hrbáček, J. & M. Hrbáčková-Esslová, 1960. Fish stock as a protective agent in the occurrence of slow-developing dwarf species and strains of the genus Daphnia. Int. Revue Hydrobiol. 45: 355–358.

Hrbáčková, M. & J. Hrbáček, 1978. The growth rate of Daphnia pulex and Daphnia pulicaria (Crustacea: Cladocera) at different food levels. Věst. Čsl. Spol. Zool. 42: 115–127.

Kořínek, V., J. Fott, J. Fuksa, J. Lellák & M. Pražáková, 1987. Carp ponds of central Europe. Managed Aquatic Ecosystems. Chapter 3: 29–62.

Straškraba, M., 1964. Preliminary results of a new method for the quantitative sorting of freshwater net plankton into main groups. Limnol. Oceanogr. 9: 268–270.

Hydrobiologia **225**: 217–227, 1991.
V. Kořínek & D. G. Frey (eds), Biology of Cladocera.

An analysis of the phototactic behaviour of *Daphnia magna* clones and their sexual descendants

Luc De Meester
Laboratory of Animal Ecology, State University of Ghent, K.L. Ledeganckstraat 35, 9000 Gent, Belgium

Key words: Cladocera, *Daphnia magna*, crosses, genetics, phototaxis, vertical migration

Abstract

Daphnia magna clones, isolated from different natural populations, were inbred and crossbred. Some aspects of the heredity of the phototactic behaviour, quantified by an index, were analysed. Heritability in the broad sense was determined for several assemblages of clones through calculation of the clonal repeatability. Heritability sensu strictu was determined by offspring-parent regressions. Frequency distributions of a phototactic index of several inbred and crossbred families are given.

The results indicate a significant contribution of additive genetic variance to the total phenotypic variance. Comparison of heritability estimates suggests an important genetic interaction component.

Some implications of these results on the variability of vertical migration patterns in natural populations are discussed.

Introduction

Phototactic behaviour is a polymorphic trait in *Daphnia magna*. Dumont *et al.* (1985) and De Meester & Dumont (1988) found repeatable differences in phototactic response of different clones isolated from natural populations. Based on the repeatability of experimental results on clones, De Meester (1989) calculated the heritability of the behavioural trait. It was found to be high under the standardised experimental conditions and in the limited population – four clones – studied. The ameiotic parthenogenesis of *Daphnia* (Hebert & Ward, 1972) enables one to compare genetically homogeneous lines without need for inbreeding. Heritability in the broad sense is defined as the ratio V_G/V_P, in which V_G is the genotypic variance, and V_P is the phenotypic variance. The genetic contribution measured by clonal repeatability is, however, likely to be overestimated by genotype-environment interactions and environmental influences caused by common environment, such as maternal effects (Falconer, 1981). The high value of heritability of phototaxis in *Daphnia magna* obtained by us previously (De Meester, 1989; $V_G/V_P = 0.83$) was therefore considered to be an upper limit, indicating a relatively high heritability of a behavioural trait. In an attempt to disentangle genetic and environmental components of the phenotypic variance, the phototactic behaviour of sexual offspring resulting from mixis within and between *Daphnia magna* clones will be analysed in this paper.

Materials and methods

Origin of the clones

We sampled several populations in Eastern and Western Flanders, Belgium. Care was taken to samples the whole water column. A few females were aselectively picked out and cloned in the laboratory. In this way, 37 clones were established (Table 1). The main parent clones for the breeding study were a subset of these clones (see Table 2).

Breeding Daphnia magna

Culture conditions were the same for all *Daphnia* clones. The animals were reared in one-liter jars filled with dechlorinated tap-water. Density was kept low: 10–30 adult females liter^{-1}. Temperature (20 ± 2 °C) and photoperiod (14 h light–10 h dark) were kept constant throughout the study. The long day photoperiod was established to reduce sexual mixis (Stross, 1969, 1987). *Scenedesmus acutus* was used as main food, at a concentration of 1×10^5 to 5×10^5 cells ml^{-1}. Small amounts of horse manure extract (filtered over 40 μm) were added to the medium at the start of a new culture. The horse manure extract seemed especially important for juvenile survival, through conditioning of the medium and supply of bacteria.

The method for obtaining sexual offspring of known origin will be described in some detail elsewhere (De Meester, in prep.). Sexual mixis was induced by changes from long day (14L/10D) to short day (10L/14D) photoperiod (Stross, 1966, 1969; Bunner & Halcrow, 1977). Ephippia were stored wet in the dark at 4 °C for 1–4 months, and were subsequently manually decapsulated and incubated in the light at 20 °C (Davison, 1969; Schwartz & Hebert, 1987). Mere induction of sexual mixis within a clone resulted in the production of intraclonal sexual offspring provided that males were present in the culture. From the genetical point of view, intraclonal sexual mixis can be considered as selfing ('clonal selfing'). We repeatedly checked for pseudosexuality through examination of ephippia produced in the absence of males (Hebert, 1981, 1987a, 1987b), but no pseudosexual clones were found. All ephippial eggs resulted from sexual mixis.

To obtain interclonal sexual offspring of known origin, i.e. offspring resulting from crosses between clones, sexual mixis was induced in a population consisting of females of one clone (the mother clone) and an adult male of another clone (the father clone). Any contamination by males produced by the mother clone was prevented by checking all cultures every two days, at which time all contaminant males, at most 48 hrs old, were discarded, and ephippia were collected. Though sperm storage is not known to occur in cladocerans (Gerritsen, 1980), we established for each cross a new female population from neonati. We did not work with genetically heterogeneous populations because there are indications that, due to differences in susceptibility to male and mictic female inducing stimuli between clones (Young, 1979; Brookfield, 1981; Carvalho & Hughes, 1983; Lynch, 1983; Korpelainen, 1986; personal observations), the assumption of random mating would not hold. Particularly in our study, assortative mating would further have been increased through partially spatial isolation.

An important assumption in the interpretation of our results is that males and females of *Daphnia magna* are genetically identical, at least with respect to the trait under consideration. Most studies indicate that males of cladocerans are indeed genetically identical to females (Banta, 1939; Hebert, 1978, 1987b; Zaffagnini, 1987).

The clones derived from sexual mixis were

Table 1. Origin and name of the clones isolated from natural populations. *n* = number of clones established from the population.

Locality	Habitat	*n*	Name
Blankaart	shallow lake	3	B_{1-3}
Citadelpark	city pond (Ghent)	6	C_{1-6}
Oostduinkerke	shallow pond	2	D_{1-2}
Mandelhoek	ditch	1	I_1
Leie	eutrophic river	5	L_{1-5}
Muinkpark	city pond (Ghent)	2	M_{1-2}
Maaltepark	city pond (Ghent)	2	N_{1-2}
Drieh. plas	pond	10	P_{1-10}
Ringvaart	channel	5	R_{1-5}
Zwin	ditch	1	Z_1

given a compound name, consisting of the name of the mother clone, followed by the name of the father clone, and a number indicating the chronological order of hatching. For instance, $P_1.P_1$-5, also written as P_1-5, is the clone that originated from the fifth hatchling of ephippia resulting from intraclonal sexual mixis in clone P_1; $R_1P_17.C_1$-2 is the second clone that originated from ephippia resulting from interclonal sexual mixis with $R_1.P_1$-7 as mother and C_1 as father clone.

Measurement of phototactic behaviour

The experimental procedure was basically the same as in De Meester (1989) and De Meester & Dumont (1989). The experimental set-up consisted of a small perspex column (25 cm height, 5 cm internal cross-section), placed in a darkened box, and illuminated from above with a 150 W fiber light source (Müster). Light intensity at the water surface was about 500 W m^{-2}. The bottom of the column was covered by black pebbles, so that light reflection was minimised. The column was externally divided in an upper compartment of 12 cm height, a lower compartment of 3 cm height, and a middle compartment of 10 cm height, and placed in a constant temperature (20 $\pm$ 2 °C) room. In each experiment 10 (sometimes 8) adult females of the same clone were followed. The test animals were placed in dechlorinated tap water for 3 to 4 hours prior to the experiment. The experimental column was also filled with dechlorinated tap water. Once placed in the experimental column, the animals were given 5 minutes dark adaptation, after which the fiber light source was lit. All experiments were carried out between 2 and 7 P.M. to obviate interferences with a circadian endogenous rhythm (Harris, 1963; Ringelberg & Servaas, 1971). Each experiment lasted 10 minutes. At one-minute intervals, the positions of the test animals were recorded. A percentage positively phototactic behaviour can be assigned to the number of animals in the upper compartment, while a percentage negatively phototactic behaviour can be defined for the lower 3 centimeters of the column. The observations of the second 5 minutes of the experimental period were averaged in the calculation of the phototactic index. This minimises the influence of the flight reaction at the start of the experiment (De Meester & Dumont, 1988), as the vertical distribution of a population in the small column stabilises within a few minutes. The phototactic index is defined as $(U - L)/(U + M + L)$, in which U, M, and L are the number of animal-observations in the upper, middle, and lower compartment, respectively. The phototactic index of a clone is defined as the average of at least 2 experiments. The replicate experiments on a given genotype (clone) were always performed with animals of another generation.

Three gross phenotypes of phototactic behaviour are apparent (De Meester & Dumont, 1988). Animals of positively phototactic clones rapidly ascend to the upper part of the experimental column, and tend to aggregate in the upper 3 centimeters. The distribution of extreme negatively phototactic animals is largely restricted to the bottom 3 centimeters of the column. Animals of intermediate phototactic behaviour tend to be more dispersed throughout the column, but are nevertheless predominantly found in the middle compartment, and clearly avoid the upper 3 centimeters of the column. These distributions are reflected in the values of the phototactic index. More specifically, a clone was called positively phototactic when the average phototactic index was more than 0.65, and when on average more than 15% of the animals were observed in the uppermost 3 centimeters. Negatively phototactic animals were defined by a phototactic index lower than -0.85, and by a high (>85) percentage of animal-observations in the bottom 3 centimeters.

Genetic analysis

Each clone is characterised by its value of the phototactic index. All data are based on experiments on females. The phototactic index of a father clone is therefore given by the behaviour of the females of that clone. This is acceptable given the genetic identity of males and females. We did not use results on the phototactic behaviour of males in the analysis, because males are not always present and are seldom abundant in a culture.

It is assumed that offspring mortality (on average, about 20% for interclonal offspring and about 40% for intraclonal offspring) was not selective with respect to the trait under consideration.

It was our aim to partition phenotypic variance into genetic and environmental components. Total phenotypic variance (V_P) has a genetic (V_G) and an environmental (V_E) component: $V_P = V_G + V_E$. It is assumed that no genotype-environment correlations or interactions are present. The genetic component itself can be partitioned into a component due to additive gene effects (V_A), and components due to interaction between alleles (dominance; V_D) and due to non-allelic interaction (epistasis; V_i): $V_G = V_A + V_D + V_I$. The environmental component can be partitioned into variance due to temporal environmental influences (V_{ES}), giving rise to variance in expression within one individual, and variance due to permanent influences, such as maternal effects (common environment; V_{EC}).

Heritability in the broad sense (V_G/V_P) was estimated through clonal repeatability (Parsons, 1973; Falconer, 1981) in several (artificial) populations. Variance components were calculated by a single classification ANOVA with unequal sample sizes, as the number of replicate experiments was not the same for all clones. The heritability estimates were compared with the value obtained by De Meester (1989) for four selected clones.

Offspring-on-parent regressions were performed to estimate the heritability, in a narrow sense, of phototaxis in *Daphnia magna*. Heritability is defined as V_A/V_P (Falconer, 1981; Mather & Jinks, 1981). As data on females were used for both parents, there should be no difference in variance between the parents, so that offspring on mid-parent regressions can be used (Falconer, 1981). Heritability estimates were obtained for both intraclonal and interclonal crosses. Assortative mating has no influence on the estimates of heritability based on regressions of offspring on mid-parent (Falconer, 1981). As crosses performed in the breeding study were not random, only offspring-to-midparent regressions could be interpret for the results of both intraclonal and interclonal mixis. The significance of the regressions was tested by analysis of the variance (Sokal & Rohlf, 1981).

Comparison of the estimates of heritability in the broad sense with those of heritability in the narrow sense was used to partition the genetic variance into an additive and an interaction component.

Maternal effects were checked for through comparison of reciprocal crosses.

The analysis was performed on a set of clones originating from different natural populations because attention was focused primarily on inter-population differences (Chapman, 1974; Arthur, 1984; Butlin & Hewitt, 1986).

Results

A total of 1667 experiments was carried out. The phototactic behaviour of 37 clones derived from natural populations, of 338 offspring clones (30 families) from intraclonal mixis, and of 189 offspring clones (26 families) derived from interclonal crosses was determined. The average number of experiments on a clone was about 3. A summary of the results for clones with 5 or more offspring clones is given in Table 2. We first calculated the heritability in the broad sense ($H_B^2 = V_G/V_P$) for all clones derived from natural populations. This should give us an idea of the genetic component of variance in nature, as compared to the limited population (4 clones) studied in De Meester (1989). Differences between clones were significant at the 0.001 level, and $h_B^2 = 0.55$ (Table 3). We also calculated the heritability of all parent clones that gave rise to 5 or more offspring clones (see Table 2). As this subset of clones was selected for different phototactic behaviour, the estimate is likely to be higher than the former. This

Table 2. Phototactic index of parent clones and their intra- and interclonal offspring. Clone: average phototactic index of the clone; Intraclonal: average phototactic index of intraclonal sexual offspring; Crosses, dam: weighted average value of offspring of interclonal crosses with the given clone as mother; Crosses, sire: weighted average value of offspring of interclonal crosses with the given clone as father. Only parent clones with more than five offspring clones are tabulated; numbers in parentheses are the number of offspring clones tested.

	Clone	Intraclonal	Crosses dam	Crosses sire
C_1	0.887	0.816 (51)	0.454 (11)	- 0.197 (72)
C_2	0.240		- 0.162 (28)	
C_3	- 0.923	- 0.993 (6)		
C_4	- 0.402	- 0.313 (28)		
R_1	0.379	0.367 (5)	0.001 (13)	- 0.145 (39)
R_2	- 0.343	- 0.419 (47)	- 0.187 (11)	
R_3	- 0.283	- 0.297 (26)	- 0.619 (9)	
P_1	- 0.430	- 0.504 (39)	- 0.182 (59)	0.028 (30)
P_2	- 0.323			- 0.168 (44)
R_2-30	- 0.028	- 0.589 (6)		
P_1-12	- 0.957	- 0.945 (15)	- 0.281 (16)	
P_1-32	- 0.566	- 0.521 (80)		
P_1-81	- 0.263	0.141 (7)	- 0.411 (9)	

Table 3. Results of the analysis of clonal repeatability on two populations: A) Clones isolated from different natural habitats (see table 1); B) clones that were used as main parents in the present breeding study (see table 2). N = number of clones; n = average number of experiments per clone.

Pop.	N	n	V_E	V_G	V_P	h^2_B	sign.
A	37	2	0.103	0.125	0.228	0.55	$p < 0.001$
B	13	2	0.091	0.250	0.341	0.73	$p < 0.005$

estimate is important, because it was calculated for a population comparable to the one for which the additive variance component was estimated. The difference between clones was significant at the 0.001 level, h^2_B was estimated to be 0.73. In a previous study (De Meester, 1989), using a different index, a comparable estimate ($h^2_B = 0.74$) was obtained for 4 clones selected for different phototactic behaviour. These broad heritability estimates all indicate that a high portion of variance in phototactic behaviour is due to gene action.

The values for the slopes of offspring-on-parent regressions are given in Table 4. The heritability estimate based on all data is high: $h^2 = 0.72$. This is due mainly to the high correlation between offspring and mid-parent values in intraclonal crosses: the heritability estimate based on values of intraclonal offspring is 0.90. The slope of the regression of offspring on mid-parent for interclonal crosses is only 0.38. The regression of both intraclonal and interclonal offspring on mid-parent values is shown in Fig. 1. On Fig. 2, the regression of intra- and interclonal offspring on mid-parent values can be compared.

We checked for maternal effects through comparison of the results of reciprocal crosses (Table 5). There is no indication for maternal effects in most of the crosses; differences among reciprocal crosses were not significant by nested ANOVA. There is a large difference in the average phototactic index of the cross $C_2.C_1$ (-0.107) as

Table 4. Results of the offspring on mid-parent regressions. b = regression slope; C.L. = 95% confidence limits of regression slope; s = significance level.

	Mid-parent
Intraclonal	b = 0.895 C.L.: 0.778–1.012 s: $p < 0.001$
Interclonal	b = 0.381 C.L.: 0.131–0.631 s: $p < 0.005$
All data	b = 0.722 C.L.: 0.604–0.839 s: $p < 0.001$

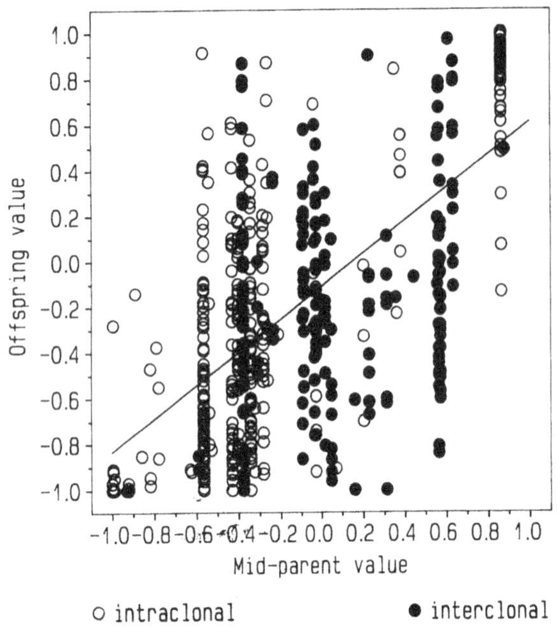

Fig. 1. Regression of the values for the phototactic index of offspring clones on the mid-parent values, for intra- and interclonal crosses. See Table 4 for statistics.

compared to $C_1.C_2$ (0.760), but it remains that 2 (out of 22) $C_2.C_1$ offspring clones have an average phototactic index larger than 0.700. Figure 3 ena-

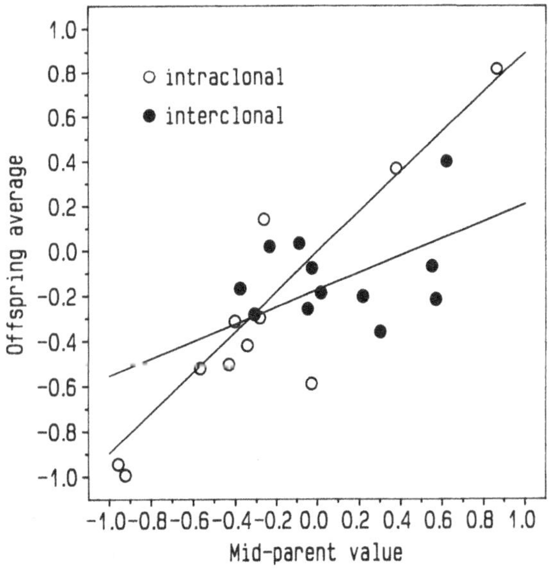

Fig. 2. Regression of the family-average value of the phototactic index of the offspring on the mid-parent value, for intra- and interclonal crosses separately. Only families with 5 or more offspring clones are presented. Regression lines based on data of all families. See Table 4 for statistics.

Table 5. Results of reciprocal crosses. (A): Average offspring value for the cross with the lesser positively phototactic parent as mother clone, ± 95% C.I.; (B): Average offspring value for the cross with the more positively phototactic parent as mother clone, ± 95% C.I. Numbers in parentheses are the number of offspring clones.

Cross	(A)	(B)
$P_1 12.R_1/R_1.P_1 12$	− 0.343 ± 0.435 (2)	− 0.220 ± 0.669 (2)
$P_1.R_1/R_1.P_1$	− 0.103 ± 0.160 (10)	− 0.044 ± 0.245 (8)
$P_1.C_1/C_1.P_1$	− 0.358 ± 0.209 (7)	− 0.309 ± 1.704 (2)
$C_2.C_1/C_1.C_2$	− 0.107 ± 0.220 (22)	0.760 (1)
$R_1.C_1/C_1.R_1$	0.259 ± 0.527 (3)	0.453 ± 0.268 (8)

bles to evaluate differences between the average offspring value of the phototactic index for a given interclonal cross and the mid-parent value, i.e. between the observed and the expected value. Differences were significant at the 5% level in 8 out of 13 crosses. All 5 crosses of the extremely positively phototactic clone C_1 with clones of intermediate phenotype show a deviation of the average offspring value towards the intermediate phenotype, suggesting genetic interaction. Crossing of the two extreme phenotypes ($C_1.P_1 12$) gives rise to offspring with, on average, an intermediate phenotype. The average value of the offspring from a cross of the negatively phototactic clone $P_1 12$ with a clone of intermediate phototactic behaviour (R_1) is almost exactly intermediate to the parent values. With the exception of 3 cases, one of which is rather striking, crosses between clones of intermediate phenotype resulted in offspring with an average value close to expectation. The following example will illustrate the point. The reciprocal crosses $R_1.P_1$ and $P_1.R_1$ gave rise to offspring with comparable phototactic behaviour, with an average value that is nearly exactly intermediate to the parental values (number of offspring: 16). A backcross of one of the offspring clones (R_1P_1-7) to the father clone P_1 resulted in an average offspring value nearly exactly intermediate to the parental values (num-

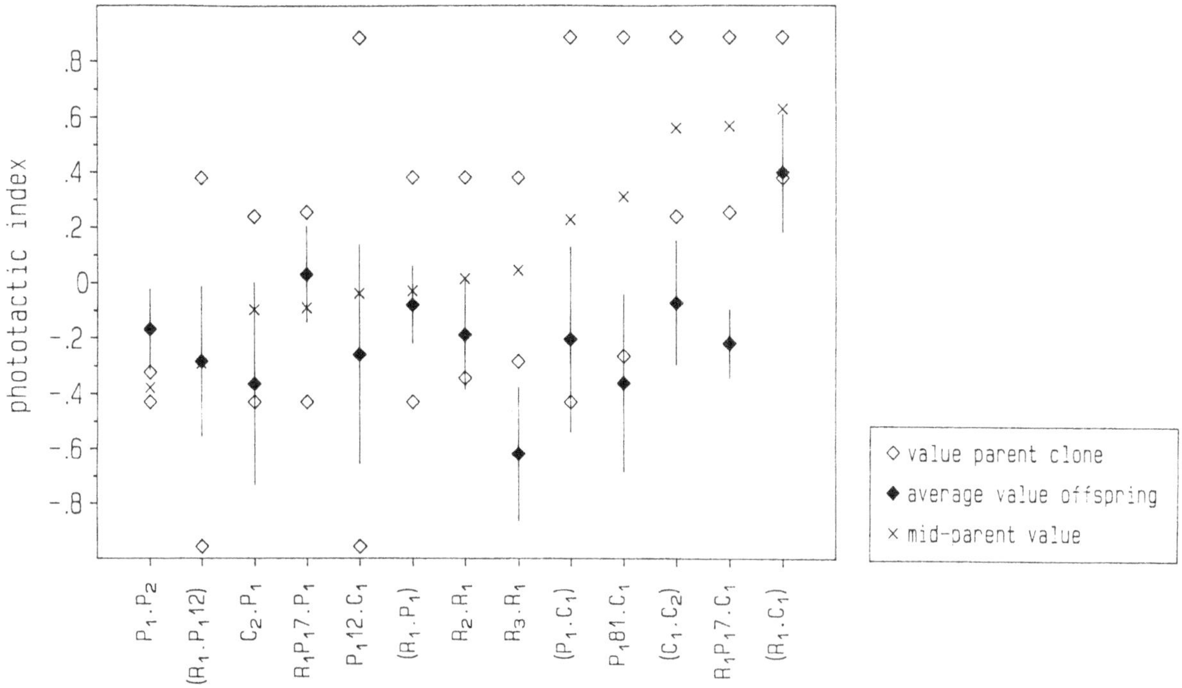

Fig. 3. Parent values, mid-parent values, and average value (±95% confidence interval) of the offspring for all crosses with 4 offspring or more. X-axis: crosses, ordered according to mid-parent value. Name of cross in parentheses indicates the combined result from reciprocal crosses.

ber of offspring: 16). Crossing the same clone with clone C_1 resulted, however, if offspring with an intermediate phenotype, the average value showing a significant deviation from the mid-parent value (number of offspring: 19; Fig. 3). Clones P_1, R_1 and R_1P_1-7 show no dominance with respect to each other, but seem dominant for the phototactic trait studied with respect to clone C_1.

As the repeatability and the regression analysis were performed on a similar population, we can assume V_P to be the same for both populations ($V_P = 0.460$; Table 3). The calculation of clonal repeatability gave $V_E = 0.122$ and $V_G = 0.338$. Based on all data, $V_A = 0.332$. Using solely the data on intraclonal offspring, $V_A = 0.412$, and for interclonal offspring, $V_A = 0.175$. This indicates that there are no dominance deviations interfering with the results on intraclonal crosses, while a genetic interaction component is suggested by the results on interclonal crosses ($V_D + V_I = 0.120$, which is about 35% of the total phenotypic variance). It is possible that the lower value of the regression slope for interclonal crosses is due to the smaller range of mid-parent values, but the

Table 6. Segregation ratios for families with 4 offspring or more. Classification of crosses according to phototactic behaviour of parents (+ +, × +, × ×, × −, − −). + = percentage of positively phototactic offspring; × = percentage of offspring with intermediate phototactic behaviour; − = percentage of offspring with negatively phototactic behaviour.

	Cross	Phototactic behaviour of offspring			
		N	+	×	−
+ +	$C_1.C_1$	52	92.2	7.8	0.0
× +	$R_1.C_1/C_1.R_1$	11	27.3	72.7	0.0
	$P_181.C_1$	7	14.3	85.7	0.0
	$P_1.C_1/C_1.P_1$	9	11.1	88.9	0.0
	$C_2.C_1/C_1.C_2$	23	0.0	100.0	0.0
	$R_1P_17.C_1$	18	0.0	100.0	0.0
× ×	$R_2.R_1$	11	0.0	100.0	0.0
	$R_1P_17.P_1$	16	0.0	100.0	0.0
	$P_1R_1/R_1.P_1$	18	0.0	100.0	0.0
	$R_2.R_2$	45	0.0	93.0	7.0
	$R_3.R_3$	22	0.0	95.5	4.5
	$C_4.C_4$	30	0.0	93.3	6.7
	$P_1.P_2$	44	0.0	88.4	11.6
	$P_1.P_1$	39	0.0	74.4	25.6
	$P_132.P_132$	78	1.3	75.6	23.1
− +	$P_112.C_1$	7	0.0	100.0	0.0
− ×	$P_112.R_1/R_1.P_112$	4	0.0	100.0	0.0
− −	$P_112.P_112$	16	0.0	6.2	93.8

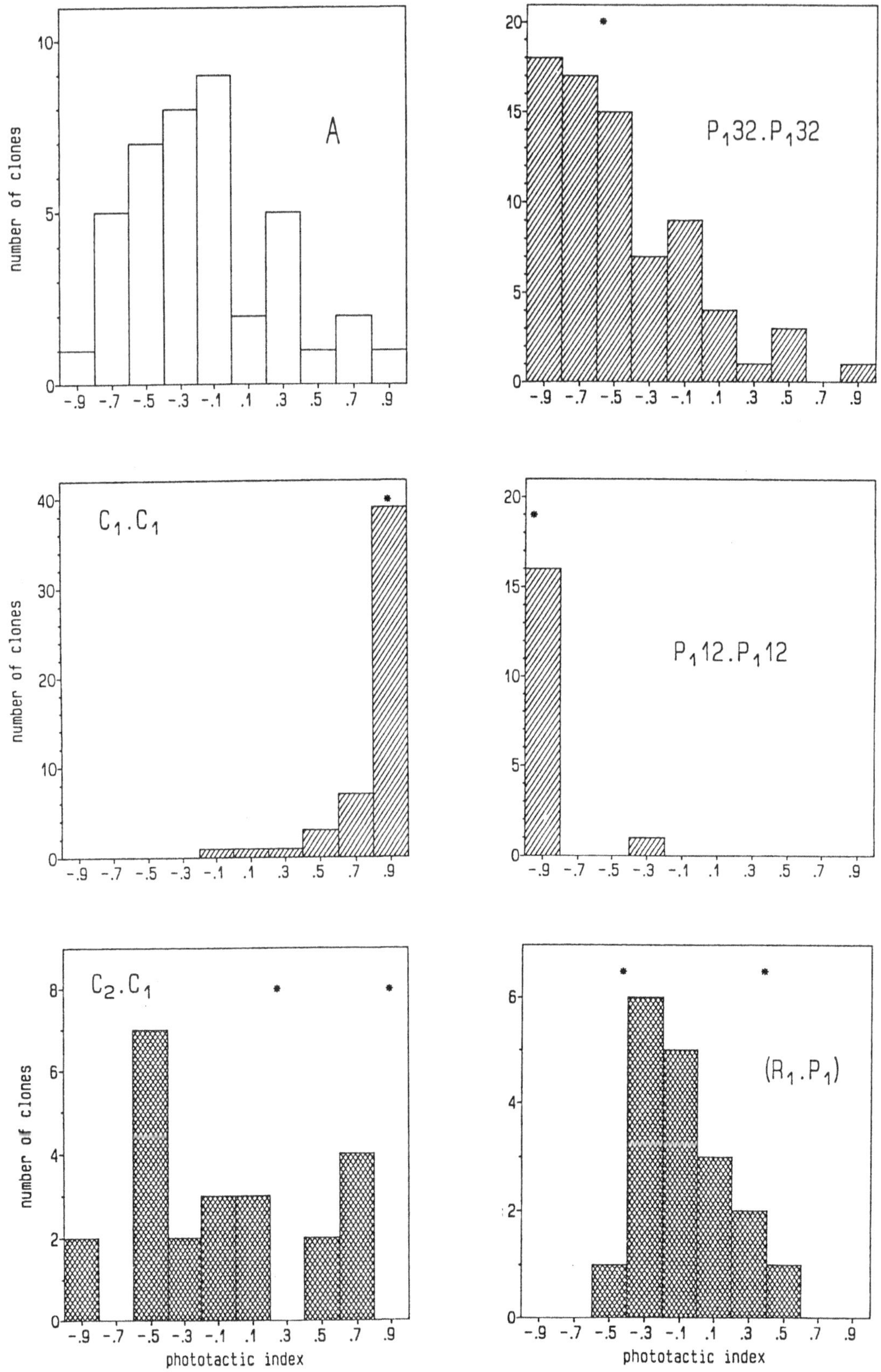

Fig. 4. Frequency distributions of the phototactic index for 37 clones isolated from natural populations (A) and for the offspring of several intraclonal ($P_1 32.P_1 32$; $C_1.C_1$; $P_1 12.P_1 12$) and interclonal ($R_1.P_1$ and $C_2.C_1$) families. The phototactic behaviour of the parent clone(s) is indicated by an asterisk.

results shown in Fig. 3 suggest that genetic interactions are a plausible alternative. The high value of the regression slope for intraclonal mixis ($b = 0.90$) is largely determined by the good correlation of the parent-offspring values for the extreme phenotypes ($P_1 12.P_1 12$, $C_3.C_3$ and $C_1.C_1$). If we accept the existence of genetic interaction, the absence of dominance deviations for the intraclonal offspring of the clones with extreme phenotype suggests that the positively phototactic clone used in the breeding study was largely homozygous for the trait under consideration.

Based on our characterisation of positively-intermediate-negatively phototactic animals, segregation ratios were calculated for all families with 5 or more offspring (Table 6). Frequency distributions of the phototactic index of the offspring from some intra- and interclonal families are shown in Fig. 4. Crosses involving genotypes with intermediate phototactic behaviour produce offspring with different ratios, according to which clone is involved. Although not conclusive, these results and the dominance deviations visualised in Fig. 3 indicate that (a) major gene effect(s) might be involved. It appears that phenotypes with intermediate phototactic behaviour can be the result of different genotypes; there is no evidence that this is so for the extreme phenotypes.

Discussion

It is our hypothesis that the variability of vertical migration patterns between and within natural populations can, at least partially, be linked to genetic differences in migration responses. Due to the major impact of light stimuli in the proces of vertical migration, it seems plausible that phototactic behaviour is one of the more important traits that characterise vertical migration. Weider (1984, 1985) reported that electrophoretically distinct *Daphnia pulex* 'clones' (clonal groups) showed different depth distributions. This is indirect evidence that a genetic component might be important in determining the vertical distribution of *Daphnia* populations. Studies on clones isolated from natural populations revealed consistent differences in phototactic behaviour (Dumont *et al.*, 1985; De Meester & Dumont, 1988). The present study substantiates these findings with the results of an analysis of the phototactic behaviour of the offspring from in- and crossbred *Daphnia* clones.

There was no incompatibility observed when crossing clones that differed in their phototactic responses, and interclonal offspring proved to be sexually fertile too. Thus, we deal with variability within one species (Hebert, 1984). The fact that crosses often resulted in offspring with average phototactic behaviour nearly intermediate to the parents implicates that males contribute genetically to the offspring (Hebert, 1987a). No pseudosexual clones were observed. About 73% of the total variability of the phototactic index of our laboratory population was found to be genetic in origin. Thus, under the standardised culture conditions and experimental procedure used in the present study, environmental and physiological variability accounted for only 27% of the total variability. All heritability estimates based on clonal repeatability are in reasonable agreement, irrespective of the group of clones they were based upon. The limitation that the estimates are valid only for the environmental circumstances in which the study was conducted remains (Sprague, 1955). Our results indicate that about 52% of the genetic variability (38% of the total phenotypic variance) is due to additive gene effects, the remainder 48% (35% of the total phenotypic variance) being caused by interactions between alleles (dominance) and/or loci (epistasis). The heritability estimate of 0.38 is to be interpreted as conservative, because of the rather low variance in mid-parent values on which the regression for interclonal crosses was based. Even this conservative estimate indicates, however, a significant and moderately high contribution of additive genetic effects to the total phenotypic variance.

The distribution of the phototactic index of clones isolated from natural populations is not significantly different from a normal distribution (see also Fig. 4a). However, as Thoday and

Thompson (1976) have shown, this need not imply a pure quantitative inheritance. The rather large non-additive genetic component and the tendency for segregation is positively-intermediate-negatively phototactic animals suggest that relatively few loci determine a positive or negative or an intermediate phenotype. At least for the intermediate phenotype, several genotypes seem to determine a comparable phenotype. The negatively and positively phototactic clones we studied were largely homozygous for the trait under consideration. The variability of the trait within the intermediate phenotype, and probably also within the extreme phenotypes, can partially be accounted for by genes with small additive effects (cf. Fig. 2: $R_1.P_1$). Even if we do not want to characterise animals as positively or negatively phototactic, or intermediate, it still holds that rather few loci or small regions of the genome that segregate as units (Thoday & Thompson, 1976), are responsible for a large part of the genetic variance.

In studies of particular natural populations, however, it is likely that the quantitative aspects of the inheritance of phototaxis will in many cases be relatively important, because the extreme phenotypes tend to be rare (e.g. De Meester, in press). Positively phototactic animals will often be selected against by visual predation. Negatively phototactic animals will often not be caught when using standard sampling equipment. The phototactic responses of *Drosophila* species has been described as a quantitative character with medium heritability (Benzer, 1967; Richmond, 1969). The response measured in some of these studies was comparable to that measured for *Daphnia* in our study. Moreover, the visual system of both organisms might be largely comparable. However, the ecological function of the trait, and therefore selective pressures on its expression, is different: in *Drosophila* phototaxis seems mainly connected with overall habitat preference (Rockwell & Seiger, 1973), while in *Daphnia* the phototactic response considered is of primary importance in vertical migration, which is a rather specialised behavioural response.

Acknowledgements

This study benefitted from (sometimes vivid) discussions with D. Ebert, K. Desender, K. Martens and Professor H.J. Dumont. A. Van Noordwijk drew attention to a scaling problem. We thank these people and an anonymous reviewer for critically reading an earlier version of the manuscript. The author is a Research Assistant with the National Fund for Scientific Research (Belgium). A grant (01.1905.85) of the State University of Ghent, and travel support from the National Fund for Scientific Research (Belgium) is acknowledged.

References

Arthur, W., 1984. Mechanisms of morphological evolution, a combined genetic, developmental and ecological approach. J. Wiley & Sons, Chichester, 275 pp.

Banta, A. M., 1939. Studies on the physiology, genetics, and evolution of some Cladocera. Publ. Carnegie Inst. 513: 285 pp.

Benzer, S., 1967. Behavioral mutants of *Drosophila* isolated by countercurrent distribution. Proc. natn. Acad. Sci. U.S.A. 58: 1112–1119.

Brookfield, J. F. Y., 1981. No evidence for frequency-dependent selection acting between clones of the water flea, *Daphnia magna*. Heredity 47: 297–315.

Bunner, H. C. & K. Halcrow, 1977. Experimental induction of the production of ephippia in *Daphnia magna* Straus (Cladocera). Crustaceana 32: 77–86.

Butlin, R. K. & G. M. Hewitt, 1986. Heritability estimates for characters under sexual selection in the grasshopper, *Chorthippus brunneus*. Anim. Behav. 34: 1256–1261.

Carvalho, G. R. & R. N. Hughes, 1983. The effect of food availability, female culture-density and photoperiod on ephippia production in *Daphnia magna* Straus (Crustacea: Cladocera). Freshwat. Biol. 13: 37–46.

Chapman, A. R. O., 1974. The genetic basis of morphological differentiation in some *Laminaria* populations. Mar. Biol. 24: 85–91.

Davison, J., 1969. Activation of the ephippial eggs of *Daphnia pulex*. J. gen. Physiol. 53: 565–575.

De Meester, L., 1989. An estimation of the heritability of phototaxis in *Daphnia magna* Straus. Oecologia 78: 142–144.

De Meester, L., in press. Evidence for intrapopulation genetic variability for phototactic behaviour in *Daphnia magna* Straus, 1820. Biol. Jb. 58.

De Meester, L. & H. J. Dumont, 1988. The genetics of phototaxis in *Daphnia magna*: Existence of three phenotypes for vertical migration among parthenogenetic females. Hydrobiologia 162: 47–55.

De Meester, L. & H. J. Dumont, 1989. Phototaxis in *Daphnia*: interaction of hunger and genotype. Limnol. Oceanogr. 34: 1322–1325.

Dumont, H. J., Y. Guisez, I. Carels & H. M. Verheye, 1985. Experimental isolation of positively and negatively phototactic phenotypes from a natural population of *Daphnia magna* Straus: A contribution to the genetics of vertical migration. Hydrobiologia 126: 121–127.

Falconer, 1981. Introduction to quantitative genetics, second edition. Longman, New York, 340 pp.

Gerritsen, J., 1980. Sex and parthenogenesis in sparse populations. Am. Nat. 115: 718–742.

Harris, J. E., 1963. The role of endogenous rhythms in vertical migration. J. mar. biol. Ass. U.K. 43: 153–166.

Hebert, P. D. N., 1978. The population biology of *Daphnia* (Crustacea, Daphniidae). Biol. Rev. 53: 387–426.

Hebert, P. D. N., 1981. Obligate asexuality in *Daphnia*. Am. Nat. 117: 784–789.

Hebert, P. D. N., 1984. Demographic implications of genetic variation in zooplankton populations. In: Wöhrmann, K. & V. Loeschcke, Population Biology and Evolution. Springer-Verlag, Berlin: 195–207.

Hebert, P. D. N., 1987a. Genotypic characteristics of the Cladocera. Hydrobiologia 145: 183–193.

Hebert, P. D. N., 1987b. Genetics of *Daphnia*. In: Peters, R. H. & R. de Bernardi (eds.), *Daphnia*, Mem. Ist. ital. Idrobiol. 45: 143–192.

Hebert, P. D. N. & R. D. Ward, 1972. Inheritance during parthenogenesis in *Daphnia magna*. Genetics 71: 639–642.

Korpelainen, H., 1986. The effects of temperature and photoperiod on life-history parameters in *Daphnia magna* (Crustacea: Cladocera). Freshwat. Biol. 16: 615–620.

Lynch, M., 1983. Ecological genetics of *Daphnia pulex*. Evolution 37: 358–374.

Mather, K. & J. L. Jinks, 1981. Biometrical genetics, third edition. Chapman and Hall, London.

Parsons, P. A., 1973. Behavioural and ecological genetics, a study in *Drosophila*. Clarendon Press, Oxford.

Richmond, R. C., 1969. Heritability of phototactic and geotactic responses in *Drosophila pseudoobscura*. Am. Nat. 103: 315–316.

geotactic responses in *Drosophila pseudoobscura*. Am. Nat. 103: 315–316.

Ringelberg, J. & H. Servaas, 1971. A circadian rhythm in *Daphnia magna*. Oecologia 6: 289–292.

Rockwell, R. F. & M. B. Seiger, 1973. Phototaxis in *Drosophila*: A critical evaluation. Am. Sci. 61: 339–345.

Schwartz, S. S. & P. D. N. Hebert, 1987. Methods for the activation of the resting eggs of *Daphnia*. Freshwat. Biol. 17: 373–379.

Sokal, R. R. & F. J. Rohlf, 1981. Biometry, second edition. Freeman, San Francisco.

Sprague, G. F., 1955. Problems in the estimation and utilization of genetic variability. In: Population genetics: the nature and causes of genetic variability in populations. Cold Spring Harbor Symp. quant. Biol. 20: 87–92.

Stross, R. G., 1966. Light and temperature requirements for diapause development and release in *Daphnia*. Ecology 47: 368–374.

Stross, R. G., 1969. Photoperiod control of diapause in *Daphnia*. III. Two-stimulus control of long-day, short-day induction. Biol. Bull. 137: 359–374.

Stross, R. G., 1987. Photoperiodism and phased growth in *Daphnia* populations: coactions in perspective. In: Peters, R. H. & R. de Bernardi (eds.), *Daphnia*, Mem. Ist. ital. Idrobiol. 45: 413–437.

Thoday, J. M. & J. N. Thompson Jr., 1976. The number of segregating genes implied by continuous variation. Genetics 46: 335–344.

Weider, L. J., 1984. Spatial heterogeneity of *Daphnia* genotypes: Vertical migration and habitat partitioning. Limnol. Oceanogr. 29: 225–235.

Weider, L. J., 1985. Spatial and temporal genetic heterogeneity in a natural *Daphnia* population. J. Plankton Res. 7: 101–123.

Young, J. P. W., 1979. Enzyme polymorphism and cyclic parthenogenesis in *Daphnia magna*. II. Heterosis following sexual reproduction. Genetics 92: 971–982.

Zaffagnini, F., 1987. Reproduction in *Daphnia*. In: Peters, R. H. & R. de Bernardi (eds.), *Daphnia*, Mem. Ist. ital. Idrobiol. 45: 245–284.

Hydrobiologia **225**: 229–243, 1991.
V. Kořínek & D. G. Frey (eds), Biology of Cladocera.

Ecological genetics of Norwegian *Daphnia*. II. Distribution of *Daphnia longispina* genotypes in relation to short-wave radiation and water colour

Anders Hobæk[1] & Hans Georg Wolf[2†]
[1] *Museum of Zoology, Dept. of Animal Ecology, University of Bergen, Muséplass 3, N-5007 Bergen, Norway;* [2] *Max-Planck-Institut für Limnologie, Abt. Ökophysiologie, Postfach 165, D-2320 Plön, Germany.* (*† Deceased*).

Key words: Daphnia longispina, ultraviolet radiation, pigments, photoprotection, genetics, clonal diversity

Abstract

In the mountain range of South Norway, transparent and melanic members of the *Daphnia longispina* complex occur. Melanic populations inhabit clearwater lakes and ponds, while transparent populations are found in ponds with slightly humic water. Mixed populations have not been detected. The distribution patterns of the two morphs are related to ambient levels of short-wave radiation, and the light absorptive properties of the inhabited waters. The pigment present is probably melanin, which is deposited in the dorsally directed parts of the carapace, head shield and antennae. Allozyme studies indicate that these distinct morphs are only remotely related, the alpine transparent group being closer to lowland, likewise transparent, populations. A melanic population found at Svalbard is closely related to the melanic mainland populations. Clonal diversity in ponds and shallow lakes is very low, in contrast to populations of deeper lakes.

Introduction

The colour of a *Daphnia* is subject to intense selection. A conspicuously coloured animal is seen from afar by visual predators, and hence transparency is favoured in the presence of planktivorous fish (Zaret, 1980). On the other hand, colour may be advantageous, for instance when high haemoglobin levels are needed (Weider & Lampert, 1985), or when pigments are necessary to protect animals from irradiation. The dosage of harmful short-wave light increases with altitude as well as latitude (Thomasson, 1956). Primary photoprotective pigments are carotenoids and melanins (Britton, 1983). Carotenoid photoprotection is well documented in copepods of high altitude lakes (Hairston, 1976, 1978, 1979). In the *D. pulex* group, clones with a melanized carapace are known from arctic areas (Brooks, 1957; Hebert & McWalter, 1983). Brownish *D. pulex* have also been studied in the Alps (Siebeck, 1978a, 1978b); in this case the pigments are probably of carotenoid nature. Thus, freshwater crustaceans with high contents of carotenoids or melanins primarily inhabit localities where irradiation levels are expectedly high.

In the mountain range of southern Norway, members of the *D. longispina* complex populate lakes and ponds. Some animals have a transparent carapace, but may have a reddish colour of varying intensity in their internal tissues, especially in lipid droplets of the ovaries and in their eggs. Other animals have a dark brown, almost black colour in the dorsal parts of their carapace; head shield and antennae. Dark specimens are known from many lakes and ponds, while the transparent ones have so far been found only in ponds. In the lowlands, transparent *D. longispina* are common in both ponds and lakes. On the assumption that the colour variants represented ecophenotypes, we undertook an electrophoretic survey of mountain populations. Preliminary results indicated that these forms were genetically quite distinct (Wolf & Hobæk, 1986). Here we present an extension of that work, including more populations from the mountains, lowlands, and the Arctic. We also examine distribution patterns of the colour morphs in the mountains, in relation to ambient radiation levels and the light absorptive properties of inhabited waters.

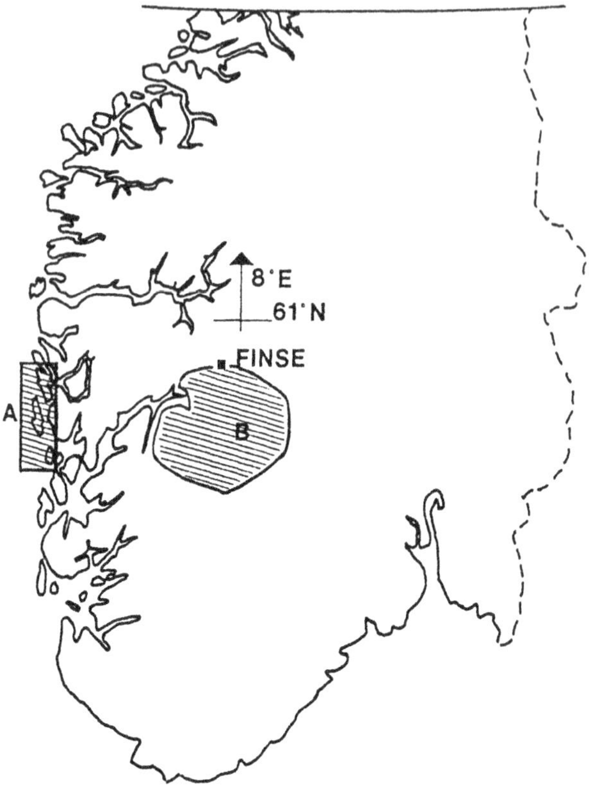

Fig. 1. Map of southern Norway, showing the location of study areas (hatched). A) Western coast around Bergen, where lowland populations were sampled. B) the Hardangervidda mountain plain, with the main study site Finse in its northern edge.

Materials and methods

Study areas

Samples were collected in the northern part of the mountain plain of Hardangervidda, southern Norway (Fig. 1). Most localities are situated within reach of the Ecology Field Station at Finse, and were sampled with varying intensity 1985–1988. For a more detailed map of the Finse area, see Wolf & Hobæk (1986). In addition, data were included from several lake and pond populations in coastal western Norway (visited 1984–1988), and from one small lake in the Kongsfjord area (79° N) on Svalbard (sampled 1987). We also included data on *D. hyalina* from Schöhsee at Plön, FRG, in an attempt to clarify the taxonomic position of Norwegian populations. Some characteristics of the Norwegian localities are given in Table 1.

Sampling, sorting

Animals were collected with a small plankton net (90 μm meshes, 20 cm diameter) thrown into the ponds and hauled ashore. In some cases we attached the net to an otter board, which permitted longer hauls. Lake samples were taken by oblique tows with a similar net. Animals were transported alive in thermos bottles. Within 12 hours after collecting, they were sorted alive into microtiter plates, single animals being placed in 10 μl of distilled water containing 0.001 M mercaptoethanol to stabilize enzymes. Samples were then frozen and stored at − 18 °C, and later transported on dry ice to Plön, FRG for electrophoretic analysis. Normally, each sample consisted of 175–200 animals, and each individual was analyzed for 5 enzymes. The sample from lake Saatevatn consisted of only 11 animals. Most mountain ponds have been analyzed repeatedly, both within one season and through several years. During the sorting procedure, animals were classified as 'melanic' or 'transparent'. This classification refers only to the appearance of

Table 1. Some characteristics of the localities investigated.

Locality	Abbr.	Region	Altitude (m)	Max depth (m)	Max length (m)	Max width (m)	Trophic state[1]	Water type[2]
Finse 0	F0	Mountain	1210	3.5	60	20	O	C
Finse 1	F1	Mountain	1230	1.5	30	10	O	C
Finse 2	F2	Mountain	1230	1.5	10	5	O	C
Finse 3	F3	Mountain	1230	1	30	15	O	C
Finse 4	F4	Mountain	1240	1	18	8	O	SH
Finse 5	F5	Mountain	1225	1	20	7	O	SH
Finse 6	F6	Mountain	1220	1	15	5	O	SH
Finse 7	F7	Mountain	1225	1	10	7	O	SH
Finse 8	F8	Mountain	1220	1	20	10	O	SH
Finse 9	F9	Mountain	1220	1	2	1	O	H
Finse 10	F10	Mountain	1220	1	6	2	O	H
Saatevatn	Saat	Mountain	1410	24	1100	650	O	C
Raggsteindalen 1	R1	Mountain	1234	4	250	70	O	C
Herdla 1	H1	West coast	2	2	12	5	M	VH
Herdla 2	H2	West coast	3	2	22	10	M	VH
Eikelandsvatn	Eik	West coast	2	17	1200	300	M	H
Kvernavatn	Kver	West coast	3	30	980	320	M	H
Myravatn	Myr	West coast	31	18	426	245	M	SH
Sarsvatn	Sars	Svalbard	60	8	1000	400	O	C

[1] O = oligotrophic; M = mesotrophic
[2] C = clear; SH = slightly humic; H = humic; VH = very humic

their carapace. Some of the transparent animals seem quite dark to the naked eye, due to internal pigments.

Electrophoresis

Prior to 1987, electrophoresis was carried out on starch gels (see Wolf [1982] for methods). Later samples were run on cellulose acetate gels, following the methods of Weider *et al.* (1987). The latter technique allowed for better resolution at certain enzyme loci. Enzymes analyzed are listed in Table 2. Based on the results of electrophoresis, a number of genotypes could be distinguished. In the following, the term 'clone' will refer to a multilocus genotype, characterized by a unique combination of alleles at the investigated loci. These 'clones' may actually consist of groups of clones, which could not be distinguished with the limited number of loci surveyed. Allele frequencies listed in Table 2 give the most recent information of each locality. In the habitats sampled repeatedly, no significant changes in allele frequencies were found. Genetic distances between populations were calculated by Nei's (1972) formula. The distance matrix was subjected to a cluster analysis, using the Fitsch-Margoliash algorithm (program Kitsch in the PHYLIP package; Felsenstein, 1987). Parameters were set to mimic UPGMA clustering, since the requirements for estimating evolutionary time with this algorithm probably cannot be met by our data.

Light measurements

To assess the light absorptive properties of the inhabited waters, samples were brought to the laboratory for absorption measurements. These

Table 2. Allele frequencies in all populations investigated. Alleles are numbered consecutively in order of increasing electrophoretic mobility. Enzymes included: GOT (glutamate-oxaloacetate-transam nase); PGI (phosphoglucose isomerase); PGM (phosphoglucomutase); MDH (malate dehydrogenase); AO (aldehyde oxidase); ME (malic enzyme); FUM (fumarase); LDH (lactate dehydrogenase); HEX (hexokinase). Populations are abbreviated as in Table 1, except Schöhsee (Sch) in Holstein, FRG.

Locus	Allele	F4	F5	F6	F7	F8	F9	F10	Kver	Eik	Myr	Sch	H1	H2	F0	F1	F2	F3	R3	Saat	Sars
GOT	1	1	1	1	1	1	1	1	0.997	1	0.880	1	1	1	0	0	0	0	0	0	0
	2	0	0	0	0	0	0	0	0.003	0	0.120	0	0	0	1	1	1	1	1	1	1
PGI	0	0	0	0	0	0	0	0	0	0.006	0	0	0	0	0	0	0	0	0	0	0
	1	0	0	0	0	0	0	0	1	0.994	1	0.985	1	1	1	1	1	1	1	0.400	1
	3	0	0	0	0	0	0	0	0	0	0	0.015	0	0	0	0	0	0	0	0.600	0
	4	1	1	1	1	1	0.940	1	0	0	0	0	0	0	0	0	0	0	0	0	0
	5	0	0	0	0	0	0.060	0	0	0	0	0	0	0	0	0	0	0	0	0	0
PGM	0	0	0	0	0	0	0	0	0	0	0.011	0	0	0	0	0	0	0	0	0	0
	1	0	0	0	0	0	0	0	0	0	0.048	0.171	0	0	1	1	1	1	1	1	1
	2	0	0	0	0	0	0.005	0	0.142	0	0.644	0.637	0.930	0.895	0	0	0	0	0	0	0
	3	0	0	0	0	0	0.010	0.060	0.278	0.403	0.191	0.181	0.030	0.016	0	0	0	0	0	0	0
	4	1	1	1	1	1	0.985	0.940	0	0	0	0	0	0	0	0	0	0	0	0	0
	5	0	0	0	0	0	0	0	0.579	0.597	0.106	0.010	0.040	0.089	0	0	0	0	0	0	0
MDH	1	0	0.020	0	0	0	0	0	0	0	0	0	0	0	0	0	0	0	0	0	–
	1.5	0	0	0.020	0	0	0	0	0	0	0	0	0	0	0	0	0	0	0	0	–
	2	0.983	0.980	0.980	1	1	0.990	1	0.962	1	1	0.990	1	1	0	0	0	0	0	0	–
	3	0.017	0	0	0	0	0.010	0	0.038	0	0	0.010	0	0	1	1	1	1	0.500	1	–
	4	0	0	0	0	0	0	0	0	0	0	0	0	0	0	0	0	0	0.500	0	–
AO	1	0	0	0.010	0.045	0	0.010	–	0.050	0	0.050	0	0	0	0	0	0	0	0	0	0
	2	1	0.990	0.990	0.950	1	0.990	–	0.950	0.581	0.950	1	1	1	0	0	0	0	0	0	0
	2.5	0	0.010	0	0.005	0	0	–	0	0.429	0	0	0	0	0	0	0	0	0	0	0
	3	0	0	0	0	0	0	–	0	0	0	0	0	0	1	1	1	1	1	1	0
	4	0	0	0	0	0	0	–	0	0	0	0	0	0	0	0	0	0	0	0	1
ME	0	0	0	0	0	0	0.010	0.130	0	0	0	0	0	0	0	0.180	0	0	0	0.170	0
	1	1	1	1	1	1	0.990	0.870	0.069	0	0	0	0	0	1	0.920	1	1	1	0.830	1
	2	0	0	0	0	0	0	0	0.931	1	1	1	1	1	0	0	0	0	0	0	0
FUM	1	0	0	0	0	0	0	0	0	0	0	0	0	0	0	0	0	0	0	0	–
	2	1	1	[illegible]	0.990	1	0.996	1	1	1	1	1	1	1	1	1	1	1	0.500	1	–
	3	0	0	0	0.010	0	0.004	0	0	0	0	0	0	0	0	0	0	0	0.500	0	–
LDH	0	0	0.010	0	0	0	0	0	0	0	–	0	0	0	0	0	0	0	0	0	0
	1	0	0	0	0	0	0	0	1	1	–	1	1	1	1	1	1	1	1	1	1
	2	1	0.990	[illegible]	1	1	1	1	0	0	–	0	0	0	0	0	0	0	0	0	0
HEX	1	0	0	0	0	0	0	–	–	–	–	–	–	–	1	1	1	1	–	1	–
	2	1	1	[illegible]	1	1	1	–	–	–	–	–	–	–	0	0	0	0	–	0	–

were conducted in 1-cm quartz cuvettes in a Shimadzu UV-160 spectrophotometer, at several wavelengths between 200 and 400 nm. The samples were not filtered, but particles were allowed to settle in the sample bottles for 24 h before measurements. We measured ambient ultraviolet (UV) light intensities with a Cole-Parmer Series 9811 radiometer. The sensor had a sensitivity peak at 365 nm, with half-bandwidth from 330 to 377 nm. To measure in water, it was necessary to glue o coverslip (0.17 mm thick) on top of the sensor. This effected a 15% reduction of light intensity in air, so a correction factor of 1.15 was applied to all field measurements.

Pigment analysis

The effects of various organic and inorganic media (acetone, 95% ethanol, 4% formalin, 40% hydrogen peroxide, 5 N sodium hydroxide) on melanic animals were registered visually. Samples conserved in formalin were used to assay melanin contents, since this fixative had no visible effect on carapace pigmentation. After washing the animals three times in water, 30–55 adult females were homogenized in an ultrasonic bath. Upon centrifugation, the residue was dissolved in 5 ml 5 N NaOH for 5 days. During this period, the tubes were heated 5 times to about 65 °C by placing them in a jar with hot tap water, and then left to cool. Finally, the samples were filtered (0.2 μm membrane filters), and absorbance of the filtrate measured at 350 nm. 5 N NaOH constituted the blank. Melanin content was assayed in animals from Kvernavatn (coastal transparents), pond F4 (alpine transparents), and pond F0 (alpine melanics). Absorbance was related to concentration by a reference curve made from synthetic melanin (Sigma No. M 8631), dissolved in 5 N NaOH. We measured the body lengths and eye diameters of all animals prior to homogenization. Length was converted to dry mass by the equation $lnW = 1.4681 + 2.8292 \cdot lnL$ (where W is dry mass in μg, and L is length in mm), to calculate the biomass from which extracts were made. This equation estimates the mass of animals without eggs or embryos (Bottrell *et al.*, 1976). To correct for melanin present in the animals' compound eyes, we calculated the total volume of eyes in each sample, as the sum of individual eye volumes calculated as spheres. Most eyes were not strictly spherical, and hence the spherical diameter was taken as the average of the longest and shortest diameter. Embryos with visible eyes were removed from mother animals. The absorption of the transparent Kvernavatn population was then used as background melanin content, but corrected for different total eye volumes in each sample. The remaining difference was taken as carapace melanin, expressed as μg melanin per mg dry body mass as well as per individual.

Results

Distribution

Both transparent and melanic *D. longispina* occur over large areas in the mountain range of South Norway. The melanic form has been found only above 900 m a.s.l. in ponds and lakes with sparse or no fish populations (Fig. 2). In several cases,

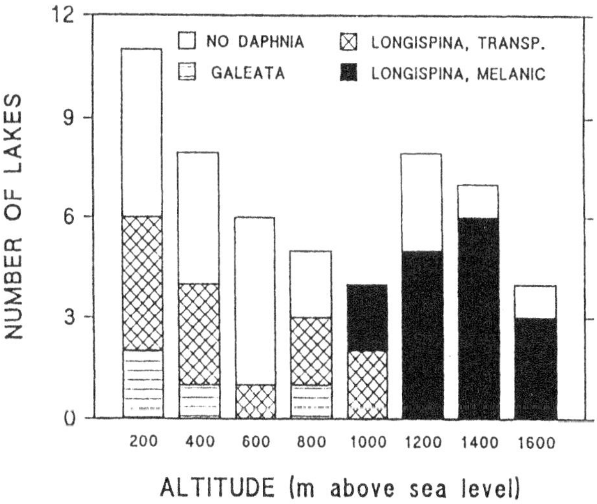

Fig. 2. The occurrence of *Daphnia* species and morphs in 53 lakes of Western Norway. Data from surveys in 13 watersheds. Lakes are sorted by altitude categories, the abcissa values giving the upper altitude limit of each interval (thus 800 designates the 601–800 m a.s.l. category). Note that only lakes, but no ponds, are included. Transparent *D. longispina* are known in ponds up to 1250 m a.s.l.

Hobæk, 1986). This is due to transcription errors in the previous paper.

The melanic Svalbard population electrophoretically resembles the melanic mainland animals. Within the entire melanic group, the average genetic distance is 0.085, (but only 0.001 within the Finse populations). The transparent mountain group consists only of Finse populations, with an average genetic distance of 0.003. Thus the two groups present at Finse are very distinct from each other, but each of them is nearly homogeneous. The transparent mountain group is more closely related to the transparent lowland group than to the melanic group (average genetic distances of 0.673 and 1.543, respectively). On the other hand, the melanic population group is genetically less distant from the lowland transparent group than the alpine transparent group, the genetic distances averaging 1.018 and 1.543, respectively.

All the pond populations are virtually monoclonal, i.e., one single clone makes up 90% or more in the population. In most cases, the dominant clones are homozygous for all loci investigated. R3 is a shallow, small lake, whose *Daphnia* population is fixed for heterozygosity at two loci, so even here there is only one clone present. By contrast, the populations of larger lakes contain a lot more variation, particularly at the PGM locus (Table 2). This was evident even in the very small sample from lake Saatevatn (3 genotypes among 11 animals). However, the rule is broken in shallow lake Sarsvatn, whose melanic population appears to be monomorphic. Lake Midnutvatn is another shallow, but large lake close to Finse. The allozyme data from its melanic population have not been included in Table 2, because we were unable to ascertain allele identities. However, even in this lake one single clone accounts for 95% of the population.

The exoskeletons of melanic animals appear completely transparent immediately after molting, as we frequently observed during the sorting of live samples. Such apparently transparent animals may give the impression of a mixed population. We have frequently included such animals in electrophoretic samples of melanic populations, and with no exception their genotypes were identical with the melanic animals.

Light and light absorption

Measurements of UV light intensity in selected ponds, as well as UV absorption in water samples, confirm the impressions from the field (Fig. 4; Table 4). Pond F4, in which there is a marked UV attenuation, is only slightly humic, while the water in F0 is very clear. Even less UV absorption occurs in water from lake Midnutvatn (Table 4). This large, shallow lake (about 4 m deep) harbors a population of melanic *D. longispina*. The light measurements in the field indicate an ambient UV intensity of up to 2 mW cm^{-2} on clear summer days in the entire water mass of transparent ponds and to many meters depth in clearwater lakes. By contrast, even in slightly humic ponds this intensity is reduced to half at 50 cm depth.

Pigment contents

Absorption curves of extracts from melanic animals resemble that of synthetic melanin, but with

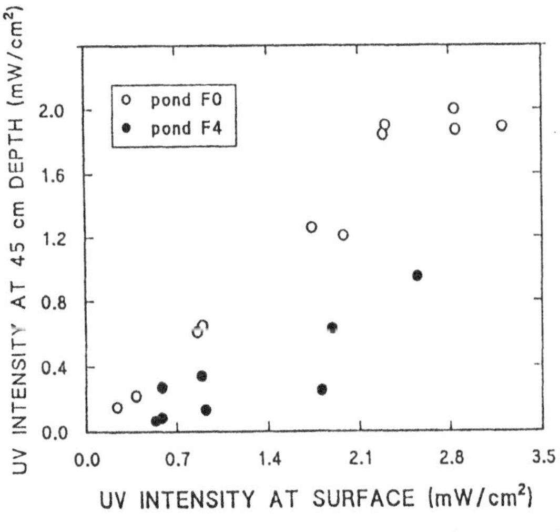

Fig. 4. UV radiation intensity above the surface (x-axis) and at 45-cm depth (y-axis) in ponds F0 and F4. Intensity was measured with a Cole-Parmer Series 9811 radiometer. The figure includes measurements from several dates in June–September 1987, under varying weather conditions.

Table 4. Absorption at UV wavelengths in surface water samples from selected localities representing a gradient of organic contents. Lake Midnutvatn (1323 m.a.s.l.) is a large, shallow lake in the vicinity of Finse. For data on other localities, see Table 1. Pigment morphs are given as melanic (M) or transparent (T).

Locality	Daphnia pigment morph	Date	Wavelength (nm)			
			400	384	254	200
Midnutvatn	M	12.8.88	0.000	-	0.005	0.008
F0	M	29.6.87	0.003	0.002	0.018	0.056
		7.7.87	0.003	0.003	0.022	0.058
		18.7.87	0.002	0.002	0.019	0.048
F3	M	3.7.87	0.008	0.009	0.052	0.067
		6.7.87	0.005	0.007	0.040	0.089
		17.7.87	0.002	0.003	0.025	0.053
		24.7.87	0.004	0.004	0.034	0.096
		5.8.87	0.004	0.004	0.024	0.062
F4	T	29.6.87	0.015	0.017	0.072	0.186
		3.7.87	0.014	0.015	0.070	0.154
		6.7.88	0.015	0.017	0.080	0.191
		17.7.87	0.010	0.011	0.070	0.183
		24.7.87	0.010	0.012	0.074	0.212
		5.8.87	0.009	0.009	0.059	0.194
		12.8.88	0.011	-	0.064	0.213
F9	T	29.6.87	0.031	0.038	0.206	0.439
H1	T	5.8.88	0.041	0.053	0.379	1.633

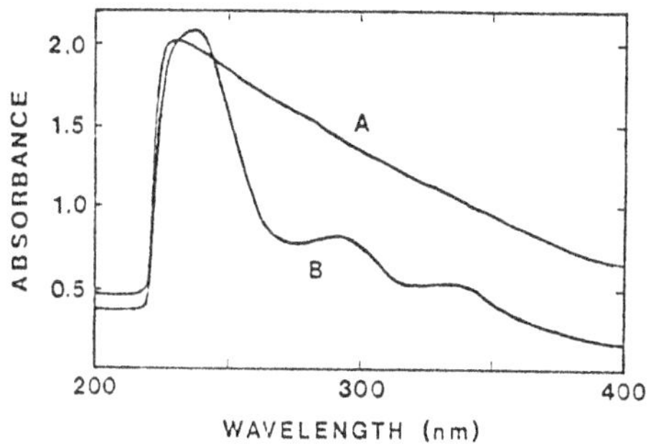

Fig. 5. Absorption spectra in the wavelength range 200–400 nm of A) synthetic melanin, and B) melanic *Daphnia longispina* extracts in 5 N NaOH.

a lower absorbance above 250 nm (Fig. 5). The chemical behaviour of the pigment was consistent with expectations (Emery, 1984). Of the chemicals tried, only NaOH was able to dissolve the pigment. Further, bleaching with H_2O_2 was complete after 20 h. Otherwise, the pigment was apparently inert, except for a noticeable reduction in colour intensity in animals killed and stored for 2 years in 95% ethanol. However, the extraction by NaOH on F0 melanic animals proceeded very slowly, and was far from complete in 5 days whether animals were homogenized or not. Hence, the estimate of melanin content of this sample is certainly too low. Table 5 lists the results of the melanin extractions. It appears that the transparent mountain population (F4) contains a substantial amount of pigment, even though this is not visible in a dissecting microscope except as a faint brownish colour along the fornix, carapace margin, and the antennae. As expected, the dark F0 animals had the highest content of melanin (2.4 μg ind^{-1}). However, expressed per body mass, both alpine groups yielded about 0.03 μg mg^{-1}.

Discussion

Measuring light intensity *in situ* involves many difficulties. The apparatus we used was not

Table 5. Contents of melanin in samples from Kvernavatn (a coastal population of transparent animals); F4 (an alpine transparant population), and F0 (an alpine melanic population). Based on absorbance measurements at 350 nm, converted to concentration by a standard curve made from synthetic melanin. Body size dimensions are the means of animals in the extraction samples.

	Body length (mm)	Body mass (μg)	Eye diameter (mm)	Melanin (μg · mg^{-1})	μg melanin · ind^{-1}	
					In eye	In carapace
Kvernavatn	1.57	16.12	0.0981	0.0022	0.0352	0
F4	2.00	31.31	0.1275	0.0312	0.0589	0.9170
F0	2.62	66.62	0.1253	0.0372	0.0569	2.4183

designed for underwater measurements, and the absolute intensities may not be quite correct. Nevertheless, the measurements are fully comparable, and our emphasis is on the relative difference between ponds and water qualities, rather than on the absolute light intensities. Actually, the results obtained fit well with Emery's (1984) measurements in arctic Canada. A better characterization of light regimes requires more advanced equipment. This seems necessary to obtain a quantitative understanding of radiation quality, dosage, and the photoprotection provided by melanin as well as carotenoids.

The estimates of melanin content presented here are preliminary, being based on methodological studies. Nonetheless, the pigment analyses confirm the assumed relation of melanin to photoprotection in alpine habitats. Direct weighing would be preferrable to biomass estimated derived from body lengths. Since the solution of visible pigments from carapace walls was incomplete in the F0 sample, it is difficult to estimate how much melanin they really deposit. After extraction, the exoskeleton remains were still clearly brownish. This may be a result of formalin fixation. Alternatively, since part of the pigment dissolved in ethanol, and some of the pigment dissolved only slowly in NaOH, more than one form of the pigment may be present. Analyses conducted on fresh material may help to resolve this problem. The ratio body mass/area increases with body length, so small animals should be expected to have higher pigment contents per unit body mass than larger ones. This has not been accounted for in our analysis, although the mean body sizes differ considerably in the samples (Table 5). Perhaps it is more correct to compare pigment contents per unit carapace area. A full analysis of pigments should consider the contents of both carotenoids and melanin, since both melanic and transparent animals sometimes have conspicuously coloured internal tissues.

Daphnia usually deposit melanin in their ephippia, so they apparently all possess the metabolic ability to synthesize the pigment. One might expect the animals to deposit melanin in their carapace only when needed, in response to ambient light conditions. In fact, this seems to be the case within the melanic group. When F0 animals are kept in artificial light without UV-radiation, they only maintain a weakly pigmented patch dorsally. Animals sampled in the same locality under ice in November and in March had a similar appearance. These observations suggest that there may be a metabolic cost to melanin production. In contrast to carotenoids, the pigment is lost with each exuvium and has to be replenished. Unless melanin production is costly, it is hard to understand why the transparent mountain animals deposit little melanin, except in their ephippia. Conceivably, this reluctance to deposit melanin could be a result of long-time selection for invisibility to escape fish predators. This would imply that the transparent and melanic mountain groups have been separated under different predation and light conditions for a long time, in accordance with the large genetic distance estimates from the allozyme study. In arctic Canada, melanic clones occur within both *D. pulex* and *D. middendorffiana*, and Hebert & McWalter (1983) therefore concluded that melanism has arisen independently several times. The distribution of transparent and melanic *D. pulex* is related to organic contents of the water (Emery, 1984), similar to the patterns in *D. longispina* reported here.

The phenomenon of stronger pigmentation in alpine than lowland animals is by no means restricted to *Daphnia* (e.g. Brehm, 1938). Planktonic copepods in the mountain study area have beautiful red and blue colours (e.g. *Cyclops scutifer*, *Heterocope saliens*, and *Arctodiaptomus laticeps*), almost certainly due to carotenoids (Hairston, 1976; Luecke & O'Brien, 1981). The phenomenon may also be involved in black or red coloration of alpine beetles (rufinism), and even in the stronger pigments of epipelagic marine copepods compared with their relatives in deeper waters (Herring, 1965). The photoprotective role of melanism is further confirmed by the peculiar pigmentation in the cladoceran genus *Scapholeberis*, which attach themselves to the water surface with their pigmented ventral edge upwards.

Although the function of melanin seems related

to short-wave radiation, some alternatives need consideration. Dodson (1984) found that melanic *D. middendorffiana* had stronger and thicker carapace walls than transparent *D. pulex*, suggesting an improved resistance towards invertebrate predators. Predation experiments with *Heterocope septentrionalis* actually demonstrated a higher survival of melanic animals (Dodson, 1984; see also Hebert & Loaring, 1980). In the Norwegian mountain localities, the abundance of invertebrate predators (*Heterocope saliens*, *Bythotrephes longimanus*, *Popyphemus pediculus*, water mites, and beetles) is in fact much higher in the humic ponds holding transparent *D. longispina* (unpublished observations). Among the predators mentioned, only *P. pediculus* and aquatic beetles have been observed in the clearwater ponds, and always at very low densities. Thus, the occurrence of invertebrate predators contradicts an important function of melanin in predator defence.

Another possible function of melanin is heat absorption. In the mountain localities, temperature obviously limits growth and reproduction, and so only a slight body temperature rise would be beneficial. It seems questionable that the animals can maintain a temperature differential to the surrounding water, considering their continuous swimming and their feeding currents. Note that the effects of heat absorption and photoprotection do not exclude each other. It would be interesting to experimentally quantify these effects. Unfortunately, the mountain animals have proved unusually sensitive to handling, and all experimental trials have failed due to excessive mortality. Emery (1984) found much higher mortality rates in transparent than melanic *D. pulex* exposed to artificial and natural short-wave radiation.

The allozyme results confirm that the melanic *D. longispina* are quite distinct from other members of the complex (Wolf & Hobæk, 1986). Within the transparent group, the situation is less clear. Obviously, allozyme data from more populations and other areas are needed to verify the reality of two distinct clusters as found here. In particular, the alpine transparent group may be expected to contain much more variation than Fig. 3 indicates, since all our populations are from one single area.

The clonal poverty of ponds and shallow lakes forms a remarkable contrast to the clonal richness of deeper lakes, and confirms the results presented by Wolf (1987). However, in pond populations of *D. magna* and *D. pulex* several clones often cohabit (Hebert, 1974; Hebert *et al.*, 1988), even in arctic habitats (Weider *et al.*, 1987). We can find no simple explanation for this difference. Another striking feature of alpine *D. longispina* pond populations is their low level of heterozygosity, in fact they are usually homozygous at all the investigated loci. Again *D. magna* and *D. pulex* populations differ, occasionally showing heterozygote excess (Hebert, 1987). In *D. pulex*, this may partly be explained by obligate parthenogenesis and polyploidy (Beaton & Hebert, 1989), but these traits are unknown in *D. magna*. The simplest explanation for these contrasts may be differences in dispersal of resting eggs between habitats. Low dispersal rates, few founders, and subsequent inbreeding seem the most likely reasons for low clonal diversity and homozygosity in *D. longispina* of alpine ponds. Why dispersal rates should be low in our study area remains a puzzle, however. Dispersal can presumably occur only by ephippia. The probability for these to be transported by birds, wind, or water would seem to depend on their ability to float on the surface. In arctic ponds at Svalbard, densely populated by *D. pulex* and/or *D. middendorffiana*, thousands of ephippia may accumulate along the shores. We have never observed anything similar in the alpine *D. longispina* populations. This difference in floating ability is probably related to the surface sculpturing of the ephippia. In the scanning electron microscope, the ephippial surface of both transparent and melanic alpine *D. longispina* appear almost smooth, having no deep depressions. By contrast, the arctic ephippia are covered with hundreds of deep, narrow pits (Hobæk, unpubl. obs.), acting as air traps which greatly increase their floating ability. This characteristic of the alpine populations is not common to all populations of the *D. longispina* complex, which vary both in their ephippial surface sculptures and their floating ability.

A shallow alpine pond constitutes a rather homogeneous habitat compared with deeper lakes. Monoclonal populations could simply be a result of clonal competition. This seems an unlikely explanation, however, as coexisting clones of *D. pulex* have been reported from very similar ponds in arctic Canada (Weider *et al.*, 1987). In fact, the high clonal diversity in that area (average 4.5 clones per pond) was in part related to a presumedly high dispersal rate by avian as well as human vectors (Weider, 1989). Furthermore, if heterosis is generally advantageous (Hebert & Crease, 1983), clonal competition should not effect an erosion of heterozygosity. Populations in alpine ponds and shallow lakes can overwinter only as resting eggs, at least during hard winters. In the 3.5 m deep pond F0, active animals have been shown to overwinter under ice. However, in winters of heavy snowfall (like 1988–89), the pond fills with snow and ice, leaving no room for active animals. These periods of obligate diapause constitute narrow population bottlenecks, suggesting that low clonal diversity is a result of random genetic drift. Different clones may differ in their reproductive response to environmental cues (Ferrari & Hebert, 1982; Korpelainen, 1986), and clones leaving few ephippia will have a high probability of extinction. This hypothesis predicts that clonal diversity is greater over a larger geographical range, and indeed the melanic populations outside the Finse area have different genotypes. Moreover, it also explains the difference observed in clonal richness between deep and shallow lakes. If dispersal rates are high, random drift will be counteracted by immigrations, and clonal composition will remain more diverse.

Conceivably, low clonal diversity can arise if one obligately asexual clone has invaded a particular area. At Finse, there seems to be only one melanic and one transparent clone, each with only minor variation within and between ponds. If the area was invaded relatively recently, asexual clones would not have diverged detectably. Since laboratory populations have been impossible to maintain, we have not been able to ascertain the reproductive system of these populations. Moreover, no genetic markers allowing the detection of allele segregation were found in the pond populations, except in population R3, which has been sampled only once. All the alpine pond populations produce males in abundance, but this fact provides no evidence for sexual reproduction, since obligate asexuals frequently retain male production (Innes & Hebert, 1988). The monoclonal conditions in our mountain ponds are completely at odds with the high clonal diversity known in obligately parthenogenetic *D. pulex* populations (Hebert *et al.*, 1988), as well as their high heterozygosity levels as discussed above. On the other hand, population R3 was fixed for heterozygosity at two loci. This state is hard to reconcile with annual recombination. Thus, the available evidence is ambiguous, and indeed both reproductive systems may occur.

The absence of transparent *D. longispina* in the clearwater localities can be explained by higher radiation levels in these localities. But why are the melanic animals absent from humic ponds? To account for this, some excluding factor must exist. This may be related to competitive exclusion by the transparents, but it is not obvious what their advantage could be. If melanin deposition is a major metabolic cost, photoprotection may imply reduced body growth and reproduction rate, resulting in inferior competitive ability in the melanic clones. However, attempts to inoculate humic ponds with melanic animals have failed, irrespective of whether transparents were initially present or not. Thus it is possible that some physiological limitation occurs. In a majority of ponds at Finse, no *Daphnia* are present, including ponds with organic contents intermediate between the clearwater and humic categories found to be inhabited. To resolve this issue, we will need an evaluation of the metabolic costs of melanin production, as well as improved methods for field experimentation in the mountain.

So far, the transparent mountain group has been observed in only ponds. This is probably a reflection of the available habitats in the mountains: larger lakes usually have clear waters.

Potentially, the transparent animals should be able to inhabit humic lakes. Published information is difficult to evaluate, since the morphological differentiation between the transparent lowland and alpine groups is difficult, and even confusion with *D. galeata* is possible. For instance, Hobæk & Raddum (1980) reported one alpine lake (765 m a.s.l.) west of Finse to harbor transparent *D. longispina*. An electrophoretic survey later suggested their affinity to *D. galeata* from Norway and Germany (Wolf & Hobæk, unpubl.), and closer inspection confirmed this conclusion (f. *microcephala* Sars, without helmets).

The dichotomy between melanic and transparent groups in the *D. longispina* complex seems to be consistent, even when populations from Svalbard and Germany are included. The genetic distance estimates suggest a specific distinction between these groups (Ayala *et al.*, 1974). This may be a premature conclusion, however. It will be necessary to assay allozyme genotypes in melanic populations from other areas (particularly from northern Scandinavia) to clarify the relatedness of our populations to previously published taxa (see below). Further, the variability within the transparent groups may be larger if populations are included from other areas. Also, the close relatedness of German *D. hyalina* with coastal *D. longispina* revives the old enigma of separating these two species. To complicate matters further, a new melanic morphotype of this complex was discovered in August 1989, further south on the Hardangervidda mountain plain. No electrophoretic data are yet available. Thus, it seems clear that a more extensive data set is required to resolve the systematic position of the electrophoretic groups detected in our material. It should be noted that morphological differences between the electrophoretic groups have been found, e.g. in body size, head and spine lengths, length of antennae and antennal setae, relative lengths of the endopodid setae of the second thoracic limb, and ephippium surface sculpture. These preliminary observations will need to be confirmed by more extensive data, and to be adjusted for allometric relations.

If the melanic group proves to be a distinct species, its proper name is difficult to decide. Several similar taxa have been published long ago, all of them later being considered synonyms or varieties of *D. longispina* or *D. hyalina* (e.g., Lilljeborg, 1901; Flößner, 1972). The oldest melanic taxon of the *D. longispina* complex is *D. longispina* v. *abbreviata*, Lilljeborg (1887). The author described it very briefly, stating nothing about its colour. However, Ekman (1904), having seen Lilljeborg's animals, identified dark brown animals from North Swedish alpine ponds with this form. Our melanic animals resemble Ekman's illustration of f. *abbreviata*, much more so than *D. longispina* f. *frigidolimnetica* Ekman, 1904. The latter taxon was raised to species level by Hrbáček *et al.* (1978), and Hrbáček (1987) seems to consider it to be the correct name for this group. In our opinion, Ekman's description suggests f. *frigidolimnetica* to be close to our alpine transparent animals, rather than the melanic ones. Later reports from Northern Sweden use the above names variously; e.g. the Latnjajaure population was considered f. *frigidolimnetica* by Ekman (1904), *D. longispina* by Pejler (1973), and *D. rosea* (sensu Sars) by Nauwerck (1978).

Sars (1890) listed a series of species and varieties from Norway. Among them is *D. lacustris* v. *alpina*, which he found in alpine lakes north of Finse. His short comments gives their colour as bluish grey to almost black. No illustrations were ever published, but two drawings found at the Sars archive at the University Library in Oslo (see Frey, 1982) indicate that this is a different taxon. In fact, his figures show a striking resemblance to the strange animals we recently found on the Hardangervidda south of Finse, having an entirely different head shape with a pronounced crest.

Next in priority comes *D. zschokkei* Stingelin, 1894, which was described from alpine lakes in Switzerland. Again, our melanic animals show a striking resemblance. Margaritora & Ferrara (1979) have since found characters to distinguish three species of the *D. longispina* complex; *D. longispina*, *D. rosea*, and *D. zschokkei*. It seems quite possible that the melanic animals from Scandinavia, Svalbard, and the Alps are the same taxon. If so, its proper name can be decided only by a redescription of the older taxa. In particular, it seems questionable whether *D. longispina* v. *abbreviata*, described from the Commander (Bering) Islands in the Pacific, really are identical with Ekman's melanic animals from Sweden. This may apply to melanic animals from Mongolia (Flößner, 1986) as well. It would appear that clarifying the systematic relation of these melanic, possibly closely related groups, offers a good starting point for a necessary taxonomic revision of the *D. longispina* complex.

Acknowledgements

A grant from the Meltzer Foundation at the University of Bergen allowed the purchase of a radiometer. The opportunity to collect on Svalbard was provided by research grants 31/86 and 4/87 from the Norwegian Polar Research Institute. Directors E. Østbye and E. Leslie permitted unlimited access to the Ecology Field Station at Finse. Cand. mag. T.C. Hestenes, Dr. J. Giske, and cand. real. T. Dale provided valuable assistance in the field, and Fr. Eva Geißler helped running the gels for electrophoresis.

References

Ayala, F. J., M. L. Tracey, L. C. Barr, J. F. McDonald & S. Perez-Salas, 1974. Genetic variation in natural populations of five *Drosophila* species and the hypothesis of the selective neutrality of protein polymorphism. Genetics 77: 343–384.

Beaton, M. J. & P. D. N. Hebert, 1989. Geographical parthenogenesis and polyploidy in *Daphnia pulex*. Am. Nat. 132: 837–845.

Bottrell, H. H., A. Duncan, Z. M. Gliwicz, E. Grygierik, A. Herzig, A. Hillbricht-Ilkowska, H. Kurasawa, P. Larsson & T. Weglenska, 1976. A review of some problems in zooplankton production studies. Norw. J. Zool. 24: 419–456.

Brehm, V., 1938. Die Rotfarbung von Hochgebirgseeorganismen. Biol. Rev. 13: 307–318.

Britton, G., 1983. The biochemistry of natural pigments. (Cambridge texts in chemistry and biochemistry). Cambridge University Press.

Brooks, J. L., 1957. The systematics of North American *Daphnia*. Mem. Connect. Acad. Arts. Sci. 13: 1–180.

Dodson, S. I., 1984. Predation of *Heterocope septentrionalis* on two species of *Daphnia*: Morphological defenses and their cost. Ecology 65: 1249–1257.

Ekman, S., 1904. Die Phyllopoden, Cladoceren und freischwebenden Copepoden der nordschwedischen Hochgebirge. Ein Beitrag zur Tiergeographie, Biologie und Systematik der arktischen, nordschwedischen und mitteleuropäischen Arten. Zool. Jb. Syst. Ökol. Geogr. Tiere 21: 1–170.

Emery, C. J., 1984. The ecological impact of near ultraviolet radiation on *Daphnia pulex*. MSc thesis, University of Windsor, Ontario.

Felsenstein, J., 1987. PHYLIP (Phylogeny Inference Package) version 3.0 Manual. Univ. of Washington.

Ferrari, D. C. & P. D. N. Hebert, 1982. The induction of sexual reproduction in *Daphnia magna*: genetic differences between arctic and temperate populations. Can J. Zool. 60: 2143–2148.

Flößner, D., 1972. Krebstiere, Crustacea. Kiemen- und Blattfüßer, Branchiopoda; Fischläuse, Branchiura. Die Tierwelt Deutschlands, 60. Teil: 1–501. G. Fischer, Jena.

Flößner, D., 1986. Beitrag zur Kenntnis der Branchiopoden- und Copepodenfauna der Mongolei. Mitt. Zool. Mus. Berl. 62: 3–40.

Frey, D. G., 1982. G. O. Sars and the Norwegian Cladocera: a continuing frustration. Hydrobiologia 96: 267–293.

Hairston, N. G., Jr., 1976. Photoprotection by carotenoid pigments in the copepod *Diaptomus nevadensis*. Proc. natn. Acad. Sci. 73: 971–974.

Hairston, N. G., Jr., 1978. Carotenoid photoprotection in *Diaptomus kenai*. Verh. int. Ver. Limnol. 20: 2541–2545.

Hairston, N. G., Jr., 1979. The adaptive significance of colour polymorphism in two species of *Diaptomus* (Copepoda). Limnol. Oceanogr. 24: 38–44.

Hebert, P. D. N., 1974. Enzyme variability in natural populations of *Daphnia magna*. III. Genotypic frequencies in intermittent populations. Genetics 77: 335–341.

Hebert, P. D. N., 1987. Genotypic characteristics of the Cladocera. Hydrobiologia 145: 183–193.

Hebert, P. D. N. & T. J. Crease, 1983. Heterosis in *Daphnia*: a reassessment. Am. Nat. 119: 427–434.

Hebert, P. D. N. & J. M. Loaring, 1980. Selective predation and the species composition of arctic ponds. Can. J. Zool. 58: 422–426.

Hebert, P. D. N. & D. B. McWalter, 1983. Cuticular pigmentation in arctic *Daphnia*: adaptive diversification of asexual lineages. Am. Nat. 122: 286–291.

Hebert, P. D. N., R. D. Ward & L. J. Weider, 1988. Clonal-diversity patterns and breeding system variation in *Daphnia pulex*, an asexual-sexual complex. Evolution 42: 147–159.

Herring, P. J., 1965. Blue pigment of a surface-living oceanic copepod. Nature 205: 103–104.

Hobæk, A. & G. G. Raddum, 1980. Zooplankton communities in acidified lakes in South Norway. SNSF project, IR 75/80. Oslo-Aas, Norway: 1–132.

Hrbáček, J., 1987. Systematics and biogeography of *Daphnia* species in the northern temperate region. Mem. Ins. ital. Idrobiol. 45: 37–76.

Hrbáček, J., V. Korínek & D. G. Frey, 1978. Cladocera. In J. Illies (ed.), Limnofauna Europaea (2. ed.). G. Fisher Verlag, Stuttgart: 189–195.

Innes, D. J. & P. D. N. Hebert, 1988. The origin and genetic basis of obligate parthenogenesis in *Daphnia pulex*. Evolution 42: 1024–1035.

Korpelainen, H., 1986. The effect of temperature and photoperiod on life history parameters of *Daphnia magna* (Crustacea: Cladocera). Freshw. Biol. 16: 615–620.

Lilljeborg, W., 1887. Contributions to the natural history of the Commander Islands. No. 9. On the Entomostraca collected by Mr. Leonhard Stejneger, on Bering Island, 1882–'83. Proc. U.S. Nat. Mus. 10: 154–156.

Lilljeborg, W., 1901. Cladocera Sueciae. Nova Acta Soc. Reg. Sc. Ups. Ser. 3 Vol. 19. 701 pp.

Luecke, C. & W. J. O'Brien, 1981. Phototoxicity and fish predation: Selective factors in color morphs of *Heterocope*. Limnol. Oceanogr. 26: 454–460.

Margaritora, F. & O. Ferrara, 1979. Contributo alla conoscenza di *Daphnia rosea* Sars (Cladocera Daphnidae) in Italia: confronti e differenze con *D. longispina* O.F. Müller e *D. zschokkei* Stingelin. Boll. Mus. Civ. St. Nat. Verona 6: 127–138.

Nauwerck, A., 1978. *Bosmina obtusirostris* im Latnjajaure. Arch. Hydrobiol. 82: 387–418.

Nei, M., 1972. Genetic distance between populations. Am. Nat. 106: 283–292.

Pejler, B., 1973. On the taxonomy of limnoplanktic *Daphnia* species in Northern Sweden. Zoon 1: 23–27.

Sars, G. O., 1890. Oversigt af Norges Crustaceer med foreløbige Bemærkninger over de nye eller mindre bekjendte Arter. II. (Branchiopoda-Ostracoda-Cirripedia). Forh. Vidensk.-Selsk. Christiania 1890 No. 1: 1–80.

Siebeck, O., 1978a. Ultraviolet tolerance of planktonic crustaceans. Ver. int. Ver. Limnol. 20: 2469–2473.

Siebeck, O., 1978b. UV-toleranz und Photoreaktivierung bei Daphnien aus biotopen verschiedener Hohenregion. Naturwissenschaften 65: 390–391.

Stingelin, T., 1894. Über zwei neue Daphnien aus dem schweizerischen Hochgebirge. Zool. Anz. 17: 378–380.

Thomasson, K., 1956. Reflections on arctic and alpine lakes. Oikos 7: 117–143.

Weider, L. J., 1989. Spatial heterogeneity and clonal structures in arctic populations of apomictic *Daphnia*. Ecology 70: 1405–1413.

Weider, L. J., M. J. Beaton & P. D. N. Hebert, 1987. Clonal diversity in high-arctic populations of *Daphnia pulex*, a polyploid apomictic complex. Evolution 41: 1335–1346.

Weider, L. J. & W. Lampert, 1985. Differential response of *Daphnia* genotypes to oxygen stress: respiration rates, hemoglobin contents and low oxygen tolerance. Oecologia 65: 487–491.

Wolf, H. G., 1982. A comparison of different electrophoretic techniques for the detection of isoenzymes in single daphniids. Arch. Hydrobiol. 95: 521–531.

Wolf, H. G., 1987. Differences in the genetic structure of pond-dwelling and lake-dwelling *Daphnia*. Verh. int. Ver. Limnol. 23: 2056–2059.

Wolf, H. G. & A. Hobæk, 1986. Ecological genetics of Norwegian *Daphnia* I. Genetic differentiation between pigmented and unpigmented alpine pond populations. Hereditas 104: 193–198.

Zaret, T. M., 1980. Predation and freshwater communities. Yale University Press, New Haven & London. 187 pp.

Hydrobiologia **225**: 245–253, 1991.
V. Kořínek & D. G. Frey (eds), Biology of Cladocera.

Differences in life-history and aging in two clonal groups of *Daphnia cucullata* Sars (Crustacea: Cladocera)

Dieter Ebert[1] & Jürgen Jacobs
Institut für Zoologie, Universität München, Seidlstr. 25, D-8000 München 2, Germany; [1]*Present address of corresponding author: Institut für Zoologie, Universität Basel, Rheinsprung 9, CH-4051 Basel, Switzerland*

Key words: *Daphnia*, life history, genotypes, PGI, aging, clones

Abstract

Life-history traits were compared in twenty-two clones of *Daphnia cucullata* from the pre-alpine lake Klostersee in southern Germany. Clones were kept under standardized conditions (18 °C; 16:8 L/D) both singly and with 25 females per glass. Clones belonged to two genotype groups distinguished by enzyme electrophoresis. 11 of the clones had the genotype 'FM' at the phosphate glucose isomerase locus and 11 the genotype 'MM'. Neonates and adults of the age class 10 to 21 days were larger in one group (FM) than in the other group (MM). The size specific growth rates within any age class depended on age and body size and differed between the genotype groups. The fecundity of young FM animals (< 16 days) was significantly higher than that of MM animals, but for older females fecundity did not differ significantly. Senescence started about 6 days earlier in FM than in MM. The data are consistent with results of a field study on *Daphnia cucullata* in the Klostersee (Jacobs, 1990).

Introduction

Correlations between life-history traits and electrophoretically marked genotype groups is a useful tool to check for potential selection in the field. This is especially true for parthenogenetically reproducing species like *Daphnia*. However, unless detailed knowledge of the clonal structure of a *Daphnia* population is available, pure field studies may lead to ambiguous results. To support correlations between genotypes and life-history traits found in field studies, one needs to conduct laboratory studies.

In a field study, Jacobs (1990), studied genotype frequencies at the phosphate glucose isomerase locus (PGI; three alleles: F, M, S) in a permanent population of *Daphnia cucullata* in the Klostersee. PGI-genotype frequencies of the two main contributing genotypes (FM + MM > 90%) changed over time. There were very few sexual animals and strong deviations from Hardy-Weinberg equilibrium, and Jacobs concluded that sexual reproduction played almost no role in the genetic structure of the population. The lake population behaved as if it was an assemblage of independent clones. A correlation between FM genotype frequency and egg ratio (number of eggs per adult female) was positive; for MM genotype frequency it was negative. Jacobs (1990) suggested that the FM genotype group produced on average more eggs than the MM group. To support this field results and to study genetic variation in life-history traits of clones from one population, we conducted a series of life-history experiments with 22 clones, 11 from each of the two main PGI genotypes from the Klostersee. The results strongly support the field study.

Material and methods

The Klostersee in south Germany (47° 58′ N., 12° 29′ E.; altitude 533 m o.S.L; 0.46 km^2; mean depth 5.88 m; maximal depth 16 m) is a fairly eutrophic, prealpine, postglacial lake (Gessner, 1953; Jacobs, 1977). The cladoceran fauna includes three pelagic species of *Daphnia* – *D. cucullata*, *D. galeata* and *D. hyalina* – which are heavily preyed upon by fish (Jacobs, 1977; Seitz, 1977). These three *Daphnia* species have been recorded in the Klostersee since 1921 (Schmidt, 1924).

Culture and experimental conditions

We collected founder animals of *Daphnia cucullata* using vertical net tows at the deepest place of the Klostersee on 3 December 1985 and 17 April 1986. In all 128 clones were obtained, 36 from the December sampling and 92 from the April sampling. We did not further distinguish between the clones from the two sampling dates. Cloning and experiments were done at 18 °C ± 1 °C, L:D 16:8 hours, 300 Lux, in water from the Klostersee that had been filtered and boiled. We used 600 ml glass beakers (height 14.5 cm, diameter 8 cm, 500 ml medium) for mass cultures and test tubes (height 11.5 cm, diameter 1.6 cm, 20 ml) for single cultures. Animals were fed daily with the green algae *Ankistrodesmus gracilis*, with 2×10^4 cells ml^{-1} day^{-1}. Because the small *Daphnia cucullata* (0.4–1.2 mm) tended to get caught at the surface film we covered the beakers with black cardboard to reduce phototactic movements toward the surface and put a few crystals of cetylalcohol on the surface to reduce surface tension (Gießler, personal communication), so that fewer individuals became attached to the water surface. Throughout all experiments the medium was renewed every third day.

Electrophoresis

Clones were typed at the phosphate glucose isomerase locus (PGI, E.C No. 5.4.1.11). Methods for starch gel electrophoresis and enzyme staining are described in Gießler (1987). Three allels of the PGI locus can be found in the Klostersee population (Jacobs, 1982). They were named according to their relative electrophoretic mobilities in the gel: F fast, M medium, S slow. From 128 laboratory clones 11 FM clones and 11 MM clones were randomly selected for the experiments. Genotypes others than FM and MM were very rare (cf. Jacobs, 1990) and were therefore not used. The diagnostic glutamate oxalacetat transaminase (E.C. No. 2.6.1.1) was typed to exclude hybrids between *Daphnia cucullata* and other *Daphnia* (Gießler, 1987). 6 hybrid clones were excluded.

Experiments

Three sets of experiments were carried out:

1. Growth and egg ratio
From each clone twenty-five neonates of the same age (±6 hours) were kept in 500 ml medium. Starting with day 7 after hatching, we measured the length and width of all individuals and recorded the number of eggs in the brood pouch daily (until day 20 and thereafter every other day). Length was measured as the distance from the base of the tail spine to the upper margin of the compound eye, width as the greatest width of the carapace perpendicular to the longest body length without tail spine. The rarely occurring degenerated eggs or embryos were not counted. Neonates that hatched during these experiments were removed and measured. Egg ratio (sum of all eggs/number of all females in culture) and mean clutch size (sum of all eggs/number of females with eggs) were calculated.

2. Birth and death rates
The experiments were identical to those of set 1, but the beakers were only checked by eye to count

offspring and dead animals: twice a day before day 20, and once a day thereafter. Dead animals and offspring were removed. For each of the 22 clones, 3–5 replicates were run (one clone had only two replicates).

3. Clutch size and hatching time

As in the previous experiments, 25 neonates were kept in beakers containing 500 ml medium. At day 8 we transferred each individual into a test tube containing 20 ml medium. Test tubes were checked three times a day for offspring until day 20, subsequently only twice a day. The time of hatching of offspring was estimated as the midpoint between the first recording of offspring and the previous control. Only the first five broods were considered. We ended up with 8–24 replicates per clone.

All experiments were performed under the same conditions and lasted for 30 days unless death of the animals terminated the experiments earlier. To reduce possible maternal effects, each experiment was started from cultures kept for at least two generations under the same environmental conditions as those in the experiments.

Data analysis

From experiment 2 we calculated life tables for each clone. The intrinsic rate of natural increase, *r*, was iteratively calculated using the Euler-Lotka equation.

Data analyses were conducted using the SAS (5.ed.,1985) computer package. Nested ANOVAs with genotype group as top level and clones as second level were used. Genotype effect was tested over clone mean squares, clones over replicates. In analyses relating to experiment 1 the data were averaged in age classes of three day length to reduce the variance of the non-continuous traits.

Table 1. Means and SE for genotype groups and results of nested ANOVA for neonate and primiparous instars. There was 1 degree of freedom for genotypes and 20 for clones for all traits. *; **; ***: significant with $p < 0.05$; <0.01; <0.001. ns: not significant.

Trait	Mean SE		Genotype F	Clones F
	FM	MM		
Neonates				
Length	0.430(0.003)	0.416(0.002)	6.71*	11.96***
Width	0.209(0.003)	0.200(0.001)	10.44**	6.69***
Primiparous				
Length	0.837(0.008)	0.818(0.006)	2.74ns	17.90***
Width	0.498(0.005)	0.480(0.003)	3.32ns	15.15***
Age	7.46 (0.16)	7.62 (0.14)	0.37ns	8.91***

Results

Body size

Table 1 shows the nested ANOVAs for various traits of neonates and primiparous instars. Clone effects were highly significant for all traits. Genotype effects were significant for neonate traits but not for primipares. The mean body length and width was significant larger for FM neonates than for MM ones.

Figure 1 shows the increase in body length and width of individuals kept in mass cultures. A nested ANOVA with age as covariable showed significant differences between both genotype groups, and for clones ($F = 4.63$, $p = 0.04$, $df = 1$

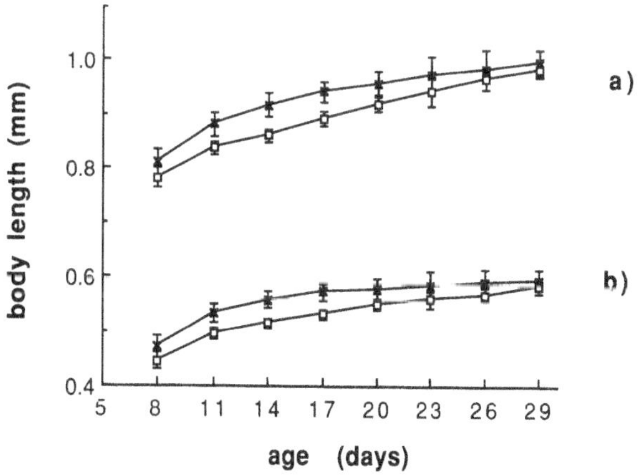

Fig. 1. Growth of (a) body length and (b) body width in two PGI-genotype groups, FM (x) and MM (□), represented in 8 age classes three days long. The graphs show only the mean values in each age class. The vertical bars indicate 95% confidence intervals.

for genotype groups; $F = 67.90$, $p < 0.0001$, $df = 20$ for clones; $F = 12190.1$, $p < 0.0001$ for age as covariable). The results for body width resembled the results for body length. Figure 1 shows clearly that the size differences were particularly large during the age interval from 10 to 21 days (compare confidence intervals in Fig. 1).

When the increments in body length in any age class were plotted against body length or against age, differences in growth pattern between the two genotype groups were more pronounced in growth relative to length than to age (Fig. 2). The same pattern was found for the width increments in each age class.

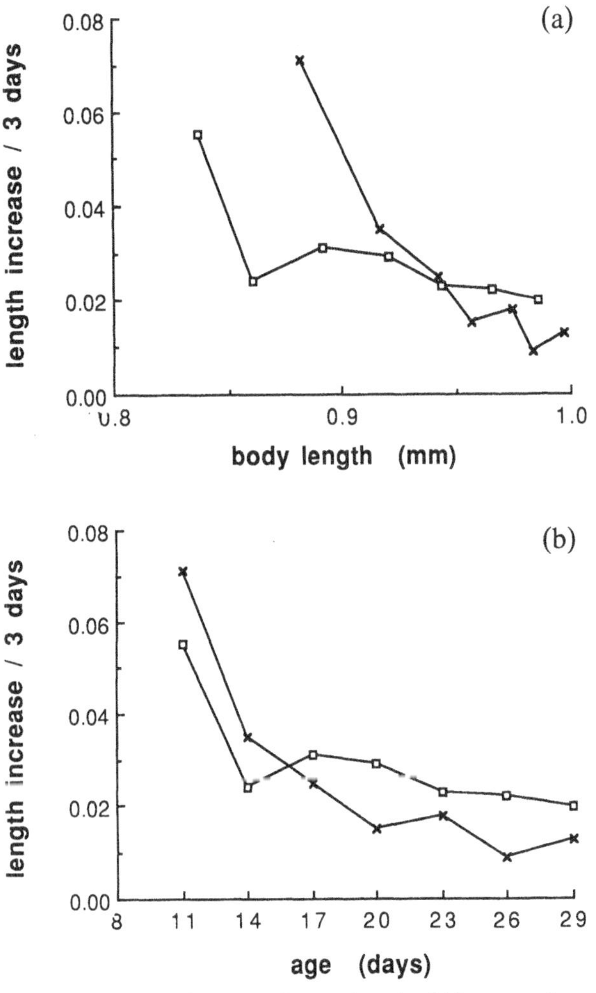

Fig. 2. Relation of increase in body length within successive age classes to (a) total body length and (b) to age, beginning with the increment of age class 7–9 days. The increase in length per age class is calculated as the difference in length between two successive age classes. FM (x), MM (□).

The width/length ratio in all animals was significantly greater for FM than for MM. This was tested with a nested ANOVA with age as a covariable ($F = 6.86$, $p = 0.01$, $df = 1$ for genotype groups; $F = 17.77$, $p < 0.0001$, $df = 20$ for clones; $F = 3178.1$, $p < 0.0001$, for covariable age). This means that throughout the first 30 days of life, FM animals were on average wider than MM animals of equal length. The width of a *Daphnia* involves both the size of its brood pouch and the size of its filtering apparatus (Jacobs, 1987). A larger filtering apparatus may increase ingestion rate and consequently the total energy uptake.

Reproduction

Egg-ratio

The egg ratio (sum of all eggs/number of all females) of eight age classes is shown in Fig. 3a. There are notable differences in age specific egg ratio. At the beginning of reproduction, FM produced more eggs than MM, but after day 16, the situation was reversed. In both genotype groups the egg ratio increased up to 14 days, and than stayed level in MM and decreased in FM clones. The low egg ratio in the first age class (7–9 day) is caused by some immature females. The mean egg ratio for the age interval from 10 to 15 days was significantly larger in FM than in MM, in the later interval (day 16–30) the egg ratios did not differ between the genotype groups. Clone effects were significant for both intervals (Table 2).

Figure 3b shows the mean clutch size (sum of all eggs/number of females with eggs) for the 8 age classes. Also here, the FM clutch size was significantly greater than the MM clutch size for the age classes 11 and 14 (Fig. 3a), but the clutch sizes for animals older than 15 days did not differ between the two groups (Table 2). Because of the different shapes of the curves for the egg ratio (Fig. 3a) and the clutch size (Fig. 3b) we plotted the distribution of clutch size in 7 age classes for FM and MM between day 10 and 30 (Fig. 4). It may be noted: (1) The age related increase in the number of females with no eggs may indicate senescence.

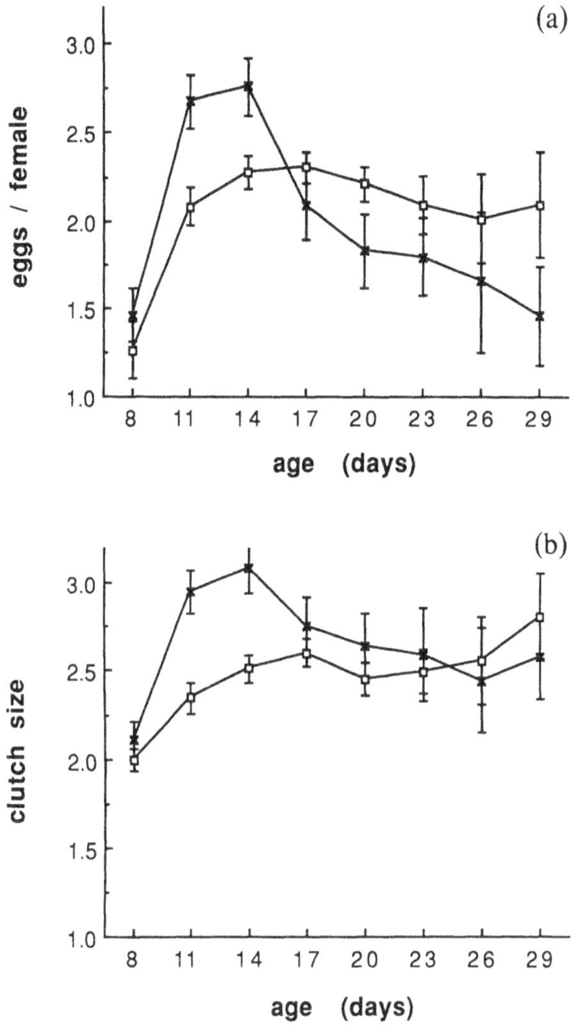

Fig. 3. Egg ratio and clutch size of 8 age classes three days long. The graph shows the mean values in each age class. (a) Egg ratio (sum of all eggs/number of all females), (b) clutch size (sum of all eggs/number of females with eggs). Vertical bars indicate one S.E. n = 33 for each point. FM (x), MM (□).

Comparing the FM and MM histograms, the percentage of animals without eggs in their brood pouch was significantly larger in the FM group for the age classes 17 (t = 3.22, df = 64, $P < 0.002$), 20 (t = 3.88, df = 45, $P < 0.001$), and 23 (t = 2.17, df = 20, $p < 0.05$) than in the MM group. (2) The distribution of clutch size with at least one egg did not show a marked shift to the left with increasing age as might be expected if average clutch size is reduced. These two points show that older daphnids tend to skip a brood between two moults, as was also described by Banta (1939) and Young

Table 2. Mean egg ratio and clutch size for two groups of age classes of the two PGI-clone groups, FM and MM. asterisks: see Table 1.

Age classes included	Mean (SE) FM	Mean (SE) MM	Genotype F	Clones F
Mass-culture experiment				
Egg-ratio				
11,14	2.72 (0.17)	2.19 (0.10)	4.50*	30.4***
17 to 29	1.78 (0.15)	2.17 (0.10)	1.06ns	17.2***
Clutch size				
11,14	3.01 (0.09)	2.43 (0.06)	7.05*	30.05***
17 to 29	2.64 (0.09)	2.59 (0.07)	0.03ns	13.34***
Single culture experiment (only clutch size)				
Clutches included				
1 and 2	2.63 (0.14)	2.39 (0.09)	1.78ns	4.34***
3,4 and 5	2.71 (0.12)	3.05 (0.09)	1.12ns	8.26***

(1979). This sign of senescence is significantly earlier in the FM than in the MM group.

Number of neonates

For the first five broods the numbers of neonates were recorded in the single animal cultures (experiment 3). Figure 5 shows the relation of number of neonates and the times when hatching took place. The figure is similar to Fig. 3a & b, however, differences between clutch sizes were not significant, neither for the two clutches before age 16 nor for the 3 clutches after age 16 (Table 2).

Differences between mass and single culture sets can in part be explained by the different methods used for fecundity estimation. The egg ratio may include eggs which were not able to develop, while neonate counts included only viable offspring.

Survival

The age dependence of survival in both genotype groups is shown in Fig. 6. Both survival curves

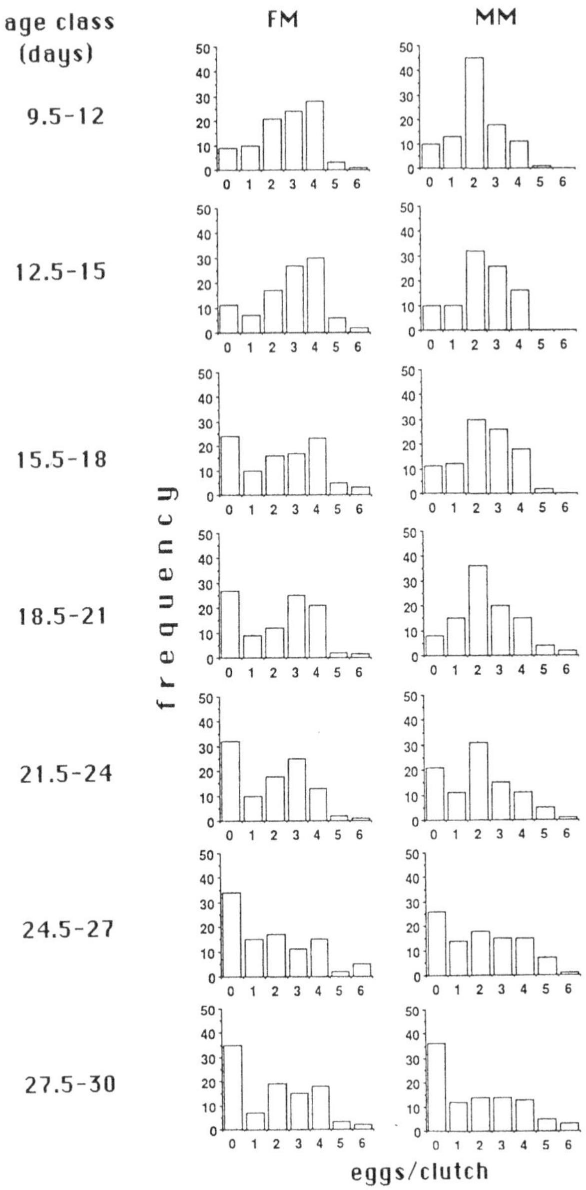

Fig. 4. Frequencies (%) of clutch sizes of the two PGI genotype groups, FM and MM, separated in 7 age classes of 3 days length.

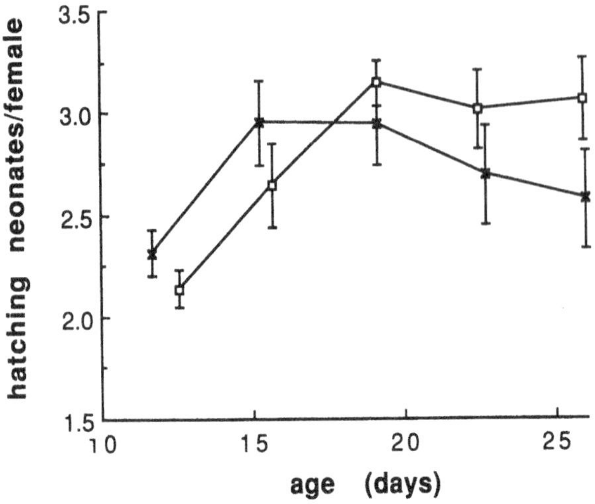

Fig. 5. Relation between number of hatching neonates and age of mother of five broods of the two PGI genotype groups, FM (x) and MM (□). Vertical bars indicate one S.E.

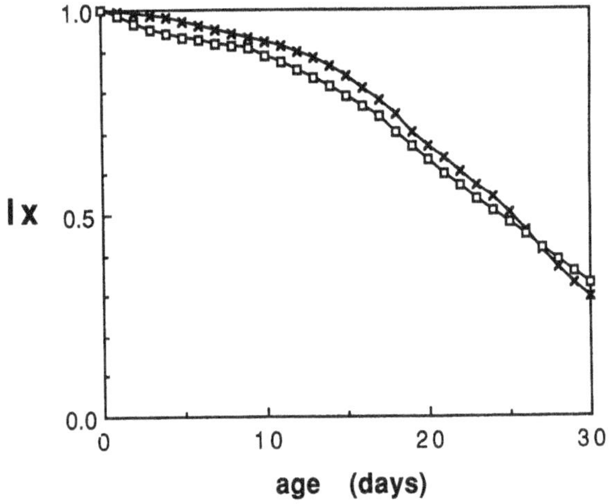

Fig. 6. Age dependent survival in two genotype groups. lx: the proportion of animals surviving to a given age. Each curve represents more than 1000 animals. FM (x), MM (□).

suggest two phases, namely a phase of high survival up to an age of about 14 days, followed by a phase of increased mortality. The curves did not differ significantly. This implies that there is no causal relationship between the differences in reproductive behaviour and survival. It is important to note that physiological survival in the laboratory is different from survival in the field.

Population growth rates

The life tables of experiment 2 were used to calculate the intrinsic rate of natural increase, *r*, according to the Euler-Lotka equation. Because *r* depends on the number of broods which a female produces before she dies, we investigated the influence of maximum age and size of mothers on

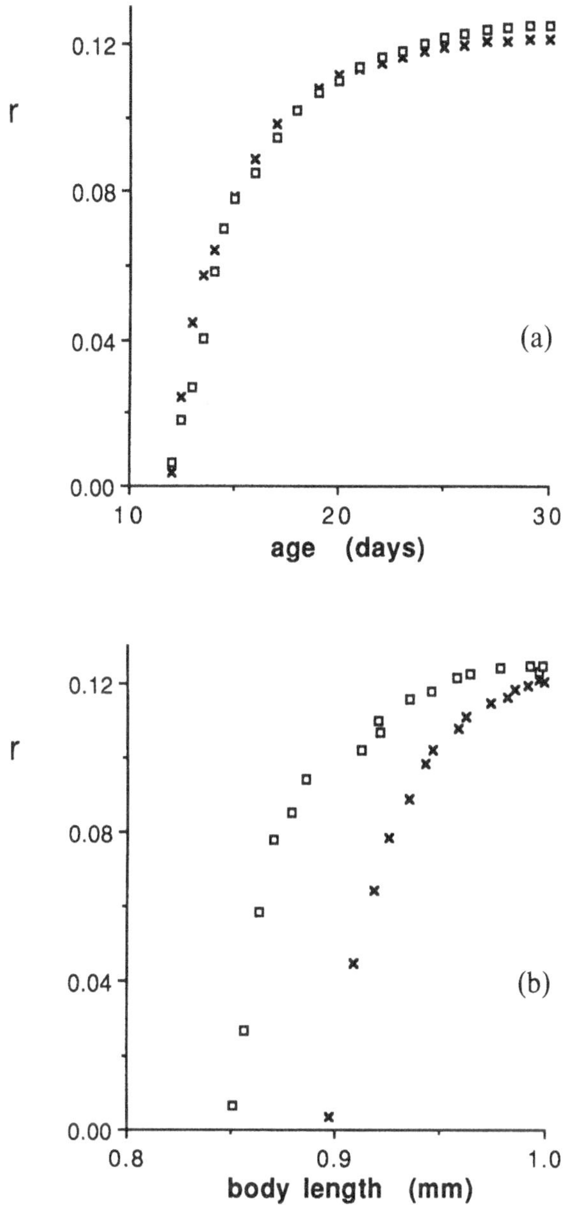

Fig. 7. Dependence of the intrinsic rate of natural increase, *r*, on (a) maximum age and (b) maximum body length at death. *r* calculated according to the Euler-Lotka equation. For further explanation, see text. FM (x), MM (□).

r (cf. Jacobs, 1978). Figure 7a & b show this relation for the two genotype groups. It may be noted: (1) If both groups are allowed to reproduce beyond 20 days, MM would outgrow FM (MM had about 4% higher *r* than FM at day 30). (2) Late broods contribute little to population growth. The major contribution to the population growth came before day 20, i.e. the first three broods. In Fig. 7a, during the interval from day 12 to 17 the FM group performed up to 35% better than the MM group for a specific maximum age. (3) The influence of maximum body size on the growth rate is pronounced (Fig. 7b). A reduction in the maximal size of the mothers from 1 to 0.9 mm would reduce *r* by 20% in MM but by 80% in FM.

Discussion

We found highly significant clone effects for most life-history traits in this study. The two electrophoretically distinguished genotype groups showed significant differences for neonate size (Table 1), growth pattern (Fig. 2) adult size (Fig. 1) and fecundity of adults younger than 16 days (Table 2, only in mass culture experiment). FM animals had a high tendency to skip broods after age 16, while in MM, this was after age 22. No differences between genotype groups was found for life-history traits of primiparae. In single cultures clutches 1 + 2 were not significant larger for the FM group, but the general reproductive pattern were the same in mass and in single cultures. Altogether, it seemed that FM individuals invested most of their resources in the early adult phase, while MM individuals spread their resources over a longer period of time. The high degree of significant clone effects in nested ANOVAs underlines the importance of using nested ANOVAs, and to test the genotype effect over the clone effect. Otherwise, clonal differences may lead to significant results of genotype effects.

In animals with indeterminate growth, fertility usually shows an increase during at least part of the lifespan (Charlesworth, 1980). Reduced fertility in spite of continued adult growth as shown for FM but not in MM (Fig. 3a, b) may be interpreted as senescence (Partridge, 1987). Comparing the reproductive strategies of FM and MM groups suggests that the cost of reproduction leads to accelerated senescence. This observation is consistent with William's (1957) hypothesis of antagonistic pleiotropy of gene action: A mutant causing a high fitness in early adulthood reduces fitness later on. It should be noted, however, that the accelerated senescence in the FM genotype group was not linked to higher physiological mortality (Fig. 6), only to reduced fecundity (Fig. 3a, b) and growth rate (Fig. 2b).

The question arises, how many clones were actually involved in our study? As we did not type other polymorphic loci we cannot answer this question directly, but Gießler (1987) showed, by typing several loci, that all of six clones of *Daphnia cucullata* from the Klostersee were electrophoretically distinct, indicating a high degree of genetic diversity. Most of the 22 clones of this study were probably genetically distinct.

Clonal differences in life-history traits of cladocerans belonging to the same population have been described by other authors (e.g. Kerfoot, 1977; Black, 1980; Loaring & Hebert, 1981; Lynch, 1983; Pace, Porter & Feig, 1984; Weider & Hebert, 1987; Carvahlo, 1987, 1988; Vanni, 1987), but there are only a few reports on the life-history of different genotype groups characterized by specific genotypes in one or more loci (e.g. Lynch, 1984; Korpelainen, 1987). Lynch (1984) found evidence for differences in life-history traits between groups with specific genotype combinations at the PGM and the PEPT-2 locus. Carvahlo (1987) found differences in the thermal tolerance of PGI marked genotypes. He also found a correlation between changes in PGI genotype frequencies and genotypic differences in parthenogenetic egg production of *Daphnia magna* in a two year field study (Carvahlo, 1988). He suggested that selection was occurring at the PGI-locus, a closely linked locus or a group of loci (Carvalho, 1988). An alternative explanation could be that PGI genotypes mark ecologically distinct clonal groups.

In a field study, a positive correlation between the frequency of FM genotypes at the PGI locus and the egg ratio was found in two bays of the Klostersee (Jacobs, 1990). The direction of FM frequency shifts corresponded with shifts in egg ratio. Jacobs suggested that, on average, the FM genotypes produced more eggs than other genotypes. Our experimental data support this hypothesis: under natural conditions of heavy size selective predation, the maximal age and size of *Daphnia cucullata* is much lower in the Klostersee than in the laboratory (Jacobs, 1977; Seitz, 1977, 1980). Therefore, if one wishes to compare laboratory data with field data, only life-history traits measured in the early adult stage, for instance during the first 15–20 days, should be taken into account. During this period there was indeed a reproductive advantage for the FM genotypes in our experiments. This supports the results of the field study (Jacobs, 1990).

There are no data on the type of selection pressure and its temporal fluctuations in the Klostersee so far. However, temporally unstable predation pressure could possibly be the reason for the coexistence of ecological distinct clonal groups (Hebert & Crease, 1980).

In closing, we want to question how important the described differences in life-histories in one environment are. The environment does not just vary in size selective predation pressure but also in food and temperature. Genetical differences in physiological responses of genotypes might compensate fitness differences between environments.

Acknowledgments

We thank A. Küppers and S. Gießler for practical help and B. Bauer, J. Bengtsson, L. Feldhaus, M. Gebhardt, S. Gießler, E. Hillesheim, A. van Noordwijk, J. Shykoff, S. Stearns and L. Yampolsky for criticizing earlier drafts of the manuscript.

References

Banta, A. M., 1939. Studies on the physiology, genetics and evolution of some cladocera. Carnegie Institution of Washington, 39: 1–285.

Black, R. W., 1980. The genetic component of cyclomorphosis in *Bosmina*. In W. C. Kerfoot (ed.), Evolution and Ecology of Zooplankton Communities, Special Symposium Vol. 3, University Press of New England, Hanover: 456–69.

Carvalho, G. R., 1987. The clonal ecology of *Daphnia magna*

(Crustacea: Cladocera). II. Thermal differentiation among seasonal clones. J. anim. Ecol. 56: 469–478.

Carvalho, G. R., 1988. Differences in the frequency and fecundity of PGI-marked genotypes in a natural population of *Daphnia magna* Straus (Crustacea: Cladocera). Funct. Ecol. 2: 453–462.

Charlesworth, B., 1980. Evolution in age-structured populations. Cambridge University Press, Cambridge.

Gessner, F., 1953. Die Limnologie des Naturschutzgebietes Seeon. Arch. Hydrobiol. 47: 553–624.

Gießler, S., 1987. Mikroevolution und Populationsgenetik im *Daphnia galeata/hyalina/cucullata* Komplex. Doctoral thesis, Universität München.

Hebert, P. D. N. & T. J. Crease, 1980. Clonal coexistence in *Daphnia pulex* Leydig: Another planktonic paradox. Science 207: 1363–1365.

Jacobs, J., 1977. Coexistence of similar zooplankton species by differential adaptation to reproduction and escape, in an environment with fluctuating food and enemy densities. II. Field data analysis of *Daphnia*. Oecologia (Berl.) 30: 313–329.

Jacobs, J., 1978. Coexistence of similar zooplankton species by differential adaptation to reproduction and escape, in an environment with fluctuating food and enemy densities. III. Laboratory Experiments. Oecologia (Berl.) 35: 35–54.

Jacobs, J., 1982. Short-term and small-scale genetic differentiation in two partially parthenogenetic pelagic species of *Daphnia*. A preliminary report. In S. D. Jayakar & L. Zonta (eds), Evolution and Genetics of Populations, University of Pavia, Pavia: 43–52.

Jacobs, J., 1987. Cyclomorphosis in *Daphnia*. Mem. Ist. ital. Idrobiol. 45: 325–352.

Jacobs, J., 1990. Microevolution in predominantly clonal populations of pelagic *Daphnia* (Crustacea: Phyllopoda): Selection, exchange, and sex. J. evol. Biol. 3: 257–282.

Kerfoot, W. C., 1977. Competition in cladoceran communities: The cost of evolving defenses against copepod predation. Ecology 58: 303–313.

Korpelainen, H., 1987. Reproductive strategies of *Daphnia magna* genotypes. Hereditas 106: 181–188.

Loaring, J. M. & P. D. N. Hebert, 1981. Ecological differences among clones of *Daphnia pulex* Leydig. Oecologia (Berl.) 51: 162–168.

Lynch, M., 1983. Ecological genetics of *Daphnia pulex*. Evolution 37: 358–374.

Lynch, M., 1984. The genetic structure of cyclic parthenogen. Evolution 38: 186–203.

Pace, M. L., K. Porter & Y. S. Feig, 1984. Life history variation within a parthenogenetic population of *Daphnia parvula* (Crustacea: Cladocera). Oecologia (Berl.) 63: 43–51.

Partridge, L., 1987. Is accelerated senescence a cost of reproduction? Funct. Ecol. 1: 317–320.

SAS Institute Inc., 1985. User's Guide: Statistics. Version 5 Edition, Cary, North Carolina.

Schmidt, C., 1924. Die Phyllopoden und Copepoden des Obinger u. Seeoner Sees. Arch. Hydrobiol. 15: 179–208.

Seitz, A., 1977. Die Bedeutung von Umweltfaktoren, Konkurrenz und Räuber-Beute-Beziehung für die Koexistenz dreier Daphnienarten. Doctoral thesis, Universität München.

Seitz, A., 1980. The coexistence of three species of *Daphnia* in the Klostersee. I. Field studies on the dynamics of reproduction. Oecologia (Berl.) 45: 117–130.

Vanni, M. J., 1987. Colonization dynamics and life history traits of seven *Daphnia pulex* genotypes. Oecologia (Berl.) 72: 263–271.

Weider, L. J. & P. D. N. Hebert, 1987. Ecological and physiological differentiation among low-arctic clones of *Daphnia pulex*. Ecology 68: 188–198.

Williams, G. C., 1957. Pleiotropy, natural selection and evolution of senescence. Evolution 11: 398–411.

Young, J. W. P., 1979. Enzyme polymorphism and cyclic parthenogenesis in *Daphnia magna*. I. Selection and clonal diversity. Genetics 92: 953–970.

Hydrobiologia **225**: 255–261, 1991.
V. Kořínek & D. G. Frey (eds), Biology of Cladocera.

The components of life-history traits variation in *Daphnia magna* population

L. Yu Yampolsky & B. A. Kalabushkin
Institute of General Genetics, USSR Academy of Sciences, Gubkin str. 3 GSP-1 V-333, Moscow, USSR

Key words: *Daphnia*, life-history, genetic variation, isozymes, heterozygosity, strategies

Abstract

The variation of six growth and reproduction traits in the parthenogenetic clones of *Daphnia magna* and its relationships with the genotypes at 5 isozyme loci were studied under controlled conditions at three food concentrations. A significant genetic component of age at maturity, growth rate and fecundity variation was detected. Nested ANOVA revealed the influences of the genotypes of the isozyme loci, that are more pronounced at the high food concentration. In some cases the genotype-environment interactions were found.

The variation of body length at 6 days and total number of offspring includes a significant heterozygosity – environment interaction. The most heterozygous clones show the least differences in life history traits between food conditions.

Life-history traits heritability, genotype-environment interactions and different reproductive strategies of the clones are discussed.

Introduction

Clonal organisms provide a remarkable possibility to study the components of life-history and fitness traits variation. Measuring the parameters of individuals of the same genotype developing under different environmental conditions one can obtain the estimates of the heritability of the quantitative traits, environmental influence and the genotype-environment interactions. *Daphnia* with their short generation time are particularly suitable for these studies.

Although the intraspecific variation of life-history traits in *Daphnia* has been studied rather intensively (Loaring & Hebert, 1981; Brookfield, 1984; Korpelainen, 1986a, b; Grechany *et al.*, 1986; Weider, 1987; Vanni, 1987), few estimates of genetically based variation component are available, as well as the estimates of the genotype-environment interaction. Most of the studies revealed significant interclonal differences in growth rate, fecundity, viability, life span, male production rate and other parameters. There are also strong evidences for the existence of the genotype-environment interactions: in many cases different clones are the most fit under different conditions and none shows the best fitness under all conditions (Carvalho, 1987; Pace *et al.*, 1984; Weider, 1985). All the mentioned studies included limited number of clones (5 to 9) that is certainly not enough to measure the genetic variation present in the population.

The only reliable heritability estimates known from literature were obtained by Lynch (1984). However in this work large number of clones were maintained under the same condition, so the genotype-environment interactions could not be measured.

The question whether the observed interclonal differences have any relationship with the genetical variation at the isozyme loci is also still unclear (Brookfield, 1984; Korpelainen, 1986a), as well as the problem of heterozygote advantages in *Daphnia*, that is important for the explanation of heterozygote excess above Hardy-Weinberg expectations observed in many *Daphnia* populations (Hebert, 1982) and for the study of heterosis in general (Berger, 1976; Zouros & Foltz, 1987).

This work is an attempt to distinguish the components of the life-history traits variation due to the genetical factors and the genotype-environment interaction and to find the relationships between the life-history traits variation and the isozyme polymorphism.

Material and methods

Daphnia clones for the experiments were obtained from the ephippia, collected in one of the ponds of Moscow Zoo in late 1987. As a whole 61 females developed from 80 ephippial eggs, but only 49 were used in the experiment, because the rest had died or didn't have enough progeny. Because some individuals died during the experiment, only 20 clones, represented at each food concentration by not less than 2 replicates, were taken for the two factors analysis. The offspring not older than 24 hours, were placed into 30 ml of artificial pond water (Hebert, Crease, 1980) (two individuals from the same clone per a vial) with three concentrations of *Chlorella* cells: 40, 80 and 160×10^3 cells per ml. *Chlorella* was added and the neonates were replaced daily; the water was changed every other day. The experiment was run under room temperature ($20 \pm 1\ ^{\circ}C$) and continuous light.

Age at first reproduction (Dj), number of eggs in the first clutch (F1), body length at maturity (L1) and at the age of 6 days (L6), spine length at the age of 6 days (SL6) and total number of progeny during first 15 days of the life (P) were measured in all the individuals.

The genotypes of all the clones at five polymorphic loci (phosphohexosoisomerase (Pgi), with 3 alleles, phosphohexosomutase (Pgm), esterase (Est), leucinaminopeptidase (Lap) and alkaline phosphotase (Alk) – all with two alleles) were analyzed by means of PAGE (Tris-EDTA-borate buffer, pH 8.6) with the staining according to the standard methods (Wolf, 1982) with minor modifications.

The data were analysed by means of nested ANOVA with clones as the factor of the lower level and genotypes at the isozyme loci (or heterozygosity level) as that of the higher level and two-way mixed model ANOVA with clones, genotypes or heterozygosity as random effect and food concentration as fixed effect. The former scheme allowed to obtain the estimation of the influence of certain loci on the background of general genetic variability, that are hardly available in bisexual organisms. More detailed information,

Table 1. Two-way mixed model ANOVAs for the influences of clones and food concentrations.

		Dj			F1		
	d.f.	MS	F	p	MS	F	p
Food	2	49.97	15.9	0.005	43.67	11.8	0.005
Clones	19	4.39	2.64	0.005**	3.22	2.03	0.025
Interact	38	3.15	1.89	0.025	3.69	2.31	0.005
Error	56	1.66			1.59		
		L1			L6		
	d.f.	MS	F	p	MS	F	p
Food	2	102.5	17.8	0.005	165.9	16.9	0.005
Clones	19	2.41	0.53	–	25.39	2.98	0.005***
Interact	38	5.76	1.26	–	9.79	1.19	–
Error	56	8.53			4.92		
		SL6			P		
	d.f.	MS	F	p	MS	F	p
Food	2	148.8	45.5	0.005	74.33	4.66	0.005
Clones	19	2.81	0.57	–	16.12	3.33	0.005*
Interact	38	3.27	0.66	–	15.93	3.29	0.005
Error	56	4.83			4.83		

See text for the traits notation. Significance of clones over the interaction and error MS pooled together: * 0.05; ** 0.01; *** 0.005.

concerning experimental design and data analysis is to be published (Yampolsky & Kalabushkin, in press).

Results

Table 1 shows that the genetical differences are significant for 4 traits of the 6 studied. For age at first reproduction, body length at 6 days and total number of offspring the differences between clones are also significant against the mean squares of the interaction and error pooled together. The portion of variation accounted for by the genetic factors estimated within each food level was higher at low and medium food than in the high food concentration. For age at first reproduction, body length at 6 days and total number of offspring these estimates reached the values of 0.46, 0.58 and 0.52, respectively at the low food; 0.66, 0.57 and 0.60 at medium food, but only 0.06, 0 and 0.44 at high food.

Nested ANOVA with the genotypes at the isozyme loci as the higher level factor and clones as the subordinate level factor revealed several cases of significant one-locus genotype influences that are pronounced only at the high food concentration (Table 2). Absence of such influence at

Table 2. Nested ANOVAs for the influence of the genotypes at isozyme looi.

Food concentration – 80,000 cells/ml		SL6							
	d.f.	MS	F	R	p				
Lap	2	19.89	5.19	0.41	0.025				
Clones	18	3.83	3.38	0.32	0.005				
Error	21	1.13		0.27					
Food concentration – 160,000 cells/ml		**F1**				**P**			
	d.f.	MS	F	R	p	MS	F	R	p
Lap	2	22.90	6.68	0.43	0.01	72.09	4.53	0.33	0.05
Clones	18	3.43	1.60	0.13	–	15.88	1.75	0.18	–
Error	21	2.14		0.44		9.07		0.49	
		P							
	d.f.	MS	F	R	p				
Pgi	5	44.74	3.25	0.28	0.05				
Clones	15	13.75	1.51	0.15	–				
Error	21	9.07		0.57					
		L1							
	d.f.	MS	F	R	p				
Est	2	32.6	7.33	0.25	0.005				
Clones	18	4.45	0.59	0	–				
Error	21	7.52		0.75					

Only the cases in which significant influence of a locus was found included. Other traits do not show any difference between the genotypes when tested over the interclonal variances. See text for the notations of the traits. The influence of the traits at the high food concentration is significant in the test of independent probabilities for multiple tests ($p < 0.05$).

Table 3. Three cases of significant genotype – food interactions from the total 30 locus-traits tests.

		F1			
	d.f.	MS	F	R	p
Food	2	32.43	5.26	0.17	–
Lap	2	0.25	0.11	0	–
Interaction	4	6.17	2.65	0.10	0.05
Error	132	2.33			
		P			
	d.f.	MS	F	R	p
Food	2	138.1	4.63	0.19	–
Est	2	5.90	0.75	0	–
Interaction	4	29.81	3.82	0.15	0.01
Error	133	7.80			
	d.f.	MS	F	R	p
Food	2	55.2	2.71	0.07	–
Pgi	5	21.8	2.94	0.01	0.025
Interaction	10	20.4	2.75	0.17	0.005
Error	124	7.42			

F1 – first clutch size; P – total number of offspring. Pooled significance level (independent probabilities test) $p < 0.05$.

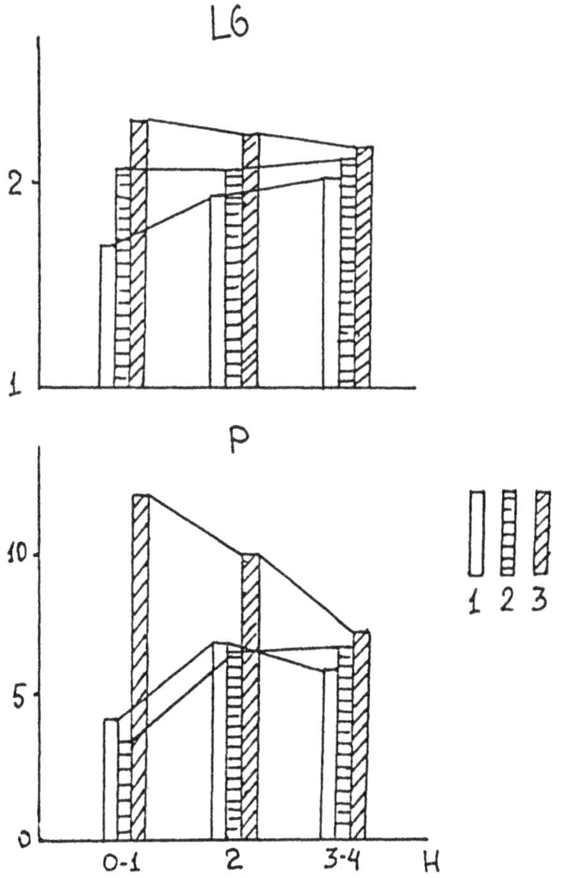

Fig. 1. Mean values of body length at 6 days (L6), mm and total number of offspring (P) of the individuals with three heterozygosity levels at three food concentrations. 1, 2 and 3 – food concentrations 40000, 80000 and 160000, respectively.

Table 4. Two-ways ANOVAs for food and heterozygosity influences.

		Dj				F1			
	d.f.	MS	F	R	p	MS	F	R	p
Food	2	41.92	3.73	0.07	–	506.94	11.08	0.29	0.025
H	2	2.45	0.36	0	–	0.30	0.02	0	–
Inter.	4	11.25	1.63	0.03	–	45.77	2.54	0.06	0.05
Error	160	6.88				18.04			
		Dj				F1			
	d.f.	MS	F	R	p	MS	F	R	p
Food	2	50.96	3.19	0.05	–	250.77	6.98	0.26	–
H	2	21.19	2.31	0.01	–	10.22	1.07	0	–
Inter.	4	15.98	1.74	0.03	–	35.91	3.77	0.10	0.01
Error	160	9.18				9.54			

Clones were grouped into 3 heterozygosity levels: 0 or 1, 2 and 3 or heterozygous loci. See text for the notations. H – heterozygosity level. Significance level for interaction influence (independent probabilities test) $p < 0.01$.

Table 5. Nested ANOVAs for the influence of the heterozygosity on the total number of offspring.

Food concentration		80,000 cells/ml				160,000 cells/ml			
	d.f.	MS	F	R	p	MS	F	R	p
H	2	34.93	4.46	0.30	0.05	64.33	3.84	0.23	0.05
Clones	18	7.83	3.69	0.40	0.005	16.74	1.84	0.23	–
Error	21	2.13		0.30		9.07		0.54	

Only the cases in which significant influence of heterozygosity was found are included. Other traits do not show any difference between the heterozygosity levels. H – heterozygosity.

other food conditions seem to indicate genotype-environment interactions. Table 3 presents the cases in which these interactions were found to be significant.

No significant differences between the heterozygosity levels were found in two-way ANOVA with food and heterozygosity as the factors, but in two cases (body length at 6 days and total offspring number) a significant portion of variance (6 and 10%) appeared to be caused by heterozygosity-food interactions (Table 4). Mean values of these two traits are shown in Fig. 1. Body length slightly decreases with heterozygosity at the high food concentration, is constant at the medium, and increases at the low one. Total offspring number decreases with heterozygosity at the high food concentration and increases at the two other food concentrations. Note, that the most heterozygous clones show the least differences in life history traits between food conditions.

As well as the individual loci, heterozygosity does not influence any trait under all conditions. Table 5 includes the cases in which nested ANOVA confirms the presence of heterozygosity effect on the background of the interclonal variation.

Discussion

Are the genetical variation components estimates obtained in this study reliable enough to be used for some conclusions concerning the heritability of the quantitative traits in *Daphnia*? The estimates for the age at maturity, body length and first clutch size, could be compared with the corresponding estimates obtained by Lynch (1984). High interclonal variation of these traits (see Table 1) is in good agreement with the Lynch's estimates of the heritability for the first parthenogenetical generations, that have values 0.57–0.76; 0.40–0.78 and 0.58–0.69, relatively. Rather high estimates of heritability in another clonal organism, *Poeciliopsis* fish were published recently by Wetherington *et al.* (1989).

One should take into account that our experimental design didn't allow to eliminate the maternal effects that could make a contribution to the interclonal variance component, since the individuals taken into the experiment had been obtained from a few mothers or from a single one and the conditions to which the mothers were exposed could slightly differ. However, the maternal effects should not affect the tests of isozyme genotypes and heterozygosity influence, because they are included into clonal differences, which are used as the error term.

It seems interesting that the effects of certain isozyme loci are more expressed in the most favorable conditions. One can suggest that this happens because the genetically determined quantitative parameters are exposed to less strong influence of the environment than under more severe conditions, when the environment has much more effects on the development of the individuals. On the basis of the data from the same experiment we showed (Yampolsky & Kalabushkin, in press) that the correlations of different traits are the highest also at the highest food concentration, that is consistent with the suggestion that better food conditions make the development of *Daphnia* quantitative traits more independent from the environment effects.

The evidence for the existence of genotype-environment and heterozygosity-environment interactions are important for the problem of clonal coexistence. Since different genotypes are most fit under different conditions, they will reach high frequencies and reproduce most intensively not simultaneously. The phenomena like this are known from nature. For example, one can suppose that the genotype-environment interactions at the esterase locus found in this study and the shifts in esterase genotypes fitness during the season in a natural population reported by Korpelainen (1987) have the same basis.

Several examples of different influence of heterozygosity on the fitness traits under different environments are known from the literature (Samollow & Soule, 1983; Singh & Green, 1984; Govindaraju & Dancig, 1987; Altukhov, 1989; Tesko *et al.*, 1990). The data obtained in this study support the idea that more heterozygous individuals are more stable and less sensitive to the deteriorating ecological conditions. As one can see from Fig. 1, the most heterozygous group of clones shows the least differences in life-history traits between the three food levels. This provides the possibility to suggest that the most heterozygous clones are the generalists, that are able to keep more or less high rate of growth and reproduction under wide variety of conditions, meanwhile the homozygous clones are the specialists, fit only to narrow sets of conditions.

The importance of these adaptive strategies in *Daphnia* populations was discussed by Lynch (1984) and Korpelainen (1987). If the relation between the adaptive strategies and heterozygosity really exists, this, perhaps, may drop new light on the genetic variation of the cyclic parthenogens.

Acknowledgements

We are very grateful to Prof. Yu. P. Altukhov and Dr. D. Ebert for help and useful criticism on the manuscript.

References

Altukhov, Yu. P., 1989. Balancing selection as the factor of allozyme polimorphism maintenance. Adv. modern biol. 107: 323–340 (In Russian).

Berger, E., 1976. Heterosis and the maintenance of enzyme polymorphism. Am. Nat. 110: 823–839.

Brookfield, J. F. Y., 1984. Measurement of the intraspecific variation in population growth rate under controlled conditions in the clonal parthenogen Daphnia magna. Genetica 63: 161–174.

Carvalho, G. R., 1987. The clonal ecology of Daphnia magna Straus (Crustacea, Cladocera). 2. Thermal differentiation among seasonal clones. J. Anim. Ecol. 56: 469–478.

Grechany, G. V., A. Ya. Nikitin & I. A. Sosunova, 1986. Hereditary determination of the dynamics of monoclonal daphnid populations. Genetika 22: 983–988 (In Russian).

Govindaraju, D. R. & B. P. Dancig, 1987. Environmental stress and the relationships among allozyme heterozygosity, biomass and biomass components in jack pine (Pinus banksiana Lamb). Genetica 74: 173–179.

Hebert, P. D. N., 1982. Clonal diversity in Cladoceran populations. In: Population Biology: Retrospect and prospect. (King, C. E. & Dawson, P. S., eds). Columbia Univ. Press, New York.

Hebert, P. D. N. & T. J. Crease, 1980. Clonal coexistance of Daphnia pulex: another planctonic paradox. Science 207: 1363–1365.

Loaring, J. M. & P. D. N. Hebert, 1981. Ecological differences among clones of Daphnia pulex Leydig. Oecologia 51: 162–168.

Lynch, M., 1984a. The limits to life history evolution in Daphnia. Evolution 38: 465–482.

Lynch, M., 1984b. The genetic structure of a cyclic parthenogen. Evolution 38: 186–203.

Korpelainen, H., 1986a. The effects of temperature and photoperiod on life-history parameters of Daphnia magna (Crustacea, Cladocera). Freshwat. biol. 16: 615–620.

Korpelainen, H., 1986b. Intraspecific variation in lifespan and fecundity under controlled conditions in the cyclic parthenogen, Daphnia magna. Arch. Hydrobiol. 105: 321–327.

Korpelainen, H., 1987. Reproductive strategies of Daphnia magna genotypes. Hereditas 106: 181–188.

Pace, M. L., K. Porter & Y. S. Feig, 1984. Life history variation within a parthenogenetical population of Daphnia parvula (Crustacea: cladocera) Oecologia 63: 43–51.

Singh, S. M. & R. N. Green, 1984. Excess of allozyme heterozygosity in marine molluscs and it's possible biological significance. Malacologia 25: 569–581.

Samollow, P. B. & M. E. Soule, 1983. A case of stress related heterozygote superiority in nature. Evolution 37: 646–649.

Tesko, W. R., M. H. Smith & J. M. Novak, 1990. Food quality, heterozygosity and fitness correlates in Peromyscus polionotus. Evolution, 44: 1318–1325.

Vanni, M. J., 1987. Colonization dynamics and life history traits of seven Daphnia pulex genotypes. Oecologia 72: 263–271.

Weider, L. J., 1985. Spatial and temporal genetic heterogeneity in a natural Daphnia population. J. Plankton Research 7: 101–123.

Weider, L. J., 1987. Life history variation among low-arctic clones of obligately parthenogenetical Daphnia pulex, a diploid-polyploid complex. Oecologia 73: 251–256.

Wetherington, J. D., 1989. Genotypic and environmental components of variation in growth and reproduction of fish hemiclones (Poeciliopsis: Poeciliidae). Evolution 43: 635–645.

Wolf, H. G., 1982. A comparison of different electrophoretic techniques for the detection of isoenzymes in single Daphnids. Arch. Hydrobiol. 95: 521–531.

Yampolsky, L. Yu., B. A. Kalabushkin. Factors of fitness traits variation in Daphnia magna. Genetika, in press. (In Russian).

Zouros, E. & D. W. Foltz, 1987. The use of allelic isozyme variation for the study of heterosis. Isozymes 13: 1–59.

Hydrobiologia **225**: 263–268, 1991.
V. Kořínek & D. G. Frey (eds), Biology of Cladocera.

Differences in demographic parameters among electrophoretic clones of *Daphnia obtusa* Kurz (Crustacea: Cladocera)

A. Bachiorri, V. Rossi & P. Menozzi
Institute of Ecology, University of Parma, 43100 Parma, Italy

Key words: Daphnia obtusa, life table, electrophoresis, clonal differences

Abstract

In an 18-month electrophoretic study, a *Daphnia obtusa* population that recently colonized Lake Orta (a large, severely polluted lake in Northern Italy) showed no genetic variation.

Populations from other localities were sampled. From clones established by single females, genotype differences were determined by two life table experiments (using Lake Orta and Lake Maggiore water as a culture medium).

As expected, in the Lake Orta water the colonizing clone (LO) had longer survival, produced more newborns and had a larger rate of increase than the other clones. In the Lake Maggiore water the LO clone still fared better than the other clones.

Introduction

Most parthenogenetic species show at least some genotypic diversity (Suomalainen *et al.*, 1976). In *Daphnia pulex* the large number of coexisting clones observed prompted Hebert and Crease (1980) to think of another 'planktonic paradox'. Determining genotypic differences through the relatively simple and inexpensive electrophoretic technique has made this kind of study easier.

Differences in ecological requirements among clones of the same species observed in nature have been confirmed by laboratory experiments: in thermal tolerance for *Daphnia magna* (Carvalho, 1987) and *Heterocypris incongruens* (Crustacea, Ostracoda) (Rossi & Menozzi, 1990), in salinity (Weider & Hebert, 1987), and in low oxygen concentration tolerance for *D. pulex* (Weider & Lampert, 1985). Differences in demographic parameters among electrophoretic clones have been analyzed by Loaring and Hebert (1981) in *D. pulex*. Association between genotype and resistance to pollution has been described in *Mytilus edulis* (Beardmore, 1980), in *Tisbe holothuriae* (Copepoda) and *Mytilus galloprovincialis* (Battaglia *et al.*, 1980).

Daphnia obtusa is a medium-size cladoceran species (0.7–2.2 mm) that commonly lives in temporary ponds (Margaritora, 1985). In 138 ponds in Yorkshire (UK) *D. obtusa* was rarely found in association with other members of the genus (Fryer, 1985). In the acidic water of the surveyed Uplands, *D. obtusa* was the only Cladoceran found. This was taken as evidence of the species' ability to colonize acidic waters (Fryer, 1985).

In an 18-month study (Bachiorri *et al.*, in press) a *Daphnia obtusa* population that recently recolonized Lake Orta (a large northern Italian lake in which severe pollution had almost completely destroyed any life forms for decades) showed constant gene frequencies in time and space. In the same study, small *D. obtusa* samples from different localities in Northern Italy revealed some genetic variability.

Estimating demographic parameters allows a quantitative comparison of the population strategies used by different clones to adapt to different environmental situations. The analysis of life tables has been used in *D. obtusa* to compare different energetic efficiencies (Manca & de Bernardi, 1987). *D. obtusa* was the material used in a classic population dynamics study (Slobodkin, 1954).

Statistical uncertainties about growth rate estimates are often ignored, due to the lack of exact algebraic expressions for estimating the variance of 'r' (Meyer *et al.*, 1986; Lensky & Service, 1982). This precludes the possibility of inferring differences among populations in the estimated parameters. Computer intensive statistical procedures have been proposed to overcome this difficulty (Meyer *et al.*, 1986), but they are rarely used by biologists.

In the present paper we report two life table experiments carried out using the Lake Orta clonal population and 4 clones from other localities. In the first, the clones were reared in Lake Orta water to confirm the genetic nature of the otherwise expected differences in tolerance to the polluted water. The second was carried out in Lake Maggiore water, used as a control condition. A Jackknife procedure was used to evaluate differences in demographic parameters among the 5 clones (Meyer *et al.*, 1986).

Materials and methods

D. obtusa laboratory clones, used in the experiments, were obtained from single females randomly selected from field samples collected in temporary ponds in Northern Italy (clones LB, MO, and CA), from Lake Orta (clone LO), and from a laboratory strain (clone P) (Fig. 1). Clones were maintained, in the laboratory, in filtered and aerated Lake Maggiore surficial water, at 20 ± 1 °C, under a 12 h-dark/12 h-light period, and fed a suspension of *Scenedesmus obliquus*, for up to 12 generations.

On each clone starch gel electrophoresis was performed, and 27 enzyme loci were analyzed

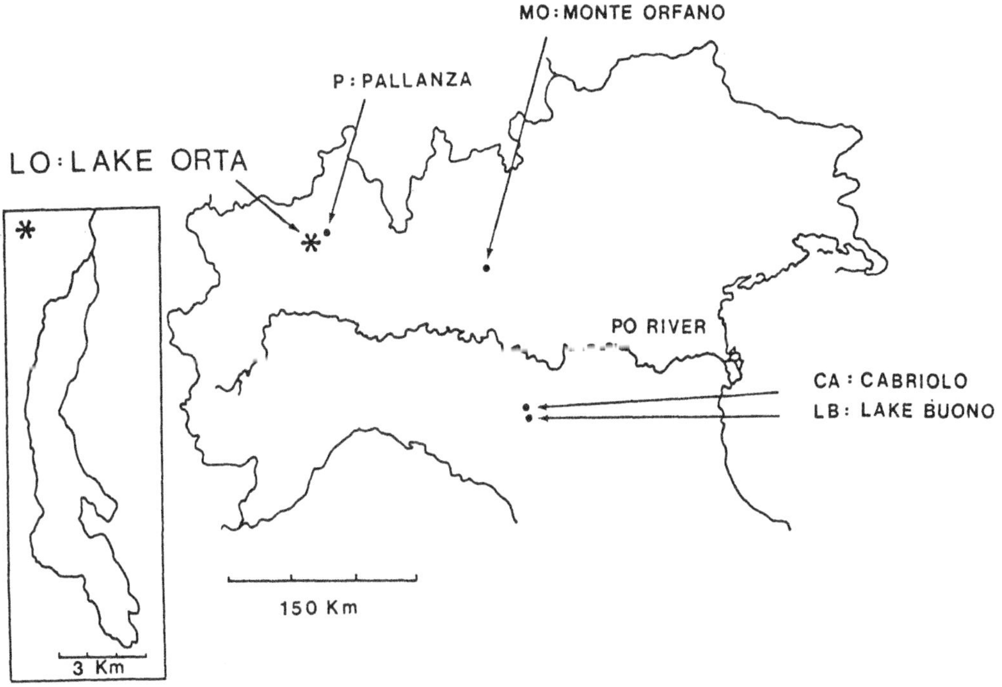

Fig. 1. Sample locations for the *D. obtusa* populations are indicated.

(Wolf, 1982; Shaw & Prasad, 1970). Electrophoretic analysis was repeated on each clone in successive generations to test the stability of electrophoretic morphs. A full report of the electrophoretic analyses will appear elsewhere (Bachiorri *et al.*, in press).

To evaluate the demographic parameters of electrophoretically distinguishable clones of *D. obtusa*, two life-table experiments were carried out under controlled laboratory conditions. For each experiment 15 newborns (12 ± 12 h old) from each of the LO, CA, MO, LB, and P laboratory clones were isolated and reared in separate 100 mL glass containers.

In experiment 1, newborns were reared in Lake Orta water. In experiment 2 the newborns were maintained in Lake Maggiore water. Both experiments were carried out at 20 ± 1 °C, under a 12 h-dark/12 h-light period at a constant food level of 40000 cell/cc. The culture medium was changed three times a week. Death events of the starting individuals were recorded daily, their neonates were counted and removed every day.

The experimental data were analyzed considering survivorship, fecundity, and growth rate. To evaluate the differences in survivorship, non parametric estimates of the survival distribution of different clones were compared (Lifetest, SAS Statistical Analysis System, 1985). We report the χ^2 statistics for the Wilcoxon test. Differences in mean life span and net reproductive rate (R_0) were evaluated by analysis of variance and Tukey pairwise comparison test. Population growth rates were calculated for each clone in the two experiments according to the equation:

$$1 = \sum_{x=0}^{N} e^{-rx} l_x m_x$$

where r = per capita rate of increase for the clone, x = age class (days; 0, 1, 2,..N), N = oldest age class in the population, l_x = probability of surviving to age x, and m_x = fecundity at age x. Iterative calculations are necessary to estimate 'r'. For the computation reported in this paper we used a program (Wildlife program, CLUP, 1989) run on a personal computer. For the statistical comparison of 'r' estimates, a measure of their variance is needed. We used the Jackknife procedure suggested by Meyer *et al.* (1986). For each clone for each experiment, 'r_{total}', the estimate for the complete data set, and 15 estimates of 'r' (r_e), were obtained omitting one replicate at a time. Fifteen Jackknife pseudovalues (r_p) were then computed as follows:

$$r_p = (15\ r_{total}) - (14\ r_e)$$

Mean, standard deviation, and mean square error were then obtained for the Jackknife pseudovalues using standard calculations.

Results and discussion

The analysis of life table experiment data showed differences in demographic parameters among clones in both experiments.

Not surprisingly, in the experiment carried out in Lake Orta water, all the demographic parameters of the Lake Orta clone were higher than for the other clones, for which Lake Orta water seemed to have a toxic effect.

Survival curves (Fig. 2) were significantly different among clones ($\chi^2 = 66.96$, $df = 4$, $p = 0.0001$); the time to 50% survival of LO clone females (36 days) was clearly longer than for the other clones (CA = 13, MO = 8, P = 10 and LB = 3 days). As expected, the mean life span of LO clone (Table 1) was also significantly higher. No differences in life span existed among the other clones.

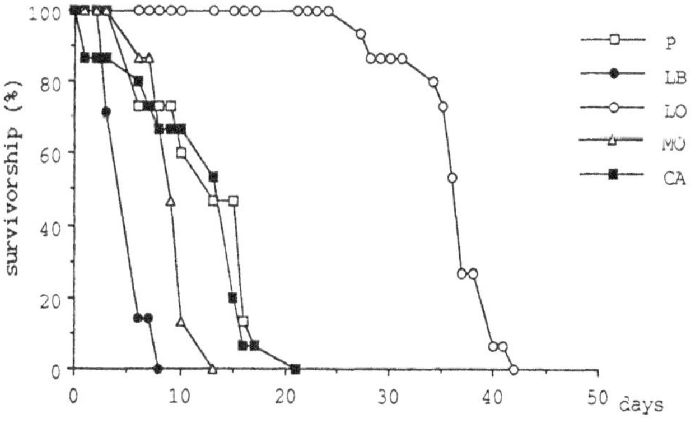

Fig. 2. Survivorship curves for the clones in experiment 1 conditions.

Table 1. Mean and standard deviation of life span (days). ANOVA, *F* statistics and *p* are reported. Letters indicate different Tukey pairwise comparison test groupings.

Clone	Experiment 1			Experiment 2		
	Mean	SD	Grouping	Mean	SD	Grouping
CA	10.73	4.77	a	16.73	3.67	a
MO	7.73	2.15	a	25.70	8.00	a b
P	10.55	5.37	a	23.67	11.21	a b
LB	3.29	1.64	b	25.60	6.51	b
LO	34.73	4.48	c	28.20	10.46	b
	$F = 140.22$ $p = 0.0001$			$F = 3.92$ $p = 0.0063$		

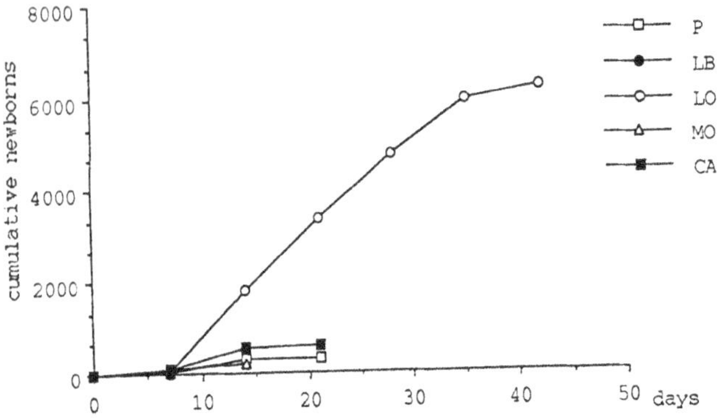

Fig. 3. Cumulative offspring production for the clones in experiment 1 conditions.

Clone LO showed high fecundity, while LB, MO, P, and CA females produced only a few individuals (Fig. 3 and Table 2). This seems to be further evidence of the toxic effect of Lake Orta water on them. In Table 3 'R_0' values, the mean number of descendants per clone, have been compared. 'R_0' value for LO clone (413.80) was by far the highest. Differences existed among the other clones that formed 2 groups, with CA and LB clones showing the highest and lowest values.

Table 2. Mean and standard deviation of neonates per daphnia per day.

Clone	N	Experiment 1		Experiment 2	
		Mean	SD	Mean	SD
CA	15	2.717	1.953	7.494	1.963
MO	15	1.638	0.975	5.439	1.992
P	15	1.454	1.572	6.891	2.022
LO	15	0.009	0.036	6.813	1.066
LO	15	11.732	1.048	9.120	1.378

The use of Jackknife allowed the estimation of a standard deviation for instantaneous growth rates (*r*). The stress produced on the organisms reared in Lake Orta water is reflected in the variation of the estimates: high mortality and low fecundity increased inter-replicate variability.

Table 3. Mean and standard deviation of reproductive rates (R_0). ANOVA, *F* statistics and *p* are reported. Letters indicate different Tukey pairwise comparison test groupings.

Clone	Experiment 1			Experiment 2		
	Mean	SD	Grouping	Mean	SD	Grouping
CA	38.87	32.24	a	131.00	40.63	a
MO	14.60	9.02	a b	151.20	60.71	a
P	21.20	23.77	a b	171.87	80.02	a
LB	0.07	0.27	b	176.00	47.17	a
LO	413.80	54.42	c	265.93	108.45	b
	$F = 497.25$ $p = 0.0001$			$F = 7.75$ $p = 0.0001$		

Table 4. Mean and standard deviation of per capita rate of increase (r). ANOVA, F statistics and p are reported. Letters indicate different Tukey pairwise comparison test groupings.

Clone	Experiment 1			Experiment 2		
	Mean	SD	Grouping	Mean	SD	Grouping
CA	0.48	0.16	a	0.55	0.09	a b
MO	0.34	0.08	b	0.49	0.04	b c
P	0.31	0.11	b	0.49	0.04	c
LB	- 0.51	0.00	c	0.51	0.02	a b c
LO	0.47	0.04	a	0.55	0.02	a
	$F = 245.01$ $p = 0.0001$			$F = 4.65$ $p = 0.0022$		

Instantaneous growth rates were significantly different among clones (Table 4): clone LO shows one of the highest values, although not statistically different from the CA clone. The apparent contradiction with the picture that emerges from the other parameters is due to the differences in life span among clones, which shortens the estimated generation time of the CA clone (R_0 is related to r ($R_0 = e^{rT}$)).

In experiment 2, Lake Maggiore water was used to study the clones' demographic parameters in non-stress-producing conditions. The average values of the parameters were comparable with results of life table studies from the literature (de Bernardi & Manca, 1982). Clones still showed differences although not as drastic as when reared in Lake Orta water. Clone LO still fared better than the other clones.

Survival curves (Fig. 4) were different among clones ($\chi^2 = 26.53$, $df = 4$, $p = 0.0001$). Differences in time to 50% survival were not as striking as in the first experiment: the highest value was for LB females (LB = 30, CA = 17, MO = 27, P = 22, and LO = 23 days). Life span was significantly different among clones (Table 1). LO clone had the highest value, but it was statistically different only from clone CA.

Clone LO is statistically different from all the others in net fecundity rate (R_0) (Table 3) although not producing as many newborns as in experiment 1 (Fig. 5 and Table 2). The other clones show R_0 values in line with those reported in the literature (de Bernardi & Manca, 1982).

Instantaneous growth rates (r) were significantly different (Table 4); LO clone showed the highest 'r' value although not different in comparison with LB and CA clones. The standard deviations, estimated by Jackknife, are much

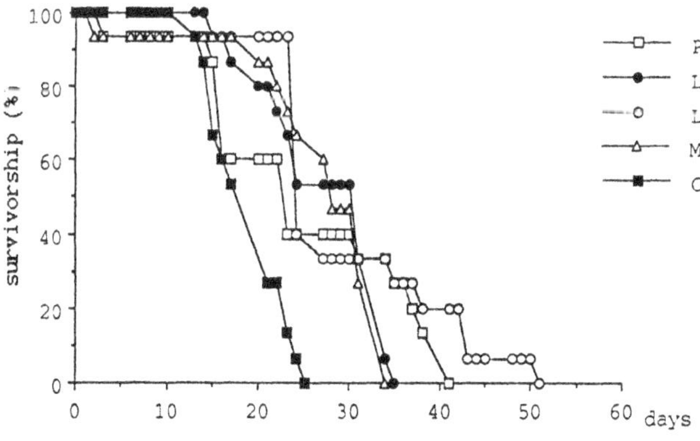

Fig. 4. Survivorship curves for the clones in experiment 2 conditions.

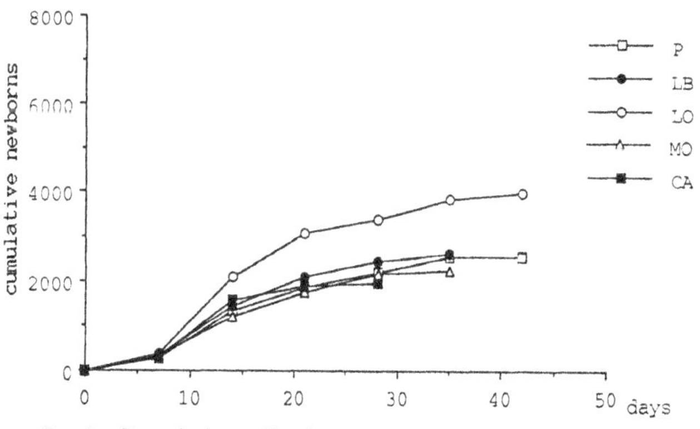

Fig. 5. Cumulative offspring production for the clones in experiment 2 conditions.

smaller and more homogeneous than in experiment 1. Averages are significantly higher than in experiment 1 for all clones, LO included.

Differences in demographic parameters calculated from life table data gave, in terms of population dynamics, a quantitative estimate of the advantage that allowed the Lake Orta clone to colonize an environment toxic for other populations of the same species. The data here presented also seem interesting for future comparison studies: the liming experiment that is currently being carried out in Lake Orta will offer a unique opportunity to follow possible changes in the genetic and demographic behaviour of this *Daphnia obtusa* population.

In the second experiment the differences in demographic parameters showed by the clones reared in non-stressful conditions might help in explaining phenomena such as the coexistence of different clones of the same species (Hebert & Crease, 1980) or the adaptability of a species to a wide range of biogeographical and environmental conditions (Fryer, 1985). Further studies would be needed to evaluate the possible effect of adaptations to low food levels in explaining the differences observed in demographic parameters in the experiment carried out in lake Maggiore water (Gliwicz & Lampert, 1990; Gorbi, pers. comm.).

Acknowledgements

We would like to thank S. Leonardi and G.M. Paris for their help in data analysis. M. Manca and C. Bonacina kindly supplied us with *D. obtusa* laboratory strain and Lake Orta samples. This research was supported by a grant from the Italian Ministero Pubblica Istruzione. A. Bachiorri and V. Rossi were students in the Ph. D. program 'Dottorato in Scienza dell'Ambiente' of the University of Parma while this research was carried out.

References

Bachiorri, A., V. Rossi, C. Bonacina, C. & P. Menozzi, 1991. Enzymatic variability of a colonizing population of *Daphnia obtusa* Kurz (Crustacea, Cladocera) in Lake Orta (Italy). Verh. Int. Ver. Limnol., in press.

Battaglia, B., P. M. Bisol & E. Rodinò, 1980. Experimental studies on some genetic effects of marine pollution. Helg. Meer 33: 587–595.

Beardmore, J. A., 1980. Genetical considerations in monitoring effects of pollution. Rapp. P.-v. Reun. Cons. int. Explor. Mer. 179: 258–266.

Carvalho, G. R., 1987. The clonal ecology of *Daphnia magna* (Crustacea – Cladocera). II Thermal differentiation among seasonal clones. J. anim. Ecol. 56: 469–478.

Clup, 1989. Software territoriale ed ambientale. Politecnico di Milano, Milano (Italy).

de Bernardi, R. & M. Manca, 1982. The consequences of life history strategies on competition between two cladocerans. Mem. Ist. ital. Idrobiol. 4: 145–161.

Fryer, G., 1985. The ecology and distribution of the genus *Daphnia* (Crustacea: Cladocera) in restricted areas: the pattern in Yorkshire. J. nat. Hist. 19: 97–128.

Gliwicz, Z. M. & W. Lampert, 1990. Food thresholds in *Daphnia* species in the absence and presence of blue-green filaments. Ecology 71: 691–702.

Hebert, P. D. N. & T. J. Crease, 1980. Clonal coexistence in *Daphnia pulex*: another planktonic paradox. Science 27: 1363–1365.

Lensky, R. E. & P. M. Service, 1982. The statistical analysis of population growth rates calculated from schedules of survivorship and fecundity. Ecology 63: 655–662.

Loaring, J. M. & P. D. N. Hebert, 1981. Ecological differences among clones of *Daphnia pulex* Leydig. Oecologia 51: 162–168.

Manca, M. & R. de Bernardi, 1987. Feeding and energy budget estimations in *Daphnia obtusa*. Hydrobiologia 145: 269–274.

Margaritora, F. G., 1985. A cura di. Fauna d'Italia 'Cladocera'. Ed. Calderini, Bologna.

Meyer, J. S., C. G. Ingersoll, L. L. McDonald & M. S. Boyce, 1986. Estimating uncertainty in population growth rates: Jackknife vs. Bootstrap techniques. Ecology 67: 1156–1166.

Rossi, V. & P. Menozzi, 1990. The clonal ecology of *Heterocypris incongruens* (Crustacea: Ostracoda). Oikos 57: 388–398.

SAS Institute Inc., 1985. SAS user' guide: statistics version 5° Edition. Cary, NC: SAS Institute Inc. North Carolina, USA.

Shaw, C. R. & R. Prasad, 1970. Starch gel electrophoresis of enzymes: a compilation of recipies. Bioch. Gen. 4: 297–320.

Slobodkin, L. B., 1954. Population dynamics in *Daphnia obtusa* Kurz. Ecological Monographs 24: 69–88.

Suomalainen, E., A. Saura & J. Lokki, 1976. Evolution of parthenogenetic insects. Evol. Biol. 9: 209–257.

Weider, L. J. & P. D. N. Hebert, 1987. Ecological and physiological differentiation among low arctic clones of *Daphnia pulex*. Ecology 68: 188–198.

Weider, L. J. & W. Lampert, 1985. Differential response of *Daphnia* genotypes to oxygen stress: respiration rates, hemoglobin content and low-oxygen tolerance. Oecologia 65: 487–491.

Wolf, H. G., 1982. A comparison of different electrophoretic techniques for detection of isoenzymes in single daphnids. Arch. Hydrobiol. 95: 521–531.

Hydrobiologia **225**: 269–280, 1991.
V. Kořínek & D. G. Frey (eds), Biology of Cladocera.

Photoperiod and temperature as inducers of gamogenesis in a dicyclic population of *Scapholeberis armata* Herrick (Crustacea: Cladocera: Daphniidae)

Dorothy B. Berner, Luat Nguyen, Steven Nguy & Shirley Burton
Department of Biology, Temple University, Philadelphia, PA 19122, USA

Key words: Cladocera, clones, Daphniidae, gamogenesis, parthenogenesis, Pennsylvania, photoperiod, temperature, *Scapholeberis*

Abstract

The induction of gamogenesis in a dicyclic population of *Scapholeberis armata* from a small pond at 40° N in Pennsylvania, USA, was studied. Animals were collected at the time of gamogenesis in mid-October (fall) 1988 and early June (spring) 1989). 10 clones were established from isolated females for each season and maintained at low density under standard conditions of 20 °C and 13.5L/10.5D, with *Chlamydomonas reinhardtii* diluted with pond water to $3 \cdot 10^5$ cells ml^{-1} as a food source. For each clone, mothers of pairs of test animals were acclimated to 4 sets of conditions: 20 °C/15.5L (spring), 15 °C/11.5L (fall), 20 °C/11.5L (control), and 15 °C/15.5L. Test animals were observed bidaily or daily for up to 37 days. Occurrence of molts and of parthenogenetic broods or ephippia was recorded. At short photoperiod, ephippial females developed, usually after one or more parthenogenetic broods, among both spring and fall animals. The response was stronger (higher % sexual females) among fall clones, and was enhanced in both groups by 15 °C water. No animals responded to long photoperiod in warm or cool water. It is likely that additional factors, such as prior photoinduction, rapid depletion of food supply, and increased inter-animal contact due to high population density are required for the induction of spring gamogenesis. Differences in the responsiveness of spring and fall clones to inducing conditions suggest some differences in genotypes of spring and fall subpopulations in this pond. It is proposed that variable responses among genotypes to factors inducing gamogenesis may be the basis for the differences and clines in the timing of gamogenesis found among ecologically and latitudinally separated populations.

Introduction

The occurrence at times of males, ephippial females, and 'resting' diapausing ephippial embryos in populations of Cladocera is familiar to all those who have studied them in the field and the laboratory. This phenomenon is frequently termed gamogenesis, but it involves several dissimilar processes. Gamogenesis occurs in males that are diploid and are themselves the product of parthenogenesis. Parthenogenetic females may switch to forming ephippia and then usually are gamogenetic, i.e. one or two haploid eggs develop. If fertilized, such eggs enter the ephippium and develop into resting embryos. However, in more northern daphniids, resting eggs may be amictic and, therefore, such ephippial females can be considered parthenogenetic (Weider *et al.*, 1987).

The timing of production of resting eggs and ephippia in a population generally can be correlated with the advent of unfavorable environmental conditions, varying with latitude, local ecology, and the past distributional history of each species (see Frey, 1982 for a brief review). The interplay of all these factors has resulted in cladocerans exhibiting monocyclic, dicyclic, or acyclic patterns of reproduction (Berg, 1931). Diapause occurs during midsummer in arctic *Daphnia middendorffiana*, a monocyclic, obligate parthenogen (Stross and Kangas, 1969). Temperate populations of *Daphnia*, especially in small ponds, may be dicyclic (Carvalho & Hughes, 1983). Frey (*loc cit*) grouped chydorid species into 'southern' and 'northern' assemblages, both monocyclic with autumnal gamogenesis. He demonstrated also that in 'northern' species whose distribution extends into the south, gamogenesis is shifted into the summer months and, in the deep south, occurs in late spring.

With cladoceran populations exhibiting apparent plasticity of reproductive patterns both geographically and ecologically, it has been a challenge to discover the environmental cues inducing the switch in any particular population from parthenogenesis to production of diapausing embryos, either sexually or asexually. In laboratory experiments, the appearance of gamogenetic individuals has been correlated with critical photoperiods and temperatures primarily, but crowding, scarcity of food, and light intensity appear to be important factors in some circumstances (Banta, 1939; Stross, 1969; Shan, 1974; Doma, 1979). Laboratory success in inducing gamogenetic individuals appears to vary with the species used, the habitat from which the population was obtained, and the season at which it was collected. Most studies have been involved with cladocerans from one region, and none has really addressed the question of how populations might have shifted their responses to seasonal cues (e.g. from one photoperiod to another) as their populations spread to more northern or southern latitudes. Shan (1974) did find significant differences in ephippial induction among geographically separated populations of the chydorid *Pleuroxus*: a northern group responded to shorter photoperiods and a southern group to longer photoperiods. A third group responded weakly to all photoperiods. Frey (1982, 1986) has proposed that the northern and southern groups may actually be unrecognized species pairs that evolved during the ice ages, shifting their timing of gamogenesis so as to accommodate to current climatic conditions. He has not proposed a mechanism for such plasticity in response to changing environmental cues.

The region in eastern North America at about 40° N is ecologically fascinating, because it is a region of overlap of northern and southern cladoceran taxa as well as other animals and plants (Frey, 1986; Berner, pers. obs.). When we discovered a late spring and fall dicyclic population of *Scapholeberis armata* in a small pond near Philadelphia, PA., it seemed an excellent opportunity to investigate the question of spring and fall induction of gamogenesis in this region of overlapping taxa. *S. armata* appears to be a widely distributed species in N. America, with a morphologically distinctive subpopulation only in the extreme southeast (Dumont & Pensaert, 1983). However, the animals in this pond could represent other, hitherto unrecognized subpopulations of this taxon. This research was designed to address the following questions:

1. Are the water temperatures and photoperiods found at the pond during the late spring and fall effective inducers of gamogenesis in this population? If not, then other factors such as crowding would have to be considered as inducers.
2. Can clones from the spring and fall respond to both spring and fall conditions? If so, this would indicate that single genotypes were perhaps capable of the kind of shift in response to differing climactic conditions proposed by Frey (1986). If not, and if clones respond solely to spring or fall conditions, then it would suggest that different genotypes within the population respond to the two sets of conditions and might therefore be reproductively isolated. This would suggest a mechanism by which species pairs might have differentiated.
3. Do spring and fall clones respond only to spring or to fall conditions? If so, then other

factors would have to be considered as inducers for the season in which temperature and photoperiod are ineffective.

Materials and methods

Animals and establishment of clones

The sexually dicyclic population of *S. armata* was found in a small (*ca.* 0.07 hectare), shallow (1 m), partially shaded, spring-fed pond on Smoke House Farm, north of Downingtown, Chester Co. PA (NW corner of Creek and Reeds Rds., 40° 3′ N, 75° 44′ W). Small, dense swarms of parthenogenetic and gamogenetic *S. armata* were first observed on 5 and 11 June 1988 in a shallow edge of the littoral zone partially protected by *Elodea* and other vegetation from fry of largemouth bass (*Micropterus salmoides*). Littoral water temperatures were 17–20 °C. During midsummer 1988, the population virtually disappeared, reappearing in early September. Although not as dense as in June, it was again strongly gamogenetic on October 15, when the water was 15 °C. pH of the water ranged from 6.9 to 7.3. Animals for experimentation were collected with a plankton net on a pole on 15 October 1988 (fall) and 6 June 1989 (spring), when gamogenesis was again occurring as in the previous year. They were transported to the laboratory in plastic bags, where single, parthenogenetic females were isolated to establish clones for experimentation. Fall clones were maintained in the laboratory for three months, and spring clones for one month, before experiments were initiated.

Culture techniques

Carboys of water for culturing *S. armata* were also collected from the pond in the spring and fall. Batches of water were filtered through Whatman 41 paper, autoclaved, and aerated before use. Vitamins were added to water in which animal were to be grown: biotin, thiamine, and B_{12} (from the stock used for algal growth medium) to give final concentrations of 0.75, 75, and 0.75 ppb, respectively. B_{12}, especially, is necessary for optimal cladoceran reproduction (Keating, 1985).

S. armata filter feeds primarily when it is attached to a substrate or, frequently, under the water's surface film. Therefore, we cultured them in broad, shallow dishes and chose *Chlamydomonas reinhardtii* Dangeard, which stays actively in suspension, as their food. Fall animals were fed *C. reinhardtii* (Carolina Biological Supply Co.) cultured with aeration in a previously autoclaved, defined medium (Keating & Dagbusan, 1984) and harvested at 4–5 days while still in the log phase of growth. 15–30 ml of algae were centrifuged at 1,000 rpm in a Model HN-S International centrifuge, resuspended in pond water, washed by recentrifugation, and resuspended to their original volume before being diluted to feed the animals. During experiments with fall animals, the algal stocks became contaminated with bacteria and were replaced with *C. reinhardtii*, strain 137 C^+, cultured in a modified Sager-Granik medium (Gealt *et al.*, 1981) with additional micronutrients (S. Dowidar & T. Punnett, pers. comm.). This alga has an NH_3 requirement, necessitating a special medium. It was later discovered that the micronutrient addition had relatively high concentrations of selenium and zinc, which possibly were toxic to and shortened the lives of the fall test animals. Spring animals were fed 137 C^+ *C. reinhardtii* grown in modified Sager-Granik medium supplemented with micronutrients from the Keating-Dagbusan formula and phosphate buffer to bring the pH to *ca.* 6.9, optimal for that alga (see Appendix I for the formula). These algae grew rapidly and were harvested 1–3 days after inoculation.

The best algal concentration to feed *C. reinhardtii* was determined in preliminary studies. Isolated *S. armata* neonates from a single fall clone were grown in vitamin-enriched pond water containing *C. reinhardtii* cultured in Keating-Dagbusan medium and diluted to concentrations of $7 \cdot 10^5$; 2, 3, 4, and $5 \cdot 10^5$; and $1.3 \cdot 10^6$ cells per ml (Table 1). From this study, a concentration of $3 \cdot 10^5$ cells ml^{-1} (equivalent to an optical density of 0.32 at 425 nm in a Bausch and Lomb spectro-

Table 1. Effects of *C. reinhardtii* concentration on *S. armata* growth and reproduction at 20C.

Concentr. cells/ml	Lifespan (days)	Total # Instars	Maximum #/brood	Total # Offspring
$7 \cdot 10^4$	16	10	7	35
$2 \cdot 10^5$	18	11	12	35
$3 \cdot 10^5$	20	12	14	93
$4 \cdot 10^5$	20	12	14	87
$5 \cdot 10^5$	18	11	14	69
$1 \cdot 10^6$	18	11	11	69

photometer 25) was chosen for the algal dilution used in this study. Similar algal concentrations are optimal for *Daphnia magna* (Porter *et al.*, 1983) and for *Ceriodaphnia* (D. Berner & J. McGeorge, unpub. data), and they have been used in other studies of cladoceran gamogenesis (Stross & Hill, 1968).

Stock clones were maintained in plastic petri dishes 35 x 10 mm, 60 x 15 mm, or in small finger bowls. If stocks were allowed to become crowded for a week or more, males and ephippial females usually appeared. Therefore, to prevent induction of gamogenesis, stock density was kept to about 2 animals ml^{-1}. Stocks in smaller dishes were subcultured into new dishes with fresh algal dilution every other day; ones in larger vessels were transferred weekly and fed a few drops of harvested algae when needed in the interim. Experimental animals were kept in 35 x 10 mm dishes and transferred every two days into ones with fresh algal dilution.

Light and temperature conditions

These conditions were based on those experienced by the animals in their natural habitat. Daylength at 40° N on June 6 was determined to be 15.5 hr and on October 15, 11.5 hr (Hoffman, 1986). In midsummer it is 13.5 hr. Water temperatures in the Smoke House Farm pond are up to 20 °C in late spring and summer and drop to 15 °C in mid-October. Therefore, 20 °C and 13.5L/10.5D were chosen as standard, non-inducing 'summer' conditions for stock clones and the beginning of experiments. There were four pairs of test conditions. Experimental 'spring' conditions were set at 20 °C and 15.5L/8.5D; 'fall' conditions were 15 °C and 11.5L/12.5D. 'Control' conditions were 20 °C with 11.5L/12.5D and 15 °C with 15.5L/8.5D. For simplicity, experimental conditions were named Warm,Long (WL); Cool,Short (CS); Warm,Short (WS); and Cool,Long (CL), respectively.

Stock clones were kept on a laboratory bench at 20 ± 2 °C. 'Warm' animals were placed in an incubator divided into 2 light-tight compartments at 20 ± 1 °C. 'Cool' animals were maintained in a constant temperature room, one group on a laboratory bench at 15 ± 1 °C and the other at 15.5 ± 1 °C in a light-tight box ventilated by two, small fans. All animals were maintained under *ca.* 1800 lumens of cool, white flourescent light at appropriate photoperiods. This intensity is far below those experienced by the animals at midday, spring and fall, even under tree shade. Each group of experimental animals was arranged in a circle under circular, 22 watt flourescent tubes so that they were evenly illuminated.

Experimental design

Isolated animals supplied with abundant food (1 animal/4 ml of diluted algae) were used to avoid possible induction of gamogenesis by inter-animal contact or food deficit (Carvalho & Hughes, 1983). Three generations descended from low density stocks of each clone were studied. Initially, one neonate from each clone was selected and isolated as a Grandmother. These were kept under the standard conditions until it was certain that they were parthenogenetic females. Four siblings from their first to forth (usually second or third) brood were isolated as Mothers, one for each of the test conditions. One mother from each clone was put into each test chamber, all initially at standard conditions. Because of asynchrony of reproduction in the grandmothers, it took two days to accumulate the mothers. Once all clones were represented, the temperature in the constant temperature room was reduced over 3 days to 15 °C. At this point, all photoperiods were reset so that each temperature regimen had one set of mothers at 11.5 hr and one at 15.5 hr of light. This shift coincided with the beginning development of the mothers' ovaries, so that all their offspring developed under the test conditions. It has been shown that the prenatal period in *Daphnia* is the most sensitive to photoperiod induction (Stross & Hill, 1968). Two sibling offspring (called 'Tests' rather than 'Daughters', because a few were males) were isolated from each mother's first to third brood and observed throughout their lifespan or until they had lived for over 30 days. This experimental design is summarized in Fig. 1. Ten clones each for fall and spring *S. armata* were used for this study. Theoretically there were twenty test animals in each of the four test conditions, but because of some deaths, most groups had a few less. Fall experimental animals were observed every other day and spring animals daily. Recorded for each female were: age, state of ovaries (with or without developing eggs), state of brood chamber (empty or with early eggs, partially developed, or eyespot embryos, and their number), if she were pre-ephippial or had a mature ephippium, if a molt or molt and shed ephippial case were present, and the number of neonates that had hatched. In spring test animals, ephippial females and ephippia were examined to see if a diapausing embryo was present. Only molts were recorded for the few males that occurred. Also recorded were pertinent facts about an animal's appearance and activity.

Results

Culture techniques

The response of six fall *S. armata* females to 5 algal dilutions of *C. reinhardtii* grown in Keating-Dagbusan defined medium (*op cit.*) are shown in Table 1. This medium has a pH of *ca.* 8, above the optimum for the alga, which grows rather slowly in it. In the modified Sager-Granick medium, with a pH of *ca.* 7, the alga grew more rapidly. Spring animals fed these algae survived somewhat longer than the above group; some spring test animals were terminated at ages of 35–37 days and 13 instars. By then, some had had broods of 15 or 16 young and a total of 85 offspring. This latter medium appears to be comparable, or slightly superior, to the Keating medium for raising algae for a cladoceran food source. We observed that the animals grown on *C. reinhardtii* did not accumulate large amounts of lipid droplets and starved rapidly if food became scarce. Perhaps the use of

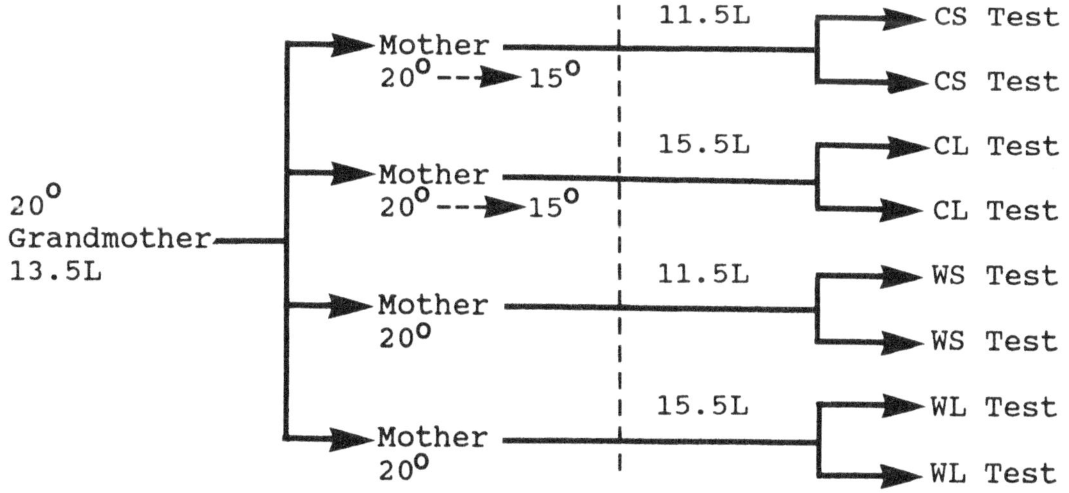

Fig. 1. Scheme for the experimental design to test 3rd generation animals of one clone for the induction of gamogenesis. The vertical dotted line indicates a shift in photoperiod.

a different alga (Cowgill *et al.*, 1986) or the addition of a diatom (Keating & Dagbusan, 1986) would promote lipid storage in *S. armata* that would help them resist starvation (Tessier *et al.*, 1983). We also found that some plastic petri dishes treated for cell culture carry a charge that causes algal cells to adhere to the surface. Animals in such dishes also tend to starve rapidly.

Induction and patterns of gamogenesis

The numbers of test animals and clones () studied under the four sets of conditions were: fall WL: 20(10), WS: 16(10), CL: 18(10), CS: 16(10); spring WL: 13(8), WS: 18(10), CL: 17(10), CS: 18(10). No ehphippial females developed among spring and fall clones under the long day

Table 2. Patterns of gamogenesis in single fall *S. armata*. (*f* = bearing young; *sf* = empty brood chamber; *E* = ephippial; *pE* = pre-ephippial).

a. Cold water/short day

	Adult instar										
Clone #	1	2	3	4	5	6	7	8	9	10	11
21	f	f	E								
5	f	f	E	pE	E						
21	f	f	E	pE	E	pE					
9	f	E	E	pE	E						
24	f	E	pE?	f	f						
14	f	E	pE	E	pE	E					
8	f	E	pE	E	f						
8	f	E	pE	E	f						
6	f	E	pE	E	pE?	sf	f				
4, 4, 5, 6, 20, 21, 12	f	E	pE	E	pE						

Total *n* tested: 16 (10 clones).

b. Warm water/short day

	Adult instar										
Clone #	1	2	3	4	5	6	7	8	9	10	11
21	f	f	f	E	sf	f	E	f	f		
24	f	f	E	pE							
9	f	f	E	pE	E						
6	f	f	E	sf	f						
12	f	f	E	E	pE						
12	f	E	sf	f	f						
5.	f	E	pE	E	f						
5	f	E	pE	E	pE						
14	f	E	pE	E	E	pE					
14	f	E	pE	E	E	sf	f	f	f		
4	E	pE	E	sf	f						

Total *n* tested: 16 (10 clones). 5 from clones 4, 8, 20, 21 and 24 remained parthenogenetic.

(15.5L/8.5D) photoperiod in either warm or cool water. The short photoperiod (11.5L/12.5D) induced different numbers of ephippial females in both spring and fall clones; larger numbers were induced at 15 °C than at 20 °C (Tables 2 and 3). Fall animals were more responsive to the inducing stimuli. 100% of the 16 CS females (all 10 clones) and 11 of 16 (69%) of the WS females, representing 8 of the 10 clones, developed ephippia. In contrast, 71% of the 18 spring CS animals, 6 of the 10 clones tested, became ephippial. Among the non-ephippial animals was 1 male. Among the spring WS animals, only 5 of 17 females (29%) (3 of 10 clones) became ephippial, and there was also 1 male in this group.

Generally, induced females became ephippial after one or two, less often three, parthenogenetic instars. In rare cases (clone 4, Table 2b; clone 142, Table 3a), the first instar was gamogenetic, and one spring CS female did not switch from parthenogenesis until her 10 th instar (clone 136, Table 3a). Often, a female did not remain gamogenetic for more than a few instars; this was true especially of the spring animals and of ones in warm water. Usually, when a female switched from parthenogenesis, she developed and shed a mature ephippial case in the following instar and one of her ovaries appeared to contain a potential diapausing egg during that instar. Among the spring test animals whose ephippia

Table 3. Patterns of gamogenesis in single spring *S. armata*. (*f* = bearing young; *sf* = empty brood chamber; *E* = ephippial; *pE* = pre-ephippial; * = ephippium containing an embryo.)

a. Cold water/short day

	Adult instar										
Clone #	1	2	3	4	5	6	7	8	9	10	11
136	f	f	f	f	f	f	f	f	f	f	f
103	f	f	f	E	E						
136	f	f	E	pE	E	pE	E	f			
106	f	f	E	pE	E*	pE	E	pE			
134	f	f	E	pE	E	pE	E				
103	f	E	pE	E	f	f	f				
134	f	E	pE	E	pE	E	sf	f	f		
112	f	E	pE	E							
112	pE	E	E	pE							
106	f	E	pE	E*	pE	E	pE				
142	f	E	pE	E	pE	E	pE	E	pE	E	
142	E	pE	pE	E	pE	E	pE	E	pE		

Total *n* tested: 18 (10 clones). 5 from clones 110 (2), 111 (2), and 133 remained pathenogenetic. 1 from clone 138 was a male.

b. Warm water/short day

	Adult instar										
Clone #	1	2	3	4	5	6	7	8	9	10	11
106	f	f	E	f	f	f	f	f	f	f	
106	f	f	E	f	f	f	f	f	f	f	f
112	f	f	E	pE	E	sf	f	f	f	f	f
112	f	f	E	E	sf	f	E*	E	E		
136	f	E	E	f	f	f	f	f	f		

Total *n* tested: 18 (10 clones). 12 from clones 103 (2), 110 (2), 111 (2), 133 (2), 134, 136, and 142 (2) remained parthenogenetic. 1 from clone 138 was a male.

were examined closely, virtually all the ephippia developed and were shed empty. If a female remained gamogenetic, a sterile, pre-ephippial instar with no egg in the brood chamber was interposed between successive ephippial instars. During such instars, the carapace has the lateral profile of an ephippial female (Dumont & Pensaert, 1983, Fig. XIX, 8), but the developing ephippium is only faintly visible, its lateral ridges are low and not prominent, and the dorsolateral carapace surface is faintly dimpled. The egg remaining in the ovary from the previous instar appears to be resorbing, while a new one develops in the opposite ovary. We did not investigate if this pre-ephippial pattern is found in females that had mated in our stock cultures. In one observed mating, however, we observed that the egg moved from the ovary into the brood chamber within five minutes of separation of the male and female. Among our test females, two clone 106, CS females (Table 3a) had eggs that moved into the ephippial chamber spontaneously. This occurred also in one WS female from clone 112 (Table 3b). We did not attempt to hatch these embryos to see if they were viable.

No males were observed among fall test animals. Among spring animals, two males (CS and CL, not siblings) from clone 138 occurred. We observed that this particular stock clone produced a large number of males. A pair of male siblings from clone 142 also occurred among the spring WL test animals. These were from the mother's first brood, whereas the clone 138 animals were from second broods.

Other life history patterns

Most of the isolated females had 3, sometimes 4, juvenile instars. During the last one, the ovaries developed eggs that then moved into the brood chamber following the next molt. In spring test animals, for which more detailed data were available, the mean number of days ($\pm$ S.D.) to hatching of the first brood varied among the 4 temperature-light conditions: CS 15 ± 2.2 ($n = 17$); CL 13.6 ± 1.9 ($n = 16$); WS 10 ± 1.2 ($n = 18$); WL 8.7 ± 1.4 ($n = 14$). Initial brood sizes ranged from 1 to 5 eggs, most often 2 or 4. Maximum brood size also varied among the conditions (ranges in (), n as above): CS 3.9 ± 1.6 (2–7); CL 5.8 ± 1.6 (3–8); WS 7.7 ± 3.2 (4–16); WL 8.6 ± 2.9 (5–18). However, in spite of differences in maturation rates and brood sizes, the mean days per instar for long-lived females were approximately the same for animals at both photoperiods at each temperature: 3.4 days for cool animals and 2.4 days for warm animals. The four males we observed lived for 20–24 days, encompassing 9–12 instars. Data were too scanty to evaluate the effect of temperature on instar length, which ranged from 1.6 to 2.7 days.

Discussion

These experiments demonstrate clearly that, for this population of *S. armata*, short photoperiod effectively induced gamogenesis in most of the fall clones and some of the spring clones. Even more clones and individuals became gamogenetic when short photoperiod was coupled with cool water. The interaction of temperature and photoperiod has been demonstrated previously in other Daphniidae. Stross & Hill (1968) found a similar enhancement of inductive photoperiod by cooler water in monocyclic *Daphnia pulex*, and they demonstrated further that the % diapausing broods in their cultures increased with decreasing photoperiod at both 19 °C and 12 °C. Carvalho & Hughes (1983) found that less than 12 hr light at 19 °C would not induce gamogenesis in clones of *Daphnia magna* but was effective at 12 °C. They also found variation in the critical photoperiod for induction, which indicated some differences among clones. It is possible that the enhanced induction of sexual females observed in our study and others is not a direct response to lower temperature per se. We suggest that it might, instead, be due to lowered filtering rate and energy assimilation at colder temperatures that would have the same inductive effect as decreased food abundance.

Hebert (1974) found as much as two months'

difference in the timing of gamogenesis in different fall genotypes of *Daphnia magna*. Thus, the much lower response of our spring clones to the inducing short photoperiod suggests some genotypic variation between that subpopulation and the one in the fall. We also observed a clonal variation in the response to temperature at the inductive short photoperiod. Some clones became ephippial at both 20 °C and 15 °C; others responded only at 15 °C. Loaring & Hebert (1981) found differences in ephippial production at two temperatures among clones that were of different genotypes. The differing responses of our clones to temperature at inducing photoperiod suggest further genotypic diversity within the population.

Of particular interest is that our spring test animals did not produce ephippial females under conditions simulating the temperature and photoperiod at which the natural population is strongly gamogenetic. This suggests that our experimental regime failed to include one or more inducing factors found in nature. The most likely of these is crowding. Banta (1939) and his collaborators and other researchers (see Hebert, 1978) have tried to assess the factors involved in crowding that lead parthenogenetic females to produce males (sex determination) or to switch to the formation of ephippia and diapause eggs. Berg (1931) demonstrated that ex-ephippial female *Daphnia cucullata* produced more ephippia in cultures of 5 females per 25 ml than of 1 per 25 ml. The latter produced more males, especially in their first broods. Doma (1979) found that males and ephippial females could be induced in populations of *Daphnia* if well-fed cultures had their algal supply cut sharply. Carvahlo & Hughes (1983) also found a negative correlation between food availability and the number of ephippia formed. Our spring population of *S. armata* similarly could have been induced by a drop in food supply following an earlier algal bloom. On the other hand, Carvahlo & Hughes (*loc cit.*) attributed ephippial formation in well-fed, crowded cultures of *Daphnia magna* to increased inter-animal contact. This could also have been a factor in our spring population, which was found in relatively dense swarms. Gamogenesis in Cladocera has been linked to swarms (Young, 1978), but Crease & Hebert (1983) suggested that swarming and gamogenesis occur independently as a result of high population density.

An important consideration in evaluating our results is that our population came from a very shallow pond, probably subject to considerable short-term fluctuation in environmental conditions. Carvahlo & Hughes (*op cit*) propose that short-term environmental cues such as food availability and population density (cueing spring gamogenesis) may be important to dicyclic *Daphnia* in small temperate ponds and that such populations are able to respond to a predictable, seasonal (fall) short photoperiod cue as well. Similar mechanisms may be operational in our population. Our experiments do not prove that the *S. armata* spring and fall clones are not influenced by long photoperiod. The spring population could have been induced by an earlier, slightly shorter photoperiod, or it could be that 15.5 h, or less, of light is a permissive period under which a second factor, such as crowding, can operate (Stross, 1969). Either is possible, since we observed that many of our stock clones became gamogenetic when they were crowded in 13.5 h light.

Berg (1931) and Stross (*loc sit.*) also observed the pattern of switches between parthenogenesis and ephippial formation that our animals exhibited. The alternation of pre-ephippial (sterile) and mature ephippial instars that we observed was first well described in *Scapholeberis* by Rammner (1929) and was observed in *Daphnia* by Stross (1969). Berg (*loc cit.*) noted the ovarian resorption of resting eggs that we observed during barren instars after the shedding of empty ephippia. He also observed the movement of some eggs into the ephippium without the stimulus of mating (Berg, *op cit*, p. 116); he found that such eggs later degenerated, whereas the ones we saw appeared to be normal at the time the ephippium was shed. If they were normal, diploid, diapausing embryos, then clones capable of this phenomenon could give rise to the obligate parthenogenic populations occurring in northern temperate and arctic regions. The male-transmitted meiosis-suppressing factor proposed by Hebert (1987) might not be involved or necessary.

Too few males were induced in these experiments for us to draw any conclusions about the roles of photoperiod and temperature in sex determination in *Scapholeberis*. However, we did confirm Rammner's (1929) observations that males live through multiple adult instars (like other Daphniidae).

The slower growth we observed of animals at cooler water temperature was to be expected as a result of lower metabolic rates. The smaller brood sizes in cooler animals may have been due to slower filtering rates and lower energy assimilation than in warm animals. At each temperature, animals in the longer photoperiod matured more rapidly, had larger broods, and tended to die younger than the ones experiencing shorter days. It is possible that light stimulates more rapid filtering, and that the long daylength animals were able to assimilate more food per day for growth and reproduction.

In summary, these experiments demonstrate that gamogenesis can be induced in spring and fall clones of a dicyclic population of *S. armata* by a short photoperiod. Cool water enhances this induction. Gamogenesis in the late spring probably involves other factors, such as induction by an earlier, somewhat shorter photoperiod, a sharp drop in food supply, or other factors associated with high population density. One or more of these factors could also be acting in concert with a permissive, long photoperiod. It would be interesting to investigate these possibilities in this population and in ones from further north and south. The fact that fewer spring clones responded to the inducing conditions suggests that there may be some genotypic differences between the spring and fall subpopulations. There is no evidence that these two subpopulations are reproductively isolated from each other.

It would be valuable to compare genotypes of clones responding to diverse inducing factors within this population and among ones from other, latitudinally separated populations. Genotypic variation allowing different clones to respond to a variety of gamogenesis-inducing factors probably is the basis for the plasticity in reproductive cycles that has enabled cladocerans to adapt to a diversity of habitats.

Acknowledgements

The authors thank Mrs. Margaret E. McCutcheon for her hospitality and permission to collect in the Smoke House Farm pond. These studies were supported, in part, by a Temple University Faculty Grant-in-Aid-of-Research to D.B.B., N.S.F. Grant No. DIR 8900929 to Temple University for Undergraduate Research, and the Temple University Program for Minority Undergraduate Summer Research.

Appendix I. Medium for Culturing *Chlamydomonas reinhardtii.*

MACRONUTRIENTS (Sager-Granick modified from Gealt, *et al.* 1981)

Compound	Stock Conc. (M)	ml/l
$FeCl_3 \cdot 6H_2O$	0.370 (fresh)	0.6
NH_4Cl	1.000	5.0
$MgSO_4 \cdot 7H_2O$	0.120	10.0
Na Citrate · $2H_2O$	0.170	10.0
$CaCl_2$	0.036	10.0
K acetate	2.200	10.0
K_2 HPO_4	1.000	0.4
Na H_2PO_4	1.000	1.4

MICRONUTRIENTS ('M' solution, from Keating & Dagbusan, 1984)

	Stock Conc. g/500ml	ml/l for 5 × 'M' stock
H_3BO_3	2.800	5.0
$MnCl_3 \cdot 4H_2O$	0.360	5.0
$LiCl_2$	0.300	5.0
RbCl	0.700	5.0
$SrCl_2 \cdot 6H_2O$	0.150	5.0
NaBr	0.032	5.0
$Na_2MoO_4 \cdot 2H_2O$	0.063	5.0
$CuCl_2 \cdot 2H_2O$	0.013	12.5
$ZnCl_2$	0.100	12.5
$Co(NO_3)_2 \cdot 6H_2O$	0.010	5.0
KI	0.003	5.0
SeO_2	0.0007	5.0
NH_4VO_3	0.0006	5.0

TO ADD LATER:

$CaCl_2 \cdot 2H_2O$	1.9 g/500 ml	10.0
$Na_2SiO_3 \cdot 5H_2O$	0.906 g/250 ml	40.0

VITAMINS (combined)

	mg/l
Thiamine-HCl	75.00
Biotin	0.75
B_{12}	1.00

To make 1 × growth medium:

$H_2O \sim 600$ ml. Add macronutrients as listed above. Add 200 ml 5 × 'M' solution. Add $CaCl_2$ and $Na_2SiO_3 \cdot 5H_2O$. Add $H_2O \rightarrow 1$ liter. Autoclave. pH should be 6.9. At time of inoculation, add 1 ml/l vitamin mixture.

References

Banta, A. M., 1939. Studies on the physiology, genetics, and evolution of some Cladocera. Paper No. 39, Department of Genetics. Carnegie Institution, Washington, D.C. x, 285 pp.

Berg, K., 1931. Studies on the genus *Daphnia* O.F. Müller with special reference to the mode of reproduction. Vidensk. Med. fra Dansk Natur. Forening. 92: 1–222.

Carvalho, G. R. & R. N. Hughes, 1983. The effect of food availability, female culture-density and photoperiod on ephippia production in *Daphnia magna* Straus (Crustacea: Cladocera). Freshwat. Biol. 13: 37–46.

Cowgill, U. M., H. W. Emmel, D. L. Hopkins, I. T. Takahashi & W. M. Parker, 1986. Variation in chemical composition, reproductive success and body weight of *Daphnia magna* in relation to diet. Int. Revue ges. Hydrobiol. 71: 79–99.

Crease, T. J. & P. D. N. Hebert, 1983. A test for the production of sexual pheromones by *Daphnia magna* (Crustacea: Cladocera). Freshwat. Biol. 13: 491–196.

Doma, S., 1979. Ephippia of *Daphnia magna* Straus – a technique for their mass production and quick revival. Hydrobiologia 67: 183–188.

Dumont, H. J. & J. Pensaert, 1983. A revision of the Scapholeberinae (Crustacea: Cladocera). Hydrobiologia 100: 3–45.

Frey, D. G., 1982. Contrasting strategies of gamogenesis in northern and southern populations of Cladocera. Ecology 63: 223–241.

Frey, D. G., 1986. The non-cosmopolitanism of chydorid Cladocera: Implications for biogeography and evolution. In Crustacean biogeography, R. H. Gore & K. L. Heck, eds. A. A. Balkema, Rotterdam: 237–256.

Gealt, M. A., J. H. Adler & W. R. Nes, 1981. The sterols and fatty acids from purified flagella of *Chlamydomonas reinhardii*. Lipids 16: 133–136.

Hebert, P. D. N., 1974. Ecological differences between genotypes in a natural population of *Daphnia magna*. Heredity 33: 327–337.

Hebert, P. D. N., 1978. The population biology of *Daphnia* (Crustacea, Daphnidae). Biol. Rev. 53: 387–426.

Hebert, P. D. N., 1987. Genotypic characteristics of the Cladocera. Hydrobiologia 145: 183–193.

Hoffman, M. S., (ed.), 1986. The world almanac and book of facts. Pharos Books: New York. 973 pp.

Keating, K. I., 1985. The influence of vitamin B_{12} deficiency on the reproduction of *Daphnia pulex* Leydig (Cladocera). J. Crust. Biol. 5: 130–136.

Keating, K. I. & B. C. Dagbusan, 1984. Effect of selenium deficiency on cuticle integrity in the Cladocera (Crustacea). Proc. Nat. Acad. Sci. USA 81: 3433–3437.

Keating, K. I. & B. C. Dagbusan, 1986. Diatoms in daphnid culture and bioassay. Envir. Toxicol. Chem. 5: 299–307.

Loaring, J. M. & P. D. N. Hebert, 1981. Ecological differences among clones of *Daphnia pulex* Leydig. Oecologia 51: 162–168.

Porter, K. G., J. D. Orcott, Jr. & J. Gerritsen, 1983. Functional response and fitness in a generalist filter feeder: *Daphnia magna* (Cladocera: Crustacea). Ecology 64: 735–742.

Rammner, W., 1929. Über periodische Erscheinungen am Cladoceren-Individuum. (Häutung, Fortpflanzung, Wachstum). Int. Rev. ges. Hydrobiol. Hydrogr. 21: 401–420.

Shan, R. K., 1974. Reproduction in laboratory stocks of *Pleuroxus* (Chydoridae, Cladocera) under the influence of photoperiod and light intensity. Int. Rev. ges. Hydrobiol. 59: 644–666.

Stross, R. G., 1969. Photoperiod control of diapause in *Daphnia*. III. Two-stimulus control of long-day, short-day induction. Biol. Bull. 137: 359–374.

Stross, R. G. & J. C. Hill, 1968. Photoperiod control of winter diapause in the fresh-water crustacean, *Daphnia*. Biol. Bull. 134: 176–198.

Stross, R. G. & D. A. Kangas, 1969. The reproductive cycle of *Daphnia* in an arctic pool. Ecology 50: 457–460.

Tessier, A. J., L. L. Henry, C. E. Goulden & M. W. Durand, 1983. Starvation in *Daphnia*: Energy reserves and reproductive allocation. Limnol. Oceanogr. 28: 667–676.

Weider, W. J., M. J. Beaton & P. D. N. Hebert, 1987. Clonal diversity in high-arctic populations of *Daphnia pulex*, a polyploid apomictic complex. Evolution 4: 1335–1346.

Young, J. P. W., 1978. Sexual swarms in *Daphnia magna*, a cyclic parthenogen. Freshwat. Biol. 8: 279–281.

Hydrobiologia **225**: 281–290, 1991.
V. Kořínek & D. G. Frey (eds), Biology of Cladocera.

Intraspecific variability in response to stimuli for male and ephippia formation in *Daphnia pulex*

Petter Larsson
Museum of Zoology, University of Bergen, Múseplass 3, N-5007 Bergen, Norway

Key words: *Daphnia pulex*, cyclical parthenogenesis, obligate parthenogenesis, male induction, ephippia induction

Abstract

Natural clones and clones from laboratory crossings of *Daphnia pulex* have been tested in their response to stimuli for male and ephippia formation. The clones are from ponds in Illinois, USA, an area with both obligate and cyclical clones of *D. pulex*. The progeny from two types of crosses were studied: 1) crosses between an obligate clone producing males and a cyclical clone producing only females, and 2) self-fertilization within a cyclical clone producing both sexes. Both the natural and the artificial clones showed great variation in response to environmental stimuli for males. Ephippia response seems more equal among the clones, but two of the artificial clones could hardly be stimulated to produce ephippia in the induction experiments. In crowded cultures, however, they produced a few. Crossings between cyclicals and obligates yielded mostly cyclical progeny. Progeny from the selfed clone had low survival and fecundity.

Introduction

During the last years, much attention has been paid to the life cycle processes in the parthenogenetic phase of the daphniid's life (Peters & Bernardi, 1987). Few, however, have evaluated the importance of sexual reproduction in the life-cycle strategy (Lynch & Gabriel, 1983; Barker & Hebert, 1986; Innes & Hebert, 1988; Hobaek & Larsson, 1990). High birth rate, growth rate and survival rate in the parthenogenetic phase are essential parameters in the success of any *Daphnia* clone, but it is evenly important to survive periods of diapause in a sufficient number. The initial number of viable ephippia hatching from a particular clone or clone-combination in spring will strongly affect the development of the population during the rest of the season.

Cladocerans have a reproductive system called cyclic parthenogenesis. When the resting eggs in the ephippia hatch, they normally become females, which reproduce by parthenogenesis for some generations. Sexual reproduction starts with the appearance of male progeny developing from parthenogenetically produced eggs. These are diploid and genetically identical with the females in the clone. The next step is production of haploid resting eggs that require fertilization to develop. The fertilized eggs are released into the protective chamber called ephippium. These eggs may go into diapause and rest for some months before they develop.

The switch to sexual reproduction is triggered by environmental cues (Mortimer, 1936; Banta, 1939; Stross & Hill, 1965, 1968) but the problem facing the various clones of cladocerans is how to cope with the environmental messages in an optimal way. When shall the sexual period start, and how many sexual individuals of each sex shall be produced?

The males have to be produced before, but close to the time other clones are producing resting eggs. Too early production of males is wasted effort. The clutches are normally unisexual, and the fitness is markedly reduced if the males do not find any sexual female to copulate with. It is also a question of numbers when the signals from the environment indicate the proper time for sexual reproduction. What is the optimal number of males to match the coming sexuality in the females, and when is it more profitable for the females to produce resting eggs instead of males?

As long as there are reasonable possibilities for growth and parthenogenetic reproduction, the population can still increase, and then more resting eggs can be produced by each clone. A too early switch to resting eggs can lead to suboptimal ephippia numbers, while a too late switch might be a catastrophe. Then the environmental conditions might be so unfavorable that the animals does not have enough resources for eggs in the ephippia.

Since the work of Weismann (1879), many attempts have been made to understand how the cladocerans are reading the environment, and a long list of factors inducing males and ephippia has been proposed (Mortimer, 1936; Banta, 1939; Stross & Hill, 1965, 1968). There seem to be two classes of environmental cues active; phenological and density dependent (Hobæk & Larsson, 1990). The photoperiod is probably the main phenological cue in the arctic and temperate zone (Stross, 1965, 1968). The nature of the density dependent cues has been less clear. Until recently (Hobaek & Larsson, 1990) all experiments on density dependent cues have been based on confining the animals in small vessels with renewal of water and food at varying intervals. This technique have made it almost impossible the differentiate between effects of food conditions, chemical crowding stimulus, and general water quality effects.

Hobaek & Larsson (1990) used a flow-through technique in their experiments on the mechanisms determining the sex of *Daphnia magna* progeny. Their findings can be summarized as follows: males were produced in up to about 50% of the clutches at good food conditions (1 & 2 mg *Scenedesmus* dw l^{-1}) when stimulated by short-day photoperiod (L : D = 8 : 16) or kept in filtered water from a tank crowded with other daphniids. When well stimulated, the females seemed to alternate between female and male clutches, achieving an approximate sex ratio of 1 : 1 in the progeny. Ephippia did not appear. Later experiments in the flow-through system, however, have indicated that this particular clone needed a combination of short days and crowded water together with low food concentrations (<0.4 mg *Scenedesmus* dw/l) before ephippia appeared (Kleiven *et al.*, submitted).

Cladocerans in various regions or localities may not respond to the environmental cues in the same way. Frey (1982) found various strategies for summer and winter diapause in southern versus northern populations of cladoceran species.

Some *Daphnia pulex* clones have abandoned sexuality and produce their resting eggs ameiotically (obligate parthenogens, Hebert, 1978, 1987a, b). Recently it has been shown that *Daphnia pulex* in ponds in North America south to the Great Lakes are exclusively obligates (Hebert & Crease, 1983, Hebert & McWalter, 1983; Lynch, 1984; Weider & Hebert, 1987a, b; Weider *et al.*, 1987; Hebert *et al.*, 1988).

In ponds just south of the Great Lakes, however, i.e. in Illinois, *D. pulex* exists in mixtures of obligate and cyclical clones (Lynch, 1984). A great variability should be expected in their respons to the environmental cues inducing the sexual periods. Hebert & Crease (1983) found that many of their obligate clones had retained the capability to produce males and thus to fertilize cyclical females. On the contrary, some cyclical clones seem unable to produce males.

The question is whether within the same region or locality, there might be different strategies among clones giving different thresholds before entering the sexual phase. It is a question both of how strong the stimuli need to be before changing the reproductive mode and how many males and ephippia should then be produced.

In this study I have used *Daphnia pulex* clones from ponds in Illinois, USA. Both natural clones and clones from forced laboratory crosses were exposed to environments with strong stimuli for male and resting egg formation to demonstrate various strategies.

Methods and materials

Animals

The animals used in the experiments were various clones of *Daphnia pulex* from ponds in the neighborhood of Champaign-Urbana, Illinois. Two clones, SWP90 and SWP96, came from the so called Smith's West Pond while CP248 came from Circle Pond. SWP90 is an obligate clone making males and SWP96 is a cyclical clone not found to make any males. F_1 progeny of SWP90 males and 96 females were called 'hybrids' and named H2, H3, H4, H6, H7; H8, H9, and H10 (Nass, 1987). H1 and H5 has gone extinct before the experiments started. H7 went extinct after the first experiment. Electrophoretic analysis proved that the F_1 individuals were true crossings between SWP90 and SWP96 (Table 1) (Nass, 1987).

Table 1. Genotypic differences among the various clones: Abrivations for enzymes: PGM = phosphoglucomutase, PGI = phosphoglucose isomerase, GOT = glutamate oxaloacetate transaminase, PEP = peptidase, LDH = lactate dehydrogenase. Abrivations for electrophoretic movements: F = fast, M = Medium speed, S = slow. The signs + and − indicate minor differences within F, M, and S. The electhrophoretic readings were done by Lisa Nass and Ken Spitze. Method described in Steiner and Joslyn (1979) and Lynch (1983).

Clones	Enzymes				
	PGM	PGI	GOT	PEP	LDH
SWP90	FF	MM +	MM	MM	FS
SWP96	MM	SS	MS	MM	SS
H2	FM	M + S			FS
H4	FM	M + S			SS
H6	FM	M + S			FS
H7	FM	M + S			FS
H8	FM	MS			FS
H9	FM	M + S			FS
H10	FM	MS			FS
CP248	MM	MM	MS −	MM	
S2	MM	MM	MS −	MM	
S4	MM	MM	S − S −	MM	
S6	MM	MM	MS −	MM	
S10	MM	MM	MM	MM	
S13	MM	MM	MS −	MM	
S15	MM	MM	MM	MM	
S16	MM	MM	MS −	MM	
S27	MM	MM	MS −	MM	

CP248 was a cyclical clone making males. Resting eggs were made by self-fertilization within the clone, and the hatchings were designated S1-30. Only S2, S4, S6, S10, S13, and S16 survived to maturity and produced female progeny. S27 produced one clutch of males and died. S7, S8, S9, S18, and S26 lived for more than 14 days but did not produce any progeny. The rest of the 'selfers' died within a few days after hatching. Electrophoretic analysis of those producing progeny (Table 1) showed that the heterozygote GOT (glutamate oxaloacetate transaminase) locus had been split in three of the eight clones, and thus verifying the occurence of sexual reproduction.

Crosses

To get sexual resting eggs, the following procedure was used. Neonate females were grown for four days in good food conditions (300 000 cells ml^{-1}). At day five the food concentration was reduced (10 000–30 000 cells ml^{-1}) and males were introduced. The males were a couple of days older than the females and had been grown separately at a good food level (300 000 cells ml^{-1}). The sex ratio was about 1 : 1. Beakers of about 250 ml were used, and there were about 20 animals in each beaker. Ephippia with eggs appeared both in the first and second clutches, but the proportion of resting eggs was very variable.

To get resting eggs hatched, the procedure of Davison (1969) and Swartz & Hebert (1987) was followed. Ephippia were kept in total darkness for

10–30 days. Then the eggs were dissected from their ephippia and rinsed three times in deionized water. The eggs were kept in deionized water under constant light (40 W neon tubes in a distance of about 20 cm) and 20 °C until hatching. Hatching appeared 4 to 40 days after the dissection. The eggs were inspected daily. Observed neonates were immediately transferred to 250 ml beakers with good food conditions.

Induction experiments

In experiments for induction of males and ephippia, the mothers of the experimental animals were kept from birth to second or third clutch at the same temperature, water type, food concentrations, and photoperiod as at the beginning of the actual experiment. Progeny from the second or third clutch were used in the experiments. The water was aged tap water prepared in large tubs. After tap water had been standing for at least ten days, *Daphnia* were introduced and maintained as a water suitability gauge. Water removed from the tubs for use, was filtered through 63 μm mesh screen and then through Whatman 1 filter paper. The water was 'crowded' in the following way: aged tap water was kept in one-gallon glass jars with a dense population (300–800 animals l^{-1}) of clone SWP96 for 24 hours before being given to the experimental animals. The same filtering procedure was used as for the crowded water. Food given to the animals was *Scenedesmus acutus*, reared in large bottles. The algae medium was replaced in the same amount as algae-suspension was taken out. The algal concentrations were estimated by using a spectrophotometer.

The experiments were run in beakers with about 35 ml water. The medium was replaced every day. All the experiments were run in incubator cabinets kept at 20 °C. The light in the incubators was from yellow-white neon tubes, and the photoperiod L : D = 8 : 16.

Experiment 1 was carried out to reveal variability among clones in their reaction to ephippia-inducing stimuli. Eggs were looked for in the ephippia from females without having had any contact with the males and from females that had such contact. SWP96, SWP90, the 'hybrids', and CP248 were given the same environmental conditions. About 40 female neonates from each of the 10 clones, born within the same 24 hours, from 10–16 different mothers of each clone, were selected for the experiment. Day 1–4: 10 animals in each beaker of 250 ml with good food conditions of 300000 cells ml^{-1}. Day 5–6: all animals from same clone (about 40) mixed in one 250-ml beaker with poor food conditions of 30000 cells ml^{-1}. The water was not changed, but food added the second day. Day 7–16: single animals were placed in plastic cups with 35 ml of medium. In each clutch there were 24 experimental animals. They were given good food conditions (300000 cells ml^{-1}). At day 11 one male of the same age was placed in each half of the cups. The animals were held untill they have produced two clutches or died.

Experiment 2 was carried out to measure differences in the clones' tendency to produce males. SWP90, SWP96, CP248, H2, H4, H6, H7, H8, H9, and H10 were used. They were all given the same environmental conditions with supposed maximum male inducing stimuli. Experimental conditions were food 300000 cells ml^{-1}, water, crowded with SWP96 (females only). 24 neonates of each clone born within the same 24 hours were placed individually in plastic cups with 35 ml water. The water was changed every day.

In experiment 3 the tendency to produce males in clones CP248, S2, S4, S6, S10, S13, S15, S16, and SWP90 was tested. The clones were given the same environmental conditions as in experiment 2, with supposed maximum male inducing stimuli.

Results

The progeny in experiment 1 yielded a high proportion of ephippia, but in various amounts among the different clones (Fig. 1). H6 and H10 did not produce any ephippia at all in this experiment, while H7 produced 36 ephippia among 39 clutches.

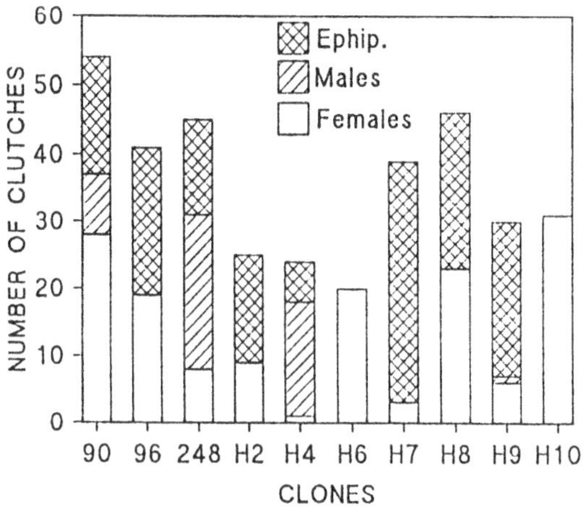

Fig. 1. Number of various clutch types among clones in experiment 1. Short day photoperiod, crowded water and a period of low food. The clones tested: SWP90, SWP96, CP248, and the 'hybrids' of SWP96 × SWP90.

Males appeared in SWP90, CP248, H4 and H9. It was surprising that H4 had 17 clutches of males among the total of 24. Just one clutch became females and 6 were ephippia.

Without having had any contact with males (Fig. 2), the SWP90 clone produced 16 ephippia with eggs from a total of 17, thus demonstrating its obligate character. H8 without males had 9 out of 22 ephippia with eggs and is thus probably an obligate clone. H2 without males had 3 out of 16 ephippia with eggs and hence is also presumed to be obligate.

SWP96 and H7 had no ephippia with eggs among 19 or more clutches without males, but had more than half of them with eggs when males were present (Fig. 3). They are obviously cyclical clones. CP248 females without males had just one ephippium with eggs among 11, and with males only 1 ephippium with eggs among 3. These results are not clear, but later crossing experiments gave separation in a heterozygote locus demonstrating that the clone is cyclical. H9 without males had one ephippium with eggs among 20, and with males it had one among 3. The clone is supposed to be cyclical. H4 produced no eggs in ephippia when it was without males while one ephippium with eggs when males were present. H4 is probably also cyclical. Among the hybrids then, only H2 and H8 seem to be obligate, but it seems as if even cyclical females can sometimes lay egg in the ephippium without any males being present. Possibly they are not viable, but it is difficult to force hatching of resting eggs, and even supposedly viable eggs might not break their diapause under the conditions imposed. Also, empty ephippia frequently appeared among both cyclicals and obligates making the judgement of cyclicals versus obligates even more difficult. Whether H6 and H10 were obligate or cyclical, could not be judged from this experiment since they did not produce ephippia there. Ephippia occurred, however, in cultures of the two clones

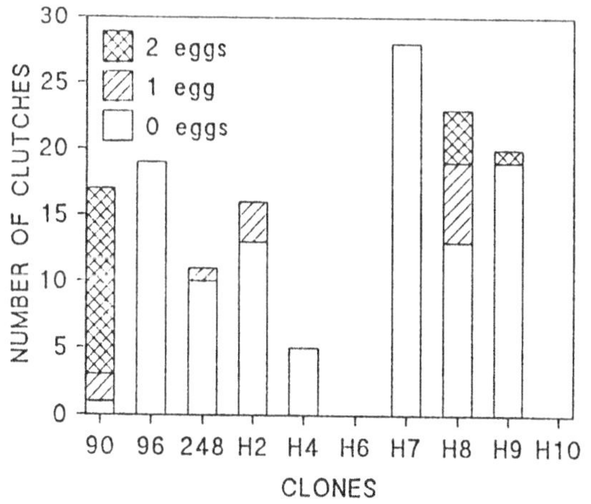

Fig. 2. Egg content of the ephippia from the clones SWP90, SWP96, CP248, and the 'hybrids' of SWP96 × SWP90 without males produced in experiment 1.

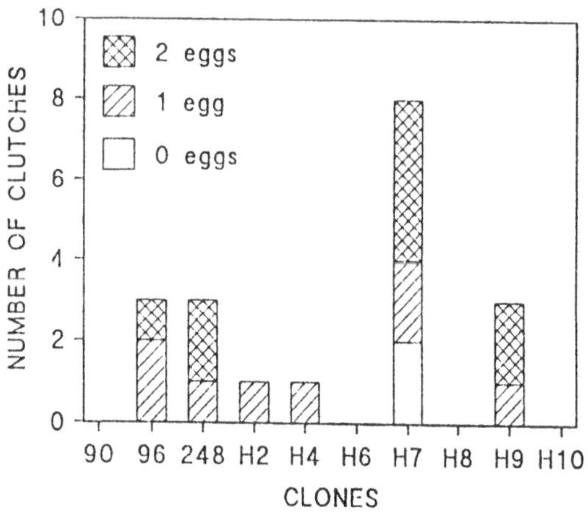

Fig. 3. Egg content in the ephippia from SWP90, SWP96, CP248, and the 'hybrids' of SWP96 × SWP90 with males produced in experiment 1.

(and in experiment 2 H6 produced one ephippium). They were all ($n_{H6} = 100$ $n_{H10} = 39$) without eggs and thus both H6 and H10 are presumed to be cyclical without males.

Experiment 2 was designed for maximum induction of males, and males appeared in clones SWP90, CP248, H2, H4, and H8 (Fig. 4). H9 produced males in experiment 1, and thus 4 out of 6 of the hybrid clones produced males. In this experiment the hybrids produced less males than the clone of their fathers, SWP90, and much less than CP248, of which 56% of the clutches had males. CP248 produced more males than any other clone and the differences were highly significant by a G-test (Tab 2). SWP90 produced fewer males than CP248, but it was still significantly different from all the others in a G-test. H4 produced more males than H2, but they were not significantly different. H2 was not significantly different from H8 or the clones with no males.

In experiment 2, also ephippia appeared in four of the clones: H2, H6, H8 and H9 (Fig. 5), but they were all empty.

In experiment 3 the CP248-selfers were compared with SWP90 and the parent clone CP248. Males appeared in all except S2 and S13. However, survival and fecundity among the selfers were very poor. S2, S4, S10 and S13 had fewer than 20 clutches from the 24 animals at the beginning of the experiment. S6 was slightly better with 34 clutches. If only the first clutches are compared (Table 3), the parent clone is significantly different from S2, S4, S6 and S13, but not from S10, S15 and S16. S6, S15 and S16 produced the highest proportion of males among the hybrids and if the three first clutches are included in the G-test, SWP90, CP248, S6, S15 and S16 could be compared (Table 4). The S15 and S16 clones were still not significantly different from the par-

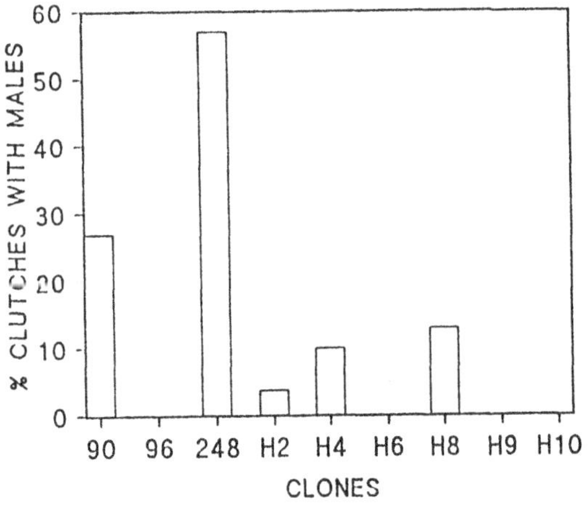

Fig. 4. Percentage of clutches with males in various clones in experiment 2. Experimental conditions: short days photoperiod, crowded water and good food conditions. The clones tested were SWP90, CP248, and the 'hybrids' of SWP96 × SWP90.

Table 2. Significance table in experiment 2. Clones SWP96, SWP90, CP248, and 'hybrids' of SWP96, and SWP90, G-tests of differences among the clones in their ability to produce males. The clones were given the same amount of male stimuli. Williams' correction is used (Sokal and Rohlf 1981. p 737). Significans: $p < 0.05$: *, $p < 0.025$: **, $p < 0.01$: ***, $p < 0.001$: ****

Clones								
	SWP96							
SWP90	****	SWP90						
CP248	****	****	CP248					
H2	*	****	****	H2				
H4	****	***	****	N.S.	H4			
H6	N.S.	****	****	N.S.	****	H6		
H8	N.S.	****	****	N.S.	**	N.S.	H8	
H9	N.S.	****	****	N.S.	***	N.S.	N.S.	H9
H10	N.S.	****	****	N.S.	****	N.S.	N.S.	N.S.

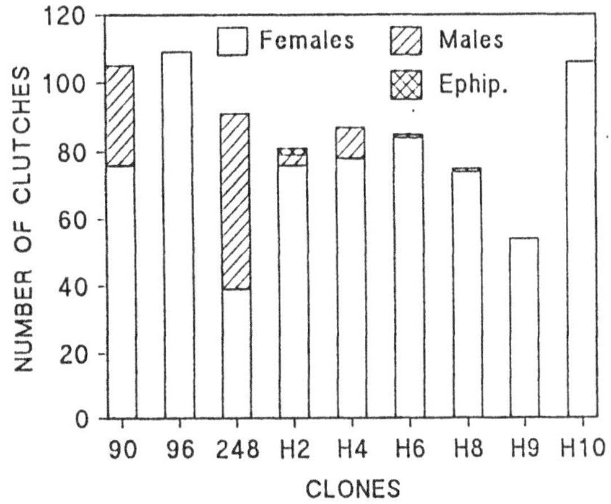

Fig. 5. Number of various clutch types among clones in experiment 2. Experimental conditions: short days photoperiod (L : D = 8 : 16), crowded water and good food conditions. The clones tested were: SWP90, CP248, and the 'hybrids' of SWP96 × SWP90.

Table 3. Significance table for experiment 3: SWP90, CP248 and the 'selfers' of CP248. G-tests of differences among clones in their ability to produce males. Only the first clutch is considered. All clutches were given the same stimuli for production of males. Williams' correction is used (Sokal and Rohlf 1981. p. 737). Significance level: $p < 0.05$: *, $p < 0.025$: **, $p < 0.01$: ***, $p < 0.001$: ****, Not significant: N.S.

Clones								
	SWP90							
CP248	***	CP248						
S2	N.S.	***	S2					
S4	N.S.	****	N.S.	S4				
S6	N.S.	****	N.S.	N.S.	S6			
S10	N.S.	N.S.	N.S.	N.S.	N.S.	S10		
S13	N.S.	***	N.S.	N.S.	N.S.	N.S.	S13	
S15	**	N.S.	***	***	***	N.S.	**	S15
S16	****	N.S.	****	****	****	N.S.	***	**

Table 4. Significance table in experiment 3. Clones: SWP90, CP248, and 'selfers' S6, S15 and S16. The three first clutches are considered. G-tests of differences among the clones in their ability to produce males. Williams' correction is used (Sokal & Rohlf, 1981. p. 737).

Clones				
	SWP90			
CP248	****	CP248		
S6	**	N.S.	S6	
S15	****	N.S.	N.S.	S15
S16	****	N.S.	N.S.	N.S.

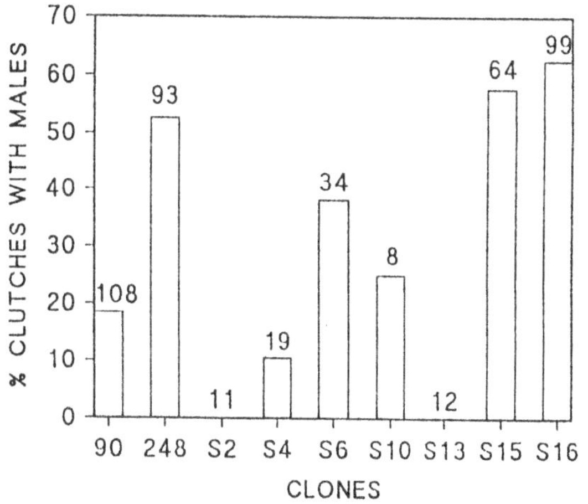

Fig. 6. Percentage of clutches with males in various clones in experiment 3. Experimental conditions: short days photoperiod (L : D = 8 : 16), crowded water and good food conditions. The clones tested were: SWP90, CP248, and 'selfers' from the crossing CP248 × CP248. Numbers on top of each bar are numbers of clutches sexed.

ent clone. When more clutches were included, more males appeared in S6, and it was not statistically different from CP248. They were all, however, significantly different from SWP90. S10 which produced 2 male clutches of a total of 8 had too few clutches to be significantly different from any other clone.

In the experiment S2, S4, and S13 produced significantly fewer males than the parent clone. Because so few clutches appeared in S2 and S13, it might still be possible that these two clones could produce males, but that they needed stronger stimuli for switching.

Survival in the experiments was very variable. The 'hybrids' in experiment 2 did it better than the selfers in experiment 3, but in both experiments CP248, SWP96, and SWP90 were better than many of the hybrids and selfers. S2 had a particularly low survival. All of the 24 experimental animals had died before 10 days after the first clutch. Also a large proportion of the selfers died before reproducing at all.

Differences in reproductivity are striking (Fig. 7 and 8). The selfers S2, S4, S6, S10 and S13 produced fewer than 10 offspring per surviving female within a period of 14 days, while S15 and S16 produced 29 and 60, respectively. The

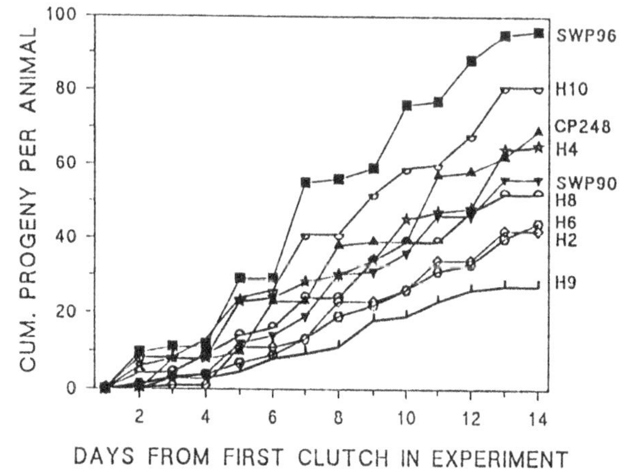

Fig. 7. Cumulative mean number of progeny per animal of clones SWP90, SWP96, CP248 and the 'hybrids' of SWP90 × SWP96: H2, H4, H6, H8, H9, H10 in experiment 2.

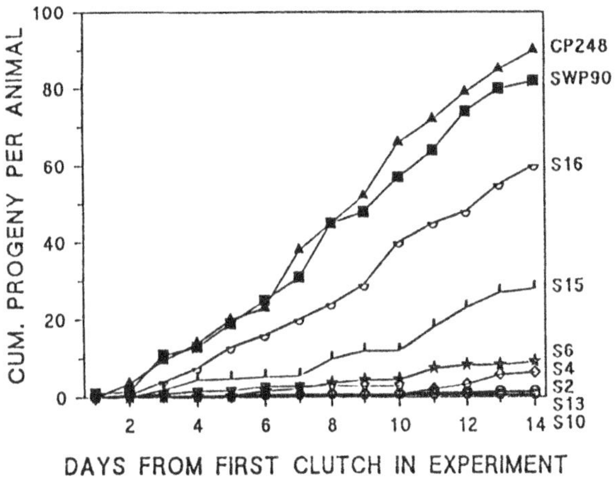

Fig. 8. Cumulative mean number of progeny per animal of the clones SWP90, CP248 and the 'selfers' S2, S4, S6, S10, S13, S15 and S16 in experiment 3.

'hybrids' produced between 29 and 81 offspring during 14 days, while the natural clones were better with 57 to 97 offspring.

Discussion

Ponds in Illinois, where these *Daphnia pulex* came from, are in a transition zone between the areas where obligate clones dominate and areas where cyclic clones are supposed to dominate (Lynch, 1984). Here the cyclical and obligate clones of *D. pulex* might compete with each other with variable success. The clones were highly variable in the production of males and ephippia by the induced stimuli.

By crossing an obligate with a cyclical clone, the offspring become both cyclical and obligate. This agrees with the same type of crossings done by Innes & Hebert (1988). The mother clone did not produce males, and the offspring clones seemed to be mostly poor male producers if any males were produced at all. The H4 clone produced more males in experiment 1 than in experiment 2. The main experimental difference between these two was the starvation period in the first experiments. This probably has affected the result. It is in agreement with Kleiven *et al.*, (submitted) who found that there was a tendency for *D. magna* to increase its male fraction above the 1 : 1 ratio at very low food concentrations (0.2 mg dw l^{-1}) in flow through-systems.

When a cyclical clone with both females and males was selfed, most of the progeny seemed to able to produce males in variable number, even though the CP248 clone produced more males than any of the other clones tested. The progeny showed very low fitness, be a sign of inbreeding depression.

When the clones were given stimuli for production of ephippia, there seemed not to be any marked difference between obligates and cyclicals. Two of the hybrids, however, did not produce ephippia under the conditions used in the experiments, and thus may have had a very different threshold for ephippia formation.

From a theoretical point of view, cyclic parthenogenesis seems to be an almost ideal reproductive system taking the advantages of both the sexual and asexual phases (Lynch & Gabriel, 1983). Most other metazoans, however, have only the sexual mode of reproduction, indicating possibly that cyclic parthenogenesis is a reproductive system that only under special circumstances is best in the long run. The fact that *D. pulex* in Northern America favors obligate asexual reproduction might further indicate that only one way of reproduction, either sexual or asexual, normally has the highest fitness. It could be that the shift from one reproduction mode to the other in some environments becomes se complicated that optimal reproduction for the cyclicals is not achieved.

In the Illinois ponds cyclical and obligate *Daphnia pulex* might compete with one another for existence, and the variation observed might possibly be a result of this. The tendency toward highly variable male production indicated that some clones specialized as 'male producers' or 'ephippia producers'. Such specialization could probably be looked upon as another type of differentiation into two sexes. Besides the question of being obligate or cyclical it could be a question of specialazing on one type of progeny.

The great diversity of reproductive strategies within the same population seems strange, and it is difficult to consider it as a stable system.

Neither the cyclic nor the obligate parthenogenesis seem to win. The appearance of strange combinations as cyclical clones without males, and clones with very high thresholds for ephippia production, could be just a result of random combinations between competing systems.

Hebert *et al.* (1989) claims that the adoption of asexuality is usually a product of internal genetic factors rather than extrinsic selection pressures. This might be right as a proximate cause, but if cyclical parthenogenesis in the long run is evolutionarily superior, there should be no reasons for the obligates to take over. Crossings between obligate males and cyclical females do not lead to only obligates clones, and although newly established obligates may have high fitness and a long life, still the selection pressure would have eliminated them in the long run. On the other hand, if the environmental cues inducing the sexual periods are poorly predictable, one year might be advantages for the obligates and another for the cyclicals and thus a balance between the two might appear.

Acknowledgements

The study was done during my sabbatical year (1987–1988) at the Department of Ecology, Ethology and Evolution, University of Illinois. Financial support was provided by a National Sciences Foundation grant BSR 86-00487 (USA) and the Norwegian Research Council for Sciences and the Humanities. I am most grateful to prof. Michael Lynch, who besides many fruitful discussions about the subject, provided me with laboratory space and access to equipment as well as various clones of *Daphnia pulex*. The 'hybrids' H2-H10 from an obligate and a cyclical clone were produced by Ms. Lisa Nass and most kindly placed at my disposal. Special thanks are given to Dr. Ken Spitze for help with the electro-phoresis, and for help to solve many practical problems during this work, and to my wife Kari and my sons Martin and Jørgen for help with practical laboratory work.

References

Banta, A. M., 1939. General characterization of Cladocera. In A. M. Banta (ed.). Studies on the physiology, genetics and evolution of some Cladocera. Publication no. 513, Carnegie Institution of Washington. Washington, D.C.: 2–12.

Barker, D. M. & P. D. N. Hebert, 1986. Secondary sex ratio of the cyclic parthenogen *Daphnia magna* (Crustacea: Cladocera) in the Canadian Arctic. Can. J. Zool. 64: 1137–1143.

Beaton, M. J. & P. D. N. Hebert, 1988. Geographical parthenogenesis and polyploidy in *Daphnia pulex*. Am. Nat. 132: 837–845.

Davison, J., 1969. Activation of the ephippial egg of *Daphnia pulex*. J. Gen. Physiol. 53: 562–575.

Frey, D. G., 1982. Contrasting strategies of gamenogenesis in northern and southern populations of Cladocera. Ecology 63: 223–241.

Hebert, P. D. N., 1978. The population biology of *Daphnia* (Crustacea, Daphniidae). Biol. Rev. 53: 387–426.

Hebert, P. D. N., 1987a. Genotypic characteristics of cyclic parthenogens and their obligately asexual derivates. *In* S. C. Stearns, (ed.). The evolution of sex and its consequences. Birkhäuser. Basel: 175–195.

Hebert, P. D. N., 1987b. Genotypic characteristics of the Cladocera. Hydrobiologia 145: 183–193.

Hebert, P. D. N. & T. J. Crease, 1983. Clonal diversity in populations of *Daphnia pulex* reproducing by obligate parthenogenesis. Heredity 51: 353–369.

Hebert, P. D. N. & D. B. McWalter, 1983. Cuticular pigmentation in arctic *Daphnia*: adaptive diversification of asexual lineages? Am. Nat. 122: 286–291.

Hebert, P. D. N., R. D. Ward & L. J. Weider, 1988. Clonal-diversity patterns and breeding system variation in *Daphnia pulex*, and asexual-sexual complex. Evolution 42: 147–159.

Hebert, P. D. N., M. J. Beaton, S. S. Schwartz & D. J. Stanton, 1989. Polyphyletic origins of asexuality in *Daphnia pulex*. I. Breeding system variation and levels of clonal diversity. Evolution 43: 1004–1015.

Hobæk, A. & P. Larsson, 1990. Sex determination in *Daphnia magna*. Ecology 71: (in press)

Innes, D. J. & P. D. N. Hebert, 1988. The origin and genetic basis of obligate parthenogenesis in *Daphnia pulex*. Evolution 42: 1024–1035.

Kleiven, O., P. Latsson & A. Hobæk (submitted). Sexual reproduction in *Daphnia magna* requires three stimuli.

Lynch, M., 1983. Ecological genetics of *Daphnia pulex*. Evolution 37: 358–374.

Lynch, M., 1984. The limits to life history evolution in *Daphnia*. Evolution 38: 465–482.

Lynch, M. & W. Gabriel, 1983. Phenotypic evolution and parthenogenesis. Am. Nat. 122: 745–764.

Mortimer, C. H., 1936. Experimentelle und cytologische Untersuchungen uber den Generationswechsel der Cladoceren. Zoologische Jahrbuch, Abteillung für allgemeine Zoologie und Physiologie 56: 323–388.

Nass, L., 1987. A study of sexual reproduction in *Daphnia pulex*. Master's thesis, Department of Ecology, ethology and evolution, University of Illinois. Unpublished.

Peters, R. H. & R. de Bernardi (ed.), 1987. Daphnia. Mem. Ist. ital. Idrobiol. 45: 1–502.

Schwartz, S. S. & P. D. N. Hebert, 1987. Methods for activation of resting eggs of *Daphnia*. Freshwat. Biol. 17: 373–379.

Sokal, R. R. & F. J. Rohlf, 1981. Biometry. Second edition. Freeman. San Francisco. 859 p.

Steiner, W. W. M. & D. J. Joslyn, 1979. Electrophoretic techniques for the genetic study of mosquitoes. Mosquito News 39: 35–53.

Stross, R. G. & J. G. Hill, 1965. Diapause induction in *Daphnia* requires two stimuli. Science 150: 1462–1464.

Stross, R. G. & J. C. Hill, 1968. Photoperiod control of winter diapause in freshwater crustacean *Daphnia*. Biol. Bul. 134: 176–198.

Weider, L. J. & P. D. N. Hebert, 1987a. Microgeographic genetic heterogeneity of melanic *Daphnia pulex* at a low-arctic site. Heredity 58: 391–399.

Weider, L. J. & P. D. N. Hebert, 1987b. Ecological and physiological differentiation among low-arctic clones of *Daphnia pulex*. Ecology 68: 188–198.

Weider, L. J., M. J. Beaton & P. D. N. Hebert, 1987. Clonal diversity in high-arctic populations of *Daphnia pulex*, a polyploid apomictic complex. Evolution 41: 1335–1346.

Weismann, A., 1876–1879. Beitrage zur Naturgeschichte der Daphnioden. Zeitschrift für wissenschaftliche Zoologie 27–33: 1–486.

Hydrobiologia **225**: 291–299, 1991.
V. Kořínek & D. G. Frey (eds), Biology of Cladocera.

Salinity tolerance and morphology of the osmoregulation organs in Cladocera with special reference to Cladocera from the Aral sea

Nick V. Aladin
Zoological Institute, Academy of Sciences, St Petersburg, 199034, USSR

Key words: salinity tolerance, osmoregulation organs, Cladocera, Aral sea

Abstract

The hyperosmotic regulation of adult Cladocera is determined mainly by the amount of salts consumed with the food and by reabsorption of salts in cells of the nuchal (neck) organ. The hypoosmotic regulation both in adults and embryos is determined mainly by excretion of salts in special epipodite cells or in cells of the nuchal (neck) organ. The salinity of the Aral sea for the last 30 years increased from 8–10‰ to 26–28‰, which led to changes in the Cladocera fauna. At present only 4 species of Cladocera inhabit the Aral sea instead of 14 species that were previously found. These changes are in agreement with osmoregulation capacities of Cladocera.

Note added in proof: Since this paper was accepted for publication, all Cladocera have disappeared from the Aral Sea. This happened when salinity reached 30–32‰. This disappearance was predicted by and agrees with earlier laboratory experiments with Aral Sea Cladocera (Aladin, 1982b).

Introduction

Very little is known about salinity tolerance and osmoregulation organs in Cladocera. Among the Cladocera more than 40 species live in brackish water. Eight species of Cladocera inhabit the oceans of the World and 5 species live in highly mineralised continental water-bodies. The present study focusses on salinity tolerance and morphology of the osmoregulation organs in Cladocera, with special reference to Cladocera from the Aral sea. It forms part of a more extensive study covering various aspects of the osmoregulatory biology of the Cladocera and summarizes the early published papers (Khlebovich & Aladin, 1976; Aladin, 1978, 1982a, 1982b, 1982c, 1983, 1984, 1987, 1988; Aladin & Andreev, 1984; Aladin & Plotnikov, 1985; Aladin & Valdivia Villar, 1987).

Materials and methods

Animals and water used for the experiments were collected from the seas: Barents, White, Baltic, Black, Azov and Japanise; from big salt lakes: Caspian and Aral; and from fresh water lakes near Leningrad and Moscow, and in the Karelian, Lithuanian, Ukrainian, Azerbaidjanian, Kazakh, and Uzbek Republics. Some Cladocera were raised in the laboratory from resting eggs. Dried mud containing these eggs were taken in Canada, USA, Peru, Argentina, Italy, Seychelles and Australia. The freezing point of the Cladocera haemolymph was measured by mean of micro-cryoscopical methods. For morphological studies of osmoregulation organs, both scanning and transmission electron micrographs were used. All the details of microcryoscopy and microscopy are given in an early paper (Aladin, 1982a).

Results and discussion

37 species of Cladocera from 10 families were studied by means of microcryoscopical methods. The results of this study are given in Fig. 1.

The Cladocera are able to osmoregulate. The hyperosmotic regulation of haemolymph (Fig. A, A_I, A_{II}) was found in freshwater and slightly brackishwater Cladocera: *Sida crystallina, Diaphanosoma brachyurum, Daphnia magna, Daphnia pulex, Daphnia longispina, Ceriodaphnia reticulata, Ceriodaphnia cornuta, Simocephalus vetulus, Bosmina longirostris, Macrothrix hirsuticornis, Chydorus sphaericus, Eurycercus glacialis, Echinisca capensis, Moina brachiata, Moina macrocopa, Polyphemus pediculus, Polyphemus exiguus, Cercopagis pengoi pengoi, Cercopagis pengoi aralensis, Apaqis cylindrata, Bythotrephes longimanus*

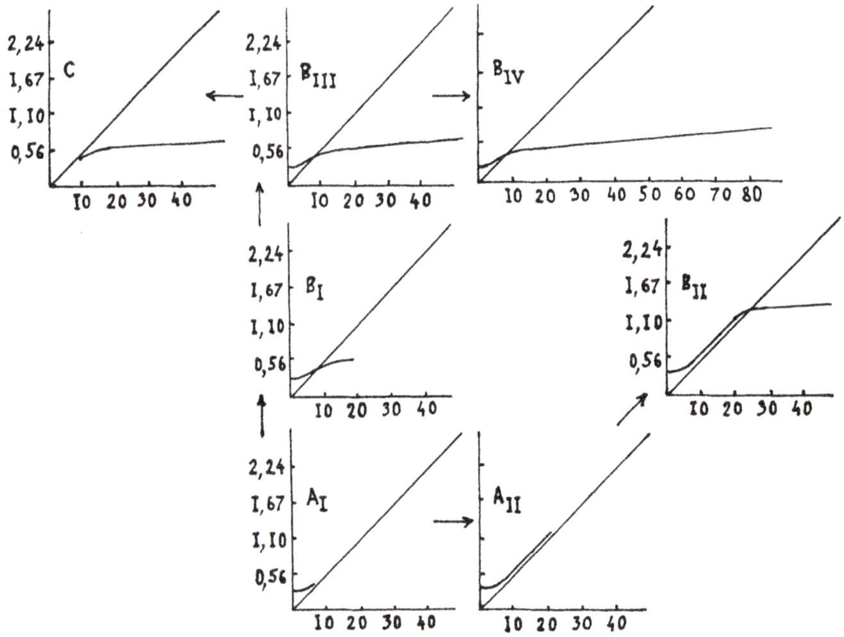

Fig. 1. Different types of osmoregulation in Cladocera.
vertical axis – freezing point (depression) of haemolymph, – °C
horizontal axis – salinity of water, ‰.
bisector line – isoosmotic line
curve line – freezing point (depression) of haemolymph
A – Hyperosmotic regulation of haemolymph

A_I – Freshwater localities: *Sida crystallina, Diaphanosoma brachyurum, Daphnia magna, Daphnia pulex, Ceriodaphnia reticulata, Ceriodaphnia cornuta, Macrothrix hirsuticornis, Chydorus sphaericus, Eurycercus glacialis, Moina brachiata, Polyphemus pediculus, Bythotrephes longimanus arctica, Leptodora kindtii.*

A_{II} – Slightly brackishwater localities: *Sida crystallina, Daphnia magna, Daphnia pulex, Daphnia longispina, Ceriodaphnia reticulata, Ceriodaphnia cornuta, Simocephalus vetulus, Bosmina longirostris, Macrothrix hirsuticornis, Chydorus sphaericus, Echinisca capensis, Moina brachiata, Moina macrocopa, Polyphemus exiguus, Cercopagis pengoi pengoi, Cercopagis pengoi aralensis, Apagis cylindrata.*

B – Combination of hyperosmotic regulation of haemolymph (at low salinity) with hypoosmotic regulation (at high salinity)

B_I – Brackishwater localities: *Podonevadne angusta, Cornigerius maeoticus hircus*

B_{II} – Highly mineralised water localities in Australia: *Daphniopsis pusilla, Daphniopsis australis*

B_{III} – Highly mineralised water localities: *Moina hutchinsoni, Evadne anonyx, Podonevadne trigona, Podonevadne camptonyx.*

B_{IV} – Extremely high mineralised water localities: *Moina mongolica*

C – Hypoosmotic regulation of haemolymph

C_I – Marine localities: *Penilia avirostris, Pleopis schmackeri, Pleopis polyphemoides, Pseudevadne tergestina, Podon leuckarti, Evadne nordmanni arctica* and *Leptodora kindtii*. The hypoosmotic regulation of haemolympha (Fig. 1, C_I) was found in the marine Cladocera: *Penilia avirostris*, *Pleopis schmackeri*, *Pleopis polyphemoides*, *Pseudevadne tergestrina*, *Podon leuckarti*, and *Evadne nordmanni*. The combination of the hyperosmotic regulation of haemolymph (at low salinity) with the hypoosmotic (at high salinity) (Fig. 1, B_I–B_{IV}) was found in Cladocera from brackish water and highly mineralised water: *Daphniopsis pusilla*, *Daphniopsis australis*, *Moina mongolica*, *Moina hutchinsoni*, *Evadne anonyx*, *Podonevadne trigona*, *Podonevadne camptonyx*, *Podonevadne angusta* and *Cornigerius maeoticus hircus*.

The hyperosmotic regulation in adults Cladocera is determined mainly by the amount of salts consumed with the food (Belyaev, 1950; Aladin, 1982c, 1984) and by reabsorption of salts in the maxillary gland (Aladin & Plotnikov, 1985). The hyperosmotic regulation in embryos of Cladocera is determined mainly by absorption of salts in cells of the nuchal (neck) organ (Aladin, 1982c, 1988; Halcrow, 1982, 1985; Aladin & Valdivia Villar, 1987). In the brood pouch the embryos cannot feed and must have the special osmoregulation organ – nuchal (Neck) organ (Fig. 2). After leaving the brood pouch the young Cladocera can feed and after the first postembryonal moulting, the nuchal (neck) organ has disappeared (Fig. 3).

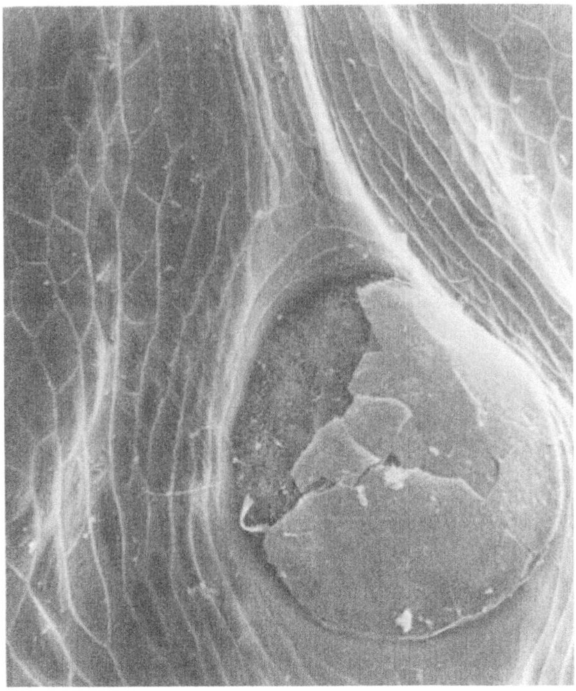

Fig. 2. Nuchal (neck) organ on the head of the first instar of *Daphnia magna*
1 – Complicated surface of cells under nuchal (neck) organ cuticle

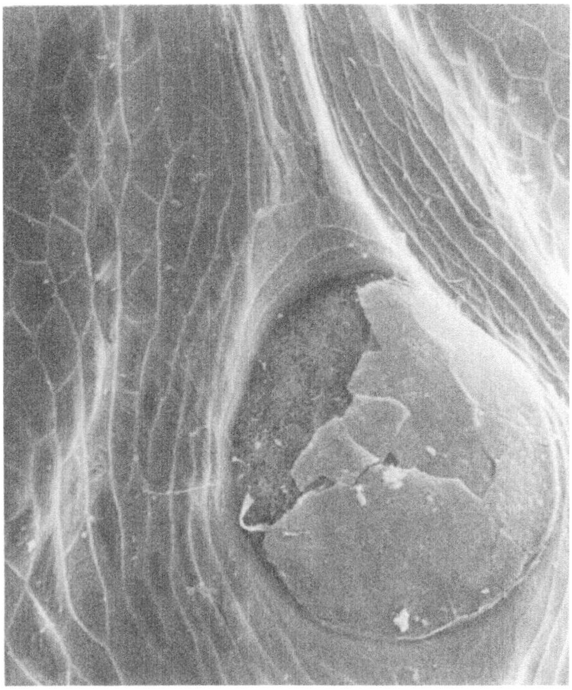

Fig. 3. Head of older juvenile *Daphnia magna* without nuchal (neck) organ

The hypoosmotic regulation both in adults and embryos is determined mainly by excretion of salts in special epipodite cells (Fig. 4) (Aladin, 1978, 1982a, 1982c, 1983, 1988) or in cells of the nuchal (neck) organ (Fig. 5) (Khlebovich & Aladin, 1976; Potts & Durning, 1980; Aladin, 1982a, 1988; Meurice & Goffinet, 1982, 1983).

The ultrastructure of the nuchal (neck) organ cells and of the special epipodite cells are very much alike. All of them have a dense cytoplasm with numerous mitochondria distributed throughout a lacunar system (Fig. 6–7). The cuticular site of the nuchal (neck) organ cells and of special epipodite cells is characterized by a high permeability to ions, in contrast to the sur-

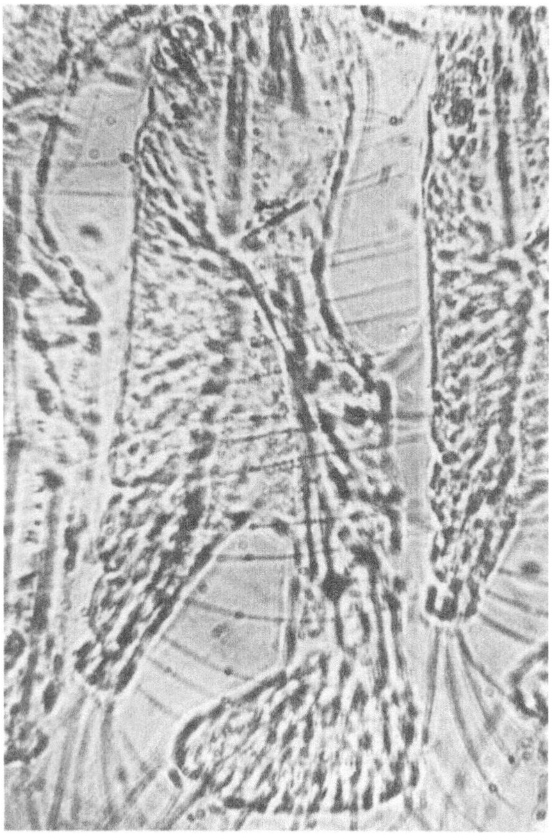

Fig. 4. Torocal appendage of *Penelia avirostris*
1 – special ion-transporting epipodite cells

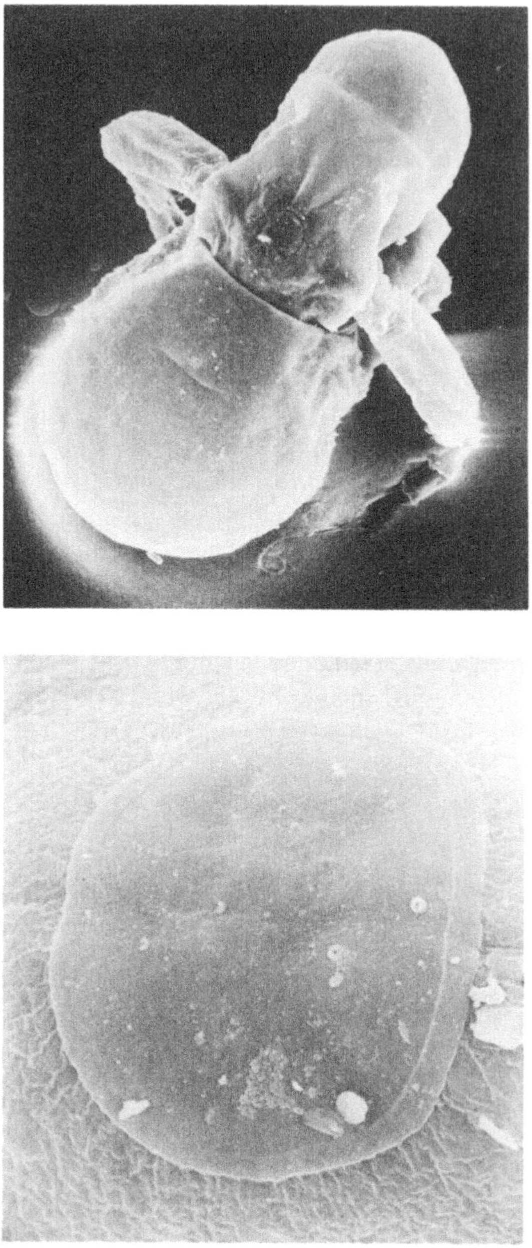

Fig. 5. Nuchal (neck) organ on the head of adult *Podon leuckarti* a – low magnification; b – high magnification
1 – nuchal (neck) organ

rounding cuticle. These cells in Cladocera exhibits structural features that are relevant to an ion-transporting and an osmoregulation function (Khlebovicn & Aladin, 1976; Aladin, 1978, 1982a, 1982c, 1983, 1988; Potts & Durning, 1980; Meurice & Goffinet, 1982, 1983; Halcrow, 1982, 1985). It is possible to compare the morphology of osmoregulating organs in Cladocera with those in the salt-tolerant anostracan *Artemia salina*. Like *Artemia salina*, the Cladocerans *Penilia avirostris*, *Moina mongolica*, *Moina hutchinsoni*, *Daphniopsis pusilla* and *Daphniopsis australis* use the special epipodite cells for osmoregulation (Copeland, 1967). Like the nauplii of *Artemia salina*, the species *Pleopis schmackeri*, *Pleopis polyphemoides*, *Pseudevadne tergestina*, *Podon leuckarti*, *Evadne nordmanni*, *Evadne anonyx*, *Podonevadne trigona*, *Podonevadne camptonyx*, *Podonevadne angusta*, *Cornigerius macoticus hircus*, and embryos of freshwater and slightly brackishwater Cladocera use the nuchal (neck) cell organ for osmoregulation (Hootman and Conte, 1975).

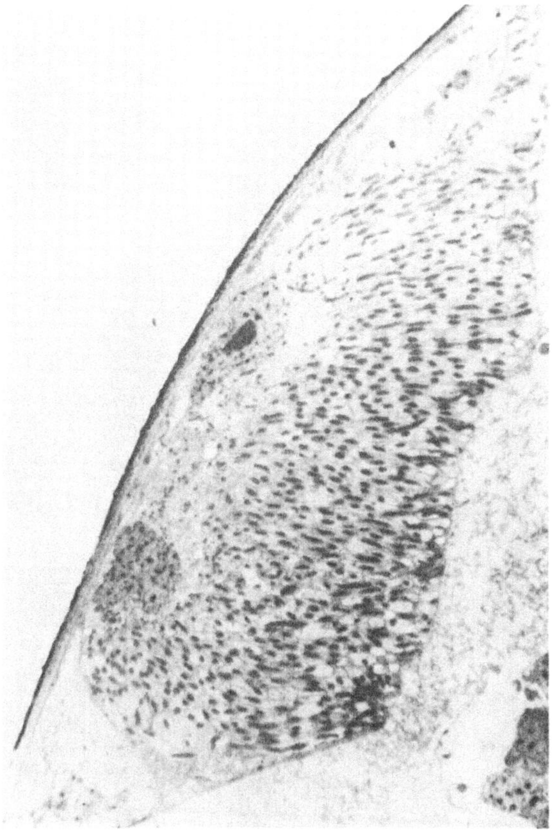

Fig. 6. Sagittal section of nuchal (neck) organ of adult *Evadne nordmanni*
1 – nucleus
2 – lacunar ventral zone
3 – cuticle
4 – haemocoelic activity

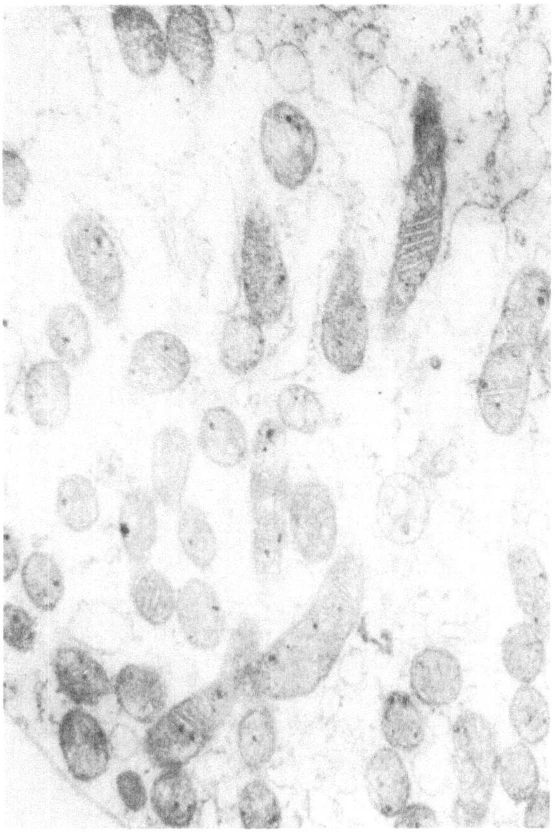

Fig. 7. The basal zone of ion-transporting cell of nuchal (neck) organ of adult *Evadne nordmanni*
1 – numerous mitochondria

The other structure that serves for osmoregulation is the closed brood pouch of *Penilia avirostris*, *Moina mongolica*, *Moina hutchinsoni*, *Pleopis schmackeri*, *Pleopis polyphemoides*, *Pseudevadne tergestina*, *Podon leuckartii*, *Evadne nordmanni*, *Evadne anonyx*, *Podonevadne camptonyx*, *Podonevadne angusta* and *Cornigerius maeoticus hircus*. When the brood pouch of these Cladocera contains developing eggs and young embryos, the osmolarity of the marsupial fluid equals that of the haemolymph (Fig. 8, 9). But when the brood pouch contains embryos in the final stages of development with a nuchal (neck) organ or epipodites, the marsupial fluid rises in concentration to that of the surrounding sea or brackishwater (Fig. 10, 11). The ability to regulate the embryonic environment and so protect the early developmental stages from extremes of osmotic stress is one of the important physiological adaptations that permit some Cladocera to inhabit the oceans and highly mineralised continental water-bodies (Khlebovich & Aladin, 1976; Aladin, 1978, 1982a, 1982, 1983).

The Aral sea, formerly the world's fourth largest lake in area, is disappearing. Between 1960 and 1989, its level dropped nearly 14 meters, its area decreased by 40%, and its volume decreased by 60% (Fig. 12) (Micklin, 1988; Micklin & Bond, 1988). Recession of the water level has resulted from reduced inflow caused primarily by withdrawals of water for irrigation. Because of desiccation of the Aral sea the salinity has in-

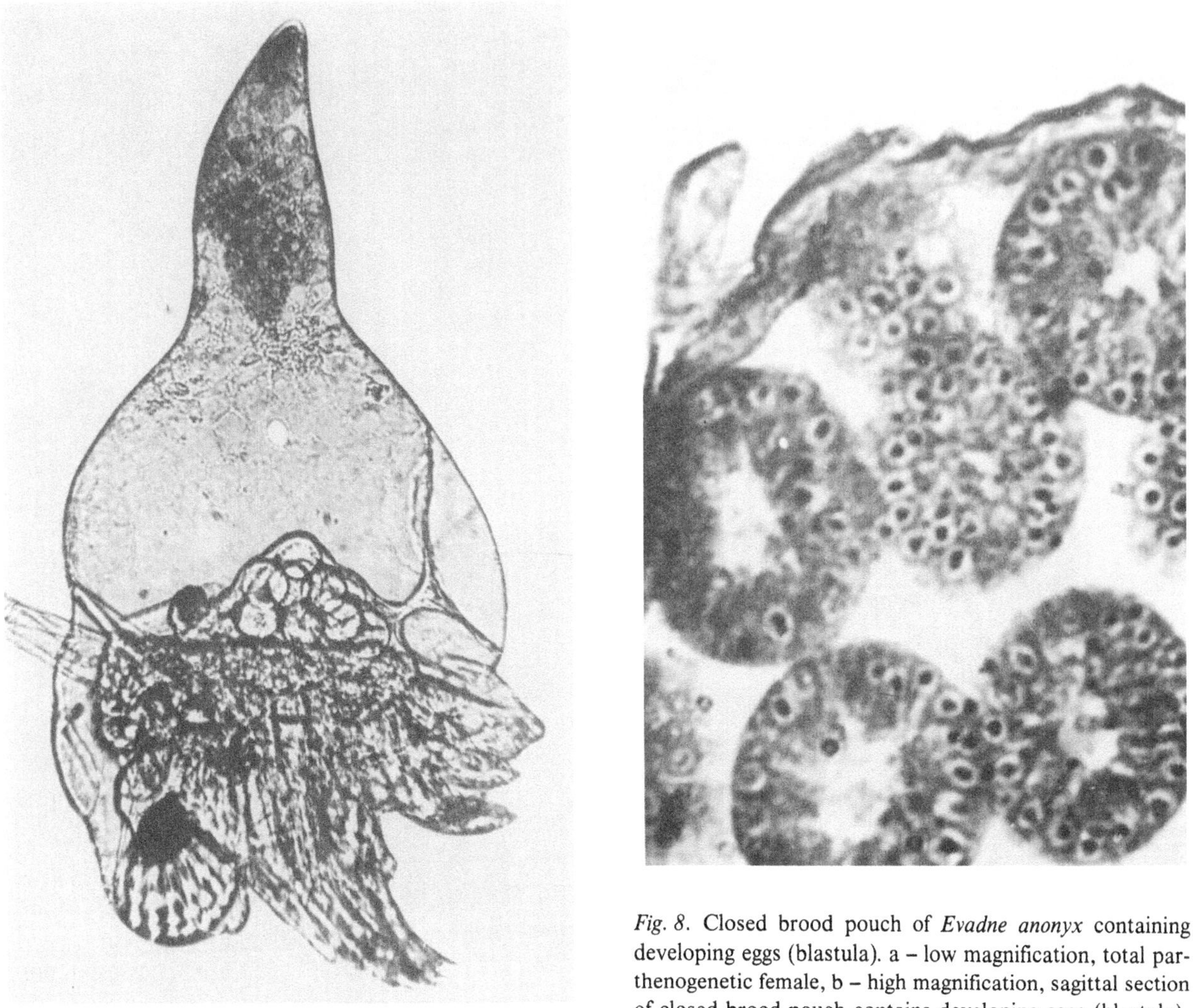

Fig. 8. Closed brood pouch of *Evadne anonyx* containing developing eggs (blastula). a – low magnification, total parthenogenetic female, b – high magnification, sagittal section of closed brood pouch contains developing eggs (blastula) 1 – developing eggs (blastula)

creased over the last 30 years from 8–10‰ to 26–28‰, and this had led to changes in flora and fauna.

Serious changes have take place in the Cladocera fauna. In the early 1960s, 14 species inhabited the Aral sea: *Diaphanosoma brachyurum, Chydorus sphaericus, Alona rectangula, Bosmina longirostris, Daphnia longispina, Ceriodaphnia reticulata, Ceriodaphnia cornuta, Ceriodaphnia pulchella, Moina mongolica, Moina micrura, Podonevadne camptonyx, Podonevadne angusta, Evadne anonyx* and *Cercopagis pengoi aralensis* (Mordukhai-Boltovskoi, 1974). Some Cladocera, such as *Alona rectangula, Ceriodaphnia reticulata, Moina mongolica, Podonevadne camptonyx, Po*donevadne angusta and *Evadne anonyx* lived in the open sea with a salinity over 8‰. Other Cladocera, such as *Diaphanosoma brachyurum, Chydorus sphaericus, Bosmina longirostris, Daphnia longispina, Ceriodaphnia cornuta, Ceriodaphnia pulchella, Moina micrura* and *Cercopagis pengoi aralensis*, inhabited the north and south parts of the Aral sea near the Syrdarya and Amudarya river deltas with a salinity less than 8‰.

The negative influence of increasing salinity on Cladocera started in 1971, when the salinity in the open Aral sea exceeded 12‰. By the middle 1970s, when the salinity in the open Aral sea exceeded 14‰, nearly all Cladocera of freshwater origin had already died. In the summer of 1980

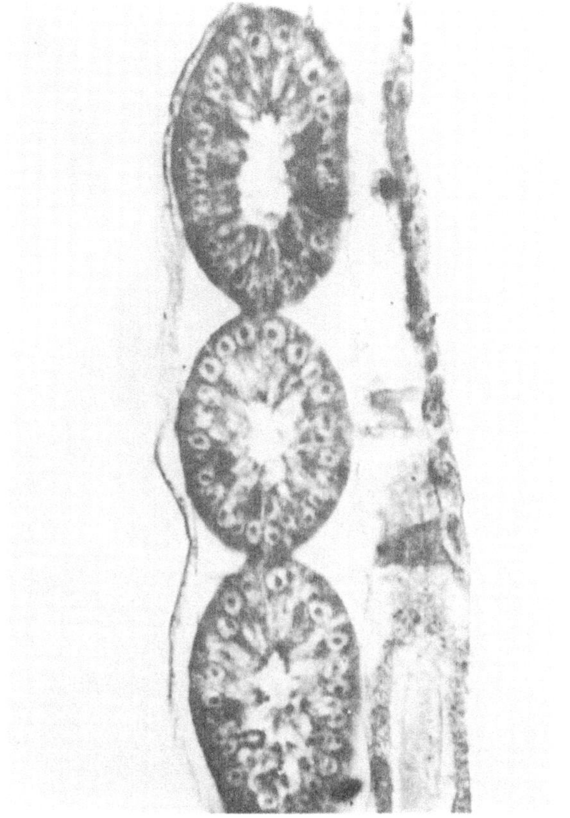

Fig. 9. Sagittal section of closed brood pouch of *Penilia avirostris* containing developing eggs (blastula)

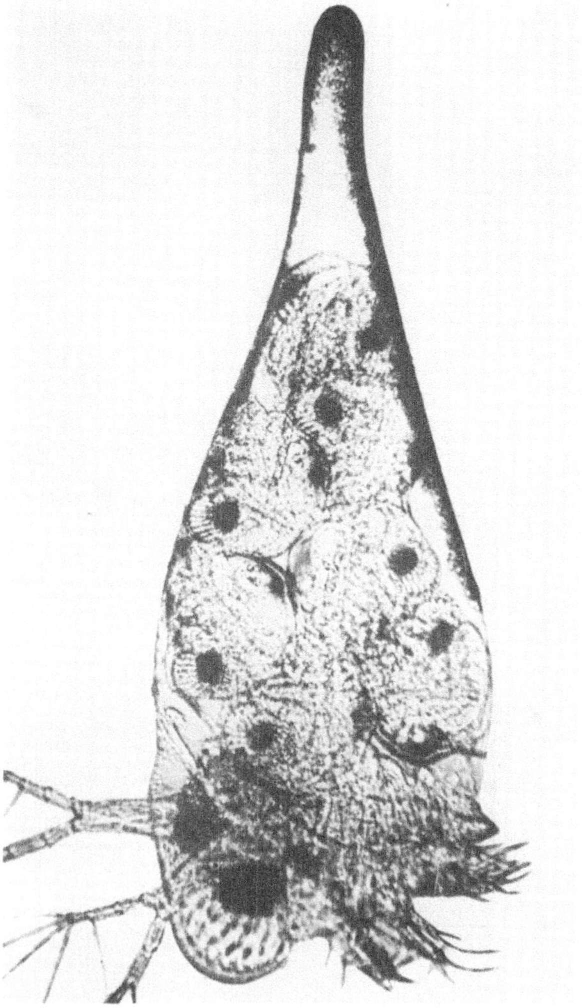

Fig. 10. Closed brood pouch of *Evadne anonyx* containing embryos at final stages of development with nuchal (neck) organ

during an expedition all over the Aral sea, only 5 species of Cladocera were found. At that time the salinity in the open Aral sea was more than 17‰. But in the north and south parts of the Aral sea near the deltas of the Syrdarya and Amudarya rivers, the salinity was about 16‰. In the open Aral sea only the podonids *Podonevadne camptonyx*, *Podonevadne angusta*, *Podonevadne trigona* and *Evadne anonyx* were found. In the southern part of the Aral sea near the Amudarya river delta *Cercopagis pengoi aralensis* was found only at two stations with a salinity of 15,92‰ (Aladin & Andreev, 1984). It is important to note that in 1980 *Podonevadne trigona* was caught in the Aral sea for the first time. Possibly this podonid became acclimatized in the Aral sea during introduction of fishes from the Caspian sea.

Regular observation during the last ten years in the north and south part of the Aral sea and in the open Aral sea around Barsakelmes Island show that now at a salinity of 26–28‰ only the 4 species of Cladocera *Podonevadne camptonyx*, *Podonevadne angusta*, *Podonevadne trigona*, and *Evadne anonyx* still live there. Until the middle 1980s when the salinity was less than 24‰, all these 4 species were numerous. But when the salinity exceeded 24‰, only *Podonevadne camptonyx* and *Evadne anonyx* remained numerous.

The serious changes that took place in the Cladocera fauna of the Aral sea during its desiccation and salinization are in agreement with

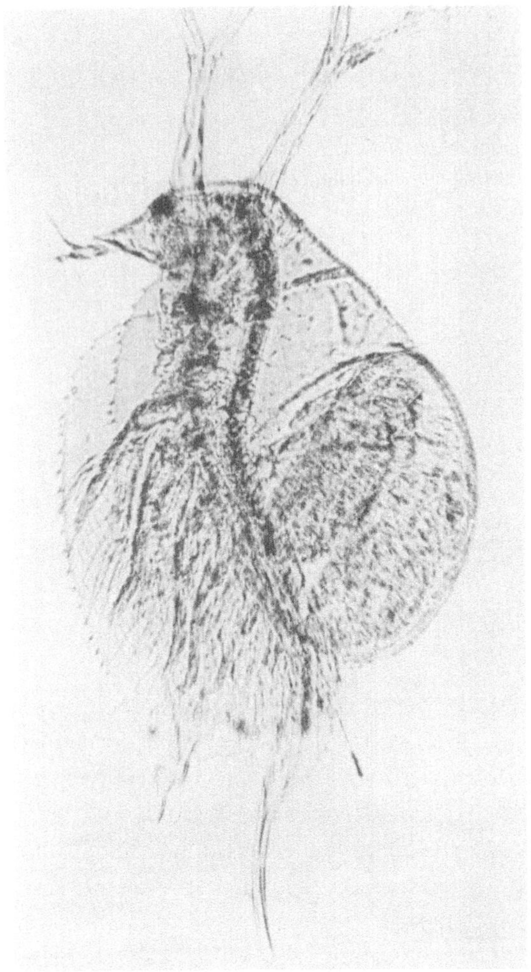

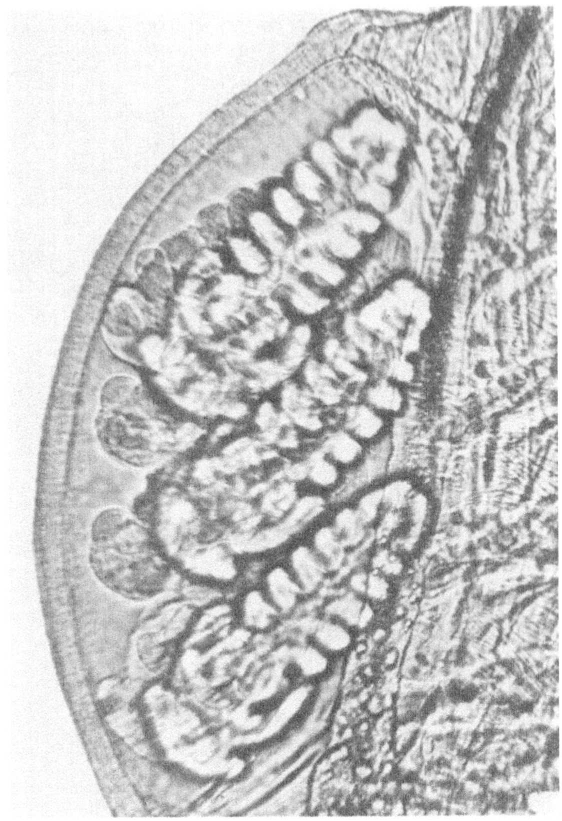

Fig. 11. Closed brood pouch of *Pinilia avirostris* containing embryos at final stages of development with epipodites. a – low magnification, partenogenetic female; b – high magnification, saggittal view on closed brood pouch containing embryos at final stages of development with epipodites.

osmoregulation capacities of Cladocera. Now only Podonidae live in the Aral sea because these Cladocera are able to hypoosmotically regulate their haemolymph by means of the nuchal (neck) organ and thus can protect the early development stages of eggs and embryos in their closed brood pouch from extremes of osmotic stress.

Other cladocerans (except *Moina mongolica*) that are not able to hypoosmotically regulate and thus cannot protect the early developmental stages of eggs and embryos died. *Moina mongolica* is able to hypoosmotically regulate by means of special epipodite cells and thus can protect early developmental stages of eggs and embryos in their closed brood pouch, but this species was last seen in the Aral see in 1974 when the salinity was about 14‰. The main reason for the disappearance of *Moina mongolica* was the acclimatization of plankton eating fishes (Mordukhai-Boltovskoi, 1974; Kortunova, 1975; Aladin & Andreev, 1984).

In the Kiziltash estuary *Moina mongolica* is found at salinities between 40,7 and 63,8‰, and reproduction occurs even at 97,14‰ (Sukhanova, 1971). In laboratory experiments *Moina mongolica* was acclimated to 88‰ of Aral sea water (Aladin, 1983). At present *Moina mongolica* lives around the Aral sea in salt lakes and former lagoons and gulfs now separated from the Aral sea. It is very possible that in the near future *Moina mongolica* will naturally reacclimatize in the open Aral sea.

Fig. 12. The Aral sea scheme
Dotted line – the shore line in 1960
Unbroken line – the shore line in 1989
Arrow 1 – Syrdarya river delta
Arrow 2 – Amudarya river delta

References

Aladin, N. V., 1978. Hypoosmotic regulation in the marine Cladocera *Penilia avirostris*. J. Comp. Biochem. Physiol. 14: 599–601 (in Russian).

Aladin, N. V., 1982a. Salinity adaptations and osmoregulation abilities of the Cladocera. Forms from open Seas and oceans. Zool. J. 61: 341–351 (in Russian).

Aladin, N. V., 1982b. Salinity adaptation and osmoregulation abilities of the Cladocera. Forms from Caspian and Aral seas. Zool. J. 61: 507–514 (in Russian).

Aladin, N. V., 1982c. Salinity adaptations and osmoregulation abilities of the Cladocera. Forms from brackish and fresh water. Zool. J. 61: 851–860 (in Russian).

Aladin, N. V., 1983. Amphiosmotic regulation in the euryhaline Cladocera *Moina mongolica* Daday. Hydrobiol. J. 19: 76–81 (in Russian).

Aladin, N. V., 1984. The influence of temperature on the osmoregulation abilities of the Branchiopoda and Ostracoda. Zool. J. 63: 1158–1163 (in Russian).

Aladin, N. V., 1987. Salinity adaptation and osmoregulation in Ostracoda and Cladocera from continental water bodies of Australia and from Seyshelles Island. Zool. J. 66: 1822–1828 (in Russian).

Aladin, N. V., 1988. Reprodactive salinity adaptations and salinity dependent features of the embryonic development in Ostracoda and Branchiopoda. Zool. J. 67: 974–982 (in Russian).

Aladin, N. V. & N. I. Andreev, 1984. The influence of Aral sea salinity on Cladocera fauna changes. Hydrobiol. J. 20: 23–28 (in Russian).

Aladin, N. V. & I. S. Plotnikov, 1985. Microcryoscopical investigation of the liguid from maxillary gland of *Daphnia magna* Straus during acclimation to different salinities. Hydrobiol. J. 21: 62–65 (in Russian).

Aladin, N. V. & R. S. Villar Valdivia, 1987. Microcryoscopical investigation of the liquid from egg and embryo of Cladocera. Hydrobiol. J. 23: 93–97 (in Russian).

Belyaev, G. M., 1950. Osmoregulation abilities of low Crustacea from inland water. Trans. allunion. Hydrobiol. Soc. 2: 194–213 (in Russian).

Copeland, D. E., 1967. A study of salt-secreting cells in the brine shrimp (*Artemia salina*). Protoplasma 63: 363–384.

Halcrow, K., 1982. Some ultrastructural features of nuchal organ of *Daphnia magna* Straus (*crustacea*: *Branchiopoda*). Can. J. Zool. 60: 1257–1264.

Halcrow, K., 1985. A note on the significance of the neck organ of *Leptodora kindtii* (Focke) (*Crustacea*, *Cladocera*). Can. J. Zool. 63: 738–740.

Hootman, S. R. & F. P. Conte, 1975. Ultrastructure of Artemia larval neck organ. J. Morphol. 145: 371–385.

Khlebovich, V. V. & N. V. Aladin, 1976. Hypotonic regulation in marine Cladocerans *Evadne nordmanni* and *Podon leuckarti*. J. Comp. Biochem. Physiol. 12: 591–592 (in Russian).

Kortunova, T. A., 1975. On changes in Aral sea zooplankton in 1959–1968 years. Zool. J. 54: 657–669 (in Russian).

Meurice, J. Cl. & G. Goffinet, 1982. Structure et fonction de l'organe nucal des Cladoceres marins gymnomeres. C. R. Acad. Sci. Paris. 29(D: 693–695.

Meurice, J. Cl. & G. Goffinet, 1983. Ultrastructural evidence of the ion-transporting role of the adult and larval neck organ of the marine gymnomeran Cladocera (*Crustacea*, *Branchiopoda*). Cell. Tissue Res. 234: 351–363.

Micklin, P. P., 1988. Desiccation of the Aral sea: A water management disaster in the Soviet Union. Science. 241: 1170–1176.

Micklin, P. P. & A. R. Bond, 1988. Reflections on environmentalism and river diversion projects. Sov. Econ. 4: 253–274.

Mordukhai-Boltovskoi, F. D., 1974. Subclass Branchiopoda. In F. D. Mordukhai-Boltovskoi (ed.) Atlas of the Aral sea invertebrates. Moscow: 112–134 (in Russian).

Potts, W. T. W. & Durning C. T., 1980. Physiological evolution in the Branchiopoda. Comp. Biochem. Physiol. 67b: 475–484.

Sukhanova, E. R., 1971. *Moina microphthalma* (Cladocera, Daphniidae) under ultrahaline conditions in the North Caucasus. Zool. J. 50: 285–287 (in Russian).

Hydrobiologia **225**: 301–307, 1991.
V. Kořínek & D. G. Frey (eds), Biology of Cladocera.

Self-maintaining plankton: pelagic Cladocera in small microcosms with lake water

Vladimir F. Matveev
A.N. Severtsov Institute of Animal Evolutionary Morphology and Ecology, USSR Academy of Sciences, Leninsky prospekt, 33, Moscow 117171, USSR

Key words: Cladocera, nutrient recycling, microcosms, density regulation

Abstract

Monospecific and mixed populations of pelagic cladocerans, *Daphnia galeata*, *Bosmina coregoni*, and *Diaphanosoma brachyurum* of Lake Glubokoe (Moscow Region) were isolated in 0.6 litre flasks with their natural medium – epilimnion water. In the laboratory they persisted for more than 100 days without external additions of food until the experiment was stopped. In three-species cultures *Diaphanosoma* was outcompeted, but in the presence of only *Bosmina* it persisted. *Daphnia* coexisted with *Bosmina* in two- and three-species cultures the whole period of observations in spite of competition. Birth and death rates of animals in microcosms depended on their density indicating that experimental conditions did not spoil natural population regulation. Demographic time lags underlay the observed oscillations in numbers. Because the time of Cladocera persistence in the cultures was comparable with lake vegetational period I conclude that summer zooplankton is capable of existing solely at the expense of nutrient recycling within epilimnion. Density regulation of cladocerans reinforces the stability of the system 'herbivore – algae', but it is weakened by the demographic time lags.

Introduction

Natural zooplankton affects phytoplankton in two ways. First, herbivore grazing may lead to the diminution of algae concentration or their complete elimination. On the other hand nutrient excretion by animals stimulates phytoplankton reproduction filling up grazing losses. The persistence of any planktonic population of Cladocera in a lake may be considered as a result of stable equilibrium in the system 'nutrients – algae – consumer – nutrients'. The studies of phosphorus recycling allow to suggest that summer primary production of lake plankton may be sustained mainly by excreted nutrients, but not by allochtonous input (Devol, 1979; Lehman, 1980; Gutelmacher, 1986). This makes one think that the system consisting of a cladoceran population and its resource may be completely autonomous. What are the mechanisms that determine its stability, that is ensure prolonged persistence? A way to study the problem can be the cultivation of Cladocera in laboratory microcosms (for review see Ringelberg & Kersting, 1978). However, one of the shortcomings of the already assumed experimental designes was an arbitrary selection of the microcosm components fulfilled on the basis of easiness of animal cultivation but not on the striving for saving the natural association of a producer and a consumer. Meanwhile, the natural association may differ completely from the artificial one by its stability due to preceding coevolution or seasonal shifts in the resource quality (Kerfoot *et al.*, 1985). On the other hand, the existing models of microcosms (Ringelberg & Kersting, 1978; Elstad, 1986; Sommer, 1988) have different compartments for food and consumer. This was designed for preventing overgrazing and consumer extinction. Indeed, complicating physical environment makes the system stable (Nixon, 1969). However, by this precluded is also the possibility of studying the natural mechanism of stability: algae and zooplankton are not separated in the epilimnion.

In the present paper an effort was made to clear up whether populations of pelagic Cladocera could persist without external food additions if they were isolated simply in flasks with a part of their natural medium. If the time of living of such populations was comparable with vegetational season one might conclude that summer plankton was really capable of existing only at the expense of internal, epilimnial nutrient recycling (*cf.* Lehman, 1980). In the latter case one should also explain what prevents the system from crash: at high consumer biomass the losses of food from grazing are not balanced by algal recruitment from nutrient excretion (Sterner, 1986). In my experiments the emphasis was given to cladocerans' population regulation, not to water chemistry as in the other studies. I used pelagic species of Lake Glubokoe (Matveev, 1986): *Daphnia galeata* Sars, *Bosmina coregoni* Baird, *Diaphanosoma brachyurum* (Lievin). They were allowed to grow in isolated cultures or in two- to three-species mixtures. By this way I tested the system's resilience to interspecific competition. Because the populations of these animals oscillated in the experiment, while the ability to avoid extinction may depend on oscillations (Karr, 1982) I also analysed the causes for such numerical variability.

Methods

0.6-litre Erlenmeyer flasks with ground-in stoppers were used for microcosms. On 16 August, 1985 before starting the experiment the flasks were filled with Lake Glubokoe water taken from epilimnion and filtered through 50 μm mesh size sieve. On 19 August 12 different-aged females of each cladoceran were put in the flasks either in isolation or together with equal quantity of individuals of one or two other species. Monospecific cultures and each combination of the two-species ones were represented by one replicate, while three-species mixtures – by three replicates. Each 1–3 days the content of flasks was filtered through special funnel with sieve (100 μm mesh size) for concentrating crustaceans. The culture water was aerated by pouring in small portions into a vessel from the height of 15 cm. Thoroughly washed flasks were filled then again with the same water where the animals were returned after census. The transfer was done with a pipette, while cladocerans and their embryos were counted in Bogorov chamber containing reserve lake water. The latter was permanently kept in the darkness in a refrigerator. In case of littering after animal transfer it was filtered through paper filter. Corpses and motionless animals with beating hearts were analysed under the microscope but afterwards were removed from the experiment to imitate their sedimentation from epilimnion. In the beginning, in the middle and at the end of cultivation 5 ml of water were taken from different flasks to determine pH. To refill the full flask volume the reserve water was used. Each time the estimated pH-values were 7.4. This indicated to the lack of oxygen limitation for animals: summer pH-values usually lie between 7.4 and 7.7 in Lake Glubokoe epilimnion. When oxygen disappears pH declines to 6.2 (Shcherbakov, 1967). The experimental flasks were kept at natural scattered light with the illumination below 1 000 lx. After day 31 (18 September) each evening the light period was prolonged up to 11:00 p.m. with a lamp yielding 200 lx to avoid animal gamogenesis. Before and after 100 day cultivation as well as on day 10, when one replicate with *Diaphanosoma* went extinct the remaining water was analysed in respect to bacterial and algal densities. Bacteria were counted on membrane filters after staining with erythrosin (Sorokin & Kadota, 1972). Algae were counted in haemocytometer after decantation and fixation with Lugol solution. For flagellates and solitary chlorococcales cell volumes were determined by approximation to the nearest geometric solid. In colonial algae only the numbers of cells were estimated. Gut contents of dead and motionless animals were analysed before removing them from the cultures. Birth rate, *b*, death rate, *d*, and the rate of population increase, *r* of Cladocera populations were estimated according to Paloheimo (1974). Time lags in responses of these variables to density were estimated then by means of cross-correlations (Matveev, 1985). In the latter case normal Pearson correlation coefficient was calculated as well as Spearman rank correlation coefficient was used to avoid underestimation of the relationships if they were nonlinear.

Results

In 7 of 9 microcosms Cladocera survived without external additions of food until the experiment was terminated – more than 100 days. In two microcosms animals died. All animals died out by unknown reasons in two-species culture (*Daphnia* + *Diaphanosoma*) on the fourth day, and in the isolated culture of *Diaphanosoma* – on the tenth day. In the first case inoculum went extinct, and in the second one the population had time to reach the density of 41 ind/flask but later declined to zero within 5 days. Before the beginning of cultivation the water contained two categories of small edible cells, *Tetraëdron minimum* and flagellates as well as colonies of two types – green *Sphaerocystis polycocca* and cyanobacteria *Coelosphaerium kuetzingianum* (Table 1). In the guts of all cladocerans during the first two weeks green cells of *Sphaerocystis* occurred occasionally, and empty covers of *Tetraëdron* were found permanently. Nevertheless, after 10 days of the experiment the density of *Tetraëdron* did not decline (Table 1: extinct culture of *Diaphanosoma*). In a month the guts were filled mainly with brown grainy mass containing occasionally *Tetraëdron*. At the end of cultivation microcosm phytoplankton was represented exclusively by small flagellates, while colonies disappeared. The densities of bacteria in the beginning, after 10 days and at the end of cultivation were of the same order of magnitude (Table 1). Flagellate density in the water of extinct culture of *Diaphanosoma* was 59 times lower than the initial one, of *Sphaerocystis* – 17 times lower. Interestingly, in other cultures flagellate density did not decline even after 100 days (Table 1).

Populations of all cladocerans in all cultures showed oscillations in numbers (Fig. 1–4). Preliminary cross-correlation analysis revealed that they were a result of demographic time lags in responses of birth and death rates to density. As a consequence, the density dependence of the rate of population increase was also delayed (Fig. 5). In all two- and three-species cultures *Daphnia* and *Bosmina* coexisted during the entire experiment. They competed because their birth and death rates were correlated to their total density (Fig. 6) (Matveev, 1983, 1985). Besides, competition was

Table 1. Algae and bacteria before the beginning, 10 days later, and after the ending of Cladocera cultivation in microcosms.

Species	Density*		
	Before	10 days later**	After
Tetraëdron minimum	$1.70 \cdot 10^4$	$1.43 \cdot 10^4$	Very scarce
Small flagellates	$1.96 \cdot 10^4$	$3.3 \cdot 10^2$	$1–2.2 \cdot 10^4$
Sphaerocystis polycocca	$4.33 \cdot 10^3$	$2.5 \cdot 10^2$	0
Coelosphaerium kuetzingianum	$1.43 \cdot 10^4$	$7.4 \cdot 10^3$	Very scarce
Bacteria	$0.98 \cdot 10^6$	$1.55 \cdot 10^6$	$0.27–1.02 \cdot 10^6$

* $\mu m^3 ml^{-1}$ – in case of *Tetraëdron* and flagellates, cells ml^{-1} – in other cases.
** Water after *Diaphanosoma* extinction in monospecific culture.

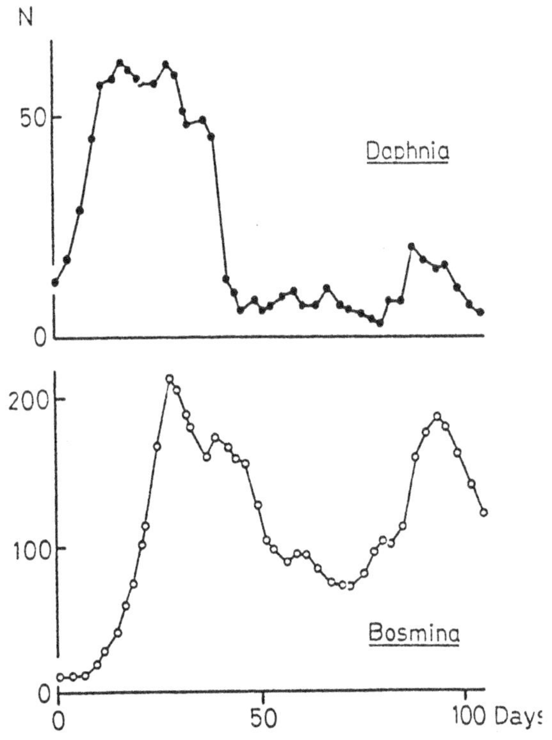

Fig. 1. Monospecific cladoceran populations. N = animal density (ind./0.6 l).

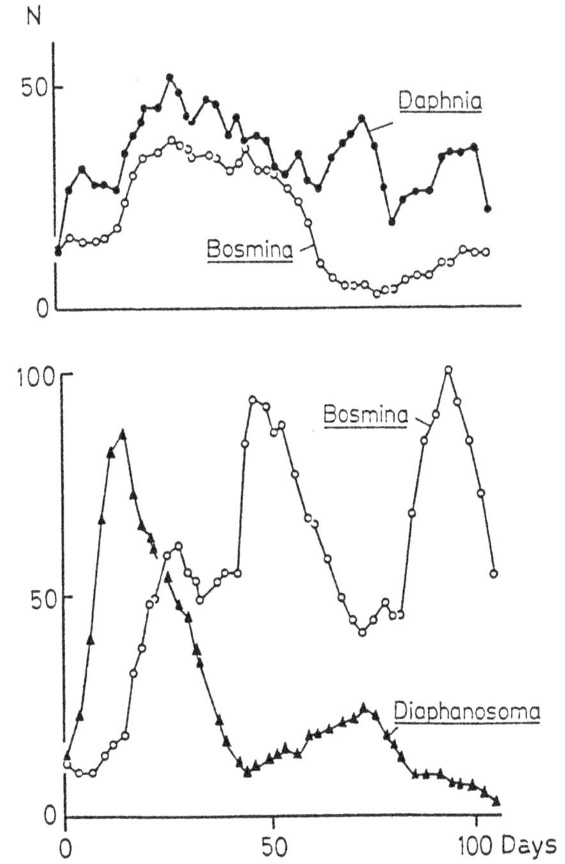

Fig. 2. Two-species cladoceran populations. N = animal density (ind./0.6 l).

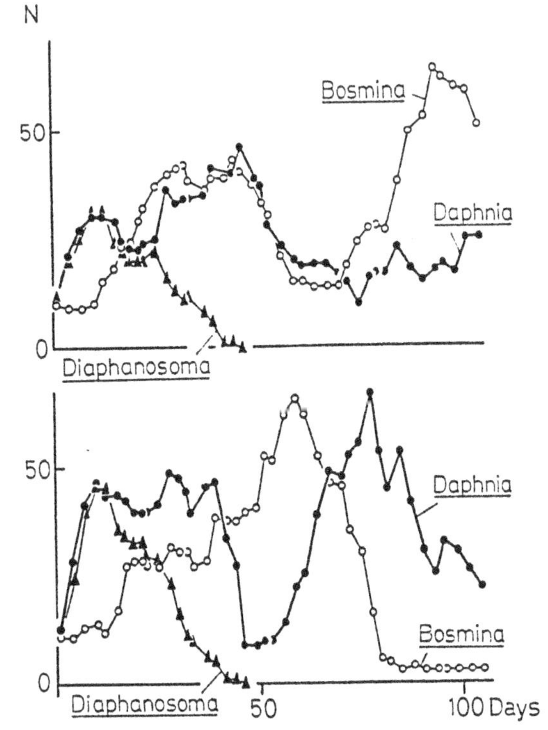

Fig. 3. Three-species cladoceran populations. N = animal density (ind./0.6 l).

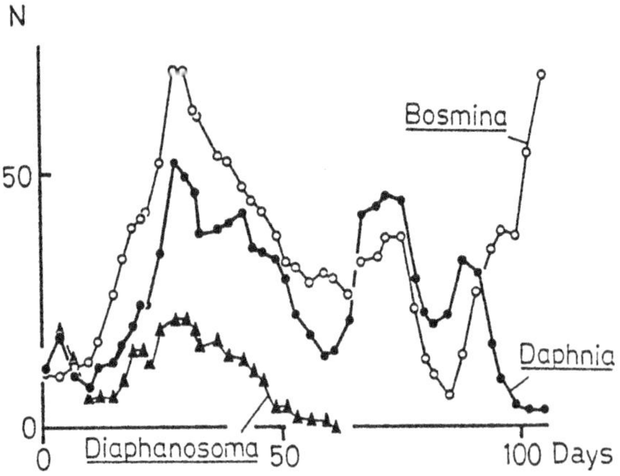

Fig. 4. Three-species cladoceran population. N = animal density (ind./0.6 l).

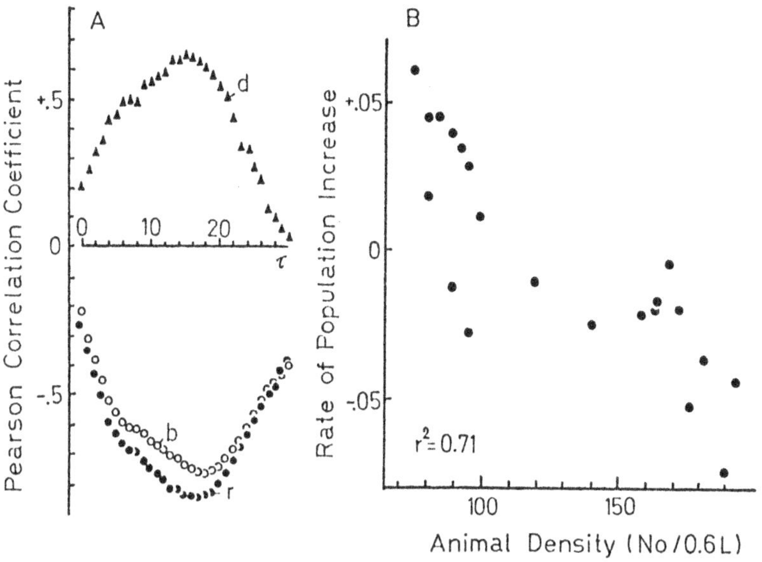

Fig. 5. A. Cross-correlations for *Bosmina* population parameters as functions of its density: d = for death rate, b = for birth rate, r = for the rate of population increase. τ = backward shift of a dependent variable in days. The time lags found: τ_{dN} = 15 days, τ_{bN} = 18 days, τ_{rN} = 17 days. B. The rate of population increase as a function of density. This plot results from cross-correlation on the left, curve 'r' (τ_{rN} = 17 days).

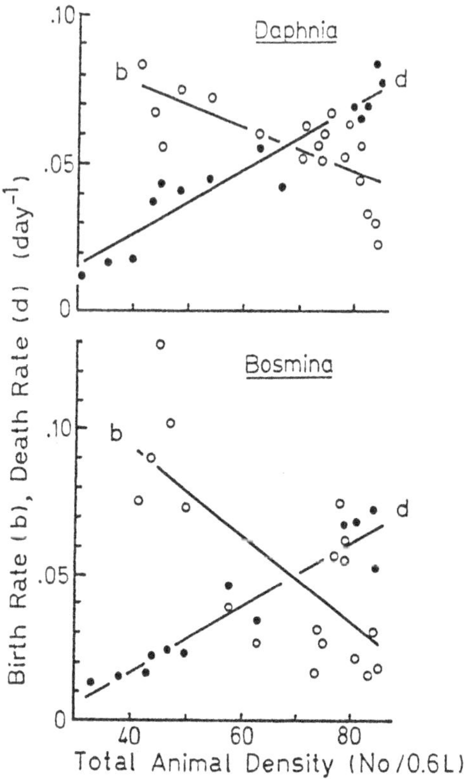

Fig. 6. Birth rate (b) and death rate (d) of cladocerans as functions of total density in two-species culture depicted on Fig. 2, above. For *Daphnia* τ_{bN} = 7 day, τ_{dN} = 14 days, for *Bosmina* τ_{bN} = 4 days, τ_{dN} = 15 days. The lines were estimated by regression analysis.

evidenced by that in the presence of other species in each mixed culture maximal density of *Bosmina* was at least three times lower than in the isolated one (about 70 *versus* 210 ind/flask, respectively). The outcome of competition was not predictable: any of the two species could dominate in numbers by the end of cultivation (Fig. 2–4). Monoculture of *Diaphanosoma* went extinct at the beginning of the experiment, nevertheless in two-species mixture within the first two weeks this species yielded a strong density peak showing high potential for population growth (Fig. 2). Thus, in the presence of the sole *Bosmina* population *Diaphanosoma* persisted. However, in all three-species mixtures it was outcompeted during 1.5–2 months (Fig. 3–4).

Discussion

Extinction risk is higher in the microcosms of small volume relative to large (Leslie & Gower, 1960; Dickerson & Robinson, 1986; Bengtsson, 1988). I used ones of the smallest volume (compare with Ringelberg & Kersting, 1978; Elstad, 1986; Sommer, 1988). In spite of this, pelagic cladocerans of Lake Glubokoe persisted in them more than 100 days until the experiment was terminated. Overgrazing and extinction did not occur even though the microcosms were unicompartmental and consumers were in permanent contact with resource. The comparison of initial and final phytoplankton composition showed that flagellates were the key food item. By the end of observations they did not change their density (Table 1). Bacteria also did not change their density. However, for averaged-sized *Daphnia* as *D. galeata* and for *Bosmina* they are probably poor food source (Geller & Müller, 1981; Pace *et al.*, 1983; Goulden & Henry, 1984; DeMott, 1985). At the same time flagellates are preferred food of Cladocera (Knisely & Geller, 1986). It is quite possible that the sharp decline in flagellate density (Table 1) was the cause of *Diaphanosoma* extinction in the monoculture.

I believe that continuous nutrient recycling was the major mechanism of Cladocera population persistence in the microcosms. Probably animal excretion was of primary importance here. The effect of bacterial mineralization was kept minimal because animal corpses were removed and flasks walls were constantly washed. Therefore, bacterial density in the cultures was relatively low all the time (Table 1). Meanwhile, algae are able to catch excreted nutrients quickly (Lehman, 1980) and reproduce on them (Sterner, 1986). Indeed, *Diaphanosoma* guts contained empty covers of *Tetraëdron* indicating that the alga was consumed and digested. Nevertheless, *Tetraëdron* density did not decline after 10 days of grazing on it (Table 1). Another mechanism promoting animal survival in microcosms was density regulation (Fig. 5B, Fig. 6). Overgrazing was prevented by slowing down of the rate of population increase which was occurring as cladoceran density grew. Probably, an increase in animal numbers gradually led to that the algal losses from grazing started exceeding their recruitment from nutrient excretion. Consequently food concentration declined. This led to a decline in animal birth rate and an increase in death rate (Fig. 6). Quite possible that such indirect density effect might be coupled with a direct one, e.g. allelopathy, lowering filtration rates of crowded animals (Helgen, 1987). In all cultures density regulation was delayed (e.g. Fig. 5A), like that, observed in open Cladocera cultures (Matveev, 1985, 1987) or in the lake (Matveev, 1976, 1983). Probably, delayed density regulation is very typical for many planktonic Cladocera.

Oscillations in animal numbers were observed in all microcosms (Fig. 1–4). Figures 5, 6 suggest that the oscillations were not a result of unsuccessful cultivation: when high mortality does not depend on density, e.g. at awkward experimental manipulations, cross-correlation curves do not have the typical form with one apex or one distinct minimum like those on Fig. 5A. I suggest that in the present study as earlier (Matveev, 1987) high time lags were the cause of oscillations. As should have been expected, closed multispecies systems proved to be competitive (Fig. 6). Nevertheless, all two-species cultures were stable. In three-species cultures *Diaphanosoma* was outcompeted either due to the effect of *Daphnia*, which was heavier than that of *Bosmina*, or due to the fact that diffuse competition is stronger than monospecific influence (Fig. 3, 4 versus Fig. 2). Bengtsson (1988) also found that in rock pool *Daphnia* probabilities of extinction were higher for three-species than for two-species associations.

Interestingly, in my cultures colonial algae disappeared by the end of observations (Table 1). The reversed course of events is usually found in lakes (Sommer *et al.*, 1986): colonial forms follow after small flagellates numerous in spring. Such plankton succession was explained by the fact that zooplankton abundant in summer eats away small cells, but is unable to suppress large colonial phytoplankton so the excreted nutrients go to the latter (Lampert, 1987). My data as well as microcosm experiments of Sommer (1988) seem not to allow accepting without reserve such simple explanation.

Acknowledgements

This work was done at the 'Lake Glubokoe' Hydrobiological Station.

References

Bengtsson, J., 1988. Life histories, interspecific competition and regional distribution of three rockpool *Daphnia* species. Acta Universitatis Upsaliensis, Comprehensive Summaries of Uppsala Dissertations from the Faculty of Science 142. Uppsala, 32 pp.

DeMott, W. R., 1985. Relations between filter mesh-size, feeding mode, and capture efficiency for cladocerans feeding on ultrafine particles. Arch. Hydrobiol. Beih. Ergebn. Limnol. 21: 125–134.

Devol, A. H., 1979. Zooplankton respiration and its relation to plankton dynamics in two lakes of contrasting trophic state. Limnol. Oceanogr. 24: 893–905.

Dickerson, J. E., Jr. & J. V. Robinson, 1986. The controlled assembly of microcosmic communities: the selective extinction hypothesis. Oecologia (Berlin) 71: 12–17.

Elstad, C. A., 1986. Population dynamics and nutrient fluxes in an aquatic microcosm. Hydrobiologia 137: 223–237.

Geller, W. & H. Müller, 1981. The filtration apparatus of Cladocera: Filter mesh-sizes and their implications on food selectivity. Oecologia (Berlin) 49: 316–321.

Goulden, C. E. & L. L. Henry, 1984. Lipid energy reserves and their role in Cladocera. In Meyers, D. G. & J. R. Strickler (eds), Trophic interactions within aquatic ecosystems. Westview Press, Boulder: 167–185.

Gutelmacher, B. L., 1986. Metabolizm planktona kak edinogo tselogo: Trophometabolicheskyie vzaimodeistviya zoo- i phytoplanktona. Nauka, Leningrad, 155 pp. (In Russian).

Helgen, J. C., 1987. Feeding rate inhibition in crowded *Daphnia pulex*. Hydrobiologia 154: 113–119.

Karr, J. R., 1982. Population variability and extinction in the avifauna of a tropical land bridge island. Ecology 63: 1975–1978.

Kerfoot, W. C., W. R. DeMott & D. L. DeAngelis, 1985. Interactions among cladocerans: Food limitation and exploitative competition. Arch. Hydrobiol. Beih. Ergebn. Limnol. 21: 431–451.

Knisely, K. & W. Geller, 1986. Selective feeding of four zooplankton species on natural lake phytoplankton. Oecologia (Berlin), 68: 86–94.

Lampert, W., 1987. Predictability in lake ecosystems: the role of biotic interactions. Ecol. Studies 61: 333–346.

Lehman, J. T., 1980. Release and cycling of nutrients between planktonic algae and herbivores. Limnol. Oceanogr. 25: 620–632.

Leslie, P. H. & J. C. Gower, 1960. The properties of a stochastic model for the predatory-prey type of interaction between two species. Biometrika 47: 219–234.

Matveev, V. F., 1976. An estimation of competition between three zooplankton (Cladocera) species in field conditions using multiple regression analysis. Zhurnal Obshey Biologii 37: 822–830. (In Russian).

Matveev, V., 1983. Estimating competition in cladocerans using data on dynamics of clutch size and population density. Int. Revue ges. Hydrobiol. 68: 785–798.

Matveev, V., 1985. Competition and population time lags in *Bosmina* (Cladocera, Crustacea). Int. Revue ges. Hydrobiol. 70: 491–508.

Matveev, V. F., 1986. Long-term changes in the community of planktonic crustaceans in Lake Glubokoe in relation to predation and competition. Hydrobiologia 141: 33–43.

Matveev, V. F., 1987. Effect of competition on the demography of planktonic cladocerans – *Daphnia* and *Diaphanosoma*. Oecologia (Berlin) 74: 468–477.

Nixon, S. W., 1969. A synthetic microcosm. Limnol. Oceanogr. 14: 142–145.

Pace, M. L., K. G. Porter & Y. S. Feig, 1983. Species- and age-specific differences in bacterial resource utilization by two co-occurring cladocerans. Ecology 64: 1145–1156.

Paloheimo, J. E., 1974. Calculation of instantaneous birth rate. Limnol. Oceanogr. 18: 692–694.

Ringelberg, J. & K. Kersting, 1978. Properties of an aquatic micro-ecosystem: I. General introduction to the prototypes. Arch. Hydrobiol. 83: 47–68.

Shcherbakov, A. P., 1967. Ozero Glubokoe. Nauka, Moscow, 379 pp. (In Russian).

Sommer, U., 1988. Phytoplankton succession in microcosm experiments under simultaneous grazing pressure and resource limitation. Limnol. Oceanogr. 33: 1037–1054.

Sommer, U., Z. M. Gliwicz, W. Lampert & A. Duncan, 19863. The PEG-model of seasonal succession of planktonic events in fresh waters. Arch. Hydrobiol. 106: 433–471.

Sorokin, Y. I. & H. Kadota, 1972. Techniques for the assessment of microbial production and decomposition in fresh waters. Blackwell Scientific Publications, Oxford, London, 112 pp.

Sterner, R. W., 1986. Herbivores' direct and indirect effects on algal populations. Science 231: 605–607.

Hydrobiologia **225**: 309–318, 1991.
V. Kořínek & D. G. Frey (eds), Biology of Cladocera.

Succession of cladocerans in a northern Italian ricefield

I. Ferrari,[1] A. Bachiorri,[2] F. G. Margaritora[3] & V. Rossi[2]
[1]*Istituto di Zoologia, Università, 44100 Ferrara, Italy*; [2]*Istituto di Ecologia, Università, 43100 Parma, Italy*; [3]*Dipartimento di Biologia Animale e dell'Uomo, Università 'La Sapienza', 00185 Roma, Italy*

Key words: ricefields, heleoplankton, cladocerans, ecological succession, diversity

Abstract

The seasonal succession of cladocerans was studied in an experimental ricefield (never treated with biocides) located on the Po River Plain (Province of Reggio Emilia) in 1982, the first year of rice cultivation, and in two later years, 1984 and 1986. A total of 26 species were identified; some of them were found in Northern Italy for the first time. The finding of *Latonopsis australis* Sars is discussed in detail. A trend of increasing specific diversity during the rice cultivation season was not recurrent every year. The most significant trends emerge when the three years are compared. The total number of species was 14 in 1982, 16 in 1984, and 20 in 1986. Diversity, too, increased strongly over the years: this trend was sustained chiefly by the increase in richness of Chydoridae species associated with the growth and decomposition of a large biomass of weeds and microalgae during the summer.

Introduction

Considerable bibliographic references are available regarding seasonal succession of heleoplankton in ricefields (Moroni, 1967; Rossi *et al.*, 1974; Pont, 1983). This area of research, however, does not seem to have been particularly reinvigorated by the consistent progress made in recent studies on production and decomposition processes and nutrient cycles in ricefield environments (Golterman, 1984; Vaquer, 1984; Minzoni, 1986; Minzoni *et al.*, 1988).

In this paper we report the results of research carried out over three complete cultivation cycles (1982, 1984, 1986) in an experimental ricefield in Northern Italy. A number of essential features of this ecosystem have already been studied: composition of the principal macrophyte associations, seasonal succession of heleoplankton, including protozoans (Ferrari *et al.*, 1984a and 1984b; Madoni, 1987; Margaritora *et al.*, 1987), variability of physical and chemical parameters in both water column and water-sediment interface (Antonietti *et al.*, 1987), decomposition processes related to nutrient cycles (Minzoni & Moroni, 1987; Viaroli, 1989). We will present here results exclusively concerning the cladoceran component of heleoplankton. First of all we will provide as complete a picture as possible of the composition of this taxocoenosis; we will then analyze the evolution of taxocoenosis structure over the years from 1982, the first year of rice cultivation, to 1986. Interest in this type of analysis seems also justified by the macroscopic characteristics of the experimental field, which was significantly different from a commonly farmed ricefield, inasmuch as it was never treated with biocides.

The study site

The experimental ricefield, named Bigliana, was located at Santa Vittoria (Reggio Emilia Province) in a reclaimed zone of the Po River Plain; it has an area of about 750 m^2. Clay made up more than 80% of the most superficial soil layers. In 1982 and 1986 rice was seeded in mid-May, just after the first submersion; the ricefield was then kept dry for about ten days between the end of May and the beginning of June and definitively dried up in mid-September. In 1984, due to rainy and cloudy weather in May, two sowings were carried out in mid-May and at the beginning of June; in the same year two other dry periods occurred in July-August due to drying up of the irrigation canal.

Weed cover increased progressively from July onwards; the dominant species were *Cyperus difformis*, *Najas minor*, *Chara braunii*, *Spirodela polyrrhyza*, *Lemna minor*, *Wolffia arrhyza*, *Typha latifolia*, *Echinocloa crus-galli*, *Schoenoplectus mucronatus*. Cover and biomass of blue-green algae and filamentous chlorophyta also tended to increase during the summer. No weeds were removed in 1982 or in 1986; in 1984 weeds were manually eliminated at the beginning of July. Rice yield was good (equivalent to about 5 tons per ha) in both 1982 and 1984. In 1986 the site took on the appearance of an abandoned field with a large weed cover; in addition, thick beds of filamentous algae were already present in June: the rice yield was fairly scanty (equivalent to less than 2 tons per ha).

Material and methods

Heleoplankton samples were collected from a permanent station (W4) in the central furrow, about 40 cm deep, at weekly intervals from May to September 1982 and during the same months in 1984 and 1986; a 3 l Ruttner bottle was used, and about 30 l of water were filtered for each sample. In 1982 and 1986 samples were also collected weekly from three additional stations (W2, W3, and W5) located among the rice plants, where average water depth was about 15 cm (Fig. 1); a half-cylindrical device (Ferrari *et al.*, 1984b) was used to collect equal volumes of water (about 10 l) at each of the three stations: the three samples were then mixed together to make up one integrated sample. Since the samples were taken to study small-bodied forms as well, rotifers in particular (Ferrari *et al.*, 1984a and 1984b), a 50 μm-mesh net was used to filter the water.

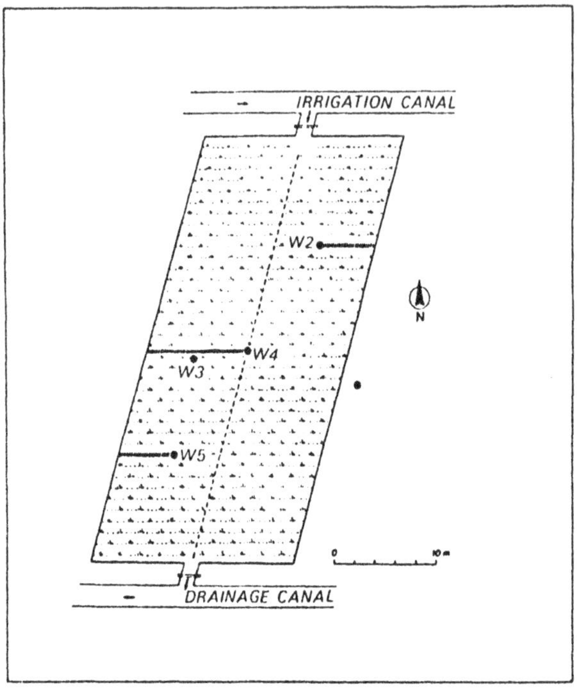

Fig. 1. Map of the experimental ricefield; heleoplankton sampling stations are indicated.

The taxocoenosis succession was studied by comparing the number of species and the Shannon-Wiener diversity index in all the samples gathered at station W4. Principal Components Analysis (PCA) was performed on the species density data for each annual series (the integrated samples from stations W2, W3, and W5 were also considered for the 1982 and 1986 series).

Results

Cladoceran species

A total of 26 species were identified (Table 1). *Moina weismanni* is a Far Eastern species not previously found in Europe; it has been described by Margaritora *et al.* (1987). A number of other

Table 1. List of cladoceran species in the experimental rice-field.

1. *Diaphanosoma brachyurum* (Liévin)
2. *Latonopsis australis* Sars
3. *Bosmina longirostris* (O. F. Müller)
4. *Ceriodaphnia reticulata* (Jurine)
5. *Ceriodaphnia laticaudata* P. E. Müller
6. *Ceriodaphnia megops* Sars
7. *Scapholeberis rammneri* Dumont & Pensaert
8. *Simocephalus vetulus* (O. F. Müller)
9. *Simocephalus serrulatus* (Koch)
10. *Daphnia curvirostris* Eylmann emend. Johnson
11. *Moina affinis* (Birge)
12. *Moina weismanni* Ishikawa
13. *Moina brachiata* (Jurine)
14. *Moina micrura* Kurz
15. *Ilyocryptus agilis* Kurz
16. *Echinisca rosea* Liévin
17. *Pleuroxus aduncus* (Jurine)
18. *Pleuroxus denticulatus* Birge
19. *Dunhevedia crassa* King
20. *Chydorus sphaericus* O. F. Müller
21. *Kurzia latissima* (Kurz)
22. *Alona rectangula* Sars
23. *Alona guttata* Sars
24. *Alona diaphana* King
25. *Oxyurella tenuicaudis* (Sars)
26. *Tretocephala ambigua* (Lilljeborg)

species were seen for the first time in Northern Italy. Margaritora in her 1985 review reported *Daphnia curvirostris* only in astatic waters in Sardinia, *Kurzia latissima* only in Sicily and in fossil samples from Monterosi Lake near Rome, *Tretocephala ambigua* only in Sardinia, *Kurzia latissima* only in Sicily and in fossil samples from Monterosi Lake near Rome, *Tretocephala ambigua* only in Sardinia, Sicily, and Apulia.

The finding of a *Latonopsis* species is particularly interesting. This species was seen in samples collected in late August and September 1986: only ten individuals consisting of young and adult females (only one of which was ephippial) were observed; no males were present. In 1932 Manfredi reported *Latonopsis occidentalis* in a ricefield in Lombardy, a northern Italian region; since that time the genus *Latonopsis* has never again been cited in Italian waterbodies.

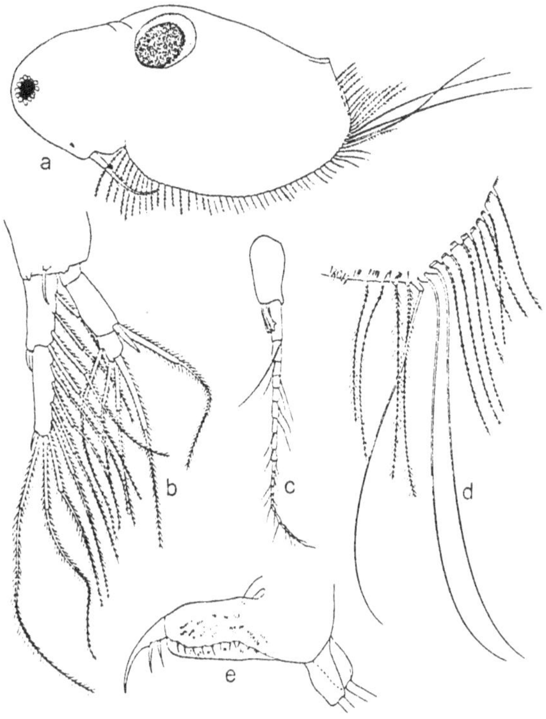

Fig. 2. Latonopsis australis Sars: a female habitus; b antenna; c antennula; d details of valve setae; e postabdomen.

Description of the Latonopsis species

Our taxonomic description is based on the criteria adopted by Korovchinsky (in Smirnov & Timms, 1983).

The adult female (Fig. 2a) is between 1.2 and 1.5 mm long. The massive body has a rectangular shape. Head length is about 1/3 of total length and a weak depression separates the head from the body. The eye is situated at the head apex and the ocellus near the antennule bases. Antennule long (Fig. 2c); the flagellum, 3 to 4 times longer than its base, is articulated and bears thin setae of different length. The antenna (Fig. 2b) is mounted on a thick base; its length is about 60% of body length; the distal edge of basipodite has a median spine and two small lateral outgrowths. Antennal setae: 4(5) – 7/0 – 1 – 4; the two ventral setae of the triarticulate branch are large and have hooked ends.

Valves bear a number (48–55) of setulated setae located on conical bases; these setae are longer on the anterior and posterior edges of the valves. The ventro-posterior corner of the valves bears three very long, naked setae mounted on extended conical bases (Fig. 2d); the length of these setae is 60–70% of body length. Groups of spinules are located on the postero-ventral and posterior margins of the valves.

The postabdomen (Fig. 2e) is relatively small, reaching about 33% of body length; its proximal part is slightly enlarged; anal teeth are small and isolated, 9–10 in number; there are numerous series of lateral setules. The claw is long and incurved; the two basal spines are about half as long as the claw.

The *Latonopsis* specimens from the Bigliana ricefield have been classified as *Latonopsis australis* Sars (syn. *L. occidentalis* Birge and

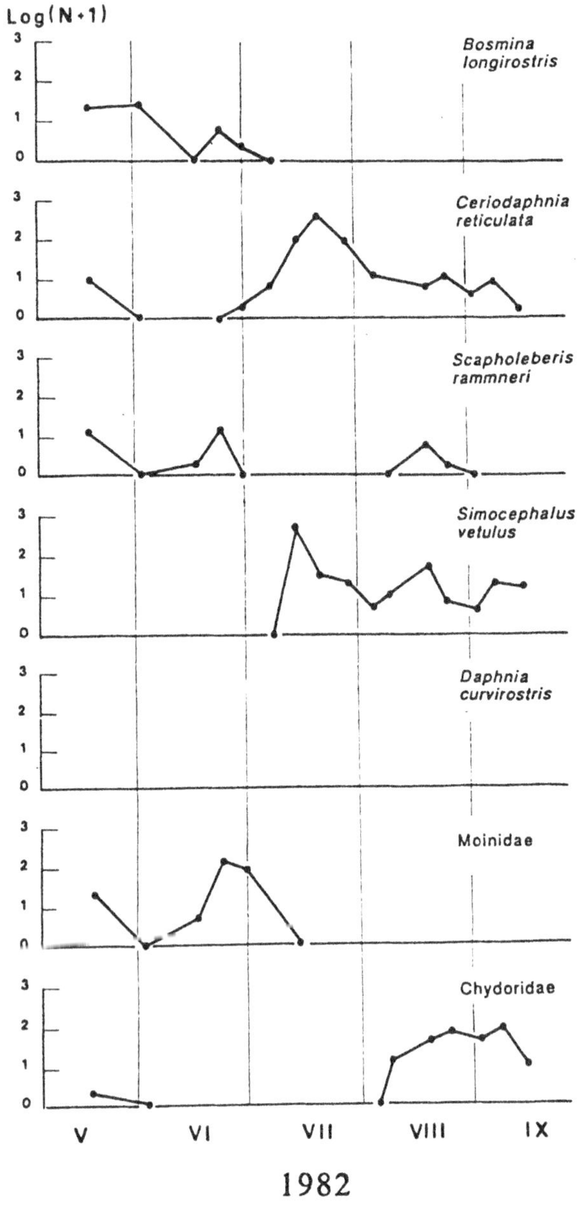

Fig. 3. Changes in density (N: individuals per liter) of the most important species (or families) of cladocerans at the W4 station in 1982.

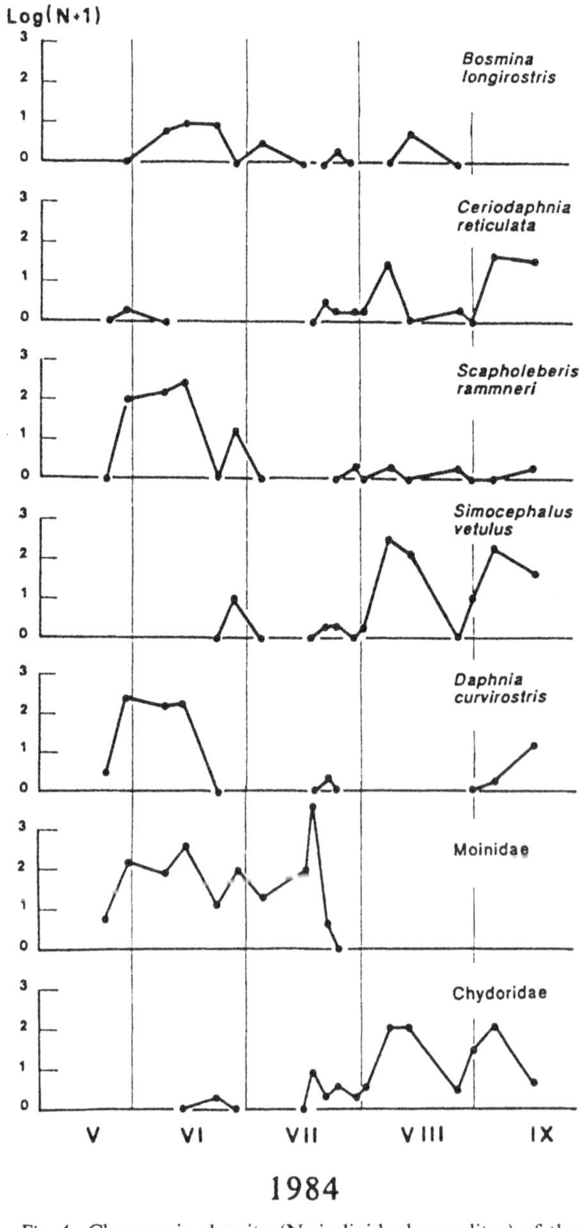

Fig. 4. Changes in density (N: individuals per liter) of the most important species (or families) of cladocerans at the W4 station in 1984.

L. breviremis Daday) because of their close affinity to the species found in Macedonian ricefields and classified as *L. australis* by Harding & Petkowski (1963). There is also good concordance with *L. australis* described by Thomas (1961).

There are, instead, fairly significant differences from the Australian type species (Korovchinsky in Smirnov & Timms, 1983). The antennula setae of the type species all have the same length, while those observed in the Bigliana ricefield specimens have varying lengths. Moreover, the basal spines of the postabdomen claw are much longer in our

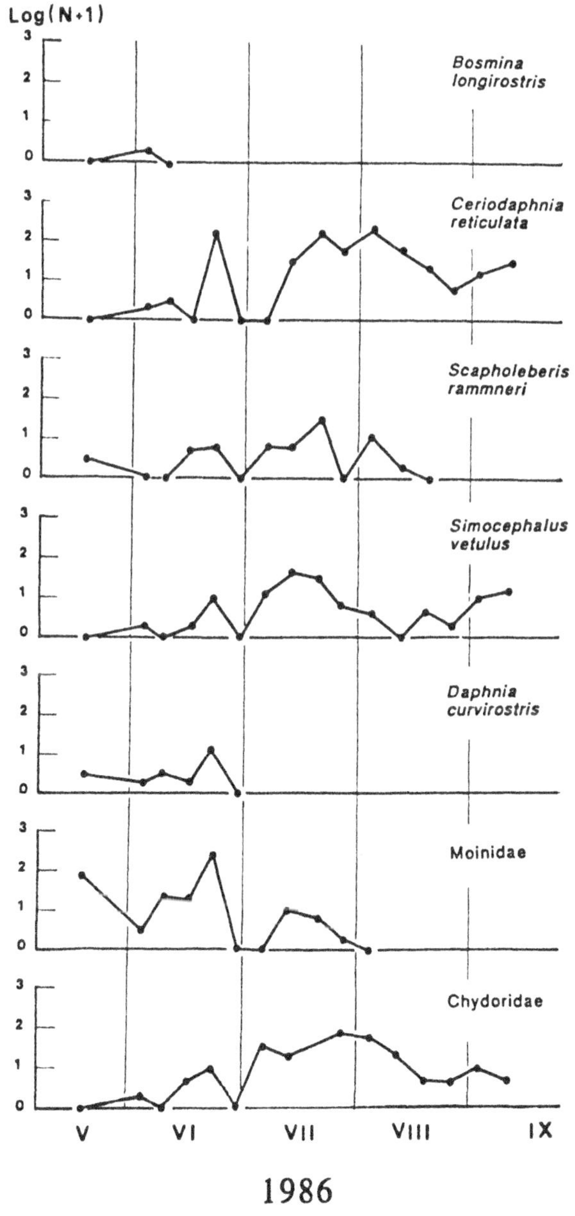

Fig. 5. Changes in density (N: individuals per liter) of the most important species (or families) of cladocerans at the W4 station in 1986.

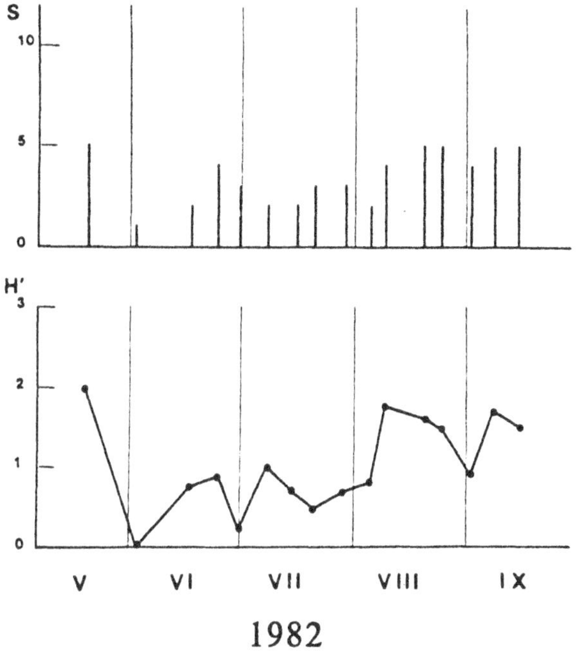

Fig. 6a.

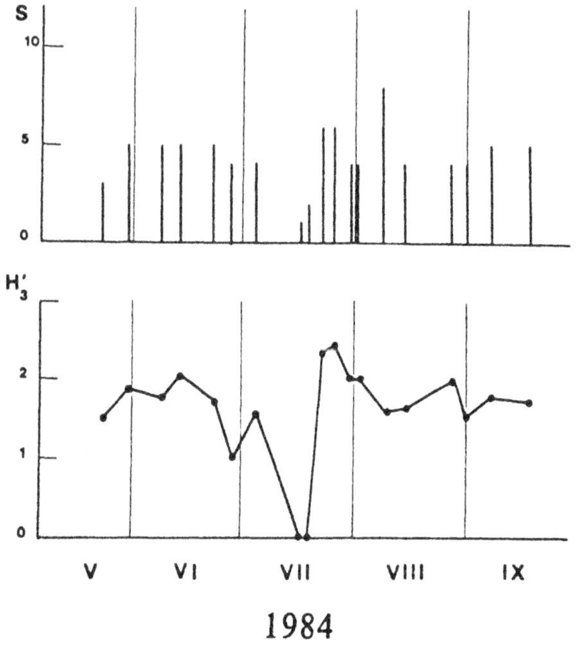

Fig. 6b.

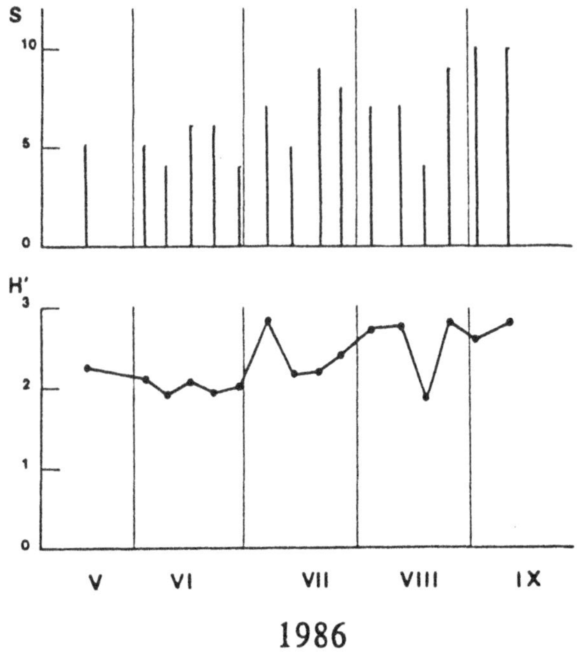

Fig. 6c.

Fig. 6. Changes in total number of species (S) and in Shannon-Wiener diversity index (H') of cladoceran taxocoenosis at the W4 station.

specimens, while the three long naked setae are quite longer in the Australian type species. Comparison with the *Latonopsis* reported by Manfredi (1932) is very difficult due to drawing inaccuracies in that paper.

Korovchinsky (1986) maintains that *Latonopsis australis* is a polytypic species; in his opinion the populations from central Asia and southern Europe probably belong to a different species, possibly *Latonopsis ferganica* Mukhamediev, 1960 or *Latonopsis hospitus* Naidenov, 1969; unfortunately, the descriptions provided by these authors are not sufficiently detailed.

Evolution of the taxocoenosis structure

Density of the most important species or families at station W4 are reported in Figs. 3, 4, and 5 for 1982, 1984, and 1986, respectively. Fig. 6 illustrates the trend of the total number of species and the Shannon-Wiener diversity index. PCA results

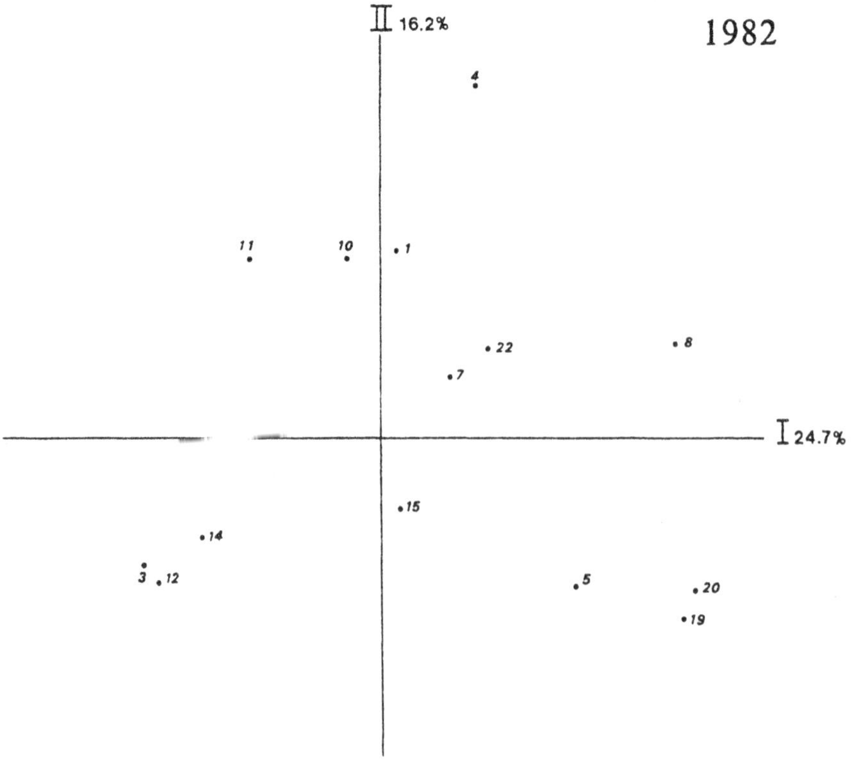

Fig. 7. PCA performed on density data of the 14 cladoceran species in the 32 samples collected in 1982 (the species are numbered as in the list on Table 1).

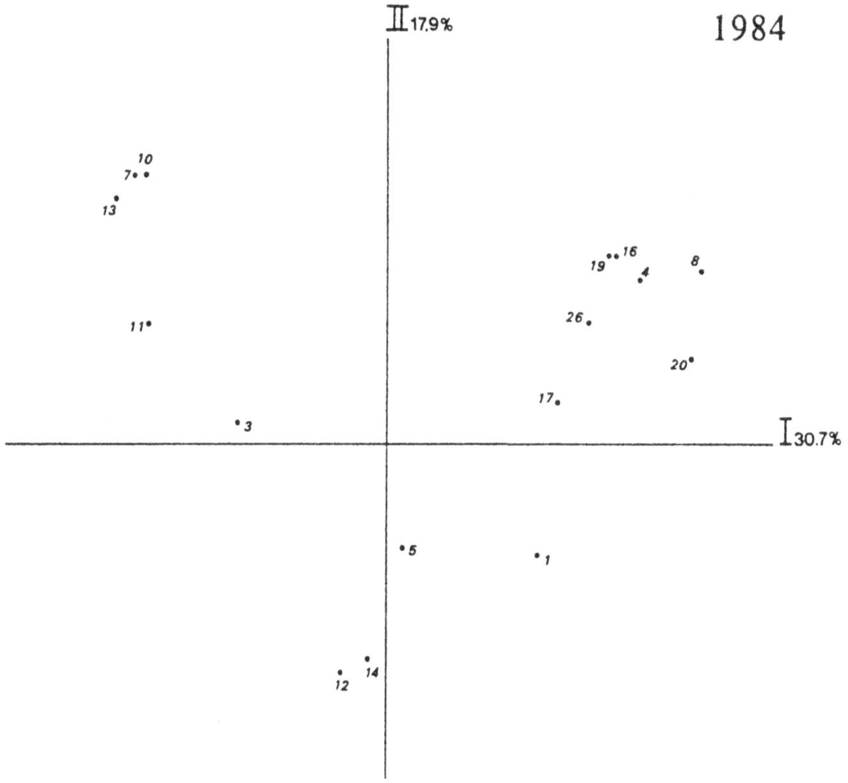

Fig. 8. PCA performed on density data of the 16 cladoceran species in the 19 samples collected in 1984 (see Fig. 7).

(Figs. 7, 8, and 9) show the species ordination on the axes of the first two components, separately for each year studied (as stated above, integrated samples collected from stations W2, W3, and W5 were also taken into account in 1982 and 1986). The salient traits of the seasonal succession in the three years may be summarized as follows.

1982 – *Bosmina longirostris*, *Moina affinis*, and *Moina weismanni* were the first species to colonize the ricefield; they were present as long as the water remained uncovered by the vegetation, i.e. until mid-July. Densities of *M. weismanni* were relatively high during the latter half of June, while *M. affinis* was quite abundant in July. Daphniidae were chiefly represented by *Ceriodaphnia reticulata* (with maximum density in July), by *Simocephalus vetulus* (with high densities up to the final drying up, especially at stations W2, W3, and W5), and by *Scapholeberis rammneri*. This species was present with much greater frequency and abundance in the 'integrated' samples than in the samples taken from the W4 station. Chydoridae (*Chydorus sphaericus* and *Dunhevedia crassa* in particular) appeared and became dominant in August–September.

1984 – In the first half of June the taxocoenosis was chiefly represented by *Moina affinis*, *Moina brachiata*, *Daphnia curvirostris*, and *Scapholeberis rammneri*; *Bosmina longirostris* was also present but with negligible densities. *Moina weismanni* then appeared transitorily, quickly outcompeted by *M. affinis*. An abrupt drop in diversity was observed in mid-July due to an exceptional density peak (more than 3700 individuals per liter) of *M. affinis*. From the end of July to September the taxocoenosis was dominated by an association made up of *Ceriodaphnia reticulata* and *Simocephalus vetulus* and four species of Chydoridae: *Chydorus sphaericus*, *Pleuroxus aduncus*, *Dunhevedia crassa*, and *Tretocephala ambigua*.

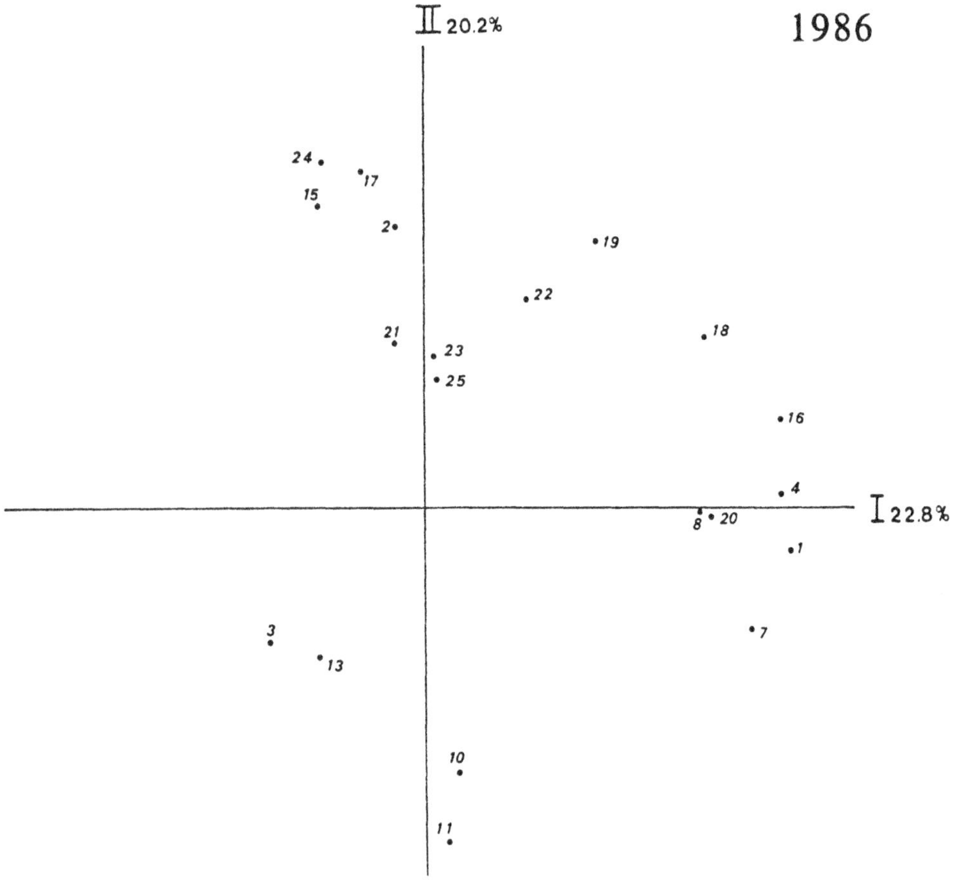

Fig. 9. PCA performed on density data of the 20 cladoceran species in the 32 samples collected in 1986 (see Fig. 7).

1986 – The spring taxocoenosis included *Moina affinis*, which was the dominant species, and *Daphnia curvirostris*. *Moina weismanni* did not appear. Abundances of both *Bosmina longirostris* and *Moina brachiata* were somewhat reduced. But in the May–June samples relatively high densities were also observed for *Ceriodaphnia reticulata*, *Simocephalus vetulus*, and *Chydorus sphaericus*, species that in 1982 and 1984 were typical of later successional phases. These species were present until the final drying up with density maxima (particularly evident in the 'integrated' samples) from the end of June to the beginning of August; the same trend was exhibited by *Diaphanosoma brachyurum*, *Scapholeberis rammneri*, and *Echinisca rosea*. There was, finally, a separate cluster of species occurring only in the last samples of the rice cultivation cycle: it principally included Chydoridae (*Pleuroxus aduncus*, *Dunhevedia crassa*, *Alona rectangula*, and, moreover, numerous other species not found in 1982 or 1984: *Pleuroxus denticulatus*, *Kurzia latissima*, *Alona guttata*, *Alona diaphana*, *Oxyurella tenuicaudis*); *Ilyocryptus agilis* and *Latonopsis australis* were also associated with this species cluster.

Conclusions

1. In our opinion, the high number of species of the Bigliana ricefield observed for the first time in northern Italy simply indicates that in the past inadequate attention has been paid to research in cladoceran distribution in this region. The finding of *Latonopsis australis* has been discussed in relation to some problems of taxonomy and biogeography: the hypothesis that this species was carried into Italy from Asia, as

is probably the case for *Moina weismanni* (Margaritora *et al.*, 1987), can be put forward.

2. It is commonly believed (Rossi *et al.*, 1974; Pont, 1983; Ferrari *et al.*, 1984a) that the seasonal evolution of cladocerans, as well as of other heleoplankton taxocoenoses, follows a pattern of characteristic successional phases: during the 'open water' period (May–June) primary production is mainly sustained by phytoplankton, and cladocerans are dominated by efficient filter feeding species of *Moina*, *Daphnia*, and *Bosmina*; subsequently, as the macrophyte cover increases, detritivorous forms (chiefly *Simocephalus* and Chydoridae) become dominant. Seasonal evolution, moreover, is said to be characterized by an increasing trend of the specific diversity from May to September. Yet, in fact, in our ricefield this pattern changed significantly each year, according to meteorological conditions and rice cultivation practices. In 1984, in particular, the ricefield was subjected to a continuous sequence of disturbances caused by the unusually frequent alternation of dry and flood periods. In that year an increasing diversity trend was not observed; on the contrary, in mid-July, after manual weeding, a sharp fall in diversity was noted. The diversity indices of the series of 'integrated' sampled collected in 1982 and 1986 were also characterized by wide seasonal fluctuations, with no recognizable significant trend. Variations in the sequence of successional phases were also notable: in 1986, for example, because of early growth of weeds and microalgae, the spring phase with dominance of euplanktonic forms was somewhat contracted.
3. Significant trends emerge when the three years are compared. The total number of species was 14 in 1982 (based on 32 samples), 16 in 1984 (19 samples), and 20 in 1986 (32 samples). At the W4 station the number of species increased from 13 in 1982 to 16 in 1984 to 19 in 1986, while the average number of species per sample increased from 3.4 to 4.4 to 6.6, respectively. In the integrated samples from the W2, W3, and W5 stations there were 13 species in 1982 and 20 in 1986. The number of Chydoridae species increased as follows: station W4: 3 in 1982, 4 in 1984, and 8 in 1986; stations W2, W3, and W5: 3 in 1982 and 9 in 1986. Diversity, too, increased strongly over the years.
4. Contrary to 1982, in 1984 and 1986 numbers of species and diversity indexes were already high at the spring flooding phase; this was probably due to the hatching of resting eggs produced in the preceding year. The increase in diversity is also very evident when values for the final phases of the three years are compared. The significant increase in diversity from the first year of rice cultivation to subsequent years in which the ricefield did not undergo a constant, heavy human manipulation is comparable to that observed in successional communities colonizing unharvested fields (Jacobs, 1975; Ferrari & Ceccherelli, 1989). In fact, our ricefield tended to resemble more and more an unharvested system. As the years passed, due to decomposition of large biomass of algae and weeds, the detritus chain prevailed overwhelmingly over the grazing chain already at the beginning of the rice growing season; at the same time the cladoceran taxocoenosis took on a high diversity structure, maintained in particular by an increased richness of detritivorous species, mainly represented by Chydoridae.

References

Antonietti, R., P. Viaroli & F. Antonietti, 1987. Changes in the physical and chemical properties of floodwater and sediments in an experimental ricefield (Reggio Emilia, Italy). Hydrobiologia 144: 83–88.

Ferrari, I., A. De Marchi, P. Menozzi, F. Minzoni & F. Piccoli, 1984a. Heleoplankton seasonal succession in an experimental ricefield in Northern Italy. Verh. Int. Ver. Limnol. 22: 1711–1716.

Ferrari, I., De Marchi A., Mazzocchi M. G., Menozzi P., Minzoni F., Piccoli F. & Moroni A., 1984b. Ricerche ecologiche a indirizzo teorico – sperimentale in un ambiente artificiale isolato: la risaia. Atti dei Convegni Lincei 62: 175–196.

Ferrari, I. & V. U. Ceccherelli, 1989. Ecological succession and evolution of freshwater ecosystems. In: O. Ravera (Ed.), Ecological Assessment of Environmental Degradation, Pollution and Recovery. Elsevier Sci. Publ. B.V., Amsterdam: 31–53.

Golterman, H. L., 1984. Sediments, modifying and equilibrating factors in the chemistry of freshwater. Verh. Int. Ver. Limnol. 22: 23–59.

Harding, J. P. & T. Petkovski, 1963. *Latonopsis australis* Sars (Cladocera) in Jugoslavia with notes on its synonymy and distribution. Crustaceana 6: 1–4.

Jacobs, J., 1975. Diversity, stability and maturity in ecosystems influenced by human activities. In Van Dobben, H. V. & R. H. Lowe-McConnell (eds), Unifying concepts in ecology, Dr. W. Junk B.V. Publ.: 187–207.

Korovchinsky, N. M., 1986. On the taxonomy and geographical distribution of the superfamily Sidoidea Baird, 1850 (Crustacea: Cladocera). Hydrobiologia 140: 243–253.

Madoni, P., 1987. Estimation of production and respiration rates by ciliated protozoa community in an experimental ricefield. Hydrobiologia 144: 113–120.

Manfredi, P., 1932. Note sulla fauna di una risaia lombarda: Rotiferi e Crostacei. Atti Soc. ital. Sc. Nat. Milano 71: 179–200.

Margaritora, F. G., 1985. Fauna d'Italia. Cladocera. Ed. Calderini Bologna, 399 pp.

Margaritora, F. G., I. Ferrari & V. Crosetti, 1987. A Far East *Moina*, *M. weismanni* Ishikawa, 1896 found in an Italian ricefield. Hydrobiologia 145: 93–103.

Minzoni, F., 1986. Studio di processi legati ai cicli dell'azoto e del fosforo nell'ecosistema di risaia. Tesi di Dottorato di Ricerca in Scienze dell'Ambiente, Università di Parma, 145 pp.

Minzoni, F. & A. Moroni, 1987. Phosphorus balance of an experimental ricefield in the Po Plain (Italy). Agric. Ecosyst. Envir. 18: 223–230.

Minzoni, F., C. Bonetto & H. L. Golterman, 1988. The nitrogen cycle in shallow water sediment systems of ricefields. Part 1: the denitrification process. Hydrobiologia 159: 189–202.

Moroni, A., 1967. Ricerche ecologiche sulle acque astatiche. Ecologia delle comunità eleoplanctoniche di risaia. Ed. Studium Parmense, Parma, 106 pp.

Mukhamediev, A. M., 1960. New species of Cladocera from ricefields of the Fergana Valley. Uzeb. biol. Zhur. 3: 35–38.

Naidenov, W., 1969. Eine neue *Latonopsis* – Art (Cladocera, Sididae) aus Bulgarien. Compt. Rend. Acad. bulg. Sci. 22: 935–938.

Pont, D., 1983. Recherches quantitatives sur l'héléoplancton des rizières de Camargue. Thèse Doct. Sci. Nat., Université de Provence (Marseille), 300 pp.

Rossi, O., A. Moroni, P. Baroni & U. Caravello, 1974. Annual evolution of the zooplankton diversity in twelve Italian ricefields. Boll. Zool. 41: 157–181.

Smirnov, N. N. & B. V. Timms, 1983. Revision of the Australian Cladocera. Rec. Aust. Mus. suppl. 1: 1–139.

Thomas, I. F., 1961. Review of the genera *Pseudosida* Herrick, 1884 and *Latonopsis* Sars, 1888 (Cladocera). Crustaceana 3: 1–8.

Vaquer, A., 1984. La production algale dans les rizières de Camargue pendant la période de submersion. Verh. Int. Ver. Limnol. 22: 1651–1654.

Viaroli, P., 1989. Alcuni aspetti della decomposizione di materiali ligno-cellulosici in ambienti acquatici temporanei. Tesi di Dottorato di Ricerca in Scienze dell'Ambiente, Università di Parma, 131 pp.

Hydrobiologia **225**: 319–323, 1991.
V. Kořínek & D. G. Frey (eds), Biology of Cladocera.

Qualitative changes in the planktonic Cladocera assemblage of a High Tatra Mountains lake during the last 100 years

Marian Vranovský
Institute of Fishery Research and Hydrobiology, Drieňová 3, Bratislava, 826 24 Czechoslovakia

Key words: Cladocera assemblage, Popradské pleso lake, changes, acidification, pollution, eutrophication, fish predation

Abstract

During the last roughly 100 years (1881–1989) some noteworthy changes have occurred in the planktonic Cladocera taxocene of the lake Popradské pleso (High Tatra Mts., Czecho-Slovakia). Primarily, in course of the twenties up to the fifties of our century *Daphnia pulicaria* and *Polyphemus pediculus*, and later also *Holopedium gibberum*, disappeared from the plankton of the lake. Conversely, approximately from the mid-seventies, *Daphnia longispina* has become a constituent of the lake plankton. In spite of lacking data on the principal parameters of the abiotic and biotic environment from the past, the author assumes that probably the most important factor responsible for the changes referred to were fluctuations in the predation pressure of fish. Besides, organic pollution might have played a definite but less important role.

Introduction

The lake Popradské pleso belongs to the most frequented and most thoroughly investigated lakes of the Slovak part of the High Tatra Mountains. As a result, data on the composition of its crustacean plankton have been available already since the 1880s. The first data on its planktonic crustaceans stem from Wierzejski (1882, 1883). In 1894 a qualitative analysis of the zooplankton of the lake was made by Daday (1896), and two decades later Minkiewicz (1914, 1917) summarized his own data and also literature data on the fauna of the Tatra lakes. Particular attention to the Cladocera of the Tatra lakes was paid by Lityński (1913, 1917). In 1928 and 1929 the zooplankton of the Popradské pleso lake and two other Tatra lakes was analysed by Kalmus (1929, 1931). Finally, from the 1950s up to the present I have my own, mostly yet unpublished, data available.

In spite of irregular intervals between examinations during the past 100 years, a comparison of the faunistic composition of planktonic Cladocera in the Popradské pleso lake shows some interesting changes.

Study lake

The Popradské pleso lake is a glacier lake. It lies 1494 m above sea level at the upper border of the timber line at the junction of the Zlomisková dolina valley with the Mengusovská dolina valley on the southern slope of the High Tatra Mts. Its area covers 6.9 ha, and the maximal depth is 17.6 m (Marček, 1971). During 190 days of the year on the average it is covered with ice. Water flows in from the Ľadový potok brook and flows out by the Krupá brook.

From 1962 until 1984 untreated waste water entered the lake from a newly constructed mountain hotel. However, a chalet, abundantly visited by tourists, had been standing there already since 1879, and its waste water was carried into the lake as well without any purification.

Data on the physical and chemical properties of water from the period before the year 1961 are lacking. During our investigations in the years 1961–1963 and 1984–1986 the pelagial water temperature at the surface did not exceed 10 °C in summer. pH varied between 5.8–7.1, and the dissolved oxygen concentration reached approximately 10 mg l^{-1} (but it dropped almost to zero in winter at the greatest depth). In the years 1984–1989 the BOD_5 values varied from 1.5 mg l^{-1} up to 4.0 mg l^{-1} O_2. Total N content varied in 1961–1963 from 0.58–1.12 mg l^{-1}, and in the years 1984–1989 between 0.04 and 0.90 mg l^{-1}. The concentration of orthophosphate P varied in the years 1961–1963 from 2 $\mu g\ l^{-1}$ up to 28 $\mu g\ l^{-1}$, and the content of total P in the years 1984–1989 from 6 $\mu g\ l^{-1}$ up to 30 $\mu g\ l^{-1}$. (Data on physical and chemical conditions are from Ertl, 1965 and Tomajka 1985, 1990).

A monospecific, autoreproducing population of the brown trout (*Salmo trutta* m. *fario*) (Dyk, 1960; Balon, 1964; Nagy & Holčík, 1988) is living in the lake, to which reference had already been in the literature in the 18th and 19th centuries (Grossinger, 1794; Wierzejski, 1883; Szontagh, 1885, 1896; Daday, 1896).

Results and discussion

What is conspicuous in the Table 1 is the lower number of Cladocera species found in the Popradské pleso lake recently as compared to the earlier accounts made by Wierzejski, Daday, and Minkiewicz. This is partly due to the fact that Minkiewicz lists not only his own findings but also summarizes the data of Wierzejski. Further, it may be due, in part, to the possibly less intensive investigation of the littoral zooplankton by the more recent authors. This may perhaps account for e.g. the missing of *Eurycercus lamellatus* and *Alonella nana* from the lists of recent authors (Kalmus, *op. cit*; Vranovský, unpubl. data; Ertl & Vranovský, 1964; and partly also Lityński, 1917). As far as *Camptocerus rectirostris* and especially *Tretocephala ambigua* are concerned (Kalmus, 1929 and 1931), the possibility of an erroneous determination may also be admitted, because the faunistic data of Kalmus cannot be relied upon fully, as found in several other instances (see Brtek, 1977, e.g.). Similarly I find Daday's (1896) report of the species *Ceriodaphnia rotunda* Sars dubious. Daday (*op. cit.*) and Minkiewicz (1914) indicate the occurrence of *Bosmina longirostris* in the lake. I assume that in both instances the datum was taken from the table in Wierzejski's (1883) paper, which is obviously erroneous and should relate instead to the Morskie Oko lake. The species *Acroperus alonoides* Hudendorff is listed from the lake only by Minkiewicz (1914, 1917) (as '*Acroperus harpae* Baird var. *frigida* Ekman'). This may indicate only that the other authors did not differentiate this species from *A. harpae* (Baird).

In contrast to the differences referred to above, some other differences, however, are real and testify to the alterations in the structure of the Cladocera taxocene in the plankton of the lake:

Primarily it is the disappearance of the species *Daphnia pulicaria* and *Polyphemus pediculus* sometime between the 1920s and the 1950s, and the disappearance of *Holopedium gibberum* in the second half of the 1960s, or in the first half of the 1970s.

As to *Daphnia pulicaria*, the Popradské pleso lake does not seem to provide a suitable biotope for this species. Namely, in other similar Tatra lakes in the forest zone *Daphnia longispina* represents the genus *Daphnia*, as a rule. However, *Daphnia pulicaria* is listed from the Popradské pleso lake by Wierzejski, Lityński, and Minkiewicz, and there is no reason to doubt the reliability of all three of these authors.

The second difference is the appearance of

Table 1. Data on the occurrence of Cladocera in the plankton of the Popradské pleso lake in the course of the years 1880–1989.

Author	Wierzejski (1883)	Daday (1896)	Minkiewicz (1914)	Minkiewicz (1917)	Lityński (1917)	Kalmus (1929, 1931)	Our data	Ertl & Vranovský (1964)	Our data	Our data	Our data	Our data	Our data	Our data	Our data
Species / Years	1880–81	1894	1909–?	1909–?	1910–15	1928–29	1959	1961–63	1976	1979	1980	1984	1985	1986	1989
Holopedium gibberum Zad.	●	●	●	●	–	–	●	●	–	–	–	–	–	–	–
Daphnia pulicaria Forbes, *em.* Hrbáček	●	–	●	●	●	–	–	–	–	–	–	–	–	–	–
Daphnia longispina O.F.M.	–	–	–	–	–	–	–	–	●	–	–	●	●	●	●
Ceriodaphnia rotunda Sars	–	●	–	–	–	–	–	–	–	–	–	–	–	–	–
Simocephalus vetulus (O.F.M.)	●	●	●	●	●	●	–	–	–	●	–	–	–	–	–
Bosmina longirostris (O.F.M.)	●*	●**	●**	–	–	–	–	–	–	–	–	–	–	–	–
Eurycercus lamellatus (O.F.M.)	●	–	●	●	●	–	–	–	–	–	–	–	–	–	–
Alonella excisa (Fischer)	●	●	●	●	–	–	–	●	●	●	–	●	–	●	–
Alonella nana (Baird)	–	●	●	●	–	●	–	–	–	–	–	–	–	–	–
Chydorus sphaericus (O.F.M.)	●	●	●	●	–	●	●	●	●	–	–	●	–	●	●
Alona quadrangularis (O.F.M.)	●	●	●	●	–	●	–	–	–	–	–	–	–	●	–
Alona rectangula Sars	–	–	–	●	–	–	–	●	–	–	–	–	–	–	–
Acroperus harpae (Baird)	●	●	●	●	–	–	–	●	●	●	●	●	●	●	●
Acroperus alonoides Hudendorff	–	–	●	●	–	–	–	–	–	–	–	–	–	–	–
Camptocercus rectirostris Schoedl.	–	–	–	–	–	●	–	–	–	–	–	–	–	–	–
Biapertura affinis (Leydig)	–	●	●	●	–	●	–	–	●	●	●	●	–	●	–
Tretocephala ambigua (Lillj.)	–	–	–	–	–	●	–	–	–	–	–	–	–	–	–
Polyphemus pediculus (L.)	●	●	●	●	–	–	–	–	–	–	–	–	–	–	–

* Wierzejski (1883) indicates the species *B. longirostris* from the Popradské pleso lake in his table, without doing so in the text. The reference in the table obviously applies to the Morskie Oko lake.
** Probably erroneous data taken over from Wierzejski's (1883) table.

Daphnia longispina in the 1970s and its establishment in the lake pelagial.

It is not easy to explain the changes referred to. As far as the abiotic factors are concerned, data are lacking on the chemical composition of the water of the lake until the 1960s. Anthropogenic acidification, however, may be ruled out as a potential cause of the mentioned changes. Acidification has eliminated the cladoceran *Daphnia pulicaria* (Stuchlík *et al.*, 1985 and Vranovský, 1985) from several High Tatra lakes lying above the tree line, and water alkalinity in the Popradské pleso lake exhibits low values at present (it drops sometimes to values near 50 μeq l^{-1}, according to Tomajka, 1990). However, the harmful effect of acidification upon the zooplankton in the higher Tatra lakes became apparent only in about the last 20 years, whereas *Daphnia pulicaria* had vanished from the Popradské pleso lake earlier. In addition, no pH decline so far manifested itself in the lake water. This factor, i.e. acidification, could not exert a detrimental effect upon *Polyphemus pediculus* and *Holopedium gibberum* either. The waste water from the mountain hotel entering the lake contains organic substances and nutrients. The BOD_5 in the lake water attains relatively high values, but the total P and total N concentrations are relatively low, and they do not appear to have increased during the last thirty years. This is obviously due to the intensive water exchange in the lake and also due to the waste water treatment installed in 1984.

I do not presume that it was the pollution and eutrophication that eliminated *Daphnia pulicaria* and *Polyphemus pediculus*, because the effects of the pollution and anthropogenic eutrophication became apparent later. Pollution, however, might have affected the survival of *Holopedium*, which vanished from the lake plankton between the mid 1960s and mid 1970s, i.e. at the time when waste water from the old chalet and the mountain hotel were being discharged into the lake without purification. It should also be mentioned that for about one season after building the new hotel, a laundry was in operation that undoubtedly increased the quantity of detergents transported into the lake.

As to the elimination of *Daphnia pulicaria* and *Polyphemus pediculus* and the establishment of *D. longispina* in the lake, I do not believe that this has been caused by abiotic factors.

Adequate data on the population density of brown trout in the Popradské pleso lake are lacking both from the past and the present. There are indications, however, that it was frequent in the lake already toward the end of the nineteenth century, i.e. at a time when *D. pulicaria* was still a constituent of the lake plankton. For example, Szontagh (1896) writes on the Popradské pleso lake in his tourist guide: '*Das Wasser... ist dunkelgrün, krystallrein und von tausend aber tausend Forellen bevölkert*'. Although actual proofs are lacking, I believe that the elimination of *D. pulicaria* and *P. pediculus* was probably brought about by an increase in trout density and thus also an increase in its predation pressure upon the zooplankton. In the case of the disappearance of *Holopedium*, pollution might have been at least as important, or even more important, than fish predation. In my opinion, the fact that *D. pulicaria* was replaced by *D. longispina* only after the mid 1970s, would indicate that recently the predation pressure of fish on cladocerans has decreased sufficiently to allow the establishment of *D. longispina*.

Conclusion

In the Cladocera taxocene of the Popradské pleso lake plankton noteworthy qualitative changes have taken place in the course of the last 100 years. Data on the principal parameters of the abiotic and biotic environment available, especially from the past, are not sufficient for definitely determining their causes. In spite of this, I assume that probably the most important factors were fluctuations in the density and the predation pressure of fish. Nevertheless, organic pollution might have played a definite, but less important role.

Acknowledgements

Some of the difficult to access older literature sources on the fauna of the Tatra lakes were willingly provided by Dr. J. Brtek (Bojnice) and Dr I. Hudec (Košice). A sample of zooplankton of 1 September 1979 was kindly placed at my disposal by Dr. F. Kubíček (Brno). Dr. H.J. Hartmann (Aubière) read the manuscript and improved its English. My cordial thanks are due to all of them.

References

Balon, E. K., 1964. Príspevok k poznaniu ichtyofauny pramennej oblasti rieky Poprad. Sborník prác o Tatranskom národnom parku 7: 140–164 (in Slovak with summaries in Russian, English, and German).

Brtek, J., 1977. Anostraca, Notostraca, Conchostraca a Calanoida Slovenska. 2. časť. Acta Rer. natur. Mus. nat. slov., Bratislava 23: 117–150 (in Slovak with a summary in German).

Daday, J., 1896. Adatok a Tátrai tavak mikrofaunájának ismeretéhez. Mathematikai és természettudományi Értesitö 14: 116–137 (in Hungarian).

Dyk, V., 1960. Pstruhy Popradského plesa. Sborník prác o Tatranskom národnom parku 4: 219–246 (in Slovak with summaries in Russian, English, and German).

Ertl, M., 1965. Chemizmus Popradského plesa. Sborník prác o Tatranskom národnom parku 8: 45–55 (in Slovak with summaries in Russian, English, and German).

Ertl, M. & M. Vranovský, 1964. Zooplanktón Popradského plesa. Biológia (Bratislava) 19: 675–689 (in Slovak with summaries in Russian and German).

Grossinger, J. B., 1794. Universa Historia physica Regni Hungariae secundum tria Regna Naturae digesta. 3. Regni Animalis. Ichthyologia, sive Historia Piscium, et Amphibiorum. Sumptibus & Typis Simonis Petri Weber, Posonii et Comaromii, 247 pp.

Kalmus, H., 1929. Untersuchung dreier Tatraseen in bezug auf ihre Fauna und auf einige für das Leben wichtige physikalische Faktoren im Oktober und November 1928. Int. Revue ges. Hydrobiol. Hydrogr. 22: 91–94.

Kalmus, H., 1931. Weitere Beiträge zur Fauna der Tatraseen. Int. Revue ges. Hydrobiol. Hydrogr. 25: 259–265.

Lityński, A., 1913. Revision der Cladocerenfauna der Tatra-Seen. 1. Teil. Daphnidae. Bull. Acad. Sci. Cracovie, Cl. Sci. Math. Natur., Juillet 1913: 566–623, Fig. 14–18.

Lityński, A., 1917. Jeziora tatrzańskie i zamieszkująca je fauna wioślarek. Sprawozdanie Komisyi fizyograficznej Akademii Umiejętności w Krakowie 51: 1–88 (in Polish).

Marček, A., 1971. Tatranské plesá podľa najnovších meraní. Zborník prác o Tatranskom národnom parku 13: 373–381 (in Slovak with summaries in Russian, German, and English).

Minkiewicz, S., 1914. Przegląd fauny jezior tatrzańskich. Sprawozdanie Komisyi fizyograficznej Akademii Umiejętności w Krakowie 48: 1–26, Tab. 1–5 (in Polish with a summary in German).

Minkiewicz, S., 1917. Skorupiaki jezior tatrzańskich. Zarys fizyograficzno-faunistyczny. Rozprawy Wydziału matematiczno-przyrodniczego Akademii Umiejętności w Krakowie 56, Sér. B: 389–447 (in Polish).

Nagy, Š. & J. Holčík, 1988. Ichtyofauna tatranských jazier. Ústav rybárstva a hydrobiológie, Bratislava, 8 pp., 2 tab. (Res. rep., in Slovak).

Stuchlík, E., Z. Stuchlíková, J. Fott, L. Růžička & J. Vrba, 1985. Vliv kyselých srážek na vody na území Tatranského národního parku. Zborník prác o Tatranskom národnom parku 26: 173–211 (in Czech with summaries in Russian, German, and English).

Szontagh, M., 1885. Képes kalauz a Tátraalji fürdökbe és a Magas Tátra hegyvidékére. Schmidt József könyvnyomdája, Igló, 313 pp. (in Hungarian).

Szontag, M., 1896. Tátraführer. Wegweiser in die Hohe Tátra und in die Bäder der Tátragegend. Singer & Wolfner, Budapest, 226 pp.

Tomajka, J., 1985. Vybrané fyzikálne a chemické ukazovatele. In M. Ertl *et al.*, Hydrocenózy jazier Tatranského národného parku. Laboratórium rybárstva a hydrobiológie, Bratislava: 67–84 (Final res. rep., in Slovak).

Tomajka, J., 1990. Vybrané chemické ukazovatele vody tatranských jazier. Ústav rybárstva a hydrobiológie, Bratislava, 11 pp., 5 tab. (Res. rep., in Slovak).

Vranovský, M., 1985. Zooplanktón jazier Tatranského národného parku. Laboratórium rybárstva a hydrobiológie, Bratislava, 42 pp. (Final res. rep., in Slovak).

Wierzejski, A., 1882. Materyjały do fauny jezior tatrzańskich. Sprawozdanie Komisyi fizyograficznej Akademii Umiejętności w Krakowie 16: 215–239 (in Polish).

Wierzejski, A., 1883. Zarys fauny stawów tatrzańskich. Pamiętnik Towarzystwa tatrzańskiego 8: 95–123 (in Polish).